INTERNATIONAL SERIES IN
NATURAL PHILOSOPHY
GENERAL EDITOR: D. TER HAAR

VOLUME 99

THEORETICAL PHYSICS AND ASTROPHYSICS

OTHER TITLES OF INTEREST

Books*

CLARK & STEPHENSON: The Historical Supernovae

ELGAROY: Solar Noise Storms

HEY: The Radio Universe, 2nd edition

LANDAU & LIFSHITZ: Electrodynamics of Continuous Media (Course of Theoretical Physics Volume 8)

MEADOWS: Stellar Evolution, 2nd edition

PACHOLCZYK: Radio Galaxies (Radiation Transfer, Dynamics, Stability and Evolution of a Synchrotron Plasmon)

REDDISH: Stellar Formation

SAHADE & WOOD: Interacting Binary Stars

Journals**

Chinese Astronomy
Vistas in Astronomy

* Available under the Pergamon textbook inspection copy scheme

** Free specimen copy available on request

THEORETICAL PHYSICS AND ASTROPHYSICS

by

V. L. GINZBURG
P. N. Lebedev Physical Institute, Academy of Sciences of the USSR, Moscow

Translated by

D. TER HAAR
Oxford University

PERGAMON PRESS
OXFORD · NEW YORK · TORONTO · SYDNEY · PARIS · FRANKFURT

U.K.	Pergamon Press Ltd., Headington Hill Hall, Oxford OX3 0BW, England
U.S.A.	Pergamon Press Inc., Maxwell House, Fairview Park, Elmsford, New York 10523, U.S.A.
CANADA	Pergamon of Canada, Suite 104, 150 Consumers Road, Willowdale, Ontario M2 J1P9, Canada
AUSTRALIA	Pergamon Press (Aust.) Pty. Ltd., P.O. Box 544, Potts Point, N.S.W. 2011, Australia
FRANCE	Pergamon Press SARL, 24 rue des Ecoles, 75240 Paris, Cedex 05, France
FEDERAL REPUBLIC OF GERMANY	Pergamon Press GmbH, 6242 Kronberg-Taunus, Pferdstrasse 1, Federal Republic of Germany

---

First edition 1979

British Library Cataloguing in Publication Data

Ginzburg, Vitaliĭ Lazarevich
Theoretical physics and astrophysics. -
(International series in natural philosophy;
vol. 99).
1. Astrophysics 2. Physics
I. Title II. Series
523.01 QB461 78-41009

ISBN 0-08-023067-9(hardcover)
ISBN 0-08-023066-0(flexicover)

In order to make this volume available as economically and as rapidly as possible the author's typescript has been reproduced in its original form. This method unfortunately has its typographical limitations but it is hoped that they in no way distract the reader.

Printed and bound at William Clowes & Sons Limited Beccles and London

CONTENTS

PREFACE TO THE ENGLISH EDITION

As any author I am glad to see my book appear in English, a language accessible to all physicists nowadays. However, it is hardly necessary to wirte a special preface to say this or to mention that in the English edition a number of errors have been corrected. The idea that such a preface might be necessary arose when I received from Professor ter Haar the preliminary list of references for the English edition and noticed the large number of references to Ginzburg. Of course, the same references also appear in the Russian edition, but there they do not appear together in one lot, and are also not so obvious in the text. The reason is that in the Russian edition, as is the usage with us, references are indicated by a number, without mentioning the name of the authors. As a result my name never appears in the text, while in the English translation one meets it very often! Such a situation is unusual for a book meant for students or, at any rate, one which is not a monograph, and it may cause some unpleasant observations. I want therefore to add the following to what I said in the preface to the Russian edition. I am, in general, not a teacher and I do not like to lecture, especially not when the same material must be presented more than once. At the same time, for the first time lectures one needs a large amount of preparation. In such a situation and if I have the possibility to choose the subject of the lectures completely freely I have given some lectures just once; the present book was written in preparing such a course. When choosing the material and wishing to share with my audience or my readers what appeared to me to be interesting and important I widely used my own papers and review articles. Of course, the result was a very one-sided picture. However, as there was no pretension of priority claims to that I had chosen the most interesting problems, there was hardly any reason for reproaching me of immodesty. A different question is whether the book is useful and deserving attention, that that is, of course, up to the reader to decide.

In conclusion I use this possibility to express my warmest thanks to Professor ter Haar for undertaking the arduous task of translation and also for a number of helpful remarks.

Moscow, April, 1978 V.L. Ginzburg

PREFACE TO THE RUSSIAN EDITION

There are many textbooks of theoretical physics among which the many-volume work of Landau and Lifshitz is the best known and most outstanding. It is impossible, however, to deal with all problems in such a course. Moreover, even the problems which are considered can usually not be looked at from different points of view. At the same time, depending on their peculiar abilities, the nature of their training, natural inclinations, and so on, different people often prefer different approaches, arguments, examples, and proofs.

A natural possibility to satisfy an existing need is, clearly, to publish different textbooks and, in particular, different supplementary textbooks which are devoted to separate problems, aspects, and methods rather than to a systematic exposition of a topic. Such supplementary texts differ in principle from the systematic ones in that the choice of material to a large extent is not predetermined. One can say the same about the style and nature of the exposition whereas in a systematic textbook one must impose very rigid restrictions with respect to conciseness, content of technical methods, unified notation, and so on. The present book is just such a supplementary text which is devoted to a few problems in theoretical physics and astrophysics. From the table of contents it is immediately clear, but we may also state it that, in general, we are dealing with problems which are in one way or other connected with electrodynamics.

In order not to violate this trend, even though it was not rigorously expressed, we left outside the limits of the book a number of problems of the general theory of relativity and of statistical physics which, in the opinion of the author, also should be the subject of supplementary courses of a similar kind.

The basis of the exposition was a lecture course for students in the physics and astrophysics departments of the Moscow Physico-Technical Institute. These lectures were not meant to replace a systematic course and had just the character of 'capita selecta' taking into account the interests of the department and not least the interests and capabilities of the author. Of course, we are not saying that the problems which the author at a particular moment in the past was occupied with are more important or more interesting than many other ones. It is simply the case that merely presenting 'more or less moved by the spirit' material with which he is well familiar, the author may perhaps hope to supplement existing texts and monographs without checking whether he is rewriting or to some extent duplicating them.

As to the nature of the exposition one should note that we are dealing here not with lectures which are written out but with a special text specially prepared for these lectures, in which rather often we have also included material which is not very suitable and in fact was not used for the lectures themselves (that is, for the oral presentation). In this respect the book is in style close to a monograph or a review article which finds reflection also in the rather large number of references to the literature. As amongst those there is a large number of references to work by the author, we emphasize that this, like the choice of material, is completely unconnected with any pretentions, but caused by the tendency, already mentioned, to touch only upon very familiar problems which were dealt with in detail in the papers referred to; moreover, a whole number of such papers were used directly in the text.

We note, finally, that the book is definitely not intended for people with a mathematical inclination — such as 'pure' theoretical physicists. The exceptionally large role played by mathematics in theoretical physics is completely unquestionable and natural, but aiming at mathematical generality and rigour is by far not always justified — one must pay for this. It is generally known, in particular, that most new physical results have been obtained by relatively simple means while the 'mathematization' occurred only in the later stages. At any rate, physics, and not mathematics is the main point of theoretical physics. An exposition of theoretical problem with a 'general physics' bias is at least as permissible as the nowadays more widely propagated tendency to mathematical perfection.

One may hope that this book will turn out to be useful for graduate students and also for post-doctoral and research workers.

In conclusion I use this opportunity to thank all who read the manuscript or parts of it and whose remarks contributed to improvements in the text.

Moscow, July, 1974 V.L. Ginzburg

Chapter I

THE HAMILTONIAN APPROACH TO ELECTRODYNAMICS

The Hamiltonian method in classical electrodynamics in vacuo.
Quantization. Photons and pseudophotons.
Does a uniform moving electron radiate?

We shall in what follows widely apply the so-called Hamiltonian method for the interpretation of a whole range of electrodynamical problems. When we use this method electrodynamics is formulated in a way which is strongly reminiscent of mechanics. The transition from classical to quantum mechanical electrodynamics is thus in the Hamiltonian framework completely analogous to the transition from classical, Newtonian mechanics to non-relativistic quantum mechanics. Nowadays much more sophisticated methods are predominant in quantum electrodynamics and in general in quantum field theory and there are strong arguments for using them. However, the use of the Hamiltonian method is still completely justified for the elucidation of a large number of physical aspects; this is, for instance, also done by Heitler (1947) in his book. Moreover, we shall in what follows apply the Hamiltonian method mainly to classical electrodynamics, both in vacuo and in a medium.

Before introducing the Hamiltonian method we shall give the main equations and relations and we shall do that in considerable detail for future convenience.

The usual form of the Maxwell equations in vacuo is:†

$$\left.\begin{aligned} \operatorname{curl} \mathbf{H} &= \frac{4\pi}{c}\rho\mathbf{v} + \frac{\partial \mathbf{E}}{\partial t}\,, \\ \operatorname{div} \mathbf{E} &= 4\pi\rho\,, \\ \operatorname{curl} \mathbf{E} &= -\frac{1}{c}\frac{\partial \mathbf{H}}{\partial t} \\ \operatorname{div} \mathbf{H} &= 0\,. \end{aligned}\right\} \qquad (1.1)$$

Here $\mathbf{H}$ is the magnetic field strength, $\mathbf{E}$ the electrical field strength, ρ the charge density, $\mathbf{v}$ the velocity of the charges, and c the velocity of light in vacuo. We assume for the sake of simplicity that there is a single point charge e, at position $\mathbf{r}_i(t)$, in the electromagnetic field. In that case

---

† We use everywhere in this book c g s units.

the charge density is given by a δ-function

$$\rho = e\,\delta(\mathbf{r} - \mathbf{r}_i(t)) \ . \tag{1.2}$$

It is well known that Eqs.(1.1) can be reduced to the equations for the electromagnetic potentials $\mathbf{A}$ and ϕ which are connected with the fields $\mathbf{E}$ and $\mathbf{H}$ by the relations

$$\mathbf{E} = -\frac{1}{c}\frac{\partial \mathbf{A}}{\partial t} - \text{grad}\,\phi \quad , \quad \mathbf{H} = \text{curl}\ \mathbf{A} \ . \tag{1.3}$$

The third and fourth of Eqs.(1.1) are automatically satisfied by virtue of (1.3), as can be verified by substitution.

From the first and second of Eqs.(1.1) we get, using (1.3) and the identity

$$\text{curl curl}\,\mathbf{A} = -\nabla^2\mathbf{A} + \text{grad div}\,\mathbf{A} \ , \tag{1.4}$$

equations for the potentials $\mathbf{A}$ and ϕ:

$$\left.\begin{aligned} \nabla^2\mathbf{A} - \frac{1}{c^2}\frac{\partial^2\mathbf{A}}{\partial t^2} - \text{grad}\left(\frac{1}{c}\frac{\partial\phi}{\partial t} + \text{div}\,\mathbf{A}\right) &= -\frac{4\pi}{c}\rho\mathbf{v}, \\ \nabla^2\phi + \frac{1}{c}\frac{\partial}{\partial t}\,\text{div}\,\mathbf{A} &= -4\pi\rho \ . \end{aligned}\right\} \tag{1.5}$$

The set of Eqs.(1.5) determines the potentials $\mathbf{A}$ and ϕ. The fields $\mathbf{E}$ and $\mathbf{H}$ can be found using Eqs.(1.3)

It is well known that the vector potential $\mathbf{A}$ and the scalar potential ϕ are not uniquely determined. Indeed, we can change to new potentials:

$$\mathbf{A}' = \mathbf{A} + \text{grad}\,\chi \quad , \quad \phi' = \phi - \frac{1}{c}\frac{\partial\chi}{\partial t} \ , \tag{1.6}$$

where χ is an arbitrary function of the coordinates and the time. This is called a gauge transformation. One can easily show that the fields $\mathbf{E}$ and $\mathbf{H}$ do not change under a gauge transformation. They can be expressed in terms of $\mathbf{A}'$ and ϕ' just as well as in terms of $\mathbf{A}$ and ϕ; one can verify this by substituting (1.6) into (1.3).

The fact that the definition of the potentials is not unique enables us to impose upon $\mathbf{A}$ and ϕ an additional condition. This condition can be chosen in such a way that the form of Eqs.(1.5) becomes as simple as possible.

For instance, we can impose as such a condition the relation:

$$\text{div}\,\mathbf{A} + \frac{1}{c}\frac{\partial\phi}{\partial t} = 0 \ . \tag{1.7}$$

This is a relativistically invariant condition which is called the Lorentz condition, and the resulting gauge is called the Lorentz gauge. It can be written in the form

$$\frac{\partial A^i}{\partial x^i} = 0 , \tag{1.7a}$$

where we have assumed summation over repeated indexes, as will be done everywhere in what follows. We use here and elsewhere in this book (apart from Chapter 12) the notation of Landau and Lifshitz (1975). We refer to that text for the definition of four-vectors, for the difference between covariant and components and for the summation convention.

One sees easily that if condition (1.7) is satisfied, the Maxwell equations take the following form:

$$\Box \mathbf{A} \equiv \left(\nabla^2 - \frac{1}{c^2}\frac{\partial^2}{\partial t^2}\right)\mathbf{A} = -\frac{4\pi}{c}\rho\mathbf{v} ,$$
$$\Box \phi \equiv \left(\nabla^2 - \frac{1}{c^2}\frac{\partial^2}{\partial t^2}\right)\phi = -4\pi\rho . \tag{1.8}$$

One should not think that the condition (1.7) and the set of Eqs.(1.8) determine $\mathbf{A}$ and ϕ completely. We can still perform a gauge transformation of the form (1.6), where in the present case χ must satisfy the homogeneous equation $\Box\chi = 0$. The fields $\mathbf{E}$ and $\mathbf{H}$ remain invariant under the transformation.

Splitting the field into longitudinal and transverse components is important, especially in the Hamiltonian framework. We split the vectors $\mathbf{E}$ and $\mathbf{H}$ into components

$$\mathbf{E} = \mathbf{E}_{\ell} + \mathbf{E}_{tr} , \quad \mathbf{H} = \mathbf{H}_{tr} , \tag{1.9}$$

where $\operatorname{div}\mathbf{E}_{tr} = 0$ and, by virtue of (1.1), $\operatorname{div}\mathbf{H}_{tr} = \operatorname{div}\mathbf{H} = 0$.

We demand that the vector potential $\mathbf{A}$ describe only the transverse field; this means that we impose on it the condition

$$\operatorname{div}\mathbf{A} = 0 , \tag{1.10}$$

instead of the additional condition (1.7). Sometimes the potential which satisfies the condition (1.10) is denoted by $\mathbf{A}_{tr}$.

If condition (1.10) is satisfied, the Eqs.(1.5) for $\mathbf{A}$ and ϕ take the form

$$\nabla^2\phi = -4\pi\rho , \tag{1.11}$$

$$\nabla^2\mathbf{A} - \frac{1}{c^2}\frac{\partial^2\mathbf{A}}{\partial t^2} = -\frac{4\pi}{c}\rho\mathbf{v} + \frac{1}{c}\operatorname{grad}\frac{\partial\phi}{\partial t} . \tag{1.12}$$

We see that we have obtained the 'static' Poisson equation for the potential ϕ. If ρ is the charge density (1.2) of a point charge, the solution is the well known one:

$$\phi = \frac{e}{|\mathbf{r} - \mathbf{r}_i(t)|} \tag{1.13}$$

where $\mathbf{r}_i(t)$ is the position of the charge at time t. The vector potential $\mathbf{A}$ now describes only the transverse field. The gauge (1.10) is called the Coulomb gauge†. The potentials $\mathbf{A}$ and ϕ are here determined apart from a gauge function $\chi(\mathbf{r},t)$ which satisfies the condition $\nabla^2\chi = 0$.

We now evaluate the energy of the electromagnetic field,

$$\mathcal{H} = \int \frac{E^2 + H^2}{8\pi} d^3\mathbf{r} \tag{1.14}$$

We substitute here the expressions for the fields $\mathbf{E}$ and $\mathbf{H}$ in the form (1.9); it is clear that in the case of the Coulomb gauge (1.10) we have

$$\mathbf{E}_{tr} = -\frac{1}{c}\frac{\partial \mathbf{A}}{\partial t}, \quad \mathbf{E}_\ell = -\operatorname{grad}\phi . \tag{1.15}$$

Substituting (1.9) and (1.15) into (1.14) we get

$$\mathcal{H} = \frac{1}{8\pi}\int (E_{tr}^2 + H^2) d^3\mathbf{r} + \frac{1}{8\pi}\int E_\ell^2 d^3\mathbf{r} + \frac{1}{4\pi}\int (\mathbf{E}_{tr}\cdot\mathbf{E}_\ell) d^3\mathbf{r}.$$

One shows easily that for a closed system, when the field 'at infinity' vanishes, the last integral equals zero. The total energy of the electromagnetic field is thus the sum of the energies of the transverse and of the longitudinal fields.

If there are several point charges in the field, the energy of the longitudinal field is simply the energy of the Coulomb interaction between the charges, that is,

$$\mathcal{H}_\ell = \frac{1}{8\pi}\int E_\ell^2 d^3\mathbf{r} = \frac{1}{2}\sum_{i,j} \frac{e_i e_j}{r_{ij}(t)} . \tag{1.16}$$

The self energy of the point charges is infinite and is here, of course, neglected. It is important to note that the longitudinal part of the electromagnetic field is not quantized. Only the transverse field is quantized (see Heitler 1947 and the discussion later in this chapter).

†It is obvious that it is possible to introduce the Coulomb gauge — and to use Eqs.(1.11) and (1.12) which are connected with it. It is therefore rather interesting that as long as thirty years ago when classical electrodynamics was already fully 'grown-up' the Hamiltonian method was usually developed on the basis of Eq.(1.8), although it led to complications (see, for instance, the first edition of Heitler's book — the best one for its time — which appeared in 1936); another indication of the unpopularity of the Coulomb gauge in the past can be seen from the fact that a paper devoted to that gauge was published as late as 1939 (Ginzburg, 1939c).

As the energy of the field of a point charge is infinite, one is often led to assume — at least in an intermediate stage — that the charge is 'smeared out' over a region of radius r_o. In that case $\mathcal{H}_\ell \sim e^2/r_o$. The electrostatic (classical) electron radius, defined by the relation $r_e = e^2/mc^2$, where e and m are the observed electron charge and mass, is equal to $r_e = 2.8 \times 10^{-13}$ cm. We shall not be concerned here with the problems connected with the electromagnetic mass of the electron — or of other particles — or whether it is a point charge, and so on.

Proceeding now along the path which will lead to the Hamiltonian formalism for electrodynamics we expand the vector potential $\mathbf{A}$ of the transverse electromagnetic field in a Fourier series,

$$\mathbf{A}(\mathbf{r},t) = \sum_\lambda q_\lambda(t)\sqrt{4\pi}\,c\;\mathbf{e}_\lambda \exp\{i(\mathbf{k}_\lambda\cdot\mathbf{r})\}\,. \tag{1.17}$$

The numerical coefficient $\sqrt{4\pi}\,c$ is a normalization factor. The polarization vector $\mathbf{e}_\lambda$ is a unit vector, that is, $e_\lambda^2 = 1$; for the sake of simplicity we assume here and henceforth that the vectors $\mathbf{e}_\lambda$ are real. In order that we may apply the expansion (1.17) we must imagine the electromagnetic field to be enclosed in a large, cubic 'box'. One can verify that the dimensions of this 'box' do not enter in any of the expressions for physically observable quantities. We shall therefore everywhere put the size of this 'box' to be equal to unity:

$$L = L^3 = 1.$$

The vector potential $\mathbf{A}$ is a real quantity; it therefore follows from the expansion (1.17) that $q_{-\lambda} = q_\lambda^*$. As the field is transverse, $(\mathbf{e}_\lambda\cdot\mathbf{k}_\lambda) = 0$, that is, the polarization vector of the harmonic with number λ of the potential is at right angles to the wavevector $\mathbf{k}_\lambda$ of this harmonic. To each direction of $\mathbf{k}_\lambda$ there correspond two vectors $\mathbf{e}_\lambda$. We should therefore introduce one more index which can take two values, or, in other words, distinguish the vectors $\mathbf{e}_{\lambda 1}$ and $\mathbf{e}_{\lambda 2}$. We shall not do this in what follows, in order to simplify the notation, but we shall sum over the polarizations, when necessary, in the final expressions; we shall assume in those cases that $(\mathbf{e}_{\lambda_1}\cdot\mathbf{e}_{\lambda_2}) = 0$.

We can realize also a different expansion of the vector potential, namely,

$$\mathbf{A} = \sum_{\lambda,i} q_{\lambda i}\,\mathbf{A}_{\lambda i}\,, \tag{1.18}$$

where the index i can take on only the values 1 and 2, while

$$\mathbf{A}_{\lambda 1} = \sqrt{8\pi}\, c\, \mathbf{e}_\lambda \cos(\mathbf{k}_\lambda \cdot \mathbf{r}) , \quad \mathbf{A}_{\lambda 2} = \sqrt{8\pi}\, c\, \mathbf{e}_\lambda \sin(\mathbf{k}_\lambda \cdot \mathbf{r}) . \tag{1.19}$$

One sees easily that the functions $\mathbf{A}_{\lambda 1}$ and $\mathbf{A}_{\lambda 2}$ in which we have expanded in (1.18), are mutually orthogonal, that is,

$$\int (\mathbf{A}_{\lambda i} \cdot \mathbf{A}_{\mu j})\, d^3\mathbf{r} = 4\pi c^2 \delta_{\lambda\mu} \delta_{ij} , \tag{1.20}$$

where the integral is over the volume of the 'box'.

We assume the field to be enclosed in a 'box' with specularly reflecting walls; the components of the wavevector $\mathbf{k}_\lambda$ must thus be integral multiples of the quantity $2\pi/L$, where L is the linear size of the 'box', that is

$$\mathbf{k}_\lambda = \left\{ \frac{2\pi}{L} n_x , \frac{2\pi}{L} n_y , \frac{2\pi}{L} n_z \right\} ;$$

n_x, n_y, n_z are here integers; the summation in (1.18) is over a hemisphere of the directions of $\mathbf{k}_\lambda$. Apparently, what we have just said contradicts the earlier statement that the size L of the 'box' is unimportant and can be put equal to 1. However, one can easily check that there is here no contradiction: for sufficiently large values of L this quantity does not occur in the final results.

It is clear from (1.18) that the transverse electromagnetic field is completely determined, if we give the set of quantities $q_{\lambda i}(t)$. The quantities $q_{\lambda i}(t)$ form a denumerable infinite set. The field is thus through (1.8) represented as a system of an infinite, though denumerable, number of degrees of freedom.

Let us see how one can express the energy of the electromagnetic field in terms of the quantities $q_{\lambda i}(t)$ which we can properly call the field coordinates. We are interested in the energy

$$\mathcal{H}_{tr} = \int \frac{E_{tr}^2 + H^2}{8\pi}\, d^3\mathbf{r} . \tag{1.21}$$

If $\mathbf{A}$ is given in the form (1.18), we can use Eqs.(1.3) and (1.15) to determine the fields $\mathbf{E}_{tr}$ and $\mathbf{H}$; we can square them and substitute them into the integral (1.21). We then get

$$\mathcal{H}_{tr} = \frac{1}{2} \sum_{\lambda, i} \left(p_{\lambda i}^2 + \omega_\lambda^2 q_{\lambda i}^2 \right) . \tag{1.22}$$

We have introduced here the notation

$$p_{\lambda i} = \dot{q}_{\lambda i} , \quad \omega_\lambda^2 = c^2 k_\lambda^2 , \tag{1.23}$$

where the dot indicates differentiation with respect to the time. In deriving Eq.(1.22) we used the orthogonality condition (1.20).

Each term in the sum (1.22) is the energy of a classical oscillator of frequency ω_λ. Therefore, (1.22) is the sum of the energies of separate oscillators, which we call the field oscillators.

If all $q_{\lambda i}(t)$ in (1.18) are known, we can determine the energy of the transverse electromagnetic field. The problem is thus reduced to the determination of the $q_{\lambda i}(t)$.

To find the equations for the $q_{\lambda i}(t)$ we substitute the expansion (1.18) into Eq.(1.12) for the transverse vector potential. Multiplying both sides of the resulting equation by $\mathbf{A}_{\lambda i}$ and integrating over the volume of the 'box' we get the following equations for the $q_{\lambda i}(t)$:

$$\ddot{q}_{\lambda i} + \omega_\lambda^2 q_{\lambda i} = \frac{e}{c}\Big(\mathbf{v}\cdot\mathbf{A}_{\lambda i}(\mathbf{r}(t))\Big) = e\sqrt{8\pi}(\mathbf{e}_\lambda\cdot\mathbf{v}(t))\begin{cases}\cos(\mathbf{k}_\lambda\cdot\mathbf{r}(t)) \,, \\ \sin(\mathbf{k}_\lambda\cdot\mathbf{r}(t)) \,. \end{cases} \tag{1.24}$$

This is the equation for an oscillator when there is an exciting force present; for $i = 1$ we choose $\cos(\mathbf{k}_\lambda\cdot\mathbf{r})$ and for $i = 2$: $\sin(\mathbf{k}_\lambda\cdot\mathbf{r})$.

We derived Eq.(1.24) in the assumption that there was a single point charge (electron with charge e; see (1.2)) moving with a velocity $\mathbf{v}(t)$ present in the field. The generalization to the case of several charges is obvious.

All relations considered here can be written completely analogously to the Hamiltonian equations of classical mechanics:

$$\dot{p} = -\frac{\partial \mathcal{H}(p,q)}{\partial q} \,, \quad \dot{q} = \frac{\partial \mathcal{H}(p,q)}{\partial p} \tag{1.25}$$

where $\mathcal{H}(p,q)$ is the Hamiltonian function of the mechanical system, and q and p are, respectively, the generalized coordinates and momenta.

Our problem now is to find such a function $\mathcal{H}(p_{\lambda i}, q_{\lambda i})$ that we can obtain equations of motion such as (1.25) from it. It is clear that Eqs.(1.24) for the $q_{\lambda i}$ for the case of a free field — a field without charges — that is, the equations

$$\ddot{q}_{\lambda i} + \omega_\lambda^2 q_{\lambda i} = 0 \,, \tag{1.26}$$

can be written in Hamiltonian form, if

$$\mathcal{H} = \mathcal{H}_{tr} = \frac{1}{2}\sum_{\lambda,i}\left(p_{\lambda i}^2 + \omega_\lambda^2 q_{\lambda i}^2\right) \tag{1.27}$$

where $\mathcal{H}_{tr}$ is the energy of the transverse electromagnetic field (1.22).

Indeed, from (1.25) and (1.27) we get

$$\dot{p}_{\lambda i} = -\frac{\partial \mathcal{H}}{\partial q_{\lambda i}} = -\omega_\lambda^2 q_{\lambda i} \; , \; \dot{q}_{\lambda i} = \frac{\partial \mathcal{H}}{\partial p_{\lambda i}} = p_{\lambda i} \; , \tag{1.28}$$

which is the same as (1.26).

The series (1.18) with the $q_{\lambda i}$ determined from (1.26) is a sum of plane electromagnetic waves propagating with the speed of light. Indeed, it follows from (1.26) that $q_{\lambda i} = C_1 \cos \omega t + C_2 \cos \omega t$, where $\omega = \omega_\lambda$. Moreover, $\omega_\lambda^2 = c^2 k_\lambda^2$ (see (1.23)) and the field thus changes as $\cos\{\omega_\lambda[t-(\mathbf{s}_\lambda \cdot \mathbf{r})/c]\}$ or $\sin\{\omega_\lambda[t-(\mathbf{s}_\lambda \cdot \mathbf{r})/c]\}$, where $\mathbf{s}_\lambda = \mathbf{k}_\lambda / k_\lambda$, $s_\lambda^2 = 1$. When there are no charges the field thus consists of plane waves moving with the velocity of light; this result is, of course, already clear from the original equations.

If the classical Hamiltonian of the electromagnetic field without charges in vacuo is the field energy, when there are charges present we must now take the interaction of them with the field into account. It is well known that in the non-relativistic case the energy of a charge in the field has the form (see, e.g., ter Haar, 1971)

$$\mathcal{H}_e = \frac{1}{2m}\left(\mathbf{p} - \frac{e}{c}\mathbf{A}\right)^2 + e\phi \; . \tag{1.29}$$

The total Hamiltonian for the system of the electromagnetic field + charged particles is thus equal to the sum of expressions (1.27) and (1.29):

$$\mathcal{H} = \frac{1}{2m}\left(\mathbf{p} - \frac{e}{c}\mathbf{A}(\mathbf{r}_i)\right)^2 + e\phi(\mathbf{r}_i) + \mathcal{H}_{tr} \; , \tag{1.30}$$

where r_i is the position of the charged particle (if there are several particles, $\mathcal{H}_e$ is the sum of expressions such as (1.29); we shall sometimes in what follows drop the index i of $\mathbf{r}_i$).

Using this Hamiltonian (in actual fact we have spoken so far all the time about a classical Hamiltonian) and (1.25) we get the following set of equations

$$\dot{p}_{\lambda i} = -\omega_\lambda^2 q_{\lambda i} + \frac{e}{mc}\left(\left[\mathbf{p} - \frac{e}{c}\mathbf{A}\right] \cdot \mathbf{A}_{\lambda i}\right) , \; \dot{q}_{\lambda i} = p_{\lambda i} \; .$$

This set can be reduced to Eq.(1.24). Indeed, as the quantity $\mathbf{r} = \partial\mathcal{H}/\partial\mathbf{p} = [\mathbf{p} - (e\mathbf{A}/c)]/m$ is simply the particle velocity $\mathbf{v} \equiv \dot{\mathbf{r}}$ we are led to the equation we have already derived earlier. We can also obtain the equation of motion for the particle in the field from the Hamiltonian (1.30); differentiating $\mathcal{H}$ with respect to $\mathbf{r}$ we find

$$\dot{\mathbf{p}} = -\frac{\partial \mathcal{H}}{\partial \mathbf{r}} = e\left\{\mathbf{E} + \frac{1}{c}[\mathbf{v} \wedge \mathbf{H}]\right\} + \frac{e}{c}\dot{\mathbf{A}} . \tag{1.31}$$

We thus get from Eq.(1.30) for the Hamiltonian both the equation of motion for the field oscillators and the equation of motion for the charged particles.

We have here considered everything in the non-relativistic approximation† in order to use the fact that non-relativistic mechanics and non-relativistic quantum mechanics are so close in form. In the relativistic quantum theory of spin-$\frac{1}{2}$ particles, where one uses the Dirac equation which does not have such an obvious classical analogue, this similarity is lost to a certain degree.

The transition from classical electrodynamics in a Hamiltonian form to quantum electrodynamics is accomplished in exactly the same way as the change from classical non-relativistic mechanics to quantum mechanics. In fact, the classical Hamiltonian for a particle with momentum $\mathbf{p}$ and position $\mathbf{r}$,

$$\mathcal{H} = \frac{p^2}{2m} + e\phi(\mathbf{r}), \tag{1.32}$$

is replaced by the Hamiltonian operator, or simply, the Hamiltonian

$$\hat{\mathcal{H}} = \frac{\hat{p}^2}{2m} + e\phi , \tag{1.33}$$

where $\hat{\mathbf{p}}$ is the operator of the particle momentum which satisfies the commutation relations ($\mathbf{r} \equiv \{x_j\}$, $j = 1, 2, 3$; note that there is no summation over j here)

$$p_j x_j - x_j p_j = -i\hbar , \tag{1.34}$$

and is given by the expression

$$\hat{\mathbf{p}} = -i\hbar\nabla . \tag{1.35}$$

If the particle is in an electromagnetic field, $\mathbf{p}$ in (1.32) is replaced by $\mathbf{p} - (e\mathbf{A}/c)$ and correspondingly $\hat{\mathbf{p}}$ in (1.33) by

$$\hat{\mathbf{p}} - \frac{e}{c}\mathbf{A} = -i\hbar\nabla - \frac{e}{c}\mathbf{A} .$$

The state of the system is determined by the wavefunction $\Psi(\mathbf{r},t)$; the change of the wavefunction with time is described by the Schrödinger equation

$$i\hbar\frac{\partial\Psi}{\partial t} = \hat{\mathcal{H}}\Psi . \tag{1.36}$$

The wavefunctions of stationary states have the form

$$\Psi_n(\mathbf{r},t) = \exp(-iE_n t/\hbar)\, \psi_n(\mathbf{r}) , \tag{1.37}$$

where the $\psi_n(\mathbf{r})$ are independent of t (n is the number of the stationary state, or its quantum number). The square of the modulus of the wavefunction of a stationary state, that is, the probability that a particle will be observed in a given point in space, is independent of the time. Substituting

---

†We are, of course, talking about the particles, as electrodynamics in vacuo (or from the quantum theoretical viewpoint the theory of spin-1 particles with zero rest mass) is always a relativistic theory.

(1.37) into (1.36) and dividing by $\exp(-iE_n t/\hbar)$ we have

$$\mathcal{H}\psi_n(\mathbf{r}) = E_n\psi_n(\mathbf{r}) . \tag{1.38}$$

Let us consider a one-dimensional harmonic oscillator with unit mass. It is well known that the Hamiltonian of such an oscillator has the form

$$\hat{\mathcal{H}} = \tfrac{1}{2}\hat{p}^2 + \tfrac{1}{2}\omega_0^2 q^2 , \tag{1.39}$$

where $q \equiv \hat{q}$ is the coordinate and ω_0 the angular frequency of the oscillator. The energy of the nth stationary state is equal to

$$E_n = \hbar\omega_0 (n + \tfrac{1}{2}) \quad ; \quad n = 0, 1, 2, \ldots, \tag{1.39a}$$

and the wavefunction has the form

$$\psi_n(q) = C_n \exp(-q^2/2q_0^2) H_n(q/q_0) , \tag{1.39b}$$

where $q_0 = \sqrt{\hbar/\omega_0}$, $H_n(x)$ is a Hermite polynomial, and C_n a normalization factor. In particular,

$$\psi_0(q) = \frac{1}{(q_0\sqrt{\pi})^{\frac{1}{2}}} \exp(-q^2/2q_0^2) . \tag{1.39c}$$

The matrix elements of the coordinate, $q_{nn'}$, and of the momentum, $p_{nn'}$, corresponding to transitions from a state with quantum number n to a state with quantum number n vanish unless $n' \neq n \pm 1$, and when $n' = n \pm 1$ equal

$$\left.\begin{aligned}
q_{n,n+1} &= \int \psi_n^* q\, \psi_{n+1} dq = \left(\frac{\hbar(n+1)}{2\omega_0}\right)^{\frac{1}{2}} , \\
q_{n,n-1} &= \left(\frac{\hbar n}{2\omega_0}\right)^{\frac{1}{2}} \\
p_{n,n+1} &= \int \psi_n^* \left(-i\hbar \frac{\partial}{\partial q}\right) \psi_{n+1}\, dq \\
&= -i\omega_0 \left(\frac{\hbar(n+1)}{2\omega_0}\right)^{\frac{1}{2}} = -i\omega_0 q_{n,n+1} , \\
p_{n,n-1} &= i\omega_0 \left(\frac{\hbar n}{2\omega_0}\right)^{\frac{1}{2}} = i\omega_0\, q_{n,n-1} .
\end{aligned}\right\} \tag{1.40}$$

We mentioned earlier that the transition from classical to quantum electrodynamics is achieved in the same way as in mechanics. The Hamiltonian of a system consisting of a particle and the field,

$$\hat{\mathcal{H}} = \hat{\mathcal{H}}_e + \hat{\mathcal{H}}_{tr} , \tag{1.41}$$

where $\mathcal{H}_e$ is the Hamiltonian for a particle in the field, which we considered earlier (see (1.29)), is replaced by the Hamiltonian operator

$$\hat{\mathcal{H}} = \hat{\mathcal{H}}_e + \hat{\mathcal{H}}_{tr} , \tag{1.42}$$

where

$$\hat{\mathcal{H}}_{tr} = \frac{1}{2} \sum_{\lambda,i} \left(\hat{p}^2_{\lambda,i} + \omega^2_\lambda q^2_{\lambda i} \right) . \tag{1.43}$$

The momentum operators $\hat{p}_{\lambda i}$ are, as in mechanics, equal to

$$\hat{p}_{\lambda i} = -i\hbar \frac{\partial}{\partial q_{\lambda i}} , \tag{1.44}$$

and satisfy the commutation relations

$$\hat{p}_{\lambda i} q_{\mu j} - q_{\mu j} \hat{p}_{\lambda i} = -i\hbar \, \delta_{\lambda\mu} \delta_{ij} . \tag{1.45}$$

If there are no charges, or if their interaction with the field, which appears in $\hat{\mathcal{H}}_e$; can be neglected, the wavefunction $\Psi(q,t)$ which describes the state of the field (q stands for the whole set of field coordinates $q_{\lambda j}$) will satisfy the equation

$$i\hbar \frac{\partial \Psi(q,t)}{\partial t} = \hat{\mathcal{H}}_{tr} \Psi(q,t) \tag{1.46}$$

The wavefunction $\psi_n(q)$ of a stationary state then satisfies the equation

$$\hat{\mathcal{H}}_{tr} \, \psi_n(q) = E_n \psi_n(q) . \tag{1.47}$$

One can easily check that E_n has the following form

$$E_n = \sum_{\lambda,i} \hbar\omega_\lambda (n_{\lambda i} + \tfrac{1}{2}) = \sum_{\lambda,i} \hbar\omega_\lambda n_{\lambda i} + \frac{1}{2} \sum_{\lambda,i} \hbar\omega_\lambda \tag{1.48}$$

The term $\frac{1}{2} \sum_{\lambda,i} \hbar\omega_\lambda$ is infinite. However, this infinity does not lead to any significant difficulties in the theory for the following reason: firstly, the physically observable quantity in this theory is not the energy itself, but the difference between energies in different states, so that the sum $\frac{1}{2} \sum_{\lambda,i} \hbar\omega_\lambda$, which is constant, does not occur in the final result. Secondly, the transition from the classical to the quantummechanical equations is not unique. One can find a method of quantizing the field equations in which this term is absent. Indeed, we shall start from the classical Hamiltonian for a single oscillator in the form

$$\mathcal{H} = \tfrac{1}{2}(p^2 + \omega^2 q^2) = \tfrac{1}{2}(p + i\omega q)(p - i\omega q) .$$

If we then change to the Hamiltonian operator, that is, replace p by $\hat{p} = -i\hbar \partial/\partial q$, we find

$$\hat{\mathcal{H}} = \tfrac{1}{2}(\hat{p}^2 + \omega^2 \hat{q}^2) - \tfrac{1}{2}\hbar\omega ;$$

this result comes about as the operators p and q do not commute. Thus, if we apply this to the field, we are led to the expression

$$E_n = \sum_{\lambda,i} \hbar\omega_\lambda n_{\lambda i} \tag{1.49}$$

for the energy of a stationary state. The wavefunction of the field has the form of a product of the wavefunctions of the separate oscillators, that is,

$$\psi_n(q) = \prod_{\lambda,i} \psi_{n\lambda i}(q_{\lambda i}) \ . \tag{1.50}$$

It is clear from Eq.(1.49) that one can treat the energy of the electromagnetic field as the energy of a set of particles with energies $\hbar\omega_\lambda$. It is often said that we have in this way at once introduced the photon concept, and that the number $n_{\lambda i}$ is the number of photons of a given kind. However, it is customary — and very reasonable — to use the term photons only for quanta of the radiation field, and in particular the field of light. In most cases one uses an even narrower definition of photons as quanta of the electromagnetic field with energy $\hbar\omega$ and momentum $\hbar\mathbf{k}$, where $k = \omega/c$. We shall in what follows call any quanta of radiation in vacuo photons, but that does not change the fact that the quanta of the transverse electromagnetic field — which we call pseudo-photons — in general do not reduce to photons. The important fact is that we do not consider here a radiation field, that is, a free electromagnetic field, which is a solution of the homogeneous field equations, but an arbitrary field, which is a solution of the inhomogeneous field equations, that is, the equations when there are currents and charges present. Such an arbitrary transverse electromagnetic field differs from a radiation field, or, in quantum language, from a collection of photons; as an example it is sufficient to mention the transverse field of a uniformly moving charge which moves in space with a velocity $v < c$. Moreover, even if the quantization (1.45) of the transverse field itself did not involve any assumptions, Eqs.(1.46) to (1.48) as well as Eqs.(1.49) and (1.50) were obtained in an inconsistent manner: we neglect the interaction between the charges and the field. If we did not do that, the field coordinates q would clearly occur both in $\hat{\mathcal{H}}_{tr}$ and in $\hat{\mathcal{H}}_e$ and as a result it would have been impossible to write the energy of the transverse field in the form (1.48) or (1.49). Nevertheless, the introduction of the pseudo-photons has some meaning, as the frequencies ω_λ in (1.49) can for the time being be assumed not to be connected with the wavevectors through the relations $\omega_\lambda^2 = c^2k_\lambda^2$. Such pseudo-photons, which are also called virtual photons, occur in the intermediate states in perturbation theory calculations (*vide infra*). In other words, the energy of the virtual photons $E_\lambda = \hbar\omega_\lambda$ and their momentum $\mathbf{p}_\lambda = \hbar\mathbf{k}_\lambda$ are not connected through the relation $E_\lambda^2 = c^2 p_\lambda^2$ $(\omega_\lambda^2 = c^2 k_\lambda^2)$ which is valid for photons with a given momentum. We shall see below that for the transverse field carried along with a moving charge we have $\omega = (\mathbf{k}.\mathbf{v})$ where $\mathbf{v}$ is the velocity of the charge. If we consider the corresponding quanta with energy $\hbar\omega$ they refer to pseudo-photons and one sometimes says that they form the 'coat' of the moving charge.

We emphasize that we do not at all wish to insist on the advisability of introducing the pseudo-photon concept, although we use this term[†]. The importance for us lies in the explication of the fact that the transverse field is in general not a collection of photons. This will be discussed below.

The pseudo-photons which occur in Eq.(1.49) do not reduce to photons of energy $\hbar\omega$ and momentum $(\hbar\omega/c)\,(\mathbf{k}/k)$ even for a radiation field. This is connected with the fact that above we used an expansion in standing waves (see (1.18) and (1.19)). Standing waves are not eigenfunctions of the momentum operator and their quantization leads to photons (we are thinking now of a pure, free, radiation field) with a vanishing momentum[††].

In order to change to the 'usual' photons with energy $\hbar\omega$ and momentum $(\hbar\omega/c)$ $(\mathbf{k}/k)$ we write the vector potential as a sum of travelling waves:

$$\mathbf{A} = \sum_\lambda \left(q_\lambda \mathbf{A}_\lambda + q_\lambda^* \mathbf{A}_\lambda^*\right) , \tag{1.52}$$

where

$$\mathbf{A}_\lambda = \sqrt{4\pi}\, c\, \mathbf{e}_\lambda \, \exp i(\mathbf{k}_\lambda \cdot \mathbf{r}) , \tag{1.53}$$

---

† The term 'pseudo-photon' is sometimes also applied in a different sense from the one used here. The method in which one approximately replaces the Fourier component of the field carried along by a moving charge by a set of photons (see, e.g. Ter-Mikaelyan, 1972) is called the pseudo-photon method.

†† To evaluate the photon momentum one can use the expression for the field momentum

$$\mathbf{G} = \frac{1}{4\pi c} \int [\mathbf{E} \wedge \mathbf{H}]\, d^3\mathbf{r} \tag{1.51}$$

and express $\mathbf{E}$ and $\mathbf{H}$ in terms of $\mathbf{A}$, that is, in terms of the quantities $p_{\lambda i}$ and $q_{\lambda i}$.

while the summation in (1.52) ism in contrast to (1.17), over half-a-sphere, that is, over half of all $\mathbf{k}_\lambda$-directions.

For the Hamiltonian of the transverse part of the field whave then

$$\mathcal{H}_{tr} = \sum_\lambda \left(p_\lambda p_\lambda^* + \omega_\lambda^2 q_\lambda q_\lambda^* \right) . \tag{1.54}$$

We shall solely consider a purely radiation field. In that case we may assume that $p_\lambda = \dot{q}_\lambda = -i\omega_\lambda q_\lambda$ (see also (1.26) and (1.28)) and we have

$$\mathcal{H}_{tr} = 2 \sum_\lambda \omega_\lambda^2 q_\lambda q_\lambda^* . \tag{1.55}$$

We note that the quantities q_λ and q_λ^* are not canonically conjugate variables, as the equations of motion in those variables do not have the form

$$q^* = -\frac{\partial \mathcal{H}}{\partial q} , \; \dot{q} = \frac{\partial \mathcal{H}}{\partial q^*} . \tag{1.56}$$

We therefore introduce new variables:

$$Q_\lambda = q_\lambda + q_\lambda^* , \quad P_\lambda = -i\omega_\lambda (q_\lambda - q_\lambda^*) . \tag{1.57}$$

We then have

$$\mathcal{H}_{tr} = \frac{1}{2} \sum_\lambda \left(P_\lambda^2 + \omega_\lambda^2 Q_\lambda^2 \right) . \tag{1.58}$$

Quantizing we get clearly Eq.(1.49) for the energy, and evaluating the momentum of the field we find

$$\mathbf{G} = \sum_\lambda \frac{\hbar\omega_\lambda}{c} n_\lambda \frac{\mathbf{k}_\lambda}{k_\lambda} , \tag{1.59}$$

that is, the momentum of a single photon is, indeed, equal to

$$\mathbf{g}_\lambda = \frac{\hbar\omega_\lambda}{c} \frac{\mathbf{k}_\lambda}{k_\lambda} \equiv \frac{\hbar\omega_\lambda}{c} \mathbf{s}_\lambda , \quad s_\lambda^2 = 1 . \tag{1.60}$$

The field and the quanta corresponding to it are characterized not only by their energy and momentum, but also by their angular momentum. The angular momentum of the electromagnetic field is classically defined as follows:

$$\mathbf{M}_{em} = \frac{1}{4\pi c} \int \left[\mathbf{r} \wedge [\mathbf{E} \wedge \mathbf{H}] \, d^3\mathbf{r} \right. . \tag{1.61}$$

An unbounded plane wave does not have a non-vanishing angular momentum along the direction of the wavevector $\mathbf{k}_\lambda$, as the Poynting vector $\mathbf{S} = c[\mathbf{E} \wedge \mathbf{H}]/4\pi$ is in the direction of $\mathbf{k}_\lambda$ in such a wave. However, for a wave in a cylindrical waveguide with perfectly conducting walls, for instance, the value of $M_{emz} \equiv M_z$, where the z-axis is along the axis of the waveguide, may be different from zero. To be concrete, for monochromatic circularly polarized waves in a waveguide we have (see Heitler, 1947 and the references cited there)

$$M_z = \pm \mathcal{H}_{tr}/\omega_\lambda \ . \tag{1.62}$$

The sign depends here on the direction in which the field rotates in the wave and $\mathcal{H}_{tr}$ is again the energy of the transverse field. After quantization (which must be done here by expanding the field in terms of the 'normal' waves in the waveguide) we have $\mathcal{H}_{tr} = \hbar\omega_\lambda n_\lambda$ and Eq.(1.62) indicates that the angular momentum of the field is the sum of the angular momenta of the field quanta, where for each quantum the value of M_z equals $\pm\hbar$. Using the nomenclature introduced earlier we can call these quanta photons (we assume the waveguide to be a vacuum and its walls to be perfectly conducting). It is, of course, not the terminology which is important, but the fact that the angular momentum of the radiation turns out to be quantized. The angular momentum — or, to be more precise, its z-component — is equal to $\pm\hbar$, that is, the spin of the photons is equal to unity. It is important to consider the problem of the angular momentum of the electromagnetic field in general and of the radiation field in particular when one expands the field in spherical waves (such waves are emitted by electrical and magnetic multipoles, including dipoles). We refer to the books by Heitler (1947) and by Berestetskii, Lifshitz and Pitaevskii (1971) for a detailed discussion of the angular momentum of the radiation field.

Let us now once again consider the complete system consisting of field and charge. In the non-relativistic approximation we have for the Hamiltonian of the system:

$$\hat{\mathcal{H}} = \frac{1}{2m}\left(\hat{\mathbf{p}} - \frac{e}{c}\mathbf{A}\right)^2 + e\phi + \hat{\mathcal{H}}_{tr} , \tag{1.63}$$

and the equation for the wavefunction takes its usual form:

$$i\hbar \frac{\partial\Psi}{\partial t} = \hat{\mathcal{H}}\Psi \ . \tag{1.64}$$

In order to be able to apply perturbation theory for solving this problem we write the Hamiltonian in the following form

$$\hat{\mathcal{H}} = \hat{\mathcal{H}}_0 + \hat{\mathcal{H}}'$$

$$\hat{\mathcal{H}}_0 = \frac{\hat{p}^2}{2m} + e\phi + \hat{\mathcal{H}}_{tr} \tag{1.65}$$

$$\hat{\mathcal{H}}' = -\frac{e}{mc}(\hat{p}\cdot\hat{\mathbf{A}}) + \frac{e^2}{2mc^2}\hat{\mathbf{A}}^2$$

where $\hat{\mathcal{H}}'$ is considered to be a perturbation.

We note that $\hat{\mathbf{p}}$ does in general not commute with $\hat{\mathbf{A}}$ and we should therefore write in the expression for $\hat{\mathcal{H}}'$: $e\{(\hat{\mathbf{p}}\cdot\hat{\mathbf{A}})+(\hat{\mathbf{A}}\cdot\hat{\mathbf{p}})\}/2mc$ rather than $e(\hat{\mathbf{p}}\cdot\hat{\mathbf{A}})/mc$, as it is just the former expression which is Hermitean. However, in the case of a transverse field, when $\operatorname{div}\mathbf{A}=0$, the terms with $(\mathbf{p}\cdot\mathbf{A})$ and $(\mathbf{A}\cdot\mathbf{p})$ are equal to one another.

The effects which are connected with the interaction of light with electrons are proportional to the 'electromagnetic interaction constant' which is the so-called 'fine structure constant'

$$\alpha = \frac{e^2}{\hbar c} = \frac{1}{137.036}\,. \tag{1.66}$$

As $\alpha \ll 1$, the interaction of an electron with the electromagnetic field is weak in a well understood sense†. We shall therefore assume that the wavefunction of a stationary state must differ little from the solution of the equation

$$\hat{\mathcal{H}}_0\psi_{n0} = E_{n0}\psi_{n0}\,, \tag{1.67}$$

which can be found, at least, in a number of cases. In particular, when there is no external electromagnetic field

$$\psi_{n0} = \exp\left[i\frac{(\mathbf{p}\cdot\mathbf{r})}{\hbar}\right]\prod_{\lambda,i}\psi_{n\lambda i}(q_{\lambda i}) \tag{1.68}$$

That the difference between the exact wavefunction and the solution of Eq.(1.67) is small is, for instance, clear from considering electrons in excited levels in the hydrogen atom. The lifetime of an electron in an excited level is of the order of 10^{-9}s and the time for completing one orbit is of the order of 10^{-15} s. If we use the uncertainty relation for the energy,

$$\Delta E\,\Delta t \sim \hbar\,, \tag{1.69}$$

† This statement refers basically to radiative effects and it does, of course, not mean that one can always consider the electromagnetic interactions to be perturbations. It is sufficient to note that the electrostatic interaction of an electron with a nucleus with charge Ze is characterized by the parameter $Ze^2/\hbar v$, where v is the electron velocity. For a hydrogen atom in its ground state $e^2/\hbar v$ 1.

we find thus that the level width $\Delta E \sim 10^{-6}$ eV, while the distance between the levels is of the order of 1 eV. The motion of the electrons in the excited levels of the hydrogen atom is thus 'quasi-stationary' and differs little from the motion which would occur, if the electrons did not interact with the radiation field. The reason for this is that, as we have already mentioned, the fine-structure constant $\alpha = e^2/\hbar c$ is much less than unity. If such a constant were of the order of unity — as is the case for the interaction between a nucleon and the meson field — the width of a level would be of the order of the distance between the levels themselves and it would be totally impossible to speak, in general, of a 'quasi-stationary' motion†.

As the functions ψ_{n0} from (1.67) form a complete set we can write Ψ in Eq. (1.64) in the form

$$\Psi = \sum_m b_m(t)\, \psi_{m0}(\mathbf{r}) \exp\left(-\frac{iE_{m0}t}{\hbar}\right) . \tag{1.70}$$

Substituting (1.70) into Eq.(1.64) with the Hamiltonian (1.65), multiplying both sides of the equation by ψ^*_{n0}, integrating over the whole of space, and using the orthonomality of the functions ψ_{n0} we get

$$\left.\begin{aligned} i\hbar \frac{db_n(t)}{dt} &= \sum_m \mathcal{H}'_{nm} b_m(t) \exp\left[\frac{i}{\hbar}(E_{n0} - E_{m0})t\right], \\ \mathcal{H}'_{nm} &= \int \psi^*_{n0}\, \hat{\mathcal{H}}'\, \psi_{m0}\, d^3\mathbf{r} \quad . \end{aligned}\right\} \tag{1.71}$$

Let us assume that at $t = 0$ we have $b_k = 1$ and $b_{n\neq k} = 0$. If we assume that the b_n with $n \neq k$ are small at all times and drop higher-order terms, we then have

$$i\hbar \frac{db_n(t)}{dt} = \mathcal{H}'_{nk} \exp\left[\frac{i}{\hbar}(E_{n0} - E_{k0})t\right] , \tag{1.72}$$

whence we easily get

$$|b_n(t)|^2 = \frac{2|\mathcal{H}'_{nk}|^2}{(E_{k0} - E_{n0})^2} \left\{1 - \cos\left[\frac{(E_{k0} - E_{n0})t}{\hbar}\right]\right\} . \tag{1.73}$$

---

† In this respect there is a considerable difference between quantum and classical theory. In classical theory relatively strong perturbations may not lead to any qualitative changes in the nature of the motion. For instance, the properties of a free oscillator in thermal equilibrium and those of an oscillator in a dense gas which strongly interacts with the oscillator are very close to one another. On the other hand, in quantum theory the properties of the motion are significantly changed when the perturbation leads to the width of a level to become of the order of the distance between levels (see Chapter 13 of the present book).

If in the first order of perturbation theory $|b_n(t)|^2 = 0$, we can use a similar procedure to find the next approximation. For instance, the matrix element is in second approximation given by

$$\mathcal{H}'^{(2)}_{nk} = \sum_{n'} \frac{\mathcal{H}'_{nn'}\mathcal{H}'_{n'k}}{E_{k0} - E_{n0}} \quad . \tag{1.74}$$

Equation (1.73) determines the probability for a transition to only a single final state with energy E_{n0}. We are usually interested in the transition to any of all possible states, that is, in the integral

$$\int |b_n(t)|^2 \, \rho(E_{n0})dE_{n0} \quad . \tag{1.75}$$

In this equation $\rho(E_{n0})dE_{n0}$ is the number of final states — which are assumed to be 'densely' distributed — in the energy range from E_{n0} to $E_{n0} + dE_{n0}$. As t tends to infinity, the integral (1.75) equals (see Eq.(1.84) or for details, Heitler's book (1947)),

$$\frac{2\pi}{\hbar} |\mathcal{H}'|^2_{E=E_{k0}} \, \rho(E_{k0})t \; , \tag{1.76}$$

and the probability for a transition per unit time is thus given by the formula

$$W = \frac{1}{t} \int |b_n(t)|^2 \, \rho(E_{n0})dE_{n0} = \frac{2\pi}{\hbar} |\mathcal{H}'|^2 \, \rho(E_{k0}) . \tag{1.77}$$

Note that the transition occurs only if there are states E_{n0} which are arbitrarily close to E_{k0}; this was reflected in (1.76). It is clear from what we have said earlier that when one evaluates the matrix elements $\mathcal{H}'_{nn'}$, one must use Eqs.(1.65), and also (1.18), (1.19), and (1.40) and understand by q the operators $q_{\lambda i}$ (for details see Heitler's book (1947)).

Using perturbation theory in such a simple or in a somewhat more complicated form enables us to find the answers to a whole set of problems in radiation theory (Heitler, 1947; Berestetskii, Lifshitz and Pitaevskii, 1971). However, a wider application of perturbation theory encounters considerable difficulties which is formally reflected in the appearance of divergent (infinite) expressions. The appearance of divergent expressions is connected with the assumption that the electron is a point particle, with the fact that the field has an infinite number of degrees of freedom, and so on. Some of those difficulties are not caused by the quantization and are classical in nature. It is sufficient to remember that the electrostatic energy of a point charge is infinite. Even in classical electrodynamics one had learned to avoid such difficulties. In particular, one used for this the mass 'renormalization' method[†]. In quantum electrodynamics one also renormalizes the charge of the particle and the situation altogether becomes more complicated. The study of the corresponding field of problems was for a long time in the centre of the attention of theoretical physics. As a result a great deal of progress was made and the infinities in quantum electrodynamics are practically made 'harmless'. A framework was developed in which one is able to find the answers to problems which may arise and, in particular, in which one can take extremely minute radiative effects into account (Heitler, 1947; Berestetskii, Lifshitz and Pitaevskii, 1971).

The present text is not concerned with any of these problems, although the material discussed a moment ago may, we hope, be useful for understanding the physical basis of quantum electrodynamics. It was only important for us in the plan for our further exposition to formulate the Hamiltonian method in classical electrodynamics and to get acquainted with the most elementary aspects of quantum electrodynamics.

It is interesting that the Hamiltonian method has in the past hardly ever been applied in classical electrodynamics; the method became popular only when one changed to quantum electrodynamics. However, as so often happens, there was a 'feedback' afterwards. To be precise, it became clear that the Hamiltonian method is very convenient also for a number of classical problems, especially when there is a medium present (see Chapters 6 and 7). Recently, when many problems turned out to be already solved, when there appeared new and more complicated problems and, when moreover, a number of powerful mathematical methods, such as the diagram technique or the Green function method, were developed and started to be widely applied, the Hamiltonian method dropped out of sight both in the quantum and in the classical radiation theory. We are, however, convinced that the Hamiltonian method nevertheless retains the advantage of clarity, simplicity, and rather large universality which makes its exposition and use very expedient, at least, for pedagogical purposes.

As an illustration we use the Hamiltonian formalism to discuss the problem of the radiation by an oscillator — a harmonically oscillating charge. To determine the field we must find the quantities $q_{\lambda i}$ in the expansion (1.18), while

---

† As far as we known the term 'mass renormalization' itself only appeared when the corresponding operations were carried out in quantum electrodynamics. This led to the assumption that the renormalization method was a product of quantum theory but this is, to say the least, incorrect (see Chapter 2 and Kramers 1944).

the equations of motion for these quantities have the form (1.24)

$$\ddot{q}_{\lambda i} + \omega_\lambda^2 q_{\lambda i} = e\sqrt{8\pi}\,(\mathbf{e}_\lambda \cdot \mathbf{v}(t)) \begin{cases} \cos(\mathbf{k}_\lambda \cdot \mathbf{r}) , \\ \sin(\mathbf{k}_\lambda \cdot \mathbf{r}) , \end{cases} \tag{1.24}$$

where $\mathbf{r} = \mathbf{r}(t)$ is the position vector of the emitting charge e; in the case of an oscillator we have

$$\mathbf{r}(t) = \mathbf{a}_0 \sin\omega_0 t \quad , \quad \dot{\mathbf{r}}(t) \equiv \mathbf{v} = \mathbf{v}_0 \cos\omega_0 t = \mathbf{a}_0\,\omega_0 \cos\omega_0 t \quad . \tag{1.78}$$

The argument $(\mathbf{k}_\lambda \cdot \mathbf{r}(t))$ in (1.24) is small, if

$$a_0 \ll \frac{1}{k_0} = \frac{\lambda_0}{2\pi} , \tag{1.79}$$

where λ_0 is the wavelength of the emitted radiation. Let us accept condition (1.79), that is, we shall assume that the amplitude of the oscillations of the charge is much smaller than the wavelength of the radiation; this is always valid for a non-relativistic oscillator, as the velocity $v_0 = \omega_0 a_0 \ll c$, while $\lambda_0 = 2\pi c/\omega_0$. In that case $(\mathbf{k}_\lambda \cdot \mathbf{r})$ in (1.24) is much less than unity and we are justified in putting $\cos(\mathbf{k}_\lambda \cdot \mathbf{r}) = 1$, $\sin(\mathbf{k}_\lambda \cdot \mathbf{r}) = 0$. Therefore $q_{\lambda 2} = 0$ and for $q_{\lambda 1}$ we get the equation

$$\ddot{q}_{\lambda 1} + \omega_\lambda^2 q_{\lambda 1} = e(\mathbf{e}_\lambda \cdot \mathbf{v}_0)\sqrt{8\pi}\cos\omega_0 t . \tag{1.80}$$

The solution of this equation which satisfies the boundary condition $q_{\lambda 1} = 0$, $\dot{q}_{\lambda 1} = 0$, when $t = 0$ has the form

$$q_{\lambda 1} = \frac{b_\lambda}{\omega_\lambda^2 - \omega_0^2}[\cos\omega_0 t - \cos\omega_\lambda t] \quad , \quad b_\lambda = e(\mathbf{e}_\lambda \cdot \mathbf{v}_\lambda)\sqrt{8\pi} . \tag{1.81}$$

Having obtained the $q_{\lambda 1}$ we have thereby completely determined the electromagnetic field and we can now evaluate all other quantities. Let us, for instance, find the energy emitted by the charge (oscillator) per unit time. To do this we must clearly use Eq.(1.22) to calculate the field energy $\mathcal{H}_{tr}$ and then find its rate of change, that is, $d\mathcal{H}_{tr}/dt$. That will just be the energy emitted by the oscillator per unit time.

For $\mathcal{H}_{tr}$ we get the expression

$$\mathcal{H}_{tr} = \sum_\lambda \left\{ b_\lambda \omega_\lambda^2 \frac{[1 - \cos(\omega_\lambda - \omega_0)t]}{(\omega_\lambda^2 - \omega_0^2)^2} + \ldots \right\} . \tag{1.82}$$

We have only written down within the braces the term which leads to an increase of $\mathcal{H}_{tr}$ with time; the other terms which are omitted in (1.82) do not contribute to the expression for the energy emitted by the oscillator per unit time for large t (we assume here that t is large).

For the evaluation of the sum in (1.82) it is convenient to change from a summation to an integration. We must then first multiply expression (1.82) by the number of field oscillators with frequencies between ω and $\omega + d\omega$; this number is equal to

$$\frac{\omega^2 d\omega d\Omega}{(2\pi c)^3}, \tag{1.83a}$$

where $d\Omega$ is an element of solid angle.

The transition from a sum to an integration is thus reduced to the substitution

$$\sum_\lambda \longrightarrow \int \frac{1}{2(2\pi c)^3} \quad \ldots \; \omega^2 \, d\omega \, d\Omega \tag{1.83b}$$

where the appearance of an extra factor $\frac{1}{2}$ is connected with the fact that we integrate over all directions, rather than over a hemisphere of **k**-directions.

We can easily integrate expressions (1.82) over ω if we use the following relation which is valid for large values of t:

$$\int_{-\infty}^{+\infty} f(\omega) \frac{[1 - \cos(\omega - \omega_0)t]}{(\omega - \omega_0)^2} d\omega = \pi f(\omega_0) t \, . \tag{1.84}$$

As a result of all these simple calculations we get an expression for the energy emitted by the oscillator per unit time into a solid angle $d\Omega$:

$$\frac{d\mathcal{H}_{tr}}{dt} = \frac{\mathcal{H}_{tr}}{t} = \frac{e^2 a_0^2 \omega_0^4}{8\pi c^3} \sin^2\theta \, d\Omega \, , \tag{1.85}$$

where θ is the angle between the direction of the oscillations $\mathbf{a}_0$ and the wavevector $\mathbf{k}_0$, where $k_0 = \omega_0/c$.

The emission which we have just determined — that is, that part of the field which increases proportional to the time t — arises when the frequency of the 'force' which occurs on the right-hand side of Eq.(1.24) is equal to the eigenfrequency of the field oscillator $\omega_\lambda = ck_\lambda$. In this respect the harmonically oscillating charge is completely typical even though in the dipole approximation (condition (1.79)) which we considered there is emission at only one frequency, ω_0. We note that in the quantum theory of radiation the situation is completely analogous in the perturbation theory framework (compare (1.82) and (1.73); for details see Heitler, 1947).

It is convenient at this stage to elucidate a few important aspects which usually are kept out of sight by discussing the somewhat rhetorical question: can a uniformly moving electron radiate?

The standard, one might say automatic, answer to that question is negative. In actual fact, however, one can make many reservations; many of them are non-trivial and have far from always been taken into consideration – which has led to paradoxes and mistakes.

Firstly, one must be precise about the frame of reference in which the electron — of course, it need not be an electron, but can be any charge, which we call 'electron' for the sake of convenience – moves uniformly, that is, with a constant velocity $\mathbf{v}$. Usually, unless a statement to the contrary is made explicitly, one is dealing with motion in inertial frames of reference. The initial field equations were written down just in such frames and we were dealing only with such frames of reference. It is clear that if the electron moves uniformly in a non-inertial frame of reference, it is accelerated relatively to an inertial system and will radiate.

Secondly, one considers uniform motion in vacuo and not in a medium. An electron moving uniformly in a medium can emit both Cherenkov and transition radiation (see Chapters 6 and 7).

Thirdly, one assumes that the electron velocity $v < c$ (the velocity of light in vacuo). Often this condition is considered to be more or less trivial, but that is not the case. The requirement of relativistic invariance does not at all lead to the condition $v < c$ and, in particular, Eqs.(1.1) are completely valid (and relativistically invariant) also when $v > c$. It is true that a particle of rest mass m can not be accelerated to a velocity $v \geq c$, as can readily be seen from the expression for the particle energy $\mathcal{E} = mc^2/(1 - v^2/c^2)^{\frac{1}{2}}$. However, this does not exclude the possibility to consider particles (tachyons) which always move with a velocity $v > c$ and an energy $\mathcal{E} = imc^2/(1 - v^2/c^2)^{\frac{1}{2}} = mc^2/(v^2/c^2 - 1)^{\frac{1}{2}}$. The difficulties which in reality arise when one considers motions with velocities $v > c$ are connected with the possible violations of the causality principle: it was just this fact and not the violation of relativistic invariance which led to the requirement that $v < c$ (see Einstein, 1907; Pauli, 1958). It is therefore probable that tachyons, which recently have been discussed extensively in the physical literature, can not exist. However, sources of radiation (even if not separate particles) moving with velocities $v > c$ all the same do exist. We shall discuss them in Chapter 8.

After this discussion we shall state the problem more precisely: can an electron moving in vacuo in an inertial frame of reference with a constant velocity $v < c$ radiate? One can give at least four arguments that an electron under such conditions does not radiate.

The first argument, and in a certain sense the most consistent one, is connected with the solution of the field Eqs.(1.1) for the case when $\mathbf{v}$ = constant. Such a solution (see, e.g., Heitler, 1947; Pauli, 1958; Landau and Lifshitz, 1975 and Chapter 3) indicates that the radiation field, that is, the field which decreases as $1/R$ and produces an energy flux at infinity, does not appear in the case under discussion; at the same time it also becomes clear that radiation would appear if $v > c$.

The second argument does not require any calculations. We can change to a frame of reference in which the electron is at rest, which can always be done when $\mathbf{v}$ = constant, $v < c$. In such a frame there is clearly no radiation: the electron is all the time at rest[†]. However, if we change from one inertial frame to another no radiation can appear, and thus there is none either for v = constant. A well known weakness of this argument is connected with the fact that it can apparently also be applied to the case $v > c$ and thus 'prove' that there is also no radiation in that case. On the other hand, a charge must in the case $v > c$ emit Cherenkov radiation even in vacuo (this is, incidentally, one of the difficulties in the theory of tachyons; see also Chapter 8). The solution of the ensuing paradox is that it is impossible to realize a frame of reference in which the electron is at rest when $v > c$.

The third argument is connected with the use of the energy and momentum conservation laws. The simplest way, although not necessary, is to use a quantum formulation for the discussion. If we consider a particle with energy $\mathcal{E} = (m^2c^4 + c^2p^2)^{\frac{1}{2}}$ and momentum $\mathbf{p}$ we can verify that the energy and momentum conservation laws do not allow such a particle to emit a photon with energy $\hbar\omega$ and momentum $\hbar\mathbf{k}$, $k = \omega/c$. In fact we shall use such an argument for discussing in Chapter 7 the conditions for radiation in a medium. Incidentally, in the framework of such an approach radiation in vacuo is only impossible when $v < c$; for tachyons with an energy $\mathcal{E} = (-m^2c^4 + c^2p^2)^{\frac{1}{2}}$ the velocity is

---

[†] In fact, it was presupposed here that the electron will be at rest at all times in some frame of reference: we need therefore give a more detailed discussion, such as, for instance, the following one. Let us assume that the electron (a free charged particle) moves with a velocity v = constant, $v < c$, and does not radiate. We can then show that such a solution is compatible with the field equations when we change over to an inertial frame of reference in which the electron is at rest, its field is the electrostatic field, and in which there is no radiation. In fact, this argument differs from the preceding one only in that the solution of the field equations for a particle at rest are simpler than for a moving particle and may be assumed to be well known.

$$v = \frac{\partial \mathcal{E}}{\partial p} = \frac{c^2 p}{\sqrt{c^2 p^2 - m^2 c^4}} > c \; ,$$

and the conservation laws do not prevent the emission of photons by a uniformly moving particle.

The fourth argument is connected with the use of the Hamiltonian method. For a uniformly moving electron $\mathbf{r}(t) = \mathbf{v}t$ and $(\mathbf{k}_\lambda \cdot \mathbf{r}) = (\mathbf{k}_\lambda \cdot \mathbf{v})t$, that is, the frequency $(\mathbf{k}_\lambda \cdot \mathbf{v})$ occurs on the right-hand side of the equations of motion for the field oscillators (see (1.24)). On the other hand, the eigenfrequency of the field oscillator in vacuo is $\omega_\lambda = ck_\lambda$ and hence, resonance can not occur when $v < c$ and thus radiation with an energy $\mathcal{H}_{tr}$ which increases with time is also impossible.

Under the provisos made (inertial frame, vacuum, velocity $v < c$) a uniformly moving electron will therefore not radiate.

The whole of this problem would not have been discussed here, probably, if what we have said were all there was. However, this is not the case. In fact, in the first period of the development of quantum electrodynamics a paradoxical statement was made that in quantum theory a uniformly moving electron nevertheless radiates. This conclusion can easily be reached as the result of simple calculations in the framework of first-order perturbation theory. Indeed, let the electron at $t = 0$ move uniformly with momentum $\mathbf{p}$ = constant and let the energy of the transverse field be equal to zero — that is, let all field oscillators be in their ground state, that is, all $n_{\lambda i} = 0$. When the interaction, say the term $\mathcal{H}' = -e(\mathbf{p} \cdot \mathbf{A})/mc$ (see (1.65)), is taken into account, the matrix elements corresponding to transitions to quantum states with $n_{\lambda i} = 1$ are non-vanishing (see (1.40)) and thus the probability $|b_n(t)|^2$ in (1.73) is also non-vanishing. It is true that the field energy will not increase with time (as $t \to \infty$) but all the same some radiation turns up. We shall not give here a detailed quantum calculation as the effect discussed here is purely classical (Ginzburg, 1939a,b). Indeed, let us state the problem exactly as before, but in classical electrodynamics in the framework of the Hamiltonian method. We shall look for a solution for the field oscillators $q_{\lambda i}$ — and thereby for the transverse field itself — for the case of a uniformly moving charge when in (1.24) $\mathbf{r} = \mathbf{v}t$ and under the conditions when at $t = 0$ all $q_{\lambda i} = 0$ and all $\dot{q}_{\lambda i} = 0$, that is the field vanishes. To simplify the discussion we shall limit ourselves to the case where $(\mathbf{k}_\lambda \cdot \mathbf{r}(t)) = (\mathbf{k}_\lambda \cdot \mathbf{v})t \ll 1$ (this is, in fact, completely unimportant); in other words, we shall assume that either the time t is rather small or that the wavelength $\lambda = 2\pi k$ is long. We can then write Eqs.(1.24) in the following form:

$$\ddot{q}_{\lambda 1} + \omega^2 q_{\lambda 2} = e\sqrt{8\pi}(\mathbf{e}_\lambda \cdot \mathbf{v}) \quad , \quad \ddot{q}_{\lambda 2} + \omega^2 q_{\lambda 2} = 0 \ . \tag{1.86}$$

The solution of these equations which satisfies the above-mentioned conditions has the form

$$q_{\lambda 1} = \frac{e\sqrt{8\pi}}{\omega_\lambda^2} (\mathbf{e}_\lambda \cdot \mathbf{v})(1 - \cos\omega_\lambda t) \quad , \quad q_{\lambda 2} = 0 \ . \tag{1.87}$$

Substituting this solution into (1.22) we get

$$\mathcal{H}_{tr} = 8\pi e^2 \sum_\lambda \frac{(\mathbf{e}_\lambda \cdot \mathbf{v})^2}{\omega_\lambda^2} (1 - \cos\omega_\lambda t) \ . \tag{1.88}$$

Changing from a sum to an integral (see (1.83)) and integrating over the angles and also over the frequency from $\omega = 0$ to some maximum value ω_{max} we have

$$\mathcal{H}_{tr} = \frac{8e^2}{3\pi c^2} (\tfrac{1}{2} v^2) \left\{\omega_{max} - \frac{\sin\omega_{max} t}{t}\right\} . \tag{1.89}$$

Exactly the same result is obtained in quantum theory. Both this fact and the absence of the quantum constant $\hbar$ in (1.89) removes any doubts as to whether there is a quantal element in the problem discussed here. The important fact is clearly the following. The electron is assumed to be moving uniformly, but at $t = 0$ we assume that there is no transverse field. On the other hand, a uniformly moving electron — if such is its motion at all times — is surrounded by its own field which it carries along and this includes a (magnetic and electric) transverse field. We assume that this field is absent at $t = 0$ and is after that, for $t > 0$, described by the field equations; this means, in fact, that the electron was at rest up to the time $t = 0$ and was accelerated instantaneously at $t = 0$ to reach its velocity $\mathbf{v}$. It is completely natural that as a result the charge radiates, firstly, while it 'accumulates' ('gets dressed by') its own accompanying field and, secondly, it emits the 'true' radiation which goes to infinity and is caused by the acceleration of the charge. This part of the transverse field — the radiation field — is in (1.87) given by the term which is proportional to $\cos\omega_\lambda t$ and corresponds to the solution of the homogeneous Eq.(1.86). From this it is already clear that we are talking about the radiation of a free field — or, in quantum language, about the emission of photons. If the interaction is switched on sufficiently slowly (adiabatically) and the electron is, physically speaking, accelerated slowly, the free field does not appear and only the entrained field is formed with its energy (for $v \ll c$)

$$\mathcal{H}_{tr} = \frac{4e^2\omega_{max}}{3\pi c^2} (\tfrac{1}{2} v^2) \ .$$

The frequency $\omega_{max} = 2\pi c/\lambda_{min}$, where λ_{min} is the shortest wavelength. For an extended charge of radius r_0 we have clearly $\lambda_{min} \sim r_0$ and $\mathcal{H}_{tr} = \frac{1}{2} m_{em} v^2$, where the electromagnetic mass $m_{em} \sim e^2/r_0 c^2$, as should be the case.

If we write down the conditions under which a uniformly moving electron does not radiate, we must thus add the requirement of stationarity or, in other words, the requirement that the motion be uniform in the whole of the range $-\infty < t < \infty$. In general such a requirement is obvious and to some extent is always understood to be satisfied, but it turned out to be somewhat hidden in the quantum calculations which we mentioned. The use of the Hamiltonian method and, chiefly, the statement of the problem which is connected with it and which is analogous to the quantal approach with a switching on of the interaction at $t = 0$ enabled us to dispose of the paradox and to make the crux of the matter completely clear. At the same time it became clear — and this is useful to bear in mind also when solving completely real problems (see Ginzburg, 1939a,b; Feinberg, 1966,1972) — that the field of a uniformly moving charge is not at all necessarily a stationary one. In other words, the charge may already have moved for some time uniformly, but the field entrained by it may still differ from the stationary field — which exists when the motion with a constant velocity has been going on for a sufficiently long time. It is particularly clear also from the example considered by us of an electron which moves uniformly for $t \geq 0$ that there exists the already mentioned difference between a free radiation field and the transverse entrained field. We need therefore not identify any arbitrary transverse electromagnetic field with a collection of photons (and it is neither strange that it became the literal norm to forget this obvious fact when the quantum theory of radiation was expounded). We note that the actual construction of quantum electrodynamics, as is already clear from the transition (quantization) from classical to quantum electrodynamics which we made earlier, is completely free from any assumption about the absence of charges and is in no way connected with any identification of a quantized transverse field with a free radiation field, that is, a collection of photons. It is therefore natural that in a consistent use of the quantum theory of radiation no incorrect results whatever are obtained. However, the fact that it is possible to ignore the difference between an arbitrary transverse field and a collection of photons is connected with the nature of the problems encountered in the quantum theory of radiation — in the overwhelming majority of cases one considers the field as $t \to \infty$, the field at infinity, and so on. However, as we mentioned, it is far from always possible to proceed in this manner (see Ginzburg, 1939a,b; Feinberg, 1966,1972).

Chapter II

RADIATION REACTION

Radiation reaction for translational motion of a charge.
Rotation of a magnetic moment (oblique magnetic rotator).

If there is a source of radiation such as a charge or an antenna, a radiation reaction will act upon it in general. The simplest and best known example of this is the motion of a point charge with a non-relativistic velocity which is described by the equation of motion

$$m\ddot{\mathbf{r}} = \mathbf{F}_0 + \frac{2e^2}{3c^3}\dddot{\mathbf{r}} \ , \qquad (2.1)$$

where $\mathbf{f} = (2e^2/3c^3)\dddot{\mathbf{r}}$ is the radiation reaction force (radiation force, or radiative friction force) and $\mathbf{F}_0$ is the external force which in the case when it is purely electromagnetic in nature has the form

$$\mathbf{F}_0 = e\,\mathbf{E}_0 + \frac{e}{c}\left[\dot{\mathbf{r}} \wedge \mathbf{H}_0\right] \ . \qquad (2.2)$$

The relativistic equation of motion including the radiation force (see, for instance, Chapter 3) goes over into (2.1) as $v \equiv |\dot{\mathbf{r}}| \to 0$ and there are no essential extra assumptions connected with deriving it from (2.1). A discussion of the conditions under which one can obtain and use (2.1) can also be related directly to the relativistic case (see Chapter 4).

The fact that one cannot apply Eq.(2.1) 'without thinking' is immediately clear if we assume the external force $\mathbf{F}_0$ to be equal to zero. The equation obtained then has not only the correct solution $\dot{\mathbf{r}} \equiv \mathbf{v} \equiv$ const. (uniform motion in the inertial frame of reference considered) but also a clearly incorrect 'self-accelerating' solution

$$\mathbf{v} = \mathbf{v}_0 \exp\left\{\frac{3mc^3}{2e^2}\,t\right\} ,$$

where the 'frequency' $\Omega_e = 3mc^3/2e^2 = 1.6 \times 10^{23}\ \mathrm{s}^{-1}$.

We can use Eq.(2.1) without any misgivings when the radiation force $\mathbf{f}$ is small compared to the external force $\mathbf{F}_0$,

$$f \ll F_0 \ . \qquad (2.3)$$

Under such conditions the force $\mathbf{f}$ plays the role of a perturbation; to a first approximation we have $m\ddot{\mathbf{r}} = \mathbf{F}_0$, and in the next approximation

$$m\ddot{\mathbf{r}} = \mathbf{F}_0 + \mathbf{f} \quad , \quad \mathbf{f} = \frac{2e^3}{3mc^3}\dot{\mathbf{E}}_0 + \frac{2e^4}{3m^2c^4}\left[\mathbf{E}_0 \wedge \mathbf{H}_0\right] . \tag{2.4}$$

For a harmonic force with angular frequency ω, when $\dot{E}_0 \sim \omega E_0$, condition (2.3) is equivalent to the requirement†

$$\frac{\lambda}{2\pi} = \frac{c}{\omega} \gg r_e = \frac{e^2}{mc^2} = 2.82 \times 10^{-13} \text{ cm} \quad , \tag{2.5}$$

$$H_0 \ll \frac{m^2c^4}{e^3} = 6 \times 10^{15} \text{ Oe} . \tag{2.6a}$$

We shall sometimes in which follows drop the factor 2π in inequalities such as (2.5), which is formally always correct when $\gg$ or $\ll$ signs are present. We note that we can write inequality (2.6a) also in the form

$$\frac{\lambda_H}{2\pi} = \frac{c}{\omega_H} \gg r_e \quad , \quad \omega_H = \frac{eH_0}{mc} = 1.76 \times 10^7 \, H_0(\text{Oe}) \text{ s}^{-1} \quad . \tag{2.6b}$$

The meaning of this condition is clear when we bear in mind that a non-relativistic charge in a constant magnetic field H_0 rotates with a frequency $\omega_H = eH_0/mc$ and emits electromagnetic waves with that frequency. The limitation (2.6) has no practical importance as even in pulsars the field scarcely exceeds a value of 10^{12} to 10^{13} Oe. We must, however, bear in mind that for relativistic particles the conditions for the applicability of the corresponding expression for the radiation force, even though obtained from (2.5) and (2.6), are quantitatively completely different due to the appearance of factors such as $\mathcal{E}/mc^2 = (1 - v^2/c^2)^{\frac{1}{2}}$ (see Chapter 4). We note also that we have so far not taken into account quantum restrictions. Condition (2.5) has therefore for electrons an essentially fictitious character, as classical considerations are adequate only up to

$$\lambda \gg \frac{\hbar}{mc} = 3.86 \times 10^{-11} \text{ cm} . \tag{2.7a}$$

This inequality can also be written in the form

$$\hbar\omega = \frac{2\pi c\hbar}{\lambda} \ll mc^2 = 5.1 \times 10^5 \text{ eV} . \tag{2.7b}$$

Condition (2.6) is replaced by the inequality $\lambda_H \gg \hbar/mc$ which is exactly the same as (2.7). The meaning of (2.7) is that one should be able to neglect the possibility of electron-positron pair creation — including such a creation in intermediate states.

---

†All numerical estimates are given for an electron ($e = 4.8 \times 10^{-10}$ esu (cgs), $m = 9.1 \times 10^{-28}$ g). The harmonic nature of $\mathbf{E}$ is used in (2.5) but is immaterial for (2.6).

However, whence follows condition (2.3) itself? We can scarcely answer this question better and more convincingly than by deriving the equation of motion from the initial equations, that is, from the field Eqs.(1.1) and the equation for an 'extended' charge,

$$m\ddot{\mathbf{r}} = e\int D(\mathbf{r}-\mathbf{r}')\,\mathbf{E}(\mathbf{r}')\,d^3\mathbf{r}' , \qquad (2.8)$$

where we have neglected for the sake of simplicity the action of the magnetic field ($v \to 0$ case), and where the charge density is $\rho(\mathbf{r}') = eD(\mathbf{r}-\mathbf{r}')$ with $\int D(\mathbf{r}-\mathbf{r}')d^3\mathbf{r}' = 1$, while $\mathbf{r}$ is the position of the centre of the charge; the main point, however, is that the field $\mathbf{E}(\mathbf{r}')$ in (2.8) is the total field, equal to the sum of the external field $\mathbf{E}_0$ and the self field of the charge itself, $\mathbf{E}'(\mathbf{r}')$. If the wavelength characterizing the external field satisfies condition (2.5), we can take $\mathbf{E}_0$ from under the integral sign in (2.8). As far as the self field $\mathbf{E}'$ is concerned, taking it into account, or, rather, eliminating it must just lead to an expression for the radiation force. The usual, rather cumbersome elimination of the field $\mathbf{E}'$ is, for instance, given by Heitler (1947); here, however, we shall accomplish these calculations by the Hamiltonian method (Belousov, 1939; Ginsburg, 1944a, 1946) which enables us to clarify a few facts which usually remain obscure and, principally, to go straightaway to the problem of the radiation reaction when a magnetic moment is in motion.

For the sake of convenience we repeat here, as we have done before and often shall do in what follows, the basic equations given in Chapter 1, but with some obvious changes in notation which are clear from (2.8) and from the explanatory remarks about that equation. We have

$$\mathbf{E}' = -\frac{1}{c}\frac{\partial \mathbf{A}}{\partial t} , \quad \mathbf{A} = \sum_{\lambda} \sqrt{8\pi}\, c\, \mathbf{e}_{\lambda}\left\{q_{\lambda 1}\cos(\mathbf{k}_{\lambda}\cdot\mathbf{r}) + q_{\lambda 2}\sin(\mathbf{k}_{\lambda}\cdot\mathbf{r})\right\} \qquad (2.9)$$

$$\left.\begin{aligned} \ddot{q}_{\lambda 1} + \omega_{\lambda}^2 q_{\lambda 1} &= \sqrt{8\pi}\, e\left(\mathbf{e}_{\lambda}\cdot\dot{\mathbf{r}}(t)\right)\int D(\mathbf{r}-\mathbf{r}')\cos(\mathbf{k}_{\lambda}\cdot\mathbf{r}')d^3\mathbf{r}' , \\ \ddot{q}_{\lambda 2} + \omega_{\lambda}^2 q_{\lambda 2} &= \sqrt{8\pi}\, e\left(\mathbf{e}_{\lambda}\cdot\dot{\mathbf{r}}(t)\right)\int D(\mathbf{r}-\mathbf{r}')\sin(\mathbf{k}_{\lambda}\cdot\mathbf{r}')d^3\mathbf{r}' . \end{aligned}\right\} \qquad (2.10)$$

The set (2.10) can be integrated in its general form. We shall, however, not be interested in the numerical coefficients which depend on the form factor D. We shall therefore simply assume that the integrals on the right-hand sides of (2.10) vanish for wavelengths $\lambda = 2\pi/k_{\lambda} < \lambda_{min} = 2\pi c/\omega_{max} \sim r_0$, where r_0 is the radius of the charge. Moreover, when $\omega_{\lambda} < \omega_{max}$ we can put

$$\int D\cos(\mathbf{k}_{\lambda}\cdot\mathbf{r}')d^3\mathbf{r}' = 1 \quad \text{and} \quad \int D\sin(\mathbf{k}_{\lambda}\cdot\mathbf{r}')\,d^3\mathbf{r}' = 0 .$$

This possibility is obvious for a charge at rest as then $\mathbf{r}' = 0$ (by virtue of what has been said r' differs from r by an amount of the order r_0 and this difference does not play a role when $\omega < \omega_{max}$). However, in the case considered of a slowly moving charge the validity of the approximation mentioned here — apart from the fact that it is corroborated by the results — is caused because the inequality $\omega_\lambda v/c \ll \omega_\lambda$ is satisfied. The fact is that when we substitute $\mathbf{r}' \approx \mathbf{r} = \mathbf{v}t$ a frequency $\omega_\lambda v/c$ appears on the right-hand side of Eqs.(2.10), while the eigenfrequency of the field oscillator is ω_λ. As a result the integral in the first of Eqs.(2.10) given below (see (2.13)) changes little when we take the time dependence of the right-hand sides of Eqs.(2.10) into account.

Proceeding as indicated we get from (2.8) to (2.10)

$$m\ddot{\mathbf{r}} = e\mathbf{E}_0 - e\sqrt{8\pi}\sum_\lambda \mathbf{e}_\lambda \dot{q}_{\lambda 1}, \tag{2.11}$$

$$\ddot{q}_{\lambda 1} + \omega_\lambda^2 q_{\lambda 1} = \sqrt{8\pi}\, e\left(\mathbf{e}_\lambda \cdot \dot{\mathbf{r}}(t)\right). \tag{2.12}$$

We have here also taken into account that in the given approximation we can put $q_{\lambda 2} = 0$ and $\mathbf{A} = \sqrt{8\pi}\, c \sum_\lambda \mathbf{e}_\lambda q_{\lambda 1}$. The solution of Eq.(2.12), under the condition that the particle surrounded by its entrained field is at $t = 0$ moving uniformly, has the form

$$q_{\lambda 1} = \frac{e\sqrt{8\pi}}{\omega_\lambda^2}\left(\mathbf{e}_\lambda \cdot \dot{\mathbf{r}}(0)\right)\cos\omega_\lambda t + \frac{e\sqrt{8\pi}}{\omega_\lambda}\int_0^t \left(\mathbf{e}_\lambda \cdot \dot{\mathbf{r}}(\tau)\right)\sin\omega_\lambda(t-\tau)\,d\tau\,. \tag{2.13}$$

Substituting (2.13) into (2.11), changing from a summation over λ to an integration over ω and the angles, and after that performing a few simple operations (integrating over the angles and transforming the integral) we have finally

$$\begin{aligned} m\ddot{\mathbf{r}} &= e\mathbf{E}_0(\mathbf{r}) - \frac{4e^2\omega_{max}}{3\pi c^3}\ddot{\mathbf{r}} + \frac{4e^2}{3\pi c^3}\ddot{\mathbf{r}}(0)\frac{\sin\omega_{max}t}{t} + \frac{4e^2}{3\pi c^3}\int_0^t\int_0^{\omega_{max}} \dddot{\mathbf{r}}(\tau)\cos\omega(t-\tau)\,d\omega\,d\tau \\ &= e\mathbf{E}_0(\mathbf{r}) - m_{em}\ddot{\mathbf{r}} + \frac{2e^2}{3c^3}\dddot{\mathbf{r}} + \frac{4e^2}{3\pi c^3}\ddot{\mathbf{r}}(0)\frac{\sin\omega_{max}t}{t} \\ &\quad + \text{terms which tend to zero as } \omega_{max} \sim c/r_0 \to \infty\,. \end{aligned} \tag{2.14}$$

The appearance of a term involving the electromagnetic mass $m_{em} = 4e^2\omega_{max}/3\pi c^3 \sim e^2/r_0c^2$ which tends to infinity as $r_0 \to 0$ indicates the necessity to renormalize the mass even in classical theory. The renormalization reduces to the fact that the total mass $m + m_{em}$ is identified with the observable particle mass†. The radiation force $\mathbf{f} = (2e^2/3c^3)\dddot{\mathbf{r}}$ is independent of r_0 and thus of

† see footnote on next page

any assumptions about the structure of the charge. Howev^r, this force is not the only reactive force. Firstly, there is the extra term proportional to $(\sin \omega_{max} t)/t$ which is important for small t. It is completely clear that it is not allowable to integrate Eq.(2.1) simply with arbitrary boundary conditions as it is inapplicable as $t \to 0$. The problems connected with the appearance of 'self-accelerated' and other incorrect solutions do therefore not occur (for details see Markov, 1946; Fradkin, 1950). Secondly, the terms of order (r_0/λ), $(r_0/\lambda)^2$, and so on, which we have not written down in (2.14) are small compared to f only if conditions such as (2.5) are satisfied which leads to the requirement (2.3). Thus, just the fact that one considers the radiation force $\mathbf{f}$ to be a perturbation allows us to use the equation of motion (2.1) without qualms; the meaning of this equation and the limits of its applicability are quite obvious.

A lot of trouble has been taken to prove that it is possible to apply Eq.(2.1) and its relativistic generalization as an exact equation where one imposes some additional conditions, for instance, to remove the 'self-accelerated' solutions. Apart from anything else it is obscure why all this is necessary. We do not know any classical problems where the radiation force (in the frame of reference in which the electron is at rest) can not be considered to be a perturbation. The idea of a classical point particle is inconsistent, and the construction of 'elementary' particles must be solved including quantum effects.≠

The analysis of the problem of the radiation reaction allows us once more to emphasize the difference between the velocity of a particle $\mathbf{v} \equiv \dot{\mathbf{r}} = \{\mathbf{p} - (e\mathbf{A}/c)\}/m$ and its generalized momentum $\mathbf{p}$ when there is a field present — or rather, when the field is taken into consideration, as there is always a transverse field when there is a moving charge. Indeed, Eqs.(2.11) to (2.14) were obtained neglecting the dependence of the vector potential $\mathbf{A}$ of the self field on the position $\mathbf{r}$. At the same time we have in the general case (see (1.31))

$$\left.\begin{aligned} \dot{\mathbf{p}} &= -\frac{\partial \mathcal{H}}{\partial \mathbf{r}} = -\,\mathrm{grad}\left\{\left(\mathbf{p} - \frac{e}{c}\int \mathbf{A}(\mathbf{r}',t)\, D(\mathbf{r}-\mathbf{r}')\, d^3\mathbf{r}'\right)^2 \Big/ 2m\right\}\,, \\ \dot{\mathbf{r}} &= \frac{\partial \mathcal{H}}{\partial \mathbf{p}} = \frac{1}{m}\left\{\mathbf{p} - \frac{e}{c}\int \mathbf{A}(\mathbf{r}',t)\, D(\mathbf{r}-\mathbf{r}')\, d^3\mathbf{r}'\right\}\,, \end{aligned}\right\} \tag{2.15}$$

† To proceed in this way is not only possible, but also necessary as the 'bare' mass m and the electromagnetic mass m_{em} nowhere appear or can be measured separately — this statement is valid in all cases where the characteristic frequency ω of the external field is very small compared to ω_{max}, as is assumed to be the case.

≠ In this connection we refer to a paper by Moniz and Sharp (1977) in which the classical Eqn.(2.1) was, in fact, obtained by taking the classical limit in a quantum-electrodynamical discussion of the radiative reaction on a charge moving with a non-relativistic velocity.

where we have neglected the action of the external field $\mathbf{F}_0$. It is clear that if $\mathbf{A}(\mathbf{r}',t)$ is independent of $\mathbf{r}'$ the reaction of the radiation field on the momentum disappears completely — that is, $\dot{\mathbf{p}}=0$, or, if the external field is taken into account, $\dot{\mathbf{p}}=\mathbf{F}_0$ — while it remains for the velocity and is described by Eq.(2.1) within the limits indicated. One cannot consider this result to be a paradox as the momentum of a charged particle is a rather complicated structure: $\mathbf{p}=m\mathbf{v}+e\mathbf{A}/c$. At the same time it is just the particle momentum which in quantum theory stands in the foreground; the fact that for a charged particle it differs from $m\mathbf{v}$ must be borne in mind when one compares classical and quantum expressions.

We now turn to the problem of the radiation reaction on a magnetic moment, that is, in the framework of the classical theory on a rigid body or on a linked set of particles which has a magnetic moment $\boldsymbol{\mu}$. Recently an analysis of this problem which arose in the discussion of models for elementary particles (Ginzburg, 1944a, 1946) became of particular interest in connection with pulsars which are in a well understood approximation just rotating magnetic dipoles or, as it is sometimes expressed, oblique magnetic rotators (Ostriker and Gunn, 1969; Michel and Goldwire, 1970; Davis and Goldstein, 1970; Ginzburg, 1971). Let us therefore consider a body (top or rotator) with an angular momentum $\mathbf{M}$ and a magnetic moment $\boldsymbol{\mu}$, while the magnetization of the rotator is $\mathfrak{m}=\boldsymbol{\mu}D(\mathbf{r})$, $\int D(\mathbf{r})\,d^3\mathbf{r}=1$; the centre of mass of the rotator is at the point $\mathbf{r}=0$ and is assumed to be fixed.

The set of equations of motion for the angular momentum $\mathbf{M}$ and the field equations are under those conditions the following ones:

$$\dot{\mathbf{M}}=[\boldsymbol{\mu}\wedge\mathbf{H}_0]+[\boldsymbol{\mu}\wedge\mathbf{H}'(\mathbf{r})]\,D(\mathbf{r})\,d^3\mathbf{r}\quad, \tag{2.16}$$

$$\mathbf{A}=-4\pi\,\mathrm{curl}\,\mathfrak{m}=4\pi[\boldsymbol{\mu}\wedge\nabla D]\;,\quad \mathbf{H}'=\mathrm{curl}\,\mathbf{A}\;, \tag{2.17}$$

where the external field $\mathbf{H}_0$ is assumed to be uniform within the limits of the rotator and the latter is for the present assumed to be uncharged. Using the expansion (2.9) for $\mathbf{A}$ we get from (2.17)

$$\left.\begin{aligned}\ddot{q}_{\lambda1}+\omega_\lambda^2 q_{\lambda1}&=-\sqrt{8\pi}\,c\int(\mathbf{e}_\lambda\cdot[\boldsymbol{\mu}\wedge D(\mathbf{r})])\cos(\mathbf{k}_\lambda\cdot\mathbf{r})\,d^3\mathbf{r}\quad,\\ &=-\sqrt{8\pi}\,c(\mathbf{e}_\lambda\cdot[\boldsymbol{\mu}\wedge\mathbf{k}_\lambda])\int D(\mathbf{r})\sin(\mathbf{k}_\lambda\cdot\mathbf{r})\,d^3\mathbf{r},\\ \ddot{q}_{\lambda2}+\omega_\lambda^2 q_{\lambda2}&=\sqrt{8\pi}\,c(\mathbf{e}_\lambda\cdot[\boldsymbol{\mu}\wedge\mathbf{k}_\lambda])\int D(\mathbf{r})\cos(\mathbf{k}_\lambda\cdot\mathbf{r})\,d^3\mathbf{r},\end{aligned}\right\} \tag{2.18}$$

We could in (2.18) integrate by parts, that is, change from ∇D to D, because the function D is assumed to be non-vanishing only in a region with dimensions of the order of the rotator radius r_0; for the same reason we can put

$$\int D \sin(\mathbf{k}_\lambda \cdot \mathbf{r})\, d^3\mathbf{r} = 0 \quad \text{and} \quad \int D \cos(\mathbf{k}_\lambda \cdot \mathbf{r})\, d^3\mathbf{r} = 1,$$

when $\omega_\lambda < \omega_{max} \sim 2\pi c/r_0$ and

$$\int D \cos(\mathbf{k}_\lambda \cdot \mathbf{r}) d^3\mathbf{r} = 0$$

when $\omega_\lambda > \omega_{max}$. Of course, it is in this way not possible to evaluate exactly the coefficients which contain r_0 and which in general depend on the form factor D.

If we proceed as indicated we may assume that $q_{\lambda 1} = 0$, while we can use for $q_{\lambda 2}$ the equation (see (2.18))

$$\ddot{q}_{\lambda 2} + \omega_\lambda^2 q_{\lambda 2} = \sqrt{8\pi}\, c\, (\mathbf{e}_\lambda \cdot [\boldsymbol{\mu}(t) \wedge \mathbf{k}])\ , \quad \omega_\lambda < \omega_{max}\,. \tag{2.19}$$

Its solution has the form

$$q_{\lambda 2} = \frac{\sqrt{8\pi}\, c}{\omega_\lambda^2} (\boldsymbol{\mu}(0) \cdot [\mathbf{k}_\lambda \wedge \mathbf{e}_\lambda]) \cos \omega_\lambda t + \frac{c\sqrt{8\pi}}{\omega_\lambda} \int_0^t (\mathbf{e}_\lambda \cdot [\boldsymbol{\mu}(t) \wedge \mathbf{k}_\lambda]) \sin \omega_\lambda (t - \tau)\, d\tau\,, \tag{2.20}$$

where we have assumed that at $t = 0$ the field $\mathbf{H}$ corresponds to the fixed magnetic moment $\boldsymbol{\mu}(0)$.

Using (2.20) we can easily find the field

$$\mathbf{H} = \text{curl}\, \mathbf{A}\ , \quad \mathbf{A} = \sqrt{8\pi}\, c \sum_\lambda \mathbf{e}_\lambda q_{\lambda 2} \sin(\mathbf{k}_\lambda \cdot \mathbf{r})\ ,$$

and substitute it into (2.16). After some simple operations similar to those mentioned for the charge we get an equation of motion $(\boldsymbol{\mu} = \boldsymbol{\mu}(t)\ , \dot{\boldsymbol{\mu}} = d\boldsymbol{\mu}/dt$, and similarly for other quantities)

$$\dot{\mathbf{M}} = [\boldsymbol{\mu} \wedge \mathbf{H}_0] - \frac{4\omega_{max}}{3\pi c^3} [\boldsymbol{\mu} \wedge \ddot{\boldsymbol{\mu}}] + \frac{2}{3c^3} [\dot{\boldsymbol{\mu}} \wedge \ddot{\boldsymbol{\mu}}] + \frac{4}{3\pi c^3} \left[\boldsymbol{\mu} \wedge \left\{ \boldsymbol{\mu}(0) \frac{\sin \omega_{max} t}{t} + \dot{\boldsymbol{\mu}}(0) \frac{d}{dt} \frac{\sin \omega_{max} t}{t} \right\} \right]$$

$$+ \text{ terms which tend to zero as } \omega_{max} \sim c/r_0 \to \infty\,. \tag{2.21}$$

If we drop the terms connected with the initial conditions and the terms which vanish as $r_0 \to 0$, the equation of motion becomes

$$\dot{\mathbf{M}} = [\boldsymbol{\mu} \wedge \mathbf{H}_0] - \mathbf{L} + \mathbf{R}\ , \quad \mathbf{L} = \frac{4\omega_{max}}{3\pi c^3} [\boldsymbol{\mu} \wedge \ddot{\boldsymbol{\mu}}]\ , \quad \mathbf{R} = \frac{2}{3c^3} [\boldsymbol{\mu} \wedge \dddot{\boldsymbol{\mu}}]\,. \tag{2.22}$$

The term $\mathbf{R}$ is the moment of the radiative friction force and is dissipative. Under stationary conditions or if one averages over time the work done by the moment of the force $\mathbf{R}$ equals the emitted energy just as in the case of the radiative friction force $\mathbf{f}$ (for details see Chapter 3). Let, for instance, the magnetic moment $\boldsymbol{\mu}$ be constant in magnitude and at right angles to the rotational axis of the rotator ($\boldsymbol{\mu} = \boldsymbol{\mu}_\perp$, $\mu_{\perp x} = \mu_{0\perp} \cos \Omega t$, $\mu_{\perp y} = \mu_{0\perp} \sin \Omega t$, with the angular velocity $\boldsymbol{\Omega}$ being along the z-axis). The radiation power is then

$$\mathcal{P} = \frac{2}{3c^3} (\ddot{\boldsymbol{\mu}})^2 = \frac{2\Omega^4 \mu_{0\perp}^2}{3c^3} \tag{2.23}$$

Under those circumstances $\mathbf{R} = 2[\boldsymbol{\mu}_\perp \wedge \ddot{\boldsymbol{\mu}}_\perp]/3c^3$ and the work done, $(\mathbf{R}\cdot\boldsymbol{\Omega})$, is just equal to (2.23). The term $\mathbf{L}$ in (2.22) is conservative and, clearly,

$$\mathbf{L} = \dot{\mathbf{M}}_m \ , \quad \mathbf{M}_m = \frac{4\omega_{max}}{3\pi c^3} [\boldsymbol{\mu} \wedge \dot{\boldsymbol{\mu}}] \ , \tag{2.24}$$

that is, $\mathbf{M}_m$ is some angular momentum of electromagnetic provenance. It follows from the derivation of Eq.(2.21), in complete analogy with the case of the charge (see (2.14)), that we must assume in Eq.(2.22) that the rotational frequency of the magnetic moment $\Omega \ll \omega_{max} \sim 2\pi c/r_0$. As to absolute magnitudes we thus have in (2.22) $L \gg \mathbf{R}$. On the other hand, the moment $\mathbf{M}_m$ can be very small in comparison with the mechanical moment $\mathbf{M}$ (for instance, this is the case for pulsars; see Ginzburg, 1971). The problem whether or not to take the terms $\mathbf{L}$ and $\mathbf{R}$ into account is thus decided by the nature of the problem. In the example just given of the calculation of the radiation by a magnetic dipole the term $\mathbf{L}$ does not play a role. If, however, one considers the scattering of electromagnetic waves by a magnetic dipole, the term $\mathbf{L}$ dominates, in contrast, over $\mathbf{R}$ (see Ginzburg, 1944a, 1946).

The appearance of the term $\mathbf{L}$ in (2.22) is somewhat unexpected as by analogy with the case of a charge we might have expected that taking the self field into account might lead to a term proportional to $\dot{\mathbf{M}}$ or $\dot{\boldsymbol{\mu}}$ imitating the contribution from them. The situation becomes clear if we consider a rotator which possesses not only a magnetic moment but also a charge with charge density $eD(\mathbf{r})$. Let us evaluate for such a rotator the electromagnetic angular momentum

$$\mathbf{M}_{em} = \frac{1}{4\pi c} \int \left[\mathbf{r} \wedge [\mathbf{E} \wedge \mathbf{H}]\right] d^3 r \ , \tag{2.25}$$

where we shall assume for the sake of simplicity that the rotator is a sphere of radius r_0, and that the field outside the sphere is that of a charge e and a magnetic dipole $\boldsymbol{\mu}$ which is at the centre of the sphere, and, finally, we put the electrical field inside the sphere equal to zero — such a model is completely realistic: it corresponds to a well conducting charged and magnetized sphere. Straightforward calculations (see Ginzburg, 1972a) then lead to the result

$$\mathbf{M}_{em} = \mathbf{M}_e + \mathbf{M}_m \; , \quad \mathbf{M}_e = \frac{2e\boldsymbol{\mu}}{3r_0c} \; , \quad \mathbf{M}_m = \frac{2}{3r_0c^2}[\boldsymbol{\mu} \wedge \dot{\boldsymbol{\mu}}] \; . \qquad (2.26)$$

The angular momentum $\mathbf{M}_m$ is the same as the one obtained earlier (see (2.24) where for complete agreement we must put $\omega_{max} = \pi c/2r_0$), while the angular momentum $\mathbf{M}_e$ is, indeed, proportional to $\boldsymbol{\mu}$, which means that in an equation such as (2.22) there appears a term proportional to $\dot{\boldsymbol{\mu}}$. If the magnetic moment $\boldsymbol{\mu}$ is proportional to the angular momentum $\mathbf{M}$, which often is the case, the electromagnetic angular momentum $\mathbf{M}_e$ will not play any part, in fact, it must be combined with $\mathbf{M}$, and the total angular momentum must be 'renormalized' and put equal to the observable value. (We have here in mind a 'point' particle; for a macroscopic rotator with $\boldsymbol{\mu} = \kappa\mathbf{M}$ we have simply

$$\mathbf{M}_e = \frac{2e\kappa}{3r_0c}\mathbf{M} \quad \text{and} \quad \mathbf{M} + \mathbf{M}_e = \left(1 + \frac{\kappa}{\kappa_e}\right)\mathbf{M} \; ,$$

where under the conditions of (2.26) $\kappa_e = \mu M_e = 3r_0c/2e \sim e/m_{em}c$, as the electromagnetic mass $m_{em} \sim e^2/r_0c^2$.) For an uncharged magnetic rotator the electromagnetic angular momentum completely reduces to $\mathbf{M}_m$ and taking this angular momentum into account may, in principle, radically alter the dynamics of the rotator (see (2.22)).

We note in conclusion that we shall encounter in what follows a few other aspects of the problem of the radiation reaction.

Chapter III

UNIFORMLY ACCELERATED CHARGE

Emission and radiative force for the uniformly accelerated motion of a charge.
Energy conservation law for charge and field.

There are in physics a few literally 'perpetual problems' which continue to be discussed in the scientific literature for decades and decades. As examples merely from the field of classical electrodynamics we can mention the problem of the electromagnetic mass and of exact solutions of the equations of motion involving the radiative friction force (see Chapter 2), the choice of the energy-momentum tensor in a medium (see Chapter 12) and the problem of the radiation and radiation reaction when a charge is moving with uniform acceleration.

The field of a uniformly accelerated charge was first considered more than sixty years ago (Born, 1909) and after that several, sometimes contradictory, statements have been published on this subject (see §32 in Pauli's book (1958) and also papers by Fulton and Rohrlich (1960), Leibovitz and Peres (1963), Rohrlich (1965), Nikishov and Ritus (1969), Kovetz and Tauber (1969), Anderson and Ryon (1969), Grandy (1970), and Ginzburg (1970c)); in the last paper we have also given a rather comprehensive list of publications, but here we shall restrict ourselves to mentioning the papers just cited.

The problem of the radiation of a uniformly accelerated charge — like other 'perpetual problems' — is not of any actual importance or, at any rate, from a practical point of view it has already for a long time been clarified to a sufficient degree. This can just be the reason why after such a long time it may still contain a few obscure points which are usually more likely to be pedagogical or methodological in nature. At the same time the neglect of such methodological problems sometimes 'takes its own revenge' and leads to misunderstandings and to the appearance of incorrect papers in very reliable scientific journals; we shall refer in fact to some examples in what follows, but we do not wish to give explicit references to incorrect publications.

The most elementary, but in fact the most important, difficulty which arises in the problem of a uniformly accelerated charge is the following one. By definition when the acceleration is uniform we have $\dot{\mathbf{v}} = \text{constant}$ and $\ddot{\mathbf{v}} \equiv \dddot{\mathbf{r}} = 0$, and hence the radiation force $\mathbf{f} = (2e^2/3c^3)\ddot{\mathbf{v}}$ vanishes (we restrict ourselves here to the non-relativistic case). On the other hand, according to the well known formula for the radiative power — which is sometimes called Larmor's formula —

$$\mathcal{P} = \frac{2e^2}{3c^3}\dot{\mathbf{v}}^2 \quad . \tag{3.1}$$

It is clear that for a uniformly accelerated motion $\mathcal{P} \neq 0$, and the following problem arises: how can a charge radiate when its radiative friction force vanishes?

In the general case the work done by the radiative friction force, $-(\mathbf{v}\cdot\mathbf{f})$, is also not equal to $\mathcal{P}$ (the minus sign is connected here with the fact that $(\mathbf{v}\cdot\mathbf{f})$ is the decrease in the particle energy, while the power $\mathcal{P}$ is positive; see also the discussion below). However, for a periodic motion the total energy is conserved on average over the time (or over a sufficiently long time interval). Indeed, $(\mathbf{v}\cdot\ddot{\mathbf{v}}) = (d/dt)(\mathbf{v}\cdot\dot{\mathbf{v}}) - \dot{\mathbf{v}}^2$ and if $(\mathbf{v}\cdot\dot{\mathbf{v}})_{t_1}^{t_1+T} = 0$ we get

$$-\int(\mathbf{v}\cdot\mathbf{f})\,dt \equiv -\frac{2e^2}{3c^3}\int(\mathbf{v}\cdot\ddot{\mathbf{v}})\,dt = \frac{2e^2}{3c^3}\int\dot{\mathbf{v}}^2\,dt \equiv \int\mathcal{P}dt \quad , \tag{3.2}$$

where the integrals over the time are taken from t_1 to $t_1 + T$, where T is the period of the motion.

In the majority of the cases the energy conservation when a time average is taken is already sufficient to remove any real contradictions. Moreover, the widely used elementary derivation of the expression for the radiative friction force $\mathbf{f}$ is just based upon a use of Eq.(3.2) and formula (3.1) for the power $\mathcal{P}$. When this method is used to find the force $\mathbf{f}$ one may suspect that the expression used, $\mathbf{f} = (2e^2/3c^3)\ddot{\mathbf{v}}$, is valid only for a periodic motion or for the wider class of motion for which we can neglect the term $(\mathbf{v}\cdot\dot{\mathbf{v}})_{t_1}^{t_1+T}$ for sufficiently long times T. But we have already seen that such a conclusion is incorrect and that one does not need to impose any such restrictions on the expression for $\mathbf{f}$ which we obtained.

One can very easily make the crux of the matter clear. However, a discussion of this problem is for us not at all an end in itself; it is more important to acquaint the reader with — or, rather, to remind him of — a number of useful formulae of the theory of radiation.

When a charge e moves in vacuo along some trajectory, the electromagnetic field is determined by the well known equations which follow from the Liénard-Wiechert potentials (see Heitler, 1947; Landau and Lifshitz, 1975)

$$\mathbf{E} = \frac{e(1 - v^2/c^2)}{[R-(\mathbf{v}\cdot\mathbf{R})/c]^3}\left(\mathbf{R} - \frac{\mathbf{v}R}{c}\right) + \frac{e}{c^2[R-(\mathbf{v}\cdot\mathbf{R})/c]^3}\left[\mathbf{R}\wedge\left[\left\{\mathbf{R} - \frac{\mathbf{v}R}{c}\right\}\wedge\dot{\mathbf{v}}\right]\right] \quad , \tag{3.3}$$

$$\mathbf{H} = \frac{1}{R}\,[\mathbf{R}\wedge\mathbf{E}]\ . \qquad (3.4)$$

The fields $\mathbf{E}$ and $\mathbf{H}$ are taken at the observation point at time t while on the right-hand side of the equation the quantities $\mathbf{R}, \mathbf{v}$, and $\dot{\mathbf{v}}$ refer to the 'emission time' $t' = t - R(t')/c$ where the vector $\mathbf{R}$ goes from the point where the charge e is situated to the observation point. Moreover, the velocity of the charge $\mathbf{v}(t') = -\,\partial\mathbf{R}(t')/\partial t'$ and $\dot{\mathbf{v}} = \partial\mathbf{v}/\partial t'$. Clearly the function $\mathbf{R}(t')$ determines the trajectory of the charge, but it is more convenient to characterize the position of the charge by the vector $\mathbf{r}(t')$ and the observation point by the vector $\mathbf{r}(t) = \mathbf{r}(t') + \mathbf{R}(t')$, whence it follows also that $\dot{\mathbf{r}} \equiv \partial\mathbf{r}/\partial t' = -\,\partial\mathbf{R}/\partial t'$.

The first term in (3.3) corresponds to the field of a charge moving with a velocity $\mathbf{v}$; this term decreases with the distance R as $1/R^2$. The second term in (3.3) decreases as $1/R$ and is the main term when $R \gg c^2(1 - v^2/c^2)/\dot{v}$; the field described by this term turns out to be transverse and is the field of an electromagnetic wave. If the charge produces such a wave field one says that it radiates. Essentially, we have here a definition and it is not only a non-trivial one, but it also needs a more precise definition. Indeed, we can consider the wave field of a charge, which decreases as $1/R$, only in the wave zone where only a single such field can exist in practice. However, we can verify the presence of a wave term (the second term in (3.3) and in the expanded form of Eq.(3.4)) also at smaller distances from the charge. In such a case, however, the complete field is not at all a radiation field which propagates with the velocity of light. For reasons which will become clear in what follows it is advisable to understand the statement 'the charge radiates' in a wider sense, namely, as the presence of a wave field, independent of whether or not another part of the field is present. We must also emphasize that when we measure the fields $\mathbf{E}$ and $\mathbf{H}$ at time t we can only reach a conclusion about the state (for instance, the acceleration) of the electron at an earlier time $t' = t - R(t')/c$.

If we consider solely the field of a single given charge, the energy flux through any closed surface surrounding the charge must be non-vanishing when there is radiation present. It is clear that the energy passing during a time $dt = [1 - (\mathbf{s}\cdot\mathbf{v})/c]\,dt'$ through an area $d\sigma = R^2 d\Omega$ in the direction of $\mathbf{s} = \mathbf{R}/R$ is equal to

$$dW_s = \frac{c}{4\pi}\Big([\mathbf{E}\wedge\mathbf{H}]\cdot\mathbf{s}\Big)\,R^2 d\Omega\, dt = \frac{e^2}{4\pi c^3}\,\frac{\big[\mathbf{s}\wedge[(\mathbf{s} - \mathbf{v}/c)\wedge\dot{\mathbf{v}}]\big]^2}{[1 - (\mathbf{s}\cdot\mathbf{v})/c]^6}\, d\Omega\, dt\ , \qquad (3.5)$$

where $d\Omega$ is an element of solid angle, while we have assumed that the field is a wave field (second term in (3.3) and (3.4)); for this reason Eq.(3.5) is in general valid only in the wave zone.

The evaluation of the total energy emitted per unit time t' gives

$$\mathcal{P} = \frac{dW}{dt'} = \frac{e}{4\ c}\int \frac{\left[\mathbf{s}\wedge[(\mathbf{s}-\mathbf{v}/c)\wedge\dot{\mathbf{v}}]\right]^2}{[1-(\mathbf{s}\cdot\mathbf{v})/c]^5}\,d\Omega$$

$$= \frac{2e^2}{3c^3}\,\frac{\dot{\mathbf{v}}^2-(\mathbf{v}\cdot\dot{\mathbf{v}})^2/c^2}{(1-v^2/c^2)^3} = -\frac{2e^2c}{3}\,w^iw_i\ . \tag{3.6}$$

Here $w^i = (w^0,\mathbf{w}) = du^i/ds$ is the four-dimensional acceleration vector of the particle (one should not confuse the unit vector $\mathbf{s}$ and the length s which enters only in the form of a differential ds)†. By virtue of the Lorentz invariance of Eq.(3.6) we can evaluate it in any inertial frame of reference. In that frame in which $\mathbf{v}=0$, Eq.(3.5) is valid for any R and the evaluation of the emitted energy and the establishment of the fact that radiation is present can thus be performed also close to the charge and not only in the wave zone. This conclusion is to some extent, of course, already clear from general considerations as the field — and, in particular, the wave field — is determined by Eqs.(3.3) and (3.4) at any distance from the charge.

The quantity $\mathcal{P} = dW/dt'$ characterizes the flux of energy through a sphere of radius R at time t, but one should emphasize that on the right-hand side we find quantities at time $t' = t - R(t')/c$ while the emitted energy also refers

---

† We use the notation of Landau and Lifshitz (1975). Here the four-velocity is

$$\frac{dx^i}{ds} = u^i \equiv (u^0,\mathbf{u}) = \left\{\frac{1}{\sqrt{1-v^2/c^2}}\ ,\ \frac{\mathbf{v}}{c\sqrt{1-v^2/c^2}}\right\},$$

$$u^iu_i = u_0^2-\mathbf{u}^2 = 1\ ,\quad ds = c\,dt\sqrt{1-v^2/c^2}\ ,$$

and

$$w^i = \frac{du^i}{ds} = \left\{\frac{(\mathbf{v}\cdot\dot{\mathbf{v}})}{c^3(1-v^2/c^2)^2}\ ,\ \frac{\dot{\mathbf{v}}}{c^2(1-v^2/c^2}+\frac{\mathbf{v}\,(\mathbf{v}\cdot\dot{\mathbf{v}})}{c^4(1-v^2/c^2)^2}\right\},$$

where $\dot{\mathbf{v}} \equiv d\mathbf{v}/dt$.

One sees easily that

$$w^iw_i = -\frac{\dot{\mathbf{v}}^2c^2(1-v^2/c^2)-(\mathbf{v}\cdot\dot{\mathbf{v}})^2}{c^6(1-v^2/c^2)^3} = -\frac{\dot{\mathbf{v}}^2-[\mathbf{v}\wedge\dot{\mathbf{v}}]^2/c^2}{c^4(1-v^2/c^2)^3}\ . \tag{3.7}$$

to a unit 'radiation time' t'. The difference between the intervals $dt = [1 - (\mathbf{s}\cdot\mathbf{v})/c]\,dt'$ and dt' is a manifestation of the Doppler effect — the pulse (train) of radiation emitted during a time dt' will have a duration dt.

If the velocity of the charge at the radiation time t is equal to zero (or, in actual fact, sufficiently small) we get for the power

$$\mathcal{P} \equiv \frac{dW}{dt'} = \frac{dW}{dt} = \frac{2e^2}{3c^3}\dot{\mathbf{v}}^2 . \tag{3.1a}$$

This expression is essentially the same as (3.1) but is written down again in somewhat more detail.

For a non-relativistic uniformly accelerated motion we have $\dot{\mathbf{v}}$ = constant. A motion in which the acceleration is constant in the comoving (eigen) frame of reference, that is, in the frame in which the particle velocity is zero, is called a relativistic uniformly accelerated motion. This means that for a uniformly accelerated motion in the comoving frame, and just in that frame, we have always $\ddot{\mathbf{v}} = 0$. In covariant form this condition can be written in the form

$$\frac{dw^i}{ds} + \alpha u^i = 0 ,$$

where α is a constant; when $\mathbf{v} = 0$, the condition written down here indeed changes to the equation $\ddot{\mathbf{v}} = 0$. Bearing in mind that $u^i u_i = 1$ and $w^i u_i \equiv (du^i/ds)u_i = 0$, we find that

$$\alpha = -\frac{dw^i}{ds}u_i = w^i w_i .$$

We are thus led to the condition which defines the uniformly accelerated motion

$$\frac{dw^i}{ds} + w^k w_k u^i = 0 ; \tag{3.8a}$$

This condition has the following form in three-dimensional notation:

$$\left(1 - \frac{v^2}{c^2}\right)\ddot{\mathbf{v}} + \frac{3}{c^2}(\mathbf{v}\cdot\dot{\mathbf{v}})\,\dot{\mathbf{v}} = 0 . \tag{3.8b}$$

When we multiply (3.8a) by w_i we see that for a uniformly accelerated motion

$$w^i w_i = -\frac{w^2}{c^4} = \text{constant} , \tag{3.9}$$

where w is the acceleration in a frame of reference in which the particle is at rest (see also (3.7)). The reverse statement is, however, incorrect: in the general case the fact that $w^i w_i$ is constant does not yet guarantee that conditions (3.8a) or (3.8b) are satisfied, while these conditions must be satisfied if we take the definition of a uniformly accelerated motion which was given earlier; our definition is an obvious one, but if one so wishes one might call a motion with a constant square of the four-dimensional acceleration, that is, one which satisfies condition (3.9), uniformly accelerated motion.

We note that when a charged particle moves in a constant and uniform electromagnetic field and if we neglect the radiation reaction condition (3.9) is immediately satisfied. Indeed, differentiating the equation of motion (for a definition of F^{ik} see Landau and Lifshitz, 1975)

$$\frac{du^i}{ds} \equiv w^i = \frac{e}{mc^2} F^{ik} u_k ,$$

with respect to s, we see that

$$\frac{dw^i}{ds} = \frac{e}{mc^2} F^{ik} w_k \quad \text{and} \quad \frac{dw^i}{ds} w_i = \tfrac{1}{2} \frac{d}{ds} (w^i w_i) = \frac{e}{mc^2} F^{ik} w_k w_i = 0 ,$$

since the electromagnetic field tensor F^{ik} is antisymmetric. We have used here not only the fact that the field is constant, that is, independent of the time, but also the fact that it is uniform, as the field is taken in the equations of motion at the point 'occupied' by the charge; thus $F^{ik} = F^{ik}(t, \mathbf{r}(t))$ and one can put the derivative of F^{ik} with respect to s equal to zero, as was done here, only when F^{ik} is independent of both t and $\mathbf{r}$.

By virtue of what we have just said we see that the motion of a charge in an arbitrary constant and uniform electromagnetic field is not always a uniformly accelerated one. The motion turns out, however, to be uniformly accelerated in the important particular case where the charge is in a constant and uniform electric field (say, in a condenser) when the velocity $\mathbf{v}$ and the acceleration $\dot{\mathbf{v}}$ are collinear, that is, when the motion is parallel to the field. We then get from (3.7) and (3.9)

$$\frac{\dot{v}}{[1 - (v^2/c^2)]^{\frac{3}{2}}} = \frac{d}{dt} \left[\frac{v}{\sqrt{1 - (v^2/c^2)}} \right] = w = \text{constant},$$

and hence

$$\ddot{v} + \frac{3 v \dot{v}^2}{c^2[1 - (v^2/c^2)]} = 0 ,$$

which for the case when $\mathbf{v}$ and $\dot{\mathbf{v}}$ are collinear is the same as condition (3.8). If we take the z-axis along the direction of the velocity $\mathbf{v}$ and if we assume, in order to get particularly simple expressions, that at $t = 0$ $z = c^2/w$ and $v = dz/dt = 0$, we have in the particular case considered here

$$
\left.\begin{aligned}
z &= c\sqrt{\frac{c^2}{w^2} + t^2}\ , \quad v = \frac{dz}{dt} = \frac{wt}{\sqrt{1 + (w^2t^2/c^2}}\ ,\\
\dot{v} &= \frac{dv}{dt} = \frac{c^3}{w^2[(c^2/w^2) + t^2]^{\frac{3}{2}}} = \frac{w}{[1 + (w^2t^2/c^2)]^{\frac{3}{2}}}\ .
\end{aligned}\right\} \tag{3.9a}
$$

As the function $z(t)$ is a hyperbola, one calls the relativistic uniformly accelerated rectilinear motion also hyperbolic motion.

The motion, of course, is hyperbolic not only in the above mentioned case of a constant and uniform electric field, collinear with the particle velocity†, but also in the corresponding gravitational field; it is only important that the equation of motion have the form

$$
\frac{d}{dt}\left[\frac{mv}{\sqrt{1 - (v^2/c^2)}}\right] = F = \text{constant}\ .
$$

It is clear from Eqs.(3.6) and (3.1a) and from what we have just said, that both in the relativistic and in the non-relativistic uniformly accelerated motion the charge radiates and that $\mathcal{P} = dW/dt' = (2e^2/3c^3)w^2$. Moreover, in a qualitative respect the radiation of a uniformly accelerated moving charge does not differ at all from that of an arbitrarily accelerated charge. This statement is true not only with respect to the calculation of the total power $\mathcal{P} = dW/dt'$, but also with respect to the spectral distribution of the radiation (Nikishov and Ritus, 1969).

The equation of motion of a charge has in the non-relativistic approximation the form (2.1) and was discussed in detail in Chapter 2. However, for the sake of convenience we write this equation once more in a somewhat different form:

$$
m\dot{\mathbf{v}} = \mathbf{F}_0 + \frac{2e^2}{3c^3}\ddot{\mathbf{v}}\ . \tag{3.10}
$$

---

† It is clear from what we have said that if the charge moves in an electric field $\mathbf{E}_0$ at an angle to it, that is, if its velocity has a component at right angles to the field, such a motion is not a uniformly accelerated one: in that case there is in the frame of reference in which the charge is at rest also a magnetic field. Nikishov and Ritus (1969) have considered the radiation of a particle when it moves in a constant and uniform electric field in an arbitrary direction.

Its relativistic generalization can be written in the form (see, for instance, Landau and Lifshitz, 1975)†

$$mc\frac{du^i}{ds} = \frac{e}{c}F_0^{ik}u_k + \frac{2e^2}{3c}\left(\frac{d^2u^i}{ds^2} + u^i\frac{du^k}{ds}\frac{du_k}{ds}\right), \tag{3.11}$$

where the external force is assumed to be the Lorentz force (F_0^{ik} is the external electromagnetic field tensor); one sometimes writes Eq.(3.11) in a different form, using the fact that $u^i(du_i/ds) = 0$, and, hence,

$$u^k\frac{d^2u_k}{ds^2} = -\frac{du^k}{ds}\frac{du_k}{ds}.$$

In the three-dimensional notation Eq.(3.11) becomes

$$\left.\begin{aligned} &\frac{d}{dt}\left[\frac{m\mathbf{v}}{\sqrt{1-(v^2/c^2)}}\right] = e\left\{\mathbf{E}_0 + \frac{1}{c}[\mathbf{v}\wedge\mathbf{H}_0]\right\} + \mathbf{f}, \\ &\mathbf{f} = \frac{2e}{3c[1-(v^2/c^2)]}\left\{\ddot{\mathbf{v}} + \dot{\mathbf{v}}\frac{3(\mathbf{v}\cdot\dot{\mathbf{v}})}{c^2[1-(v^2/c^2)]}\right. \\ &\qquad\left. + \frac{\mathbf{v}}{c^2[1-(v^2/c^2)]}\left((\mathbf{v}\cdot\dot{\mathbf{v}}) + \frac{3(\mathbf{v}\cdot\dot{\mathbf{v}})^2}{c^2[1-(v^2/c^2)]}\right)\right\}. \end{aligned}\right\} \tag{3.12}$$

It is clear from (3.8), (3.11), and (3.12) that the radiation force for a uniformly accelerated and, in particular, for a hyperbolic motion vanishes.

We note that the limitations of the applicability of the equation of motion (2.1) — or (3.10) — which we discussed in Chapter 2 lead, of course, to the impossibility of using the relativistic Eqs.(3.11) or (3.12) without any provisos. The corresponding limitations are, however, not at all connected with the problem of a uniformly accelerated charge which we are considering just now.

What points which lack clarity and what paradoxes arise here ?

The first lack of clarity has already been mentioned: the presence of radiation notwithstanding the vanishing of the radiative friction force. The second lack of clarity which is discussed for the case of a uniformly accelerated

---

† The expression for the radiation force,

$$g^i = \frac{2e^2}{3c}\left(\frac{d^2u^i}{ds^2} + u^i\frac{du^k}{ds}\frac{du_k}{ds}\right)$$

has the property that it reduces to the force $\mathbf{f} = (2e^2/3c^2)\ddot{\mathbf{v}}$ in the comoving frame of reference (see the earlier discussion of the condition (3.8a)).

charge is connected with the application of the equivalence principle for the motion of a charge in the uniform gravitational field (see Ginzburg, 1970c). The third difficulty arises when one attempts to describe for all t and z the field of a charge which always (for $-\infty < t < \infty$) moves with uniform acceleration. In particular, Leibovitz and Peres (1963) conclude with the statement: "We are therefore led to the conclusion that the Maxwell equations are incompatible with the existence of a single charge uniformly accelerated at all times". This conclusion may turn out to be fully valid as the total emitted energy during hyperbolic motion which is unbounded in time is infinite, and as $t \to \pm\infty$ the kinetic energy of the charge is also infinite (the velocity of the charge is equal to c). However, we do not need to look for an appropriate solution in the case of any real physical statement of the problem in which the particle moves with uniform acceleration only during a finite time interval. For instance, if we are dealing with motion in a uniform and constant electric field (in a concrete case, in a condenser) the charge moves in the condenser for $t_1' < t' < t_2'$, while for $t' < t_1'$ and $t' > t_2'$ its velocity is, say, constant (we remind ourselves that such a motion in a condenser is one with a uniform acceleration and is hyperbolic only actually in the case when the field vector and the particle velocity are parallel). If we take this fact into account we need not doubt that one can find the solution for the field in the form of retarded potentials. The fact that the field of a charge which shows a uniformly accelerated motion or, in particular, a hyperbolic motion is a very specific one is clear also from the example of going over to the wave zone. We have already mentioned that the field decreases as $1/R$ (wave zone) for $R \gg c^2(1-v^2/c^2)/\dot{v}$. However, such a zone does not exist at all for hyperbolic motion at a fixed observation time t. Indeed, for fixed t the distance $R = c(t-t') \to \infty$, as $t' \to -\infty$. However, as $t' \to -\infty$ for a particle performing a hyperbolic motion (see Eq. (3.9a)) we have

$$1 - \frac{v^2}{c^2} = \frac{1}{1 + (w^2/c^2)t'^2} \approx \frac{c^2}{w^2 t'^2} , \quad \dot{v}(t') \approx \frac{c^3}{w^2 t'^3}$$

whence $(c^2/\dot{v})(1 - v^2/c^2) \approx ct'$. It is thus clear that for hyperbolic motion and for a fixed $t = t' + R/c$, and for $R \to \infty$, the given condition $R \gg (c^2/\dot{v})(1 - v^2/c^2)$ cannot be satisfied. Hereby the well known arbitrariness of the concept of the energy emitted by the charge is clear: one must state about what value of t or t' one is speaking.

Nevertheless the situation is completely well defined for a motion which has uniform acceleration during a finite time interval. For a given time of observation t and a known law of motion for the charge we can find $R(t')$ and the time of radiation t'. If the value of t' lies within the interval (t'_1, t'_2), when the acceleration of the charge is uniform, one can say that at that time t' no radiation force was acting upon the charge and at the same time the charge radiated: the flux of energy through a sphere of radius $R(t')$ at time $t = t' + R/c$ was different from zero.

We are thus back at the first paradox: the presence of radiation in the absence of the radiative force. We shall concentrate our attention on this as we can thereby more explicitly clarify what the content is of the energy conservation law in electrodynamics and how it is connected with the calculation of the emitted energy and the work done by the radiative friction forces.

To determine the energy emitted by the charge or the radiation intensity at a give surface one evaluates the Poynting vector $\mathbf{S} = c[\mathbf{E} \wedge \mathbf{H}]/4\pi$ far from the charge and, if we are dealing with the loss of energy by the charge, we find the flux of this vector through a closed surface. This was just the way in which we found, in particular, the standard Eqs.(3.6) and (3.1). Of course, we cannot restrict the use of such formulae as they are only valid in vacuo. If, on the other hand, the charge moves in a medium we get, generally speaking, completely different results. It is sufficient to state that even a uniformly moving charge can radiate in a medium — this is just observed in the case of Cherenkov or transition radiation (see Chapters 6 and 7). The evaluation of the Poynting vector and its flux through a surface remains nonetheless a completely correct method for determining the energy emitted also when a charge moves in a medium (to be exact, in a medium without spatial dispersion, as when such dispersion is present the energy flux density does not reduce to the Poynting vector; see Chapter 10). One can calculate the energy losses of a charge or the emitted energy also in two different ways: through determining the time-derivative of the energy in the field, $(d/dt)\int[(\mathbf{E}\cdot\mathbf{D}) + H^2]d^3\mathbf{r}/8\pi$ or by finding the work done, $e(\mathbf{v}\cdot\mathbf{E}') = (\mathbf{v}\cdot\mathbf{f})$, by the charge against the field produced by the charge itself; in other words, one calculates the work done by the radiative friction force $\mathbf{f}$ which, when there is a medium present, is, of course, no longer given by Eqs. (3.10) or (3.11). For the often encountered case — specified later on — all three methods mentioned here lead to the same result; as one of the many examples we mention the evaluation of the energy of the Cherenkov radiation†. However, in general the total energy flux, the change in the field energy, and the work done by the radiation force are not equal to one another. Forgetting this fact has led, for instance, to errors in the theory of synchrotron emission for the case when particles are moving along spiral (non-circular) orbits (see Chapter 5).

The paradox arising in connection with the radiation of a uniformly accelerated charge is also connected with illegitimate identification of the energy flux with the work done by the radiation force.

In a well known way (see, for example, Chapters 10 and 12) one gets from the equations of the electromagnetic field the relation (Poynting's theorem)

$$\frac{d}{dt}\left(\frac{E^2+H^2}{8\pi}\right) = -(\mathbf{j}\cdot\mathbf{E}) - \operatorname{div}\mathbf{S} \quad , \quad \mathbf{S} = \frac{c}{4\pi}[\mathbf{E}\wedge\mathbf{H}] \ . \tag{3.13}$$

Here and henceforth in this chapter we restrict ourselves to the vacuum case and we shall consider the motion of a single point charge for which $\mathbf{j} = e\mathbf{v}\delta(\mathbf{r}-\mathbf{r}_e(t))$. After integrating (3.13) over some volume V, bounded by the surface σ, we have

$$\frac{d\mathcal{H}_{em}}{dt} = -e(\mathbf{v}\cdot\mathbf{E}) - \oint S_n d^2\sigma \quad , \quad \mathcal{H}_{em} = \int \frac{E^2+H^2}{8\pi} d^3\mathbf{r} \quad . \tag{3.14}$$

On the other hand, we get from the equations of motion (3.12)

$$\frac{d\mathcal{E}}{dt} = e(\mathbf{v}\cdot\mathbf{E}_0) + (\mathbf{v}\cdot\mathbf{f}) \qquad , \quad \mathcal{E} \equiv \mathcal{H}_k = \frac{mc^2}{\sqrt{1-v^2/c^2}} \quad . \tag{3.15}$$

In (3.14) we have appropriately the total field $\mathbf{E} = \mathbf{E}_0 + \mathbf{E}'$, where $\mathbf{E}'$ is the field of the charge itself; at the point where the charge is $e\mathbf{E}' = \mathbf{f}$, and hence in (3.14) we have $e(\mathbf{v}\cdot\mathbf{E}) = e(\mathbf{v}\cdot\mathbf{E}_0) + (\mathbf{v}\cdot\mathbf{f})$. We are thus, as one should expect, led from (3.14) and (3.15) to the conservation law

$$\frac{d(\mathcal{H}_{em}+\mathcal{E})}{dt} = -\oint S_n d^2\sigma \ . \tag{3.16}$$

The field energy $\mathcal{H}_{em}$ includes the energy of the external fields $\mathbf{E}_0$ and $\mathbf{H}_0$, for instance, the energy of the field in a condenser through which the charge

---

† In Tamm and Frank's original paper (1937) the energy flux was calculated, Ginzburg (1939a,b) determined the rate of change of the field energy (see also Chapter 6), and Landau and Lifshitz (1960), for example, found the work done by the radiative friction force which corresponds to the Cherenkov radiation.

considered travels and is accelerated. We shall therefore assume, exclusively for the simplification of the problem, that the charge is accelerated by some external field of non-electromagnetic nature; the effects of this field are neglected in (3.11) — and also in Eqs.(3.15) and (3.16).

The conservation law (3.14) then becomes

$$\frac{d\mathcal{H}_{em}}{dt} = -(\mathbf{v}\cdot\mathbf{f}) - \oint S_n d^2\sigma \qquad (3.17)$$

where $\mathcal{H}_{em}$ is the energy of the field of the charge (as stated, we have assumed that all other electromagnetic fields are absent); we emphasize that everywhere in (3.13) to (3.17) a single time — the time of observation t — occurs as the argument which has not been written out explicitly.

Equation (3.17) which has a completely clear meaning shows that the work done by the radiation force, $(\mathbf{v}\cdot\mathbf{f})$, the change in the field energy, $d\mathcal{H}_{em}/dt$, and the total energy flux $\oint S_n d^2\sigma$ are connected through a single relation and that they are, in general, far from equal to one another†. If, however, we consider a stationary motion with $d\mathcal{H}_{em}/dt = 0$, we have $-(\mathbf{v}\cdot\mathbf{f}) = \oint S_n d^2\sigma$. By moving the surface σ to infinity, when $\oint S_n d^2\sigma = 0$, we can further evaluate the energy $\mathcal{H}_{em}$ in the whole of space. In that case $d\mathcal{H}_{em}/dt = -(\mathbf{v}\cdot\mathbf{f})$. What we have said here explains why we can, say, determine the energy losses of the particle $(\mathbf{v}\cdot\mathbf{f})$ in the stationary regime by evaluating $\oint S_n d^2\sigma$ or $d\mathcal{H}_{em}/dt$.

It is difficult to realize stationary radiation in the exact sense of the word (as an example of a stationary process we may mention Cherenkov radiation), and one is usually dealing with a periodic process when the energy of the field in a fixed volume satisfies the relation $\mathcal{H}_{em}(t_1) = \mathcal{H}_{em}(t_1 + T)$. This is just the situation in the case, for instance, of a fixed oscillator or of synchrotron radiation by a charge moving in a circular orbit (it is important here that the radiating charge returns to the same point after the period T). For a periodic process

$$\int_{t_1}^{t_1+T} \left(\mathbf{v}(t)\cdot\mathbf{f}(t)\right) dt = -\int_{t_1}^{t_1+T} \oint S_n(t)\, d^2\sigma\, dt . \qquad (3.18)$$

---

† In order not to complicate the exposition we do not consider the signs, that is, we consider the absolute magnitudes of the quantities $(\mathbf{v}\cdot\mathbf{f})$, $d\mathcal{H}_{em}/dt$, and $\oint S_n d^2\sigma$.

It is clear that the fact that the time of observation t and the time of emission t' are not the same is here unimportant, as for a periodic process the choice of the time t' does not play a role. If, on the other hand, we consider a motion for which the field energy $\mathcal{H}_{em}(t < t_1) = \mathcal{H}_{em}(t > t_2) = \mathcal{H}^{(0)}_{em}$, Eq.(3.18) will again be valid, but with $t_1 + T$ replaced by any time $t > t_2$. This is just the situation — or at any rate practically the situation — in the case of a charge which is 'reflected' by the electric field in a condenser (one assumes that for $t < t'_1 \leq t_1$ or $t > t_2 \geq t'_2$ the velocity of the charge is constant). We must only bear in mind that the energy $\mathcal{H}_{em}(t)$ depends on the volume V bounded by the surface σ — the time t_1 may be assumed to be the time when the charge enters the condenser t'_1, but the time t_2 must be larger than the time t'_2 when the particle leaves the condenser, as the radiation field must be able to leave the volume V.

We saw (see (3.2)) that in the non-relativistic case the radiative friction force **f** satisfies this requirement: we remind ourselves that in (3.2) by definition $\mathcal{P} = \oint S_n\, d^2\sigma$.

In the relativistic case we can (after a few elementary substitutions) write down the time component of Eq.(3.11)

$$\frac{d}{dt'}\left[\frac{mc^2}{\sqrt{1 - v^2/c^2}}\right] = e(\mathbf{v}\cdot\mathbf{E}_0) + \frac{2e^2}{3}\left(\frac{dw^0}{ds} + c\,w^i w_i\right). \tag{3.19}$$

Using Eqs.(3.6) and (3.12) we can write Eq.(3.19) in the form

$$\left.\begin{aligned} \frac{d}{dt'}\left[\frac{mc^2}{\sqrt{1 - v^2/c^2}}\right] &= e(\mathbf{v}\cdot\mathbf{E}_0) + (\mathbf{v}\cdot\mathbf{f}) = e(\mathbf{v}\cdot\mathbf{E}_0) + \frac{2e^2}{3}\frac{dw}{dt'} - \mathcal{P}, \\ w^0 &= \frac{(\mathbf{v}\cdot\dot{\mathbf{v}})}{c^3(1-v^2/c^2)^2}, \quad \mathcal{P} = \frac{dW}{dt'} = -\tfrac{2}{3}e^2 c\, w^i w_i . \end{aligned}\right\} \tag{3.20}$$

The time is denoted by t' in (3.19) and (3.20); this time characterizes the motion of the charge and is the time of emission when we consider the radiation. Meanwhile, in (3.17) and in the original Eqs.(3.13) and (3.14) there occurs a single time t for the charges and the field. The radiation power $\mathcal{P} = dW/dt'$ differs therefore from $dW/dt = \oint S_n(t)\, d^2\sigma$.

The charge, entering the condenser parallel to the braking field emits at all times t' while it stays in the field (as we have stated, $t'_1 \leq t' \leq t'_2$) electromagnetic waves, and

$$\mathcal{P} = \frac{dW}{dt} = \frac{2e^2}{3c^3}w^2 = \frac{2e^4E_0^2}{3m^2c^3} = \text{constant}.$$

This means that at a sufficiently large distance $R(t')$ from the charge at time $t = t' + R(t')/c$ a radiation field can be observed with the appropriate value of the energy flux. No radiation force acts on the charge for $t' < t_1'$ and $t' > t_2'$, and it moves according to the law

$$\frac{d}{dt}\left[\frac{m\mathbf{v}}{\sqrt{1 - v^2/c^2}}\right] = \mathbf{F}_0 = e\mathbf{E}_0 .$$

However, at the times t_1' and t_2' the friction force acts upon the charge and the work done by that force over the whole interval of the accelerated motion will be

$$\int_{t_1}^{t_1} (\mathbf{v}\cdot\mathbf{f})\, dt' = -\int_{t_1}^{t_2} \mathcal{P}\, dt' = -\frac{2e^2}{3c^3} w^2 (t_2' - t_1') ,$$

that is, exactly equal to the energy emitted.

From what we have said it is clear that the vanishing of the radiation force during the uniformly accelerated motion is in no sense paradoxical, notwithstanding the presence of radiation. Indeed, the non-vanishing total energy flux through a surface surrounding the charge while the radiation force equals zero is exactly equal to the decrease in the field energy in the volume enclosed by that surface. In the general case, however, all three quantities $d\mathcal{H}_{em}/dt$, $(\mathbf{v}\cdot\mathbf{f})$, and $\oint S_n\, d^2\sigma$ (see Eq.(3.17)) are different from zero. There are no grounds for expecting that the work done by the radiation force, $(\mathbf{v}\cdot\mathbf{f})$, and the energy flux $dW/dt = \oint S_n\, d^2\sigma$ or the flux $dW/dt' = \mathcal{P}$ are necessarily equal, especially as the force is applied to the charge, while the flux is calculated through a sphere of radius R. In complete agreement with the spirit of field theory the energy flux through a surface is directly determined by the field near the surface, and not by the field on the trajectory of the charge which is inside the surface.

It may seem that the explanations given here are too detailed. However, we have done this as, for instance, in the detailed paper by Fulton and Rohrlich (1960; see also Rohrlich, 1965) which is specially devoted to the radiation of a uniformly accelerated charge, the conservation law (3.17) is not used at all. Instead, as in a number of other papers, the concept of an 'acceleration energy' is introduced.

$$Q = \frac{2e^2 w^0}{3} = \frac{2e^2}{3c^3}\, \frac{(\mathbf{v}\cdot\dot{\mathbf{v}})}{(1 - v^2/c^2)^2} .$$

It is clear from (3.30) that $(\mathbf{v}\cdot\mathbf{f}) = dQ/dt' - \mathcal{P}$ and one can write Eqs. (3.15), (3.19), and (3.20) in the form

$$\left.\begin{aligned} &\frac{d\mathcal{E}}{dt'} - (\mathbf{v}\cdot\mathbf{f}) = \frac{d\mathcal{E}}{dt'} - \left(\frac{dQ}{dt'} - \mathcal{P}\right) = e(\mathbf{v}\cdot\mathbf{E}_0)\ , \\ &\mathcal{E} = \frac{mc^2}{\sqrt{1-v^2/c^2}}\ , \quad Q = \frac{2e^2}{3c^3}\ \frac{(\mathbf{v}\cdot\dot{\mathbf{v}})}{(1-v^2/c^2)^2}\ . \end{aligned}\right\} \tag{3.21}$$

The quantity Q is sometimes interpreted as part of the 'internal energy of a charged particle' and sometimes is assumed to be part of the energy of the field which immediately surrounds the particle, without contributing to its electromagnetic mass. From this point of view we may assume, when the radiation force vanishes, that the emitted energy $\mathcal{P}$ derives from the 'acceleration energy' Q or from the 'internal energy' $\mathcal{E} - Q$. If we assume that Q is part of the field energy, the radiation energy $\mathcal{P}$ derives from the field energy. Formally the latter is completely correct as $\mathcal{P} = dW/dt'$ is the flux (per unit time t') of field energy through some surface surrounding the charge.

It appears, however, to us that the introduction of an 'acceleration energy' or an 'internal energy' of the charge not only does not add to the understanding of the energy balance, but, rather, even complicates the problem. The charge possesses merely the energy $\mathcal{E} = mc^2/(1-v^2/c^2)^{\frac{1}{2}}$; the splitting of the radiation force $\mathbf{f}$ or the work done by that force $(\mathbf{v}\cdot\mathbf{f})$ into two, or any other number of, parts is, of course, not unique, and already for that reason can not have any particular meaning. To be more precise, if one could ascribe some meaning to it, this would be possible only in connection with an identification of part of the work $(\mathbf{v}\cdot\mathbf{f})$ with the expression for $\mathcal{P}$ which is determined from independent considerations. Therefore, for a discussion of the problem of the energy balance during uniformly accelerated motion and of the radiation by a charge we can not see the need for special starting points or for going beyond the limits of the conservation laws (3.15) and (3.17). Of course, it can also be convenient and natural to write the work $(\mathbf{v}\cdot\mathbf{f})$ as a sum of two terms (see (3.21)), but there is no need to give these terms any new meaning.

Chapter IV

RADIATION OF A MOVING PARTICLE

Motion and radiation in an ondulator. Motion in a magnetic field. Radiation reaction and limits of applicability of the classical theory. Radiative (magneto-brems) losses when charged particles move in a magnetic field.

The radiation of a non-relativistic particle ($v \ll c$) differs very appreciably and even in a qualitative manner from that of relativistic particles ($v \sim c$). We shall in what follows repeatedly encounter the corresponding peculiarities and it is therefore expedient to remind ourselves of them.

If a charged particle moves in vacuo (only this case will be discussed in the present and the next chapter) it emits radiation only when it is accelerated, and in the non-relativistic case, when the velocity $v \ll c = 3 \times 10^{10}$ cm/s, the radiation has most often a dipole character. To be more precise, the intensity of the higher multipole radiation is proportional to an extra factor of order $(v/c)^{2n} \sim (a/\lambda)^{2n}$, where a is the size of the emitting system and $\lambda \equiv \lambda_0 = 2\pi c/\omega = cT$ is the wavelength of the radiation, $T \sim a/v$ is the characteristic period or quasi-period for the particle motion and we have $n = 1$ for quadrupole radiation, $n = 2$ for octupole radiation, and so on. As an example, quadrupole radiation is thus usually important only if the dipole moment of the system is zero or anomalously small†. For a dipole (oscillator) with a moment $\mathbf{p} = e\mathbf{r}$ which only changes in magnitude, the electric field in the wave zone changes according to $E \propto \sin\theta$, and the intensity (energy flux per unit solid angle $d\Omega$) as

$$I = \frac{dW_s}{d\Omega\, dt} = \frac{(\ddot{p})^2}{4\pi c^3} \sin^2\theta \quad , \tag{4.1}$$

where θ is the angle between $\mathbf{p}$ and the wavevector $\mathbf{k}$ (Fig.4.1); for a harmonic

---

† Of course, this is only the simplest possibility. Quadrupole radiation can also dominate over the dipole radiation when the frequency of the variation in the quadrupole moment is larger than the frequency of the variation in the dipole moment. The different angular dependence (or, as one says, polar diagram) of the dipole and the quadrupole radiations may also turn out to be important.

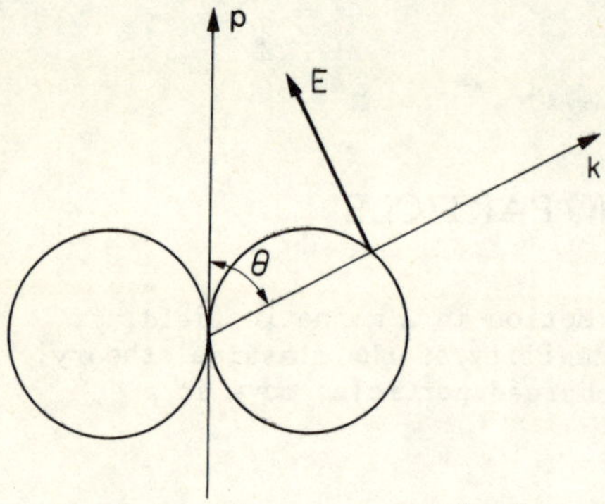

Fig.4.1. The electric field strength of a fixed dipole as function of the angle θ between the axis **p** and the wavevector **k**.

motion, when $\mathbf{p} = e\mathbf{a}_0 \sin\omega_0 t$, Eq.(4.1) changes to Eq.(1.85), in which there was also a time averaging.

One often calls the radiation of a non-relativistic electron when it moves in a magnetic field cyclotron radiation†. The frequency of this radiation — which is called dipole radiation — is, of course, equal to the frequency of rotation of the electron in the field H_0, that is,

$$\omega_H = \frac{eH_0}{mc} = 1.76\times 10^7\, H_0(\text{Oe})\,\text{s}^{-1} \quad . \tag{4.2}$$

In the simplest case of circular motion, when the velocity along the field $v_z = 0$, the radius of the orbit is equal to

$$r_H = \frac{v}{\omega_H} = \frac{mc^2}{eH_0}\frac{v}{c} = \frac{\lambda_H}{2\pi}\frac{v}{c} \ . \tag{4.3}$$

It is clear that we have always $2\pi r_H/\lambda_H \ll 1$ and $v/c \ll 1$, and this guarantees the validity of the dipole approximation; $\lambda_H = 2\pi c/\omega_H$ is the wavelength of the cyclotron radiation.

The radiation when a charge executes a non-relativistic circular motion in a magnetic field is the same as for two mutually perpendicular harmonic oscillators which are shifted in phase by $\frac{1}{2}\pi$ or, which is the same, for a constant electric dipole with dipole moment $e\mathbf{r}_H$ at right angles to the magnetic field, rotating with a frequency ω_H. The intensity of the cyclotron radiation, averaged over one period, when a charge moves along a circle is

$$I = \frac{dW_s}{d\Omega\, dt} = \frac{e^2\omega_H^4 r_H^2}{8\pi c^3}\,(1+\cos^2\alpha) \ , \tag{4.1a}$$

where α is the angle between $\mathbf{H}_0$ and $\mathbf{k}_0$ while $d\Omega = \sin\alpha\, d\alpha\, d\phi$; we show in Fig.4.2 the polar diagram for this case.

---

† There is as yet no well established terminology for the emission of particles moving in a magnetic field. It seems to us that if we take into account the terminology which has already been used it is convenient to call such radiation in the general case magneto-brems radiation, and to call it cyclotron and synchrotron radiation for non-relativistic and relativistic particles, respectively. In other words, in that terminology cyclotron and synchrotron radiation are limiting, particular cases of magneto-brems radiation in the non-relativistic and relativistic cases, respectively.

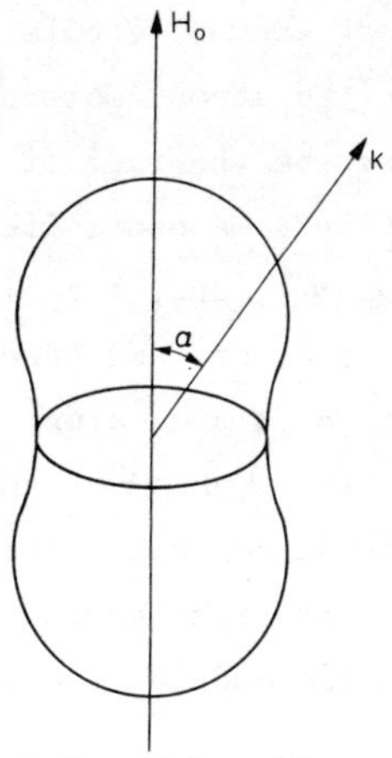

Fig.4.2. Polar diagram for cyclotron radiation, that is, the intensity of the cyclotron radiation as function of the angle α between the magnetic field vector $\mathbf{H}_0$ and the wavevector $\mathbf{k}$.

For a spiral motion, as long as the velocity component parallel to the field $v_z \equiv v_H = (\mathbf{v}\cdot\mathbf{H}_0)/H_0 \ll c$, the intensity distribution is qualitatively little different from the one discussed earlier, as long as we are not speaking about asymmetries; its nature will become clear from what follows.

Relativistic, or more correctly ultrarelativistic particles (and just this case we shall normally call relativistic in what follows), for which

$$\xi \equiv \frac{1}{\gamma} \equiv \sqrt{1 - \frac{v^2}{c^2}} = \frac{mc^2}{\mathcal{E}} \ll 1 \qquad (4.4)$$

radiate already completely differently (we have here introduced apart from ξ also the notation $1/\gamma$ as one encounters it often in the literature. In this case the dipole radiation does in general not dominate, and the simplest way to elucidate the nature of the radiation is to use the formulae for changing from one inertial system of coordinates to another. To be concrete, let the radiation have a dipole character in the coordinate system, in which the particle at a given time is at rest or moves with a non-relativistic velocity, and let it occur with a frequency ω_{00}. In the laboratory frame of reference, in which the emitter as a whole moves with a velocity v the frequency is given by the well known formula for the Doppler effect (see, for instance, Landau and Lifshitz, 1975)

$$\omega(\theta) = \frac{\omega_{00}\sqrt{1 - v^2/c^2}}{1 - (v/c)\cos\theta} \ . \qquad (4.5)$$

It is important to emphasize that the angle θ between $\mathbf{v}$ and $\mathbf{k}$ is measured here in the laboratory frame. When (4.4) is satisfied we have

$$\omega(0) = \omega_{00}\sqrt{\frac{1 + v/c}{1 - v/c}} \approx 2\omega_{00}\frac{\mathcal{E}}{mc^2} \equiv 2\gamma\,\omega_{00} \ , \qquad (4.6)$$

and the frequency $\omega(\theta)$ is large compared to ω_{00} for angles

$$\theta \leqslant \xi \equiv \frac{1}{\gamma} \equiv \frac{mc^2}{\mathcal{E}} \ll 1 \ . \qquad (4.7)$$

If, however, $\theta > \xi$, the frequency of the radiation decreases rather steeply with increasing angle θ; incidentally, we shall elucidate in Chapter 5 the obvious meaning and, one might say, the content of the Doppler effect.

Radiation which is analogous to the radiation of a fast moving dipole is realized in a great number of cases: for fast moving excited atoms, molecules, or nuclei (we do not dwell here upon the necessity to describe the radiating system itself quantum mechanically), when a charge moves in a magnetic field at a very small 'pinch angle' (the angle χ between $\mathbf{H}$ and $\mathbf{v}$), and, finally, for the motion in various 'ondulators'. By an ondulator we understand here a device which provides for a periodic motion of a charge on a path L along a trajectory which is nearly a straight line. In an electric ondulator the motion of a particle is the same as, for instance, in a condenser under the action of a uniform electric field $\mathbf{E} = \mathbf{E}_0 \cos \omega_0 t$ which is at right angles to the unperturbed (large) particle velocity $\mathbf{v}_0$. In a magnetic ondulator there is an inhomogeneous static magnetic field with a spatial period ℓ which leads to oscillations of the particle with angular frequency $\omega_0 \equiv 2\pi c/\lambda_0 = 2\pi v_0/\ell$ — such a case is realized in practice when a particle moves successively over magnets with NS NS NS ... pole arrangements, where N and S are, respectively, the North and South poles.

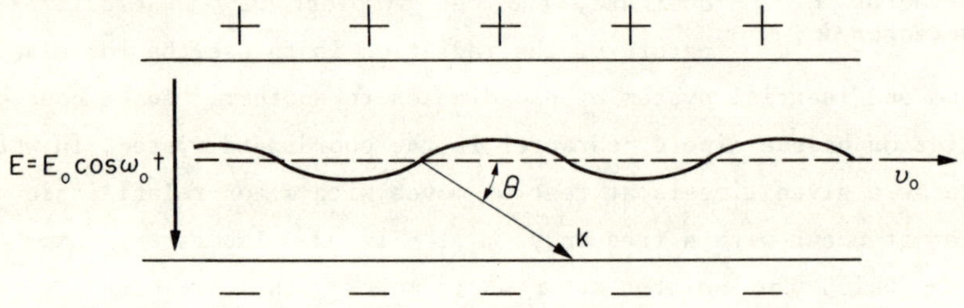

Fig.4.3. Motion of a particle in an electric ondulator

The equation of motion of a particle in an electric ondulator (Fig.4.3) is

$$\frac{d}{dt}\left[\frac{m\mathbf{v}}{\sqrt{1-v^2/c^2}}\right] = e\mathbf{E}_0 \cos \omega_0 t \ , \tag{4.8}$$

where the velocity $\mathbf{v} = \mathbf{v}_0 + \mathbf{v}'$, $v' \ll v_0$, v_0 = constant, $(\mathbf{v}_0 \cdot \mathbf{E}_0) = 0$. To a good approximation we can therefore write†

$$\left.\begin{aligned} &\frac{\mathcal{E}}{c^2}\frac{d\mathbf{v}}{dt} \equiv \frac{\mathcal{E}}{c^2}\frac{d^2\mathbf{r}_\perp}{dt^2} = e\mathbf{E}_0 \cos\omega_0 t \quad ; \quad \mathcal{E} = \frac{mc^2}{\sqrt{1-v_0^2/c^2}} \ ; \\ &\mathbf{r}_{\|} = \mathbf{v}_0 t \ , \quad \mathbf{r}_\perp = \mathbf{a}_0 \cos\omega_0 t \quad ; \quad \mathbf{a}_0 = -\frac{e\mathbf{E}_0}{m\omega_0^2}\frac{mc^2}{\mathcal{E}} \ , \end{aligned}\right\} \tag{4.9}$$

where $\mathcal{E}/c^2$ plays the role of the 'transverse' mass.

---

† We consider here the approximation which allows us to calculate the radiation in the dipole approximation considered below; Alferov, Bashmakov & Bessonov (1975) have given a much more complete theory of radiation in an ondulator.

The dipole moment which appears as the result of the field **E** equals $\mathbf{p} = e\mathbf{r}_\perp$. The emission is that of a moving dipole — with $\gamma = \mathcal{E}/mc^2 \gg 1$ —, provided

$$a_0 \ll \frac{\lambda_0}{\pi\gamma} = \frac{c}{\gamma\omega_0} , \tag{4.10}$$

that is, provided

$$eE_0\lambda_0 \ll 2\pi mc^2 . \tag{4.10a}$$

Satisfying this condition necessarily guarantees that the velocity $v' \sim a_0\omega_0$ is small compared to the velocity of light c — indeed, $v' \sim (eE_0\lambda_0/2\pi\mathcal{E})\,c$. The wavelength λ_0/γ occurs in (4.10) as in the system of coordinates moving with the average particle velocity $v_0 \approx c$ the length of an element of periodicity in the ondulator is $\ell' = \ell/\gamma \approx \lambda_0/\gamma$ while the amplitude of the acillations is, as before, equal to a_0.

The emitted frequency equals

$$\omega(\theta) = \frac{\omega_0}{1 - (v/c)\cos\theta} , \quad \lambda(\theta) = \frac{2\pi c}{\omega(\theta)} , \tag{4.11}$$

where we have dropped the index of v, as we shall also do in what follows. The difference between the first of Eqs.(4.11) and (4.5) is connected with the fact that in (4.8) and (4.9) the frequency ω_0 is measured in the laboratory frame whereas in (4.5) ω_{00} is measured in the rest frame. In the relativistic case we have according to (4.11)

$$\omega(0) = \frac{\omega_0}{1 - v/c} \approx 2\omega_0\left(\frac{\mathcal{E}}{mc^2}\right)^2 \equiv 2\omega_0\gamma^2 , \quad \frac{\mathcal{E}}{mc^2} \gg 1 \tag{4.12}$$

which differs from (4.6) by an extra factor $\mathcal{E}/mc^2$.

We can consider an ondulator as a device to transform the frequency ω_0 of the external field to the frequency $\omega(\theta)$, and huge 'transformation coefficients' are fully attainable. For instance, for an energy $\mathcal{E} \sim 5$ GeV, which is normal in present-day electron accelerators, the factor $\gamma^2 \sim 10^8$, as $mc^2 = 5.1 \times 10^5$eV. Therefore, for $\omega_0 \sim 10^{10}\,s^{-1}$ ($\lambda_0 \sim 20$ cm, radio-band), the frequency $\omega(0) \sim 10^{18}\,s^{-1}$ ($\lambda(0) \sim 10^{-7}$ cm $= 10$ Å, X-rays). It is clear that one can use an ondulator in principle as a generator for radiation in a frequency range where other methods are insufficiently effective, and also that one can apply the emission arising in an ondulator to detect particles which are flying past (see papers by Ginzburg (1947, 1972b,c) and by Alferov et al. (1975) and the literature cited there); new possibilities arise when we introduce into the ondulator some transparent medium — we shall discuss this in Chapter 6.

The expressions for the field strengths and the radiation intensity contain, like Eqs.(4.5) and (4.11) for the Doppler effect, the characteristic denominator $1-(v/c)\cos\theta$ to some power or other. We restrict ourselves here to referring to what are, perhaps, the most important expressions in radiation theory, namely, Eqs.(3.3) and (3.4) for the field of a point charge or the original expressions for the Liénard-Wiechert potentials (see Landau and Lifshitz, 1975, §63):

$$\mathbf{A} = \frac{e\mathbf{v}}{cR[1-(v/c)\cos\theta']} \ , \quad \phi = \frac{e}{cR[1-(v/c)\cos\theta']} \ , \tag{4.13}$$

where θ' is the angle between the velocity $\mathbf{v}$ and the radius vector $\mathbf{R}$, taken from the position of the charge to the point of observation; clearly, in the wave zone the angle θ' is the same as the angle θ between $\mathbf{v}$ and $\mathbf{k}$. Finally, there occurs in Eq.(3.5) for $dW_s = I\,d\Omega\,dt$ a factor $[1-(v/c)\cos\theta]^6$ in the denominator. From all this it is clear that in the relativistic case the emission is mainly in the forward direction— it is concentrated mainly within the range of angles $\theta \leqslant \xi \equiv 1/\gamma \equiv mc^2/\mathcal{E}$. In quantum (or, rather, corpuscular) language this result is particularly obvious: photons with the largest energy $\hbar\omega$ are emitted just in the forward direction. The polar diagram of the emission of a dipole moving along some trajectory with a velocity comparable to the velocity of light, c, is shown in Fig.4.4.

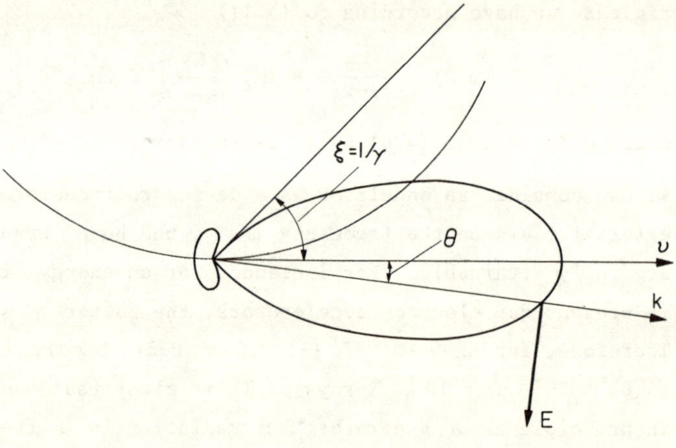

Fig.4.4. Polar diagram for the emission of a dipole.

We show the distribution of the electric field in a plane through the axis of the dipole, as a function of the angle θ between the translational velocity $\mathbf{v}$ and the wavevector $\mathbf{k}$. the dipole is moving at right angles to its axis. The field distribution is shown for the case $v = \frac{2}{3}c$.

When a particle is moving in an ondulator, and also in many other cases, one is usually interested in the total energy emitted along the whole of its path (along the length L of the ondulator), while the observation is at distances $R \gg L$, that is, far from the emitter. In that case we are interested in the quantity (see (3.5))

$$\frac{dU_s}{d\Omega} = \frac{c}{4\pi}\int([\mathbf{E}\wedge\mathbf{H}]\cdot\mathbf{s})R^2\,dt = \int\frac{dW_s}{d\Omega\,dt}\left[1-\frac{(\mathbf{s}\cdot\mathbf{v})}{c}\right]dt'$$

$$= \frac{e^2}{4\pi c^3}\int\frac{[\mathbf{s}\wedge[(\mathbf{s}-\mathbf{v}/c)\wedge\dot{\mathbf{v}}]]^2}{[1-(\mathbf{s}\cdot\mathbf{v})c]^5}\,dt'$$

$$= \frac{e^2}{4\pi c^3}\int\left\{\frac{2(\mathbf{s}\cdot\dot{\mathbf{v}})(\mathbf{v}\cdot\dot{\mathbf{v}})}{c[1-(\mathbf{s}\cdot\mathbf{v})/c]^4}+\frac{\dot{\mathbf{v}}^2}{[1-(\mathbf{s}\cdot\mathbf{v})/c]^3}-\frac{(1-v^2/c^2)(\mathbf{s}\cdot\dot{\mathbf{v}})^2}{[1-(\mathbf{s}\cdot\mathbf{v})/c]^5}\right\}dt'\,. \tag{4.14}$$

Apart from the notation, this formula is the same as Eq.(73.11) in Landau and Lifshitz's book (1975) and follows, as we have said, directly from Eq.(3.5). For an ondulator $(\mathbf{v}\cdot\dot{\mathbf{v}}) = 0$, $\dot{\mathbf{v}} = -\omega_0^2\mathbf{r}$, $(\mathbf{s}\cdot\mathbf{v}) = v\cos\theta$. Moreover, in (4.9) there occurs just the 'source time' t', because of its meaning, even though we have omitted the prime here. The result of integrating over t' reduces to replacing $\cos^2\omega_0 t'$ by $\frac{1}{2}$ and multiplying by the time the particle spends in the ondulator $T = L/v_0 \equiv L/v$. As a result we get

$$\frac{dU_s}{d\Omega} = \frac{e^2\omega_0^4 a_0^2 L\,\{[1-(v/c)\cos\theta]^2-(1-v^2/c^2)\sin^2\theta\,\cos^2\phi\}}{8\pi c^3 v\,[1-(v/c)\cos\theta]^5}$$

$$= \frac{e^4 E_0^2 L\{[1-(v/c)\cos\theta]^2-(mc^2/\mathcal{E})^2\sin^2\theta\,\cos^2\phi\}}{8\pi c^3 v\, m^2[1-(v/c)\cos\theta]^5}\left(\frac{mc^2}{\mathcal{E}}\right)^2 , \tag{4.15}$$

where ϕ is the angle between the field $\mathbf{E}_0$ and the projection of $\mathbf{s} = \mathbf{k}/k$ onto the plane at right angles to $\mathbf{v}$.

In the relativistic — or actually, the ultra-relativistic — case (4.4) when $1/(1-v/c) \approx 2(\mathcal{E}/mc^2)^2$, we have

$$\frac{dU_s}{d\Omega} \approx \frac{e^4 E_0^2 L\,(mc^2/\mathcal{E})^2}{8\pi m^2 c^4[1-(v/c)\cos\theta]^3} ,$$

$$U = \int\frac{dU_s}{d\Omega}\,d\Omega \approx \frac{1}{3}\left(\frac{e^2}{mc^2}\right)^2\left(\frac{\mathcal{E}}{mc^2}\right)^2 E_0^2 L \;\; ; \;\; \frac{\mathcal{E}}{mc^2} \gg 1\,. \tag{4.16}$$

The radiation is then, as we mentioned earlier and as is clear from (4.15), concentrated within angles $\theta \sim mc^2/\mathcal{E}$ and has a characteristic frequency $\omega \sim \omega(0) \sim \omega_0\,(\mathcal{E}/mc^2)^2$. The total emitted energy is proportional to the factor $e^4\mathcal{E}^2/m^4 = e^4c^4/m^2(1-v^2/c^2)$, that is, it depends for a given charge both on the total energy $\mathcal{E}$ and on the rest mass m — or on m and the velocity v. At the same time the acceleration depends merely on $\mathcal{E}$ (see (4.9)).

We shall in Chapter 5 also discuss the radiation by a charge which moves with relativistic velocity in a magnetic field. Now, however, we shall discuss the radiation force acting on relativistic particles.

We have already in Chapter 3 derived the relativistic equation of motion including the radiation force (see (3.11), (3.12), (3.19) and (3.20)). It is, however, convenient, as in the non-relativistic case, to consider at once the approximate nature of the formula for the radiation force, expressing it in terms of the field strength. This was done by Landau and Lifzhitz (1975, §76), but it is convenient to do it again here, starting from the equation of motion

$$mc\,\frac{du^i}{ds} = \frac{e}{c}F^{ik}u_k + g^i \;, \quad g^i = \frac{2e^2}{3c}\left(\frac{d^2u^i}{ds^2} - u^iu^k\frac{d^2u_k}{ds^2}\right) . \tag{4.17}$$

To a first approximation

$$mc\,\frac{du^i}{ds} = \frac{e}{c}F^{ik}u_k \;; \quad \frac{d^2u^i}{ds^2} = \frac{e}{mc^2}\frac{\partial F^{ik}}{dx^\ell}u_ku^\ell + \frac{e^2}{m^2c^4}F^{ik}F_{k\ell}u^\ell . \tag{4.18}$$

Substituting d^2u^i/ds^2 from (4.18) into the expression for g^i we can write (4.17) in the form

$$mc\,\frac{du^i}{ds} = \frac{e}{c}F^{ik}u_k + \frac{2e^3}{3mc^3}\left\{\frac{\partial F^{ik}}{\partial x^\ell}u_ku^\ell - \frac{e}{mc^2}F^{i\ell}F_{k\ell}u^k + \frac{e}{mc^2}\left(F_{k\ell}u^\ell\right)\left(F^{km}u_m\right)u^i\right\} \tag{4.19}$$

We can in the ultra-relativistic case everywhere except in expressions such as $\gamma = \{1 - v^2/c^2\}^{-\frac{1}{2}}$ put $v = c$. The main radiative term in (4.19) is thus the last one and as $v \to c$ we can write the three-dimensional equation of motion in the form†

† For the sake of convenience we remind ourselves here of the connection between any four-dimensional force g^i and the corresponding three-dimensional force $\mathbf{f}_i$. We have

$$g^i = (g^0, \mathbf{g}) = \left\{\frac{(\mathbf{f}\cdot\mathbf{v})}{c^2\sqrt{(1 - v^2/c^2)}} \;, \; \frac{\mathbf{f}}{c\sqrt{(1 - v^2/c^2)}}\right\}$$

where

$$\frac{dp^i}{ds} = mc\,\frac{du^i}{ds} = g^i \;, \quad \frac{d\mathbf{p}}{dt} = \mathbf{f} \;,$$

since

$$u^i = \left\{\frac{1}{\sqrt{(1 - v^2/c^2)}} \;, \; \frac{\mathbf{v}}{c\sqrt{(1 - v^2/c^2)}}\right\} , \quad ds = c\,dt\,\sqrt{(1 - v^2/c^2)}$$

$$\left.\begin{aligned} \frac{d\mathbf{p}}{dt} &= \frac{d}{dt}\frac{m\mathbf{v}}{\sqrt{(1-v^2/c^2)}} = e\left(\mathbf{E}+\frac{1}{c}[\mathbf{v}\wedge\mathbf{H}]\right)+\mathbf{f} \\ \mathbf{f} &= \frac{2e^4}{3m^2c^5}\left(F_{k\ell}u^{\ell}\right)\left(F^{km}u_m\right)\mathbf{v} = -\frac{2e^2}{3m^2c^4}\frac{(E_y-H_z)^2+(E_z+H_y)^2}{1-v^2/c^2}\frac{\mathbf{v}}{v}, \\ \gamma &= \frac{1}{\sqrt{(1-v^2/c^2)}} \gg 1, \end{aligned}\right\} \quad (4.20)$$

where we have in the last expression for $\mathbf{f}$ chosen the x-axis along the direction of $\mathbf{v}$ in order to write down explicitly the components of the field. The fact that for $\gamma \gg 1$ the force $\mathbf{f}$ is parallel to $\mathbf{v}$ is also clear from Eq. (3.12) but that equation is, however, less convenient than (4.20).

We now want to make a few remarks which are important and are sometimes not taken into consideration. In the relativistic case the radiation is mainly in the forward direction, in the direction of the velocity. The recoil, or reaction, must thus be in the backward direction, as directed radiation carries momentum. This is also demonstrated by the fact that the force $\mathbf{f}$ is antiparallel to $\mathbf{v}$. It seems that from this it also follows that the radiation reaction will lead to a decrease of the velocity components in all directions. Such a conclusion is, however, in general false.

Let us as an example consider the motion in a constant and uniform magnetic field $\mathbf{H}_0$. As we shall in what follows repeatedly be interested in this case we shall dwell upon it in somewhat greater detail for the case of a particle of charge Ze and mass M.

The equation of motion then has the form (we neglect for the time being the reaction of the radiation)†

$$\frac{d}{dt}\left[\frac{M\mathbf{v}}{\sqrt{1-v^2/c^2}}\right] = \frac{Ze}{c}[\mathbf{v}\wedge\mathbf{H}_0]. \quad (4.21)$$

One can easily integrate this equation and it then becomes clear that the particle moves along a spiral: its velocity is $\mathbf{v} = \mathbf{v}_{\parallel} + \mathbf{v}_{\perp}$, where $\mathbf{v}_{\parallel}$ = constant

† For application of Eq. (4.21) this proviso is, strictly speaking, insufficient and we must also note that we shall also assume the charge Ze and the mass M to be constant. Of course, such an assumption is practically always implied, but it is not necessarily true, as for ions the charge and mass are not constant quantities when we take into account nuclear fission and/or the 'stripping off' of atomic electrons,

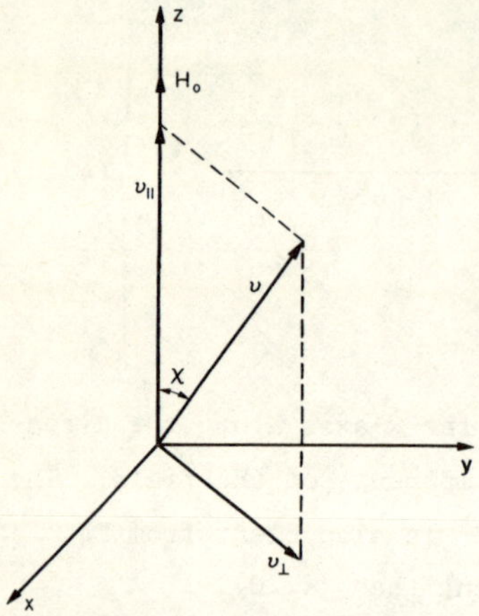

Fig. 4.5 The quantities $\mathbf{v}_\parallel$ $\mathbf{v}_\perp$, and χ

is the constant velocity along the field (along which we take the z-axis), while $\mathbf{v}_\perp$ is the velocity component in the xy-plane, that is, at right angles to $\mathbf{H}_0$ (Fig.4.5); we have

$$v_{\perp,x} = v_\perp \cos\omega_H^* t\ ,\ v_{\perp,y} = -v_\perp \sin\omega_H^* t\ ,\ v_\parallel = \text{constant},$$

$$v_\parallel = v\cos\chi\ ,\quad v_\perp = v\sin\chi\ ,\ v^2 = v_\perp^2 + v_\parallel^2\ , \tag{4.22}$$

$$\omega_H^* = \frac{ZeH_0^2}{Mc}\frac{Mc^2}{\mathcal{E}} = \omega_H\frac{Mc^2}{\mathcal{E}}\ ,\quad r_H = \frac{v\sin\chi}{\omega_H^*}\ .$$

Here χ is the angle between $\mathbf{v}$ and $\mathbf{H}_0$, and r_H the radius of the circle which the projection of the radius vector of the particle describes in the xy-plane, we shall, as usual, call this radius the radius of curvature, but it should not be confused with r_H^*, the radius of curvature of the trajectory in space of the particle which is equal to

$$r_H^* = \frac{r_H^2 + (v_\parallel/\omega_H^*)^2}{r_H} = \frac{v}{\omega_H^*\sin\chi}\ . \tag{4.23}$$

Unless we make a statement to the contrary we shall in what follows understand by e the absolute magnitude of the electron charge and take the frequencies ω_H and ω_H^* to be positive. In the ultra-relativistic case we have for electrons

$$r_H = \frac{\mathcal{E}\sin\chi}{eH_0} = \frac{\mathcal{E}\,(\text{eV})\sin\chi}{300\,H_0} \qquad \mathcal{E} \gg mc^2 = 5.1\times10^5\,\text{eV}\ . \tag{4.24}$$

We express the field H_0 always in Oersted (Gauss) and r_H — like other lengths — in cm. It is, however, often more convenient to express the energy $\mathcal{E}$ in eV, which we shall express as $\mathcal{E}(\text{eV}) = \mathcal{E}/1.6\times10^{-12}$, where $\mathcal{E}$ is in erg. The gyro-frequency ω_H^* is for a given field H_0 in the relativistic case lower than the non-relativistic gyro-frequency ω_H for an obvious reason: by virtue of the increase in the 'dynamic' mass $\mathcal{E}/c^2 = M/\sqrt{(1-v^2/c^2)}$. For electrons

$$\omega_H^* = \omega_H\frac{mc^2}{\mathcal{E}} = 1.76\times10^7\ H_0\frac{mc^2}{\mathcal{E}}\ \text{s}^{-1}\ . \tag{4.25}$$

Let us now turn to the radiation reaction. As the reaction force (4.20) is antiparallel to $\mathbf{v}$, it seems, as we mentioned earlier, that both velocity components $v_\perp$ and $v_\parallel$ must decrease when a charge moves in a magnetic field. Meanwhile it is clear that in the approximation (4.19) considered here (and only such an approximation is justified; see Chapter 2 and what follows) and when we take into account the radiation force in a constant and uniform magnetic field, $v_{\parallel} = \text{constant}$. Indeed, as in a constant magnetic field $v_{\parallel} = \text{constant}$ or $dv_{\parallel}/ds = 0$, when we neglect the radiation force, when we iterate, changing from (4.17) to (4.19), we can verify that as before $dv_{\parallel}/ds = 0$. The apparent contradiction with Eq.(4.20) is resolved when we bear in mind that that equation itself is valid only as $v \to c$, and that we can use it to consider the change in the momentum $\mathbf{p} = m\mathbf{v}/\sqrt{(1 - v^2/c^2)}$, but not the change in the velocity. The momentum of the particle, indeed, decreases under the action of the radiation, and this is true for both $\mathbf{p}_{\parallel}$ and $\mathbf{p}_{\perp}$.

After a sufficiently lone time $\mathbf{p}_{\perp} \to 0$ and

$$\mathcal{E} = \frac{mc}{\sqrt{1 - v_{\parallel}^2/c^2}} \;, \quad p_{\parallel} = \frac{mv_{\parallel}}{\sqrt{1 - v_{\parallel}^2/c^2}} \;, \quad p_{\perp} \to 0 \;, \quad t \to \infty \;. \tag{4.26}$$

From this it is clear that the particle remains relativistic, as $t \to \infty$, only for very small initial 'pinch angles' χ_0, so that $v_{\parallel} = v \cos\chi_0 \sim c$.

We saw in Chapter 2 that in the non-relativistic case the radiation reaction must be considered to be a perturbation — it must be small compared to other forces. The transition to the relativistic case and, in particular, the actual transition from Eq.(4.17) to (4.19) may impel us to reach a similar conclusion. However, such a conclusion about the necessity to require that the radiation force be small compared to the Lorentz force (in particular, in Eq.(4.20)) would be erroneous. Indeed, the relevant conditions are (2.5) and (2.6) which were obtained in Chapter 2 and which guarantee that the radiation force is small in the frame of reference in which the particle at a given time is at rest. In the laboratory frame in which the particle velocity is v condition (2.6a) becomes

$$H_0 \ll \frac{m^2c^4}{e^3}\frac{mc^2}{\mathcal{E}} = 6 \times 10^{15}\,\frac{mc^2}{\mathcal{E}}\ \text{Oe} \;, \tag{4.27}$$

where H_0 is the field in the laboratory frame (this field is in the rest frame of the particle larger in order of magnitude by a factor $\gamma = 1/\sqrt{1 - v^2/c^2}$, which leads to (4.27) rather than (2.6a); we are dealing here essentially with the Lorentz force component at right angles to the velocity $\mathbf{v}$, while the radiation force is for $\mathcal{E} \gg mc^2$ in the direction of $-\mathbf{v}$ and hence, like the component of $\mathbf{E}$ along $\mathbf{v}$, the same in the laboratory frame as in the rest frame of the particle).

Physically the situation becomes clearer if we frame our discussion in terms of the wavelengths or of the frequencies of the radiation, which is analogous to the transition from (2.6a) to the equivalent condition (2.6b). If $\omega = 2\pi c/\lambda$ is the frequency of the radiation in the laboratory frame, in the rest frame of the particle we have due to the Doppler effect $\omega_{00} \sim \omega/\gamma = \omega mc^2/\mathcal{E}$ and $\lambda_{00} \sim \lambda\,\mathcal{E}/mc^2$ (see (4.6); we take here for ω the maximum frequency $\omega(0)$). Condition (2.5) thus becomes

$$\lambda_{00} \sim \lambda\mathcal{E}/mc^2 \gg r_e ,$$

or

$$\lambda = \frac{2\pi c}{\omega} \gg r_e \frac{mc^2}{\mathcal{E}} = \frac{e^2}{mc^2}\frac{mc^2}{\mathcal{E}} , \quad \omega \ll \frac{c}{r_e}\frac{\mathcal{E}}{mc^2} = \frac{c\mathcal{E}}{e^2} . \qquad (4.28)$$

One obtains the same result by requiring that the wavelength λ be larger than the 'size' of the electron $r_e mc^2/\mathcal{E}$ in the laboratory frame; we have here the relativistic 'compression' of a moving object.

Let us now apply condition (4.28) to the case of emission in a magnetic field. We shall show in Chapter 5 that the characteristic frequency of the radiation in that case is $\omega \sim (eH_0/mc)(\mathcal{E}/mc^2)^2$. Substituting such a frequency into (4.28) also leads to the condition (4.27).

If we remain within the framework of classical theory, we can thus take the action of the radiation force into account, using Eqs.(4.17), (4.19), and (4.20) and the conditions (4.27) and (4.28). We have already mentioned in Chapter 2 that, if we take quantum effects into account (see (2.7)), we can use the classical theory only for wavelengths $\lambda \gg \hbar/mc = r_e/\alpha$, $\alpha = e^2/\hbar c \approx 1/137$. However, this inequality refers only to the rest frame. In the laboratory frame the condition for the applicability of the classical theory is obtained from (4.28) by replacing r_e by $\hbar/mc$, that is, it has the form

$$\lambda \gg \frac{\hbar}{mc}\frac{mc^2}{\mathcal{E}} , \quad \hbar\omega \ll \mathcal{E} . \qquad (4.29)$$

This inequality (the two expressions are in fact equivalent), expressed in terms of ω, is completely obvious: if the particle emits a photon of energy $\hbar\omega$ which is comparable with the particle energy $\mathcal{E}$, the classical approach clearly becomes inapplicable; it is sufficient to say that in a classical approach emission of waves with frequencies $\omega \gg \mathcal{E}/\hbar$ are not excluded which, however, violates the energy conservation law†.

---

† See footnote on the opposite page

For the case of motion in a magnetic field and of synchrotron radiation the quantal restriction on the field and the energy, which guarantees that one can use the classical theory, is

$$H_0 \ll \frac{e^2}{\hbar}\frac{m^2c^4}{e^3}\frac{mc^2}{\mathcal{E}} = \frac{m^2c^3}{e\hbar}\frac{mc^2}{\mathcal{E}} = 4.4\times10^{13}\,\frac{mc^2}{\mathcal{E}}\,. \tag{4.30}$$

We obtain this restriction by applying condition (4.29) to the synchrotron radiation with a characteristic frequency $\omega \sim (eH_0/mc)(\mathcal{E}/mc^2)^2$.

The characteristic 'quantum' field $m^2c^3/e\hbar$ has the following meaning: an electric field $E_0 \sim m^2c^3/e\hbar$ along a path length $\hbar/mc$ performs on the charge e an amount of work of the order of mc^2. Physically it follows from this that in such (and, of course, in stronger) fields it is already possible to produce electron-positron pairs for which one needs an energy of at least $2mc^2 \approx 1$ MeV. A particle of energy $\mathcal{E}$ can produce pairs already in fields $F_0 \gtrsim (m^2c^3/e\hbar)(mc^2/\mathcal{E})$, as in its rest frame the field reaches at once the critical value $m^2c^3/e\hbar$ (for details about quantum effects in strong fields see Nikishov and Ritus (1970), Ritus (1972a,b), and the literature cited in these papers).

If we now turn to Eq.(4.20), we see that the radiative reaction force is small compared to the Lorentz force, provided (we assume that $H_\perp \sim H_0$, and so on; see also (4.44) below and the remarks following inequality (4.27))

$$F_0 \ll \frac{m^2c^4}{e^3}\left(\frac{mc^2}{\mathcal{E}}\right)^2 = 6\times10^{15}\left(\frac{mc^2}{\mathcal{E}}\right)^2 , \quad F_0 \sim E_0 \text{ or } H_0\,. \tag{4.31}$$

By virtue of the extra factor $mc^2/\mathcal{E}$ which occurs here, as compared to the criteria (4.27) and (4.29), it may well turn out that the radiative reaction force is even large compared to the Lorentz force, and the classical Eq.(4.20) is moreover applicable; condition (4.30) is satisfied. The range of fields for which such a situation occurs is clearly determined by the inequalities

$$\frac{m^2c^4}{e^3}\left(\frac{mc^2}{\mathcal{E}}\right)^2 \ll H_0 \ll \frac{m^2c^3}{e\hbar}\frac{mc^2}{\mathcal{E}} \sim 4\times10^{13}\,\frac{mc^2}{\mathcal{E}}\,. \tag{4.32}$$

---

† In actual fact the situation is somewhat more complicated. If the motion of the particle is given, as is often assumed in classical theory, it may in principle emit at arbitrarily high frequencies — the reaction of the radiation (recoil) is compensated by the external forces which secure the motion with the given parameters. However, a free particle with energy $\mathcal{E}$ can, of course, never emit photons with energies $\hbar\omega > \mathcal{E}$.

For fixed H_0 condition (4.32) can also be written as

$$\sqrt{\frac{m^2c^4}{e^3H_0}} \ll \frac{\mathcal{E}}{mc^2} \ll \frac{m^2c^3}{e\hbar H_0} \sim \frac{4\times 10^{13}}{H_0} . \tag{4.33}$$

For fields of the order of 10^9 to 10^{13} Oe (at the surface of pulsars) the upper bound to the parameter $\mathcal{E}/mc^2$ (that is, the restriction to 'classical' behaviour) is rather strong and has a completely realistic value. Moreover, the reaction force dominates over the Lorentz force for energies

$$\frac{\mathcal{E}}{mc^2} \gg \sqrt{\frac{m^2c^4}{m^3H_0}} \sim \frac{10^8}{\sqrt{H_0}} . \tag{4.34}$$

It is clear from the derivation that the relative smallness of the reaction force in the rest frame of the particle is not retained in the laboratory frame because the reaction force and the Lorentz force depend differently on the energy as, generally speaking, they have different directions. The derivation of Eq.(4.19) from (4.17) remains correct also in the case (4.34) as it was derived in an invariant manner and is thus valid when the components of one four-vector (the vector g^i) are small compared to those of the other four-vector (the vector $(e/c)F^{ik}u_k$) in any frame of reference (we have repeated here the appropriate explanation from §76 of Landau and Lifshitz's book (1975)).

Expression (4.20) for the reaction force is very convenient for finding radiation losses, that is, energy losses due to radiation. For instance, in a constant magnetic field the radiation losses are

$$\mathcal{R}(\mathcal{E}) = \frac{2e^4}{3m^2c^3} H_\perp^2 \left(\frac{\mathcal{E}}{mc^2}\right)^2 , \quad \frac{\mathcal{E}}{mc^2} \gg 1 , \tag{4.35}$$

where $\mathbf{H}_\perp \equiv \mathbf{H}_{0\perp}$ is the projection of the field $\mathbf{H}_0$ onto the plane at right angles to $\mathbf{v}$. For an arbitrary velocity v for a particle of charge Ze and mass M we have

$$\mathcal{R} = \frac{2(Ze)^4 H_\perp^2 v^2}{3M^2c^5(1-v^2/c^2)} = \frac{2(Ze)^4 H_\perp^2}{3M^2c^3} \left\{\left(\frac{\mathcal{E}}{Mc^2}\right)^2 - 1\right\} . \tag{4.36}$$

The most direct way to obtain this expression exists in using Eq.(4.19). One often for the determination of the radiation losses calculates the emitted energy, but we noted already in Chapter 3 a well known limitation to such a procedure, and it may lead to confusion (see Ginzburg, Sazonov, and Syrovatskii, 1968; Ginzburg and Syrovatskii, 1969; and the discussion in Chapter 5 of the peculiarities of synchrotron radiation for spiral motion). However, an evaluation of the emitted energy leads in the case of motion along a circle in a magnetic field (see Landau and Lifshitz, 1975, §74) to the correct Eq.(4.36) with $H_\perp = H_0$; the generalization of this formula to the case of spiral motion can be directly performed in the above mentioned manner, but we can, in general, give also a different reason for the resulting substitution $H_0 \to H_\perp \equiv H_{0\perp}$. In fact, there are no losses when $H_\perp = 0$, we can assume that the losses are known when $H_\perp = H_0$, and, finally, when $\mathcal{E}/mc^2 \gg 1$, (4.35) is valid; therefore it is rather obvious to have Eq.(4.36) as it gives the correct result in three limiting cases.

When we neglect the radiation force the energy $\mathcal{E} = Mc^2/\sqrt{(1 - v^2/c^2)}$ is conserved during motion in a magnetic field; to prove this it is sufficient to multiply Eq.(4.21) scalarly by $\mathbf{v}$. When we take the radiation force into account we have thus

$$\frac{d\mathcal{E}}{dt} = - \mathcal{R} , \tag{4.37}$$

or, in the ultra-relativistic case

$$\frac{d\mathcal{E}}{dt} = -\frac{2(Ze)^4 H_\perp^2}{3M^2c^3}\left(\frac{\mathcal{E}}{Mc^2}\right)^2 = -0.98 \times 10^{-3} H_\perp^2 \left(\frac{Z^2 m}{M}\right)^2 \left(\frac{\mathcal{E}}{Mc^2}\right)^2 \text{ eV/s} . \tag{4.38}$$

For electrons $(Z = 1,\ M = m)$

$$\frac{d\mathcal{E}}{dt} = -\frac{2}{3}\frac{e^2 H_\perp^4}{m^2c^3}\left(\frac{\mathcal{E}}{mc^2}\right)^2 = - 0.98 \times 10^{-3} H_\perp^2 \left(\frac{\mathcal{E}}{mc^2}\right)^2 \text{ eV/s}$$

$$= - 1.58 \times 10^{-15} H_\perp \left(\frac{\mathcal{E}}{mc^2}\right)^2 \text{ erg/s} . \tag{4.39}$$

We can write this equation also in the following form

$$\frac{d\mathcal{E}}{dt} = -\beta\,\mathcal{E}^2 , \quad \beta = \frac{2e^4 H_\perp^2}{3m^4c^7} = 1.95 \times 10^{-9} \frac{H_\perp^2}{mc^2} \text{erg}^{-1} \text{ s}^{-1} . \tag{4.40}$$

Hence we have

$$\mathcal{E}(t) = \frac{\mathcal{E}_0}{1 + \beta\,\mathcal{E}_0 t} , \quad \mathcal{E}_0 = \mathcal{E}(0) . \tag{4.41}$$

The energy of the electron is thus reduced by a factor two after a time

$$T_m = \frac{1}{\beta\,\mathcal{E}_0} = \frac{5.1 \times 10^8}{H_\perp^2}\frac{mc^2}{\mathcal{E}_0} \text{ s} . \tag{4.42}$$

It is further clear from (4.41) that for any initial energy $\mathcal{E}_0$ the energy of electron at time t cannot exceed the value

$$\mathcal{E}_{max}(t) = \frac{1}{\beta t} = \frac{5.1 \times 10^8}{H_\perp^2 t} mc^2 = \frac{2.6 \times 10^{14}}{H_\perp^2 t} \text{ eV} , \tag{4.43}$$

where t is measured in seconds.

The result (4.43), that is, the existence of a maximum (limiting) energy $\mathcal{E}_{max}$ for a charge moving in a magnetic field can be generalized to the case of a non-uniform field (see Landau and Lifshitz, 1975, §76). As far as we know, the first discussion of the problem of the limiting energy was for the case of cosmic rays reaching the Earth (Pomeranchuk, 1939). In that case the particle must traverse the terrestrial magnetic field $H \sim 0.2$ to 0.5 Oe, that is, traverse a path length $L \sim R_{\oplus} \sim 10^9$ cm (the Earth's radius $R_{\oplus} \approx 6360$ km), for which it needs a time $t \sim R_{\oplus}/c \sim 3 \times 10^{-2}$ s. As a result we get for electrons $\mathcal{E}_{max} \sim 10^{17}$ eV (we have put in (4.43) $H_{\perp} \sim H_0 \sim 0.2$ Oe, which corresponds to the particle being incident in the equatorial region; Pomeranchuk's calculations (1939) led to a value $\mathcal{E}_{max} = 4 \times 10^{17}$ eV for incidence at the Earth's equator).

We have written Eqs. (4.38) to (4.43) in various forms and with various numerical coefficients as they are widely used, in particular in astrophysics (see Chapter 15).

When a particle moves in a constant and uniform magnetic field, the radiation force is small compared to the Lorentz force, if during a single revolution, that is, during a period $T = 2\pi/\omega_H^* = (2\pi Mc/ZeH_0)(\mathcal{E}/Mc^2)$, the loss of energy $\Delta\mathcal{E}$ by the particle is small compared to $\mathcal{E}$. We are thus led to the condition (for $M = m$, $Z = 1$)

$$\frac{\mathcal{E}}{Mc^2} \ll \sqrt{\frac{m^2c^4H_0}{e^3H_{\perp}^2}} \sim 10^8\sqrt{\frac{H_0}{H_{\perp}^2}} = \frac{10^8}{\sqrt{H_0\sin^2\chi}} . \qquad (4.44)$$

As one should have expected, this condition is the same as the one obtained earlier, but it makes it somewhat more precise (see (4.34), where we considered the opposite inequality and put $H_{\perp} \sim H_0$). In the cosmos inequality (4.44) is often well satisfied. For instance, in the interstellar field $H_0\sin^2\chi \sim 10^{-6}$ Oe and (4.44) is satisfied for electrons with energies $\mathcal{E} \ll 10^{17}$ eV. However, when $H_0 \sim 10^4$ Oe (sunspots, magnetic stars, accelerators) we are led to a much more rigid requirement $\mathcal{E} \ll 10^{12}$ eV; however, in the cases mentioned just now one does not meet with electrons with energies $\mathcal{E} \gg 10^{11}$ eV in practice. In magnetic white dwarfs ($H_0 \sim 10^7$ to 10^8 Oe) and in pulsars ($H_0 \sim 10^9$ to 10^{13} Oe) inequality (4.44) is already violated for relatively low energies. On the other hand, at the same time, the larger the losses, the more difficult it is to accelerate particles to high energies and, generally speaking, the fewer such particles there will be. In practice therefore the condition (4.44) that the radiation force be small is violated only in exceptional cases. In such cases, the particle does, of course, not move along a circle or a spiral, but along a curve with a fast decreasing radius (see papers by Shen (1970); Suvorov and Chugunov (1973), and Lubart (1974) for some calculations for that case). At the same time the synchrotron radiation does not change its nature (intensity, spectral structure, polarization) even when condition (4.44) is violated as only a small part of the trajectory is responsible for emission with $\gamma = \mathcal{E}/mc^2 \gg 1$, and it would be changed only if the condition for classical behaviour (see (4.33))

$$\frac{\mathcal{E}}{mc^2} \ll \frac{m^2c^3}{e\hbar H_0} \sim \frac{4 \times 10^{13}}{H_0} \quad , \tag{4.45}$$

were not satisfied.

We shall clarify what has been said here in Chapter 5. We shall everywhere in what follows assume that condition (4.45) is satisfied†, and normally, for the sake of simplicity, we shall also assume that condition (4.44) is valid, although it is not used in calculations for the majority of cases.

It is clear from Eq.(4.38) for the losses that in a given field $H_\perp$ and at the same energy $\mathcal{E}$ protons emit less than electrons by a factor $(M/m)^4 \sim 10^{13}$. Therefore the synchrotron (magneto-brems) losses for protons, and also for other nuclei, usually do not play any role. However, in very strong fields when electrons with high energies 'do not survive', proton synchrotron emission may in principle turn out to be important. We note that a charged particle moving in a magnetic field (and, in general, being accelerated) emits not only electromagnetic waves, but also quanta of all those fields with which it interacts. For example, for all charged particles there is magneto-brems gravitational radiation. Furthermore, protons must in a magnetic field emit π^+- and π^0-mesons ($p \to n + \pi^+$, $p \to p + \pi^0$ processes, where n is a neutron), and also positrons and neutrinos ($p \to n + e^+ + \nu$ process, where ν is a neutrino). However, the intensity of non-electromagnetic synchrotron radiation is usually under real conditions negligibly small (Ginzburg and Zharkov, 1965) and it does not play a role by itself. Nevertheless, from a methodological point of

---

† When inequality (4.45) is not satisfied we are in the quantum region and we must use the methods of quantum electrodynamics for our discussion (Nikishov and Ritus, 1970; Ritus, 1972a,b). We have already remarked that characteristic for the quantum region is the electron-positron pair production, while for yet larger fields or energies there is also the creation of pairs of other kinds of particles (mesons, baryons). Undoubtedly, the quantum region (strong fields, high energies) is of very great interest, and this interest will only increase in connection with the discovery of pulsars, the attaining of stronger fields in the laboratory, and the appearance of new accelerators.

view and bearing in mind that there may always be new possibilities (which one should never forget) one should remember the non-electromagnetic synchrotron radiation. Incidentally, even though we apply the term synchrotron (or, in general, magneto-brems) only to the radiation by particles moving in a magnetic field, in the literature the term synchrotron radiation is also used in a wider sense, for instance, in application to the emission of gravitational, electromagnetic, or other waves (say, waves of a scalar field) when relativistic particles move in a strong gravitational field (Chitre and Price, 1972; Doroshkevich, Novikov and Polnarev, 1973). In all these cases the radiation has some common features with the usual electromagnetic synchrotron emission, but on the whole its character (intensity as function of energy, polar diagram, and so on) is appreciably changed depending on the kind of accelerating and emitted fields. Remembering that it is 'impossible to encompass the unbounded' we shall in Chapter 5 explicitly talk about the synchrotron radiation by electrons or, to spell it out, about the emission of electromagnetic waves by ultrarelativistic electrons moving in a constant and uniform magnetic field.

Chapter V

SYNCHROTRON RADIATION

Peculiarities of the synchrotron radiation.
Some applications of the theory of synchrotron radiation in astrophysics.
Limits of applicability of the theory.

The character and, in particular, the spectrum of the synchrotron radiation depends very strongly on the ratio of the angle θ between the wavevector of the radiation, $\mathbf{k}$, and the particle velocity $\mathbf{v}$, and the angle χ between $\mathbf{v}$ and the external magnetic field $\mathbf{H}_0$ (we restrict our discussion here to motion in a magnetic field, although, in fact, we are dealing with a very general result; see, for instance, Landau and Lifshitz, 1975, §77). The fact of the matter is that the radiation is mainly concentrated within angles $\theta \sim mc^2/\mathcal{E} \ll 1$ and, if the angle $\chi \leqslant mc^2/\mathcal{E}$, in a given direction $\theta \leqslant mc^2/\mathcal{E}$ the radiation is 'accumulated' from the whole trajectory or, at any rate, from a significant part of it. If, however,

$$\chi \gg \xi = \frac{mc^2}{\mathcal{E}} \quad , \quad \frac{mc^2}{\mathcal{E}} \ll 1 \tag{5.1}$$

the synchrotron radiation gets to the observer only from a small section of the trajectory (for details see below).

The radiation is completely analogous to that in an ondulator when

$$\chi \leqslant \xi = \frac{mc^2}{\mathcal{E}} \quad , \quad \frac{mc^2}{\mathcal{E}} \ll 1 \tag{5.2}$$

as in that case (see (4.22))

$$r_H = \frac{v \sin\chi}{\omega_H^*} \approx \frac{c\chi}{\omega_H^*} \leqslant \frac{c}{\omega_H} = \frac{\lambda_H}{2\pi} \quad , \quad \omega_H = \frac{eH}{mc} \quad , \quad \omega_H^* = \omega_H \frac{mc^2}{\mathcal{E}} \quad ; \tag{5.3}$$

as only the external field occurs in our expressions we drop here and henceforth the index of $\mathbf{H}_0$.

To be more precise, the radiation is similar to that of a fast moving dipole (oscillator) with a moment $er_H = ec\chi/\omega_H^* = \mathcal{E}\chi/H \ll mc^2/H$, provided (see (4.10))

$$\chi \ll \frac{mc^2}{\mathcal{E}} \quad , \quad r_H \ll \frac{\lambda_H}{2\pi} \ . \tag{5.4}$$

We shall not discuss this case here in detail, as the situation was explained qualitatively in Chapter 4 for details see Landau and Lifshitz, 1975, §77 and Germantsev and Ginzburg 1952).

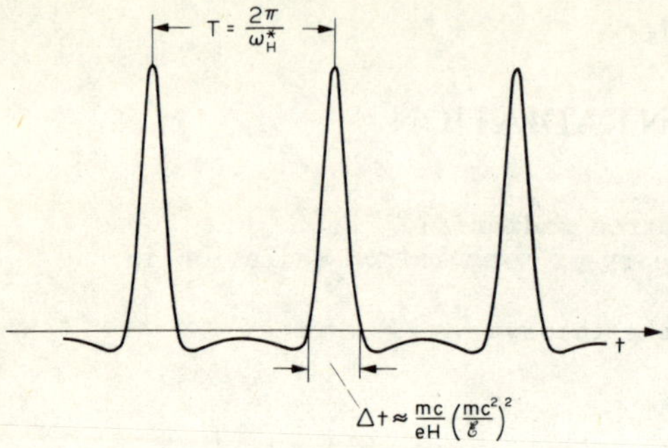

Fig.5.1. Time-dependence of the electric field in the wave zone for a particle moving along a circle in the magnetic field.

This picture is obtained if the field of a fast moving dipole (see Fig.4.4) is rotated with an angular velocity ω_H^*.

As to the radiation when condition (5.1) is satisfied, it is expedient to start with the important special case of motion along a circle, when $\chi = \pi/2$.

The whole of the radiation is then concentrated close to the plane of the orbit within angles $\theta \leqslant mc^2/\mathcal{E}$. At large distances from the orbit, the 'observer', whom we assume to be in a position in or close to the plane of the orbit within an angle $\theta \leqslant mc^2/\mathcal{E}$, registers radiation pulses which follow one another after time intervals equal to the period of the rotation of the charge

$$T = \frac{2\pi}{\omega_H^*} = \frac{2\pi mc}{eH} \frac{\mathcal{E}}{mc^2} \ . \tag{5.5}$$

One can easily explain the shape of the radiation pulses (Fig.5.1) by considering the electric field of a fast moving oscillator (dipole; see Fig.4.4), which swings round with respect to the observer due to the rotation of the particle in the magnetic field (the acceleration vector which corresponds to the dipole axis is all the time at right angles to the field **H** and rotates around it with the frequency ω_H^*). The length of each pulse is

$$\Delta t \sim \frac{r_H \xi}{c}\left(\frac{mc^2}{\mathcal{E}}\right)^2 \approx \frac{mc}{eH}\left(\frac{mc^2}{\mathcal{E}}\right)^2 , \tag{5.6}$$

where $r_H = v/\omega_H^* \approx \mathcal{E}/eH$ is the radius of curvature of the particle trajectory, while the factor $(mc^2/\mathcal{E})^2$ appears as a result of the Doppler effect. Indeed, within the angular range $\xi = mc^2/\mathcal{E}$ the electron moves in the direction of the observer during a time $\Delta t' \sim r_H \xi/c \approx mc/eH$. During that time the electron traverses a distance $v\Delta t'$ (this is just the Doppler effect). As a result the observed length of the pulse is of the order of $(c-v)\Delta t'$ and for its duration we have

$$\Delta t = \Delta t' \left(1 - \frac{v}{c}\right) \sim \frac{1}{2}\Delta t' \left(\frac{mc^2}{\mathcal{E}}\right)^2 ,$$

which is equivalent to (5.6).

The spectrum of radiation which is in the form of pulses which repeat at time intervals $T = 2\pi/\omega_H^*$ will, clearly, consist of overtones of the frequency ω_H^*. In actual fact, however, we may assume the spectrum in the region of the high harmonics to be continuous, as $T \gg \Delta t$, and the maximum of the spectrum corresponds to the frequency

$$\omega_m \sim \frac{1}{\Delta t} \sim \frac{eH}{mc}\left(\frac{\mathcal{E}}{mc^2}\right)^2 . \qquad (5.7)$$

It is important here that the field of the radiation changes sign (see Fig.5.1). This is just the reason that there is a maximum in the spectrum (vide infra). The effective width of the spectrum of the radiation is also of the order of ω_m and we can therefore estimate the average spectral density of the synchrotron radiation by dividing the total power of this radiation $\mathcal{P} = \mathcal{R} = (2e^4H^2/3m^2c^3)(\mathcal{E}/mc^2)^2$ (see Eq.(4.39))† by ω_m. As a result we find

$$\bar{p} \sim \frac{\mathcal{P}(\mathcal{E})}{\omega_m} \sim \frac{e^3H}{mc^2} . \qquad (5.8)$$

One of the characteristic features of synchrotron (and in general of magneto-brems) radiation is its polarization. The predominant direction of the electric vector in the emitted waves lies in the same plane as the direction of the acceleration and the line of sight (the vector **k**). As the direction of the acceleration changes all the time when a particle moves in a magnetic field, the waves will in general be elliptically polarized. Indeed, if the oscillator (see Fig.4.4) moves toward the observer, the polarization of the radiation propagating in the direction of the velocity of the translational motion does not change. Hence it is clear that the magneto-brems radiation of a single electron is in the general case elliptically polarized and the electric field **E** in the wave is a maximum in the plane through the direction

† For motion along a circle — and in general when the emitter on the whole does not approach the separate source — the power of the radiation $\mathcal{P}$ and the radiation losses $\mathcal{R}$ are equal to one another (see Chapter 3 and the remarks made there).

of the acceleration. This means that the predominant direction of the field **E** in the wave is perpendicular to the projection of the magnetic field on the plane of the figure (as usual we understand by the plane of the figure the plane perpendicular to the line of sight).

Let us now consider the emission when the motion is along a spiral (but with condition (5.1) satisfied). With respect to each separate pulse the situation is the same here as for motion along a circle, with the field H replaced by its component $H_\perp = \sin\chi$ at right angles to the velocity. Indeed, we now have for the length of the pulse

$$\Delta t \sim \frac{r_H^* \xi}{c}\left(\frac{mc^2}{\mathcal{E}}\right)^2 \approx \frac{mc}{eH_\perp}\left(\frac{mc^2}{\mathcal{E}}\right)^2 , \qquad (5.9)$$

where $r_H^* = v/(\omega_H^* \sin\chi) \approx \mathcal{E}/eH_\perp$ is the radius of curvature of curvature of the spatial trajectory of the particle (see (4.23)). We have born in mind in (5.9) that the electron moves in the direction of the observer within the angular range $\xi = mc^2/\mathcal{E}$ during a time $\Delta t' \sim r_H^* \xi/c \approx mc/eH_\perp$; the transition from $\Delta t'$ to Δt must be done in the same way as for the circular motion.

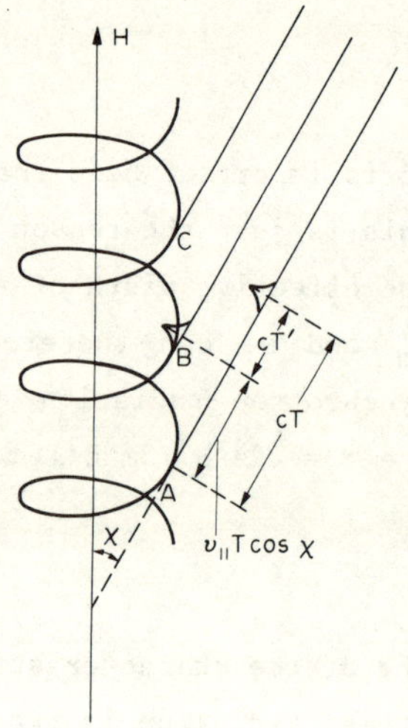

Fig.5.2. Emission during spiral motion. The time between pulses T' differs from the period $T = (2\pi mc/eH)(\mathcal{E}/mc^2)$ due to the Doppler effect.

Moreover, whereas for the circular motion the pulses follow one another after a period T (see (5.5)), for spiral motion the radiation pulses follow one another after a time T' which differs from T due to the Doppler effect.

One can use Fig.5.2 to find the time T' in an elementary way. For a given observer flashes of radiation occur when the electron is at the points A,B,C, ... (for the sake of simplicity we assume here and in what follows that the radiation is strictly 'needle-shaped'). In other words, just at those points the electron 'looks at' the observer. The interval between the times when the electron passes the points A and B is, of course, equal to the period $T = 2\pi/\omega_H^*$. The distance between the points A and B is equal to $v_{\|} T = vT\cos\chi$, where χ is the angle between **v** and **H**, and the pulse emitted at the point A during that time traverses a path length cT. It is clear from Fig.5.2 that the pulse emitted at the point B arrives at the observer lagging behind the first pulse by an amount

$$T' = T\left(1 - \frac{v_{\|}\cos\chi}{c}\right) = T\left(1 - \frac{v\cos^2\chi}{c}\right) \approx T\sin^2\chi = \frac{2\pi}{\omega_H^*}\sin^2\chi\,, \qquad (5.10)$$

where we used the fact, when changing to the penultimate expression, that the whole argument is for the limiting case as $v \to c$. We remember again that the picture used where the radiation reaches the observer in the form of separate pulses is applicable only when $\chi \gg \xi = mc^2/\mathcal{E}$. In fact, however, an expression such as $T' = T(1 - v_{\|}\cos\chi)$ has a general character and its appearance is not necessarily linked with the assumption about the 'needle' nature of the radiation or with the possibility to divide it into separate pulses (for details see Ginzburg, Sazonov, and Syrovatskii, 1968; Ginzburg and Syrovatskii, 1969).

The radiation spectrum of an ultra-relativistic electron consists thus in the wave zone of harmonics of the frequency

$$\Omega_H = \frac{2\pi}{T'} = \frac{\omega_H^*}{\sin^2\chi}\,. \qquad (5.11)$$

By itself this fact is not very important when we bear in mind that in all cases of interest to us the harmonics are not resolved and we are dealing with a continuous spectrum. However, the change in the interval between the pulses shows up not only in the spectrum, but also in all characteristics of the radiation field, in particular in its intensity registered at the point of observation. Indeed, let the electron lose through radiation in each revolution (over a time $T = 2\pi/\omega_H^*$) an energy $\Delta\mathcal{E} = \mathfrak{R}T$, where $\mathfrak{R} = (2e^4H_\perp^2/3m^2c^3)(\mathcal{E}/mc^2)^2$ as is clear from (4.37) and (4.39). By virtue of what we have said earlier it is then clear that this energy reaches an 'observer' which is positioned on a fixed sphere at a distance R from the electron over a time T' and, hence, the average observed power of the radiation (total energy flux) will be equal to

$$\mathfrak{P} = \frac{\Delta\mathcal{E}}{T'} = \frac{\mathfrak{R}T}{T'} = \frac{\mathfrak{R}}{\sin^2\chi}\,. \qquad (5.12)$$

At first sight it might look as if we have here a contradiction of the energy conservation law. The electron loses an energy $\mathfrak{R}$ per unit time. The whole of this energy changes to radiation and must, surely, be equal to the total flux of radiation through the sphere under consideration. One often proceeds as follows: one evaluates the radiative losses suffered by the particle and equates them to the total flux of radiation. In a stationary case and for an emitter with a fixed centre of mass one can, indeed, proceed in this way. However, in general, as we reminded ourselves in Chapter 3, the work performed by the emitter per unit time (the power of the losses $\mathcal{R}$) is equal to the total flux through a surface plus the change in the field energy $(d/dt)\int[(E^2+H^2)/8\pi]d^3\mathbf{r}$ in the volume enclosed in that surface. In the case in which we are interested the region of space occupied by the radiation and lying between the moving electron and the surface which is fixed in space on which the observation is performed decreases all the time. At the same time the energy enclosed in that region also decreases and hence the power of the observed radiation $\mathcal{P}$ is larger than the power of the losses $\mathcal{R}$. All the same, in a number of papers the power of the losses $\mathcal{R}$ has been used when going over to spectral quantities for the intensity. Such an approach can of course not lead to a correct expression for the intensity of the radiation fixed at some non-moving surface, if the motion of the emitter is taken into account. However, if the radiating particles are in a fixed volume (e.g., the shell of a supernova) or, to be more precise, if the distribution function of the radiating particles does not change with time, the intensity of the radiation of the assembly of particles is the same as the spectral power of the losses. This conclusion is evident from the energy conservation law and is, of course, confirmed by direct calculations (see Ginzburg, Sazonov, and Syrovatskii, 1968; Ginzburg and Syrovatskii, 1969, and below).

We have assumed that the fact that the whole of this essentially entirely elementary problem had not been elucidated for such a long time and had led to the use of formulae which were either not completely or not always correct†, justifies such a detailed exposition and, in fact, a repeated return to the appropriate remarks.

The elementary considerations and formulae given here provide us with a picture which is qualitatively completely clear and they also enable us to give estimates for the characteristic features, such as the intensity, spectrum, and polarization, of the synchrotron radiation for concrete cases. For obtaining quantitative formulae we need, on the other hand, rather cumbersome calculations. These calculations are performed on the basis of the well known formulae for the retarded potentials and can be found, for instance, in the paper

---

† This refers, in particular, to several formulae given by Ginzburg and Syrovatskii (1964a, 1966a). Fortunately, in that case these formulae were applied only to cases (or in conditions) where the corresponding differences between the expressions for the intensity can be neglected (see below and Ginzburg, Sazonov, and Syrovatskii, 1968; and Ginzburg and Syrovatskii, 1969).

by Ginzburg, Sazonov, and Syrovatskii (1968; see also Ginzburg and Syrovatskii 1969); for the case of circular motion, that is, when $\sin\chi = 1$, a number of calculations can be found in §75 of Landau and Lifshitz's book (1975)†. We shall therefore restrict outselves here to giving some final results and also to discussing their application in astrophysics. Bearing in mind such applications we digress somewhat from the nature of exposition which we have adopted on the whole in the present book; that is, in agreement with the work by Ginzburg and Syrovatskii (1964a, 1966a) we shall give a rather large number of formulae which are, in fact, auxiliary in nature or convenient for calculations. Of course, readers who are not interested in applications of the synchrotron theory can omit these details. The next part of the present chapter (up to and including Eq.(5.66)) is thus in a well understood sense auxiliary in nature.

We can expand the field of the synchrotron emission of a particle in a Fourier series in the overtones of the frequency $\Omega_H = 2\pi/T' = \omega_H^*/\sin^2\chi$, as the motion has a period T'. In other words, the radiation field at large distances from the charge can be written in the form

$$\mathbf{E} = \mathrm{Re}\sum_{n=1}^{\infty} \mathbf{E}_n \exp\left[i\omega_n\left(\frac{R}{c} - t\right)\right], \quad \omega_n = n\Omega_H = \frac{\omega_H^*}{\sin^2\chi}, \quad n = 0,1,2,3\ \ldots \tag{5.13}$$

In the ultra-relativistic case considered we have for an electron, up to terms of order $\xi^3 = (mc^2/\mathcal{E})^3$

$$\mathbf{E}_n = \frac{2e\omega_H^*}{\sqrt{3}\pi cr}\frac{n}{\sin^5\chi}\{(\xi^2+\psi^2)K_{\frac{2}{3}}(g_n)\,\boldsymbol{\ell}_1 + i\psi(\xi^2+\psi^2)^{\frac{1}{2}}K_{\frac{1}{3}}(g_n)\,\boldsymbol{\ell}_2\}, \tag{5.14}$$
$$\psi \ll 1, \quad \xi = \frac{mc^2}{\mathcal{E}} \ll 1,$$

where $\psi = \chi - \alpha$ is the difference between the angles between $\mathbf{v}$ and $\mathbf{H}$ and between $\mathbf{k}$ and $\mathbf{H}$; the angle ψ is, clearly, the angular distance between the generatrixes of the cone described by the vectors $\mathbf{v}$ and the direction of the wavevector $\mathbf{k}$ (if the vectors $\mathbf{v}$, $\mathbf{k}$ and $\mathbf{H}$ lie in one plane, as may happen at some appropriate time, the angle ψ has the same absolute magnitude as the angle θ). In (5.14) e is the absolute magnitude of the charge (the electron charge) and, by definition $\omega_H^* > 0$. For a positively charged particle (positron)

---

† It is necessary to remind the reader about the use of different notations. In the above-mentioned papers by Ginzburg et al. the angle between $\mathbf{v}$ and $\mathbf{H}$ was denoted by θ, but here it is denoted by χ; while θ here denotes the angle between $\mathbf{v}$ and $\mathbf{k}$.

the amplitude of the field is the complex conjugate with respect to $\mathbf{E}_n$ according to (5.14) which corresponds to the opposite sense of rotation of the electric vector. Further, in (5.14) $\boldsymbol{\ell}_1$ and $\boldsymbol{\ell}_2$ are two mutually orthogonal unit vectors in the plane of the figure, and $\boldsymbol{\ell}_2$ is directed along $\mathbf{H}_\perp$ while $\boldsymbol{\ell}_1 = [\boldsymbol{\ell}_2 \wedge \mathbf{k}]/k$ (see Fig.5.3). Finally, $K_{\frac{1}{3}}(g_n)$ and $K_{\frac{2}{3}}(g_n)$ are Bessel functions of the second kind of the imaginary argument g_n (Macdonald functions; see, for instance, Gradshteyn and Ryzhik, 1965, §§8.4 and 8.5)

$$g_n = \frac{n}{3\sin^3\chi}(\xi^2 + \psi^2)^{\frac{3}{2}} = \frac{\nu}{2\nu_c}\left(1 + \frac{\psi^2}{\xi^2}\right)^{\frac{3}{2}} . \quad (5.15)$$

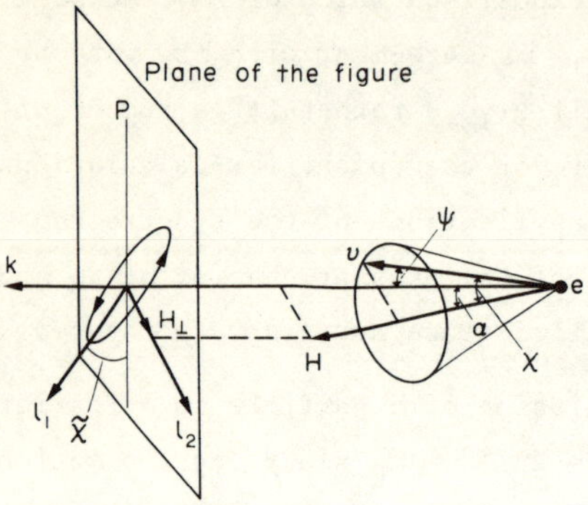

Fig.5.3. The ellipse of the oscillations of the electric vector in the wave emitted by a particle moving in a magnetic field.

The charge is assumed to be negative (electron); for a positively charged particle (positron) the direction of rotation is opposite to the one shown. P is the plane of the figure (the plane perpendicular to the direction of the emission, or, what amounts to the same, to the direction to the observer), $\boldsymbol{\ell}_1$ and $\boldsymbol{\ell}_2$ are two mutually orthogonal unit vectors in the plane of the plane of the figure, with $\boldsymbol{\ell}_2$ directed along $\mathbf{H}_\perp$ – the projection of the magnetic field $\mathbf{H}$ onto the plane of the figure

In the last of Eqs.(5.15) we changed from the number of the harmonic to the frequency $\nu = \omega/2\pi = n\omega_H^*/2\pi \sin^2\chi$ and introduced the notation

$$\nu_c = \frac{3\,\omega_H^* \sin\chi}{4\pi\xi^3} = \frac{3\,e\,H_\perp}{4\,\pi\,mc}\left(\frac{\mathcal{E}}{mc^2}\right)^2 . \quad (5.16)$$

The presence of i in front of the second term in the braces in Eq.(5.14) corresponds to the elliptical polarization of the radiation. One of the axes of the oscillations of the electric vector is directed along $\mathbf{H}_\perp$, and the other, the major axis, is perpendicular to $\mathbf{H}_\perp$. The ratio of the axes which we denote by $\tan\beta$ is by virtue of (5.14) equal to

$$\tan\beta = \frac{\psi K_{\frac{1}{3}}(g_n)}{(\xi^2 + \psi^2)^{\frac{1}{2}} K_{\frac{2}{3}}(g_n)} . \quad (5.17)$$

When $\psi > 0$ the rotation is left-handed (for an observer counter-clockwise), and when $\psi < 0$ it is right-handed, where we take the angle ψ to be positive, if $\chi > \alpha$, that is, if the vector $\mathbf{k}$ lies outside the velocity cone.

The polarization degenerates to become linear only when $\psi = 0$, that is, when the wavevector lies strictly on the surface of the velocity cone. For large ψ the polarization tends to become circular, as for large values of the argument

$K_{\frac{2}{3}}(x) \approx K_{\frac{1}{3}}(x) \approx (\pi/2x)^{\frac{1}{2}} e^{-x}$; however, the intensity of the radiation then becomes negligibly small (see Fig.5.4 below).

The radiation field can be characterized by the 'polarization tensor of the radiation' which by definition equals

$$\tilde{p}_{\alpha\beta}(n) = \frac{c}{8\pi} E_{n,\alpha} E^*_{n,\beta} , \tag{5.18}$$

where $\alpha,\beta = 1,2$ and $E_{n,\alpha}$ are the components of the electric vector occurring in (5.13) and (5.14), while the energy flux density (Poynting vector), averaged over a period, in the nth harmonic is equal to

$$\tilde{p}_n = \mathrm{Tr}\, \tilde{p}_{\alpha\beta}(n) \equiv \tilde{p}_{11} + \tilde{p}_{22} = \frac{c}{8\pi} |\mathbf{E}_n|^2 . \tag{5.19}$$

In the range of high harmonics the radiation spectrum is practically continuous, $\nu = \omega/2\pi = n\omega^*_H/2\pi \sin^2\chi$ and it is convenient to introduce instead of $\tilde{p}_n$ the 'spectral density of the polarization tensor'

$$\tilde{p}_{\alpha\beta}(\nu) = \tilde{p}_{\alpha\beta}(n) \frac{dn}{d\nu} = \frac{2\pi \sin^2\chi}{\omega^*_H} \tilde{p}_{\alpha\beta}(n) . \tag{5.20}$$

We introduce for the field of an ultra-relativistic electron the functions (here $g_\nu = g_n$; see (5.15)

$$\tilde{p}^{(1)}_\nu \equiv \tilde{p}_{11}(\nu) = \frac{3 e^2 \omega^*_H}{4\pi^2 R^2 c \xi^2 \sin^2\chi} \left(\frac{\nu}{\nu_c}\right)^2 \left(1 + \frac{\psi^2}{\xi^2}\right)^2 K^2_{\frac{2}{3}}(g_n) , \tag{5.21}$$

$$\tilde{p}^{(2)}_\nu \equiv \tilde{p}_{22}(\nu) = \frac{3 e^2 \omega^*_H}{4\pi^2 R^2 c \xi^2 \sin^2\chi} \left(\frac{\nu}{\nu_c}\right)^2 \left(\frac{\psi}{\xi}\right)^2 \left(1 + \frac{\psi^2}{\xi^2}\right) K^2_{\frac{1}{3}}(g_\nu) , \tag{5.22}$$

$$\tilde{p}_{12}(\nu) = \tilde{p}_{21}(\nu) = -i \frac{3 e^2 \omega^*_H}{4\pi^2 R^2 c \xi^2 \sin^2\chi} \left(\frac{\nu}{\nu_c}\right)^2 \left(1 + \frac{\psi^2}{\xi^2}\right)^{\frac{3}{2}} \frac{\psi}{\xi} K_{\frac{1}{3}}(g_\nu) K_{\frac{2}{3}}(g_\nu) . \tag{5.23}$$

Clearly, $\tilde{p}^{(1)}_\nu d\nu$ is the radiative flux in the frequency range $d\nu$ with the electric vector in the wave directed along $\boldsymbol{\ell}_1$; similarly the direction 2 is characterized by the vector $\boldsymbol{\ell}_2$. The spectral density of the radiative flux for both polarizations is $\tilde{p}_\nu = \tilde{p}^{(1)}_\nu + \tilde{p}^{(2)}_\nu$.

If we use Eqs.(5.21) and (5.22) to calculate the total radiative energy flux through a fixed surface, that is, evaluate the integral of the current density over all frequencies and directions, it turns out, in accordance with (5.12), to be equal to $\mathcal{P} = \mathcal{R}/\sin^2\chi$, where $\mathcal{R}$ are the energy losses caused by the synchrotron radiation by an ultra-relativistic electron (see (4.37) and (4.39)); for the sake of convenience we give once more the expression for $\mathcal{R}$:

$$\mathcal{R} = -\frac{d\mathcal{E}}{dt} = \frac{2e^4 H_\perp^2}{3m^2c^3}\left(\frac{\mathcal{E}}{mc^2}\right)^2 = \frac{2}{3}\left(\frac{e^2}{mc^2}\right)^2 cH_\perp^2\left(\frac{\mathcal{E}}{mc^2}\right)^2 . \qquad (5.24)$$

As we have already emphasized, the difference between $\mathcal{P}$ and $\mathcal{R}$ is connected with the fact that, when $\chi \neq \pi/2$, the radiation field, averaged over a period of rotation of the electron, is not stationary. Indeed, when the motion of the electron is helical, the electron approaches the observer and the centre of its orbit remains fixed only when $\chi = \frac{1}{2}\pi$ (circular motion). By virtue of the energy conservation law the total energy flux through a fixed surface σ is equal to

$$\mathcal{P} = \int_\sigma \tilde{p}_\nu \, d\nu \, d\sigma = \mathcal{R} - \frac{d}{dt}\int \frac{E^2 + H^2}{8\pi} d^3\mathbf{r} . \qquad (5.25)$$

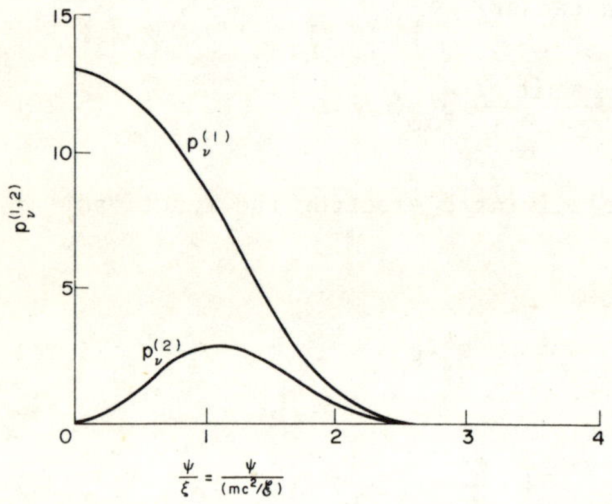

Fig.5.4. Angular distribution of the synchrotron radiation fluxes of a separate electron corresponding to the two main directions of polarization. $p_\nu^{(1)}$ is the flux with the polarization at right angles to the projection of the magnetic field onto the plane of the figure, $p_\nu^{(2)}$ and the flux with the polarization along that projection with $\nu/\nu_c = 0.29$. We have taken as the scale unit along the vertical axis the coefficient

$$(3e^3H/4\pi^2R^2mc^2\xi)(\nu/\nu_c)^2$$

in Eqs.(5.26) and (5.27). The angle $\psi = 0$ corresponds to the direction of the instantaneous electron velocity.

When the electron approaches the 'observer' (that is, the surface σ) the field energy localized between the emitter and the surface σ changes — and this explains the difference between $\mathcal{P}$ and $\mathcal{R}$.

In most applications of the theory of synchrotron (magneto-brems) radiation, one is first of all dealing not with separate particles, but with their totality. Secondly, this ensemble is usually stationary or, at any rate, changes very slowly as a whole. To be concrete, if we are dealing with relativistic electrons in a supernova shell or in other nebulae, the length of the emitting region along the line of sight can be taken to be unchanged during the time necessary for the electromagnetic wave to traverse that length. In other words, the velocity of the shell (of the boundaries of the nebula, ...) V_r along the line of sight is small compared to the velocity of light. If we neglect terms of order V_r/c the volume of the shell can be assumed to be constant and it is clear from the energy conservation law as an average over all emitting particles in the shell $\mathcal{P} = \mathcal{R}$ (see (5.25)), as we already mentioned. A detailed calculation (Ginzburg, Sazonov, and Syrovatskii, 1968; Ginzburg and Syrovatskii, 1969) confirms this conclusion, of course. Bearing this in mind we can use for stationary emitters the quantities

$$p_\nu^{(1)} = \tilde{p}_\nu^{(1)} \sin^2\chi = \frac{3}{4\pi^2 R^2} \frac{e^3 H}{mc^2 \xi} \left(\frac{\nu}{\nu_c}\right)^2 \left(1 + \frac{\psi^2}{\xi^2}\right)^2 K_{\frac{2}{3}}^2(g_\nu) \quad , \tag{5.26}$$

$$p_\nu^{(2)} = \tilde{p}_\nu^{(2)} \sin^2\chi = \frac{3}{4\pi^2 R^2} \frac{e^3 H}{mc^2 \xi} \left(\frac{\nu}{\nu_c}\right)^2 \left(\frac{\psi}{\xi}\right)^2 \left(1 + \frac{\psi^2}{\xi^2}\right) K_{\frac{1}{3}}^2(g_\nu), \tag{5.27}$$

instead of the quantities $\tilde{p}_\nu^{(1)}$ and $\tilde{p}_\nu^{(2)}$; as before

$$g_\nu = \frac{\nu}{2\nu_c}\left(1 + \frac{\psi^2}{\xi^2}\right)^{\frac{3}{2}} \quad , \quad \nu_c = \frac{3\,e\,H_\perp}{4\,\pi\,mc}\left(\frac{\mathcal{E}}{mc^2}\right)^2 \quad . \tag{5.15a}$$

We show in Fig.5.4 the angular distribution of the radiation fluxes $p_\nu^{(1)}$ and $p_\nu^{(2)}$. We have chosen as scale unit along the vertical axis the coefficient $(3\,e^3\,H/4\,\pi^2\,R^2\,mc^2\,\xi)(\nu/\nu_c)^2$ in Eqs,(5.26) and (5.27). The curves are drawn for $\nu/\nu_c = 0.29$ which, as we shall see below, corresponds to the maximum in the frequency spectrum of the global emission (in all directions) of an electron. Fig.5.4 shows that in the range of small angles ψ the main contribution to the radiation comes from oscillations with the electric field directed at right angles to the projection $\mathbf{H}_\perp$ of the magnetic field onto the plane of the figure, that is, in that range $p_\nu^{(1)} \gg p_\nu^{(2)}$.

From the above it is clear that we should use Eqs.(5.21) to (5.23) rather than Eqs.(5.26) and (5.27) for synchrotron radiation sources which move with relativistic velocities. In that case, if the source as a whole moves with a velocity V_r along the line of sight in the direction of the 'observer', the intensity of the radiation is increased by a factor $(1 - V_r/c)^{-1}$ as compared to a fixed source with the same electron distribution function (see Ginzburg, Sazonov, and Syrovatskii, 1968; Ginzburg and Syrovatskii, 1969). Recently one has begun to ascertain that synchrotron radiation sources under cosmic conditions may have relativistic velocities. For instance, in the case of explosions of galactic nuclei leading to the formation of radio-galaxies, the emitting 'radio-clouds' probably move in a number of cases with velocities comparable to the velocity of light. It is possible that shells, jets, and outbursts moving with relativistic velocities also exist in the case of other objects, in the first place in quasars. For that reason there is undoubtedly interest not only in stationary, but also in non-stationary (relativistic) synchrotron, and in general magneto-brems radiation sources (Ginzburg, Sazonov, and Syrovatskii, 1968; Ginzburg and Syrovatskii, 1969; Rees, 1967; Ryle and Longair, 1967; Ozernoy and Sazonov, 1969). In what follows we shall, however, concentrate our attention on stationary (or more precisely, quasi-stationary) sources for which we can use Eqs.(5.26) and (5.27) in the range (5.1).

Before going any further, we shall consider the quantities which characterize radiation as this problem is usually not touched upon in electrodynamics textbooks.

Any radiation flux is characterized not only by its frequency dependence, but also, in general, by four independent parameters, for instance, the position of the principal axes of the polarization ellipse, the intensities along the two principal directions, and the sense of rotation of the electric vector. It is, however, convenient to use for these parameters the Stokes parameters (see, for instance, Ginzburg and Syrovatskii, 1966a; Gardner and Whiteoak, 1966; Chandrasekhar, 1960; Shurcliff, 1962). For the radiation by a single particle these parameters I_e, Q_e, U_e, and V_e can be expressed in terms of the radiation flux densities with respect to the two main directions of polarization $p_\nu^{(1)}$ and $p_\nu^{(2)}$, and also in terms of $\tan\beta$, the ratio of the minor to the major axis of the ellipse of the oscillations of the electric vector (see (5.17)), and the angle $\tilde{\chi}$ between some arbitrary fixed direction in the plane of the figure and the major axis of this ellipse (that is, the direction at right angles to the projection of $\mathbf{H}$ onto the plane of the figure)†. The corresponding equations are the following ones:

$$\left.\begin{aligned} I_e &= p_\nu^{(1)} + p_\nu^{(2)} , & Q_e &= (p_\nu^{(1)} - p_\nu^{(2)})\cos 2\tilde{\chi} , \\ U_e &= (p_\nu^{(1)} - p_\nu^{(2)})\sin 2\tilde{\chi} , & V_e &= (p_\nu^{(1)} - p_\nu^{(2)})\tan 2\beta . \end{aligned}\right\} \quad (5.28)$$

Like $p_\nu^{(1)}$ and $p_\nu^{(2)}$, the Stokes parameters have the dimensions of an energy flux density per unit frequency range; the index e indicates that these parameters refer to the radiation of a single electron.

---

† The angle $\tilde{\chi}$ is reckoned clockwise and is clearly defined in the interval $0 \le \tilde{\chi} < \pi$. The notation $\tilde{\chi}$ is introduced here in order not to confuse the angle $\tilde{\chi}$ with the angle χ between $\mathbf{v}$ and $\mathbf{H}$.

The Stokes parameters possess two important advantages: they can be directly measured and they are additive for independent (incoherent) radiation fluxes, that is, radiation fluxes with random phases over which one can average (Ginzburg and Syrovatskii, 1966a). Experimentally the Stokes parameters can be determined by the usual methods for studying polarized radiation (Shurcliff 1962), namely, by means of introducing a phase difference ε between one of the projections of the electric vector oscillations in the wave (for instance, along the direction s_1 in Fig.5.5) and the other projection, the one onto the perpendicular direction (s_2 in Fig.5.5).

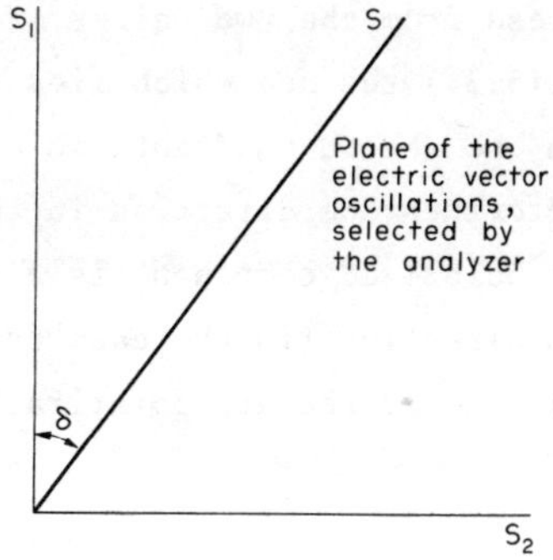

Fig.5.5. Definition of the Stokes parameters.

We introduce an additional retarding phase ε to the direction s_2 relative to the oscillations in the perpendicular direction s_1. The angle δ determines the plane of the position of the analyzer. The measured radiation flux is directed towards the observer.

The subsequent analysis reduces to establishing the dependence of the intensity of the resulting radiation on the position of the analyzer which selects the projection of the oscillations onto some arbitrary direction s (see Fig.5.5). If we denote the angle in the plane of the figure between s_1 and s by δ, the intensity of the radiation which leaves the analyzer will be the following function of ε and δ (see, for instance, Chandrasekhar, 1960):

$$I_e(\varepsilon,\delta) = \tfrac{1}{2}\{I_e + Q_e \cos 2\delta + (U_e \cos\varepsilon - V_e \sin\varepsilon)\sin 2\delta\} \ . \qquad (5.29)$$

By choosing the retarding phases ε and the position of the analyzer δ appropriately we can measure the values of all the Stokes parameters.

We note that the first Stokes parameter I_e determines the total radiative flux density (or the intensity in the case of spatially distributed sources; vide infra) while the degree of polarization Π and the angle $\tilde{\chi}$ are given by the equations

$$\Pi = \frac{\sqrt{(Q_e^2 + U_e^2 + V_e^2)}}{I_e} \qquad (5.30)$$

and

$$\tan 2\tilde{\chi} = \frac{U_e}{Q_e} \ . \qquad (5.31)$$

We choose from the two values of the angle $\tilde{\chi}$ $(0 \le \tilde{\chi} < \pi)$ which are determined by Eq.(5.31) the one which lies in the first quadrant, when $U_e > 0$, and the one in the second quadrant, when $U_e < 0$. By definition the angle $\tilde{\chi}$ then characterizes the direction in the plane of the figure in which the intensity of the polarized component is a maximum, and is reckoned clockwise from a chosen direction (in the case considered from the direction $\mathbf{s}_1$). If there is no elliptic (circular) polarization $V_e = 0$ and

$$\Pi = \frac{I_{max} - I_{min}}{I_{max} + I_{min}} .$$

Let us now consider the emission by a system of independently moving particles. Let $N(\mathcal{E}, \mathbf{R}, \boldsymbol{\tau}) d\mathcal{E}\, d^3\mathbf{R}\, d^2\Omega_\tau$ be the number of particles in a volume element $d^3\mathbf{R} = R^2\, dR\, d^2\Omega$ with an energy within the range $\mathcal{E}$, $\mathcal{E} + d\mathcal{E}$, and velocities within the solid angle $d^2\Omega_\tau$ around the direction $\boldsymbol{\tau}$. As in the conditions considered the emission from separated electrons is incoherent so that the Stokes parameters are additive, the intensity of the radiation from such a system in the direction $\mathbf{k}$ of observation is equal to†

$$I_\nu \equiv I(\nu, \mathbf{k}) = \int I_e(\nu, \mathcal{E}, \mathbf{R}, \chi, \psi)\, N(\mathcal{E}, \mathbf{R}, \boldsymbol{\tau})\, d\mathcal{E}\, d^2\Omega_\tau\, R^2\, dR . \tag{5.32}$$

Here $I_e(\nu, \mathcal{E}, \mathbf{R}, \chi, \psi)$ is given by the first of Eqs.(5.28) and the integration over dR is along the line of sight in the direction $\mathbf{k}$. One can similarly give expressions for the other Stokes parameters.

We emphasize that in contrast to the Stokes parameters (5.28) for the emission by a single electron which have the dimensions of the spectral radiative energy flux density, Eq.(5.32) determines the intensity of the radiation, that is, the energy flux through unit area at right angles to the direction of observation, per unit solid angle and unit frequency range. The intensity of radiation is measured in radio-astronomy in units $W\,m^{-2}\,Hz^{-1}\,sterad^{-1} = 10^3\,erg\,cm^{-2}\,s^{-1}\,Hz^{-1}\,sterad^{-1}$. One often uses then as the unit of flux the so-called 'flux unit' = 1 J (Jansky) which equals $10^{-26}\,W\,m^{-2}\,Hz^{-1} = 10^{-23}\,erg\,cm^{-2}\,s^{-1}\,Hz^{-1}$.

If the source (radiating system of electrons) has small angular dimensions, one uses as the experimentally measured quantity (as in the case of a single particle) the spectral radiation flux density

---

† We understand in what follows by the direction of observation (the direction of the line of sight) the direction of the wavevector $\mathbf{k}$, that is, the direction in which the observed radiation arrives.

$$\Phi_\nu = \int I_\nu \, d^2\Omega = \int I_e(\nu, \mathcal{E}, \mathbf{R}, \chi, \psi) \, N(\mathcal{E}, \mathbf{R}, \boldsymbol{\tau}) \, d\mathcal{E} \, d^2\Omega_\tau \, d^3\mathbf{R} \,, \tag{5.33}$$

where $d^3\mathbf{R} = R^2 dR\, d^2\Omega$ and the integration is over the whole volume of the source.

When we apply (5.32) and (5.33) and similarly expressions for the other Stokes parameters to synchrotron radiation we can integrate in the general form over $d^2\Omega_\tau$ for an arbitrary electron distribution $N(\mathcal{E}, \mathbf{R}, \tau)$. Indeed, the integrand is non-vanishing practically only in a narrow range of angles $\Delta\psi \sim mc^2/\mathcal{E}$, and for the integration over $d^2\Omega_\tau$ the only important contribution will thus come from a narrow circular sector $\Delta\Omega_\tau = 2\pi \sin\alpha\, \Delta\psi$, where $\alpha = \chi - \psi \approx \chi$ is the angle between the direction of observation $\mathbf{k}$ and the magnetic field $\mathbf{H}$.† In the limits of small solid angle $\Delta\Omega_\tau$ the distribution of the electrons is practically unchanged over directions and one can put $N(\mathcal{E}, \mathbf{R}, \tau) \approx N(\mathcal{E}, \mathbf{R}, \mathbf{k})$, where $\mathbf{k}$ is the direction of the emission (the direction along the line of sight from the source to the observer), and the integration over ψ can be extended to the whole range from $-\infty$ to $+\infty$. Using the relations (Westfold, 1959; Trubnikov, 1958)

$$\left.\begin{aligned}
\int_{-\infty}^{+\infty} p_\nu^{(1)} d\psi &= \frac{\sqrt{3}\, e^3 H}{2\pi m c^2 R^2} \, \frac{\nu}{2\nu_c} \left[\int_{\nu/\nu_c}^{\infty} K_{\frac{5}{3}}(\eta)\, d\eta + K_{\frac{2}{3}}\left(\frac{\nu}{\nu_c}\right) \right], \\
\int_{-\infty}^{+\infty} p_\nu^{(2)} d\psi &= \frac{\sqrt{3}\, e^3 H}{2\pi m c^2 R^2} \, \frac{\nu}{2\nu_c} \left[\int_{\nu/\nu_c}^{\infty} K_{\frac{5}{3}}(\eta)\, d\eta - K_{\frac{2}{3}}\left(\frac{\nu}{\nu_c}\right) \right],
\end{aligned}\right\} \tag{5.34}$$

we then get from (5.28) and (5.32)

$$I_\nu = I(\nu, \mathbf{k}) = \frac{\sqrt{3}\, e^3}{mc^2} \int d\mathcal{E}\, dR\, N(\mathcal{E}, \mathbf{R}, \mathbf{k})\, H \sin\chi \, \frac{\nu}{\nu_c} \int_{\nu/\nu_c}^{\infty} K_{\frac{5}{3}}(\eta)\, d\eta \,. \tag{5.35}$$

In the general case the field strength $\mathbf{H}$, the angle $\alpha \approx \chi$ between $\mathbf{H}$ and $\mathbf{k}$, and also the particle density (concentration) $N(\mathcal{E}, \mathbf{R}, \mathbf{k})$ will in this expression depend on $\mathbf{R}$.

---

† We shall not distinguish in what follows between the angles α and χ; this is clearly admissable as an ultra-relativistic particle emits practically solely in the direction of its motion.

We can similarly give expressions for the other Stokes parameters, for instance

$$Q(\nu,\mathbf{k}) = \frac{\sqrt{3}\,e^3}{mc^2} \, d\mathcal{E}\, dR\, N(\mathcal{E}, \mathbf{R}, \mathbf{k})\, H \sin\chi \, \cos 2\tilde{\chi} \, \frac{\nu}{\nu_c} K_{\frac{2}{3}}\left(\frac{\nu}{\nu_c}\right). \tag{5.36}$$

The Stokes parameter $U(\nu,\mathbf{k})$ differs from $Q(\nu,\mathbf{k})$ only in that $\cos 2\tilde{\chi}$ in the integrand in Eq.(5.36) must be replaced by $\sin 2\tilde{\chi}$. As to the parameter $V(\nu,\mathbf{k})$ which characterizes the presence of elliptic polarization of the radiation, in the ultra-relativistic approximation considered here it turns out to vanish. Indeed, it follows easily from (5.17) and (5.28) that

$$V_e \sim 2\,\frac{\psi}{\xi}\left(1 + \frac{\psi^2}{\xi^2}\right)^{\frac{3}{2}} K_{\frac{1}{3}}(g_\nu)\, K_{\frac{2}{3}}(g_\nu) \,.$$

As this function is an odd function its integral over all ψ vanishes and, hence, $V(\nu,\mathbf{k}) = 0$. The radiation from a system of electrons thus turns out to be linearly polarized. This result is valid up to terms of order $mc^2/\mathcal{E}$ and it can easily be understood, if we remember that the sign of ψ determines the sense of rotation of the electric vector in the wave emitted by a single electron. As the power of the radiation (see (5.21) or (5.26) and (5.27)) is independent of the sign of ψ while the distribution of the particles over directions of motion in the limit of very small angles $|\psi| \leqslant mc^2/\mathcal{E}$ is practically constant, the contributions to the radiation in a given direction from particles with positive and negative ψ will be the same, and the polarization will be linear.

An appreciable elliptic polarization could in the ultra-relativistic case occur only if the velocity distribution of the electrons is strongly anisotropic. For this it is necessary that the distribution changes considerably within the limits of a very small angle $|\psi| \sim mc^2/\mathcal{E}$, and moreover, just in the direction of observation. If, moreover, we take into account possible fluctuations in the directions of the magnetic field, it is clear that one needs very special conditions for the realization of such a possibility (pulsars are of particular interest in this respect).

We now give expressions for the intensity and polarization of the radiation in some concrete cases which are of particular importance for astronomical applications.

If all electrons have the same energy (mono-energetic spectrum) and the magnetic field is uniform, it follows from (5.35) that the intensity is equal to

$$I_1(\mathbf{k}) = \frac{\sqrt{3}\,e^3}{mc^2} N_e(\mathbf{k})\, H \sin\chi \,\frac{\nu}{\nu_c} \int_{\nu/\nu_c}^{\infty} K_{\frac{5}{3}}(\eta)\, d\eta \equiv N_e(\mathbf{k})\, p(\nu) \,, \qquad (5.37)$$

where $N_e(\mathbf{k}) = \int N_e(\mathbf{R},\mathbf{k})\, dR$ is the number of electrons along the line of sight with velocities in the direction towards the observer, per unit solid angle.

The degree of polarization can be seen from (5.30) and (5.36) to equal

$$\Pi = \frac{K_{\frac{2}{3}}(\nu/\nu_c)}{\int_{\nu/\nu_c}^{\infty} K_{\frac{5}{3}}(\eta)\, d\eta} = \begin{cases} \frac{1}{2} \,, & \text{when } \nu \ll \nu_c \,, \\ 1 - \frac{2}{3}\frac{\nu}{\nu_c} \,, & \text{when } \nu \gg \nu_c \,. \end{cases} \qquad (5.38)$$

As in the approximation considered the integration over the angular distribution of the electrons is equivalent to an integration of the radiative power of a single electron over all directions, Eq.(5.37) differs only by the factor $N_e(\mathbf{k})$ from the spectral distribution of the power of the total radiation (in all directions) of a single electron:

$$p(\nu) = \sqrt{3}\,\frac{e^3 H \sin\chi}{mc^2}\,\frac{\nu}{\nu_c}\int_{\nu/\nu_c}^{\infty} K_{\frac{5}{3}}(\eta)\, d\eta = \sqrt{3}\,\frac{e^3 H_\perp}{mc^2}\, F\!\left(\frac{\nu}{\nu_c}\right) . \qquad (5.39)$$

$$F\!\left(\frac{\nu}{\nu_c}\right) = \frac{\nu}{\nu_c}\int_{\nu/\nu_c}^{\infty} K_{\frac{5}{3}}(\eta)\, d\eta$$

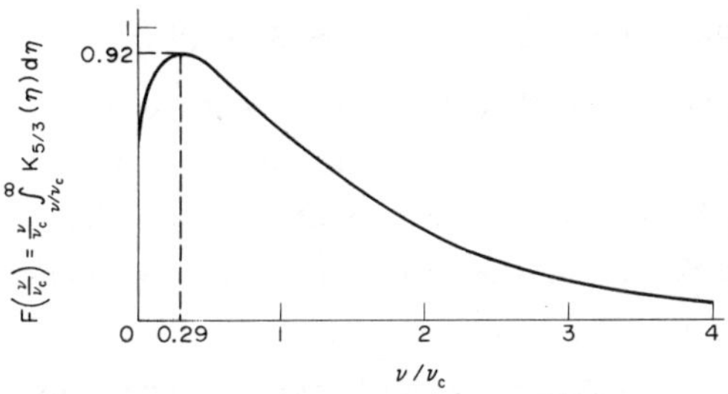

Fig.5.6. Spectral distribution of the power of the total radiation (in all directions) of a charged particle moving in a magnetic field (see (5.39)).

We show in Fig.5.6 the function $F(x) = x\int_x^{\infty} K_{\frac{5}{3}}(\eta)\, d\eta$ which reflects the spectral distribution of the emitted power, and its values together with the values of the function $F_p(x) = x K_{\frac{2}{3}}(x)$ are given in one of the appendixes of the book by Ginzburg and Syrovatskii (1964a) (it is clear from (5.38) that the polarization $\Pi = F_p(x)/F(x)$). We note that the maximum in the spectrum of the synchrotron emission of a single electron occurs at the frequency

$$\nu_m \approx 0.29\,\nu_c = 0.07\,\frac{eH_\perp}{mc}\left(\frac{\mathcal{E}}{mc^2}\right)^2 = 1.3\times 10^6\, H_\perp \left(\frac{\mathcal{E}}{mc^2}\right)^2 \qquad (5.40a)$$

$$= 1.8\times 10^{18}\, H_\perp (\mathcal{E}(\mathrm{erg}))^2 = 4.6\times 10^{-6}\, H_\perp (\mathcal{E}(\mathrm{eV}))^2 .$$

The frequency ν is here in Hz .

For the maximum frequency (5.40a) the spectral density of the power of the total radiation of a single electron is equal to

$$p_m = p(\nu_m = 0.29\,\nu_c) \approx 1.6\,\frac{e^3 H_\perp}{mc^2} = 2.16\times 10^{-22}\,H_\perp\ \mathrm{erg\ s^{-1}\,Hz^{-1}} \qquad (5.40b)$$

If we forget about the numerical coefficient, that is, look only at order of magnitude estimates we can easily obtain this last relation as follows. It is clear from qualitative considerations or from Fig.5.6 that the width of the radiation spectrum of an electron is $\Delta\nu \sim \nu_c \sim (eH_\perp/mc)(\mathcal{E}/mc^2)^2$. The total power of the synchrotron radiation, taking into account the above remarks, may be put equal to the losses (5.24). It is clear that the average spectral density of the power of the radiation is

$$\overline{p(\nu)} \sim \frac{P(\mathcal{E})}{\Delta\nu} \sim \frac{P(\mathcal{E})}{\nu_c} \sim \frac{e^3 H_\perp}{mc^2} \;,$$

which is the same as (5.8); we have repeated the estimate here for the sake of convenience.

The energy spectrum of the electrons along the line of sight can often be approximated in a narrow energy range by a power-law function of the form

$$N_e(\mathcal{E}, \mathbf{k})\,d\mathcal{E} = K_e(\mathbf{k})\,\mathcal{E}^{-\gamma}\,d\mathcal{E}\ , \quad \mathcal{E}_1 \le \mathcal{E} \le \mathcal{E}_2 . \qquad (5.41)$$

Here $N_e(\mathcal{E}, \mathbf{k})$ is the number of electrons along the line of sight moving in the direction of the observer per unit solid angle and unit energy range.

Such an approximation usually happens to be applicable for the electrons which are responsible for cosmic radio-emission within a rather wide range of energies, and one can often take the limits $\mathcal{E}_1$ and $\mathcal{E}_2$ of the spectrum such that electrons with energies $\mathcal{E} < \mathcal{E}_1$ and $\mathcal{E} > \mathcal{E}_2$ are unimportant for the frequency range of the radiation which is of interest to us. We can under these assumptions use in the integrals (5.35) and (5.36) the function (5.41) for all energies and therefore (when $\gamma > \frac{1}{3}$) apply the relations

$$\left.\begin{aligned}
&\int_0^\infty \mathcal{E}^{-\gamma}\,\frac{\nu}{\nu_c}\,K_{\frac{2}{3}}\!\left(\frac{\nu}{\nu_c}\right) d\mathcal{E} = \tfrac{1}{4}\,\Gamma\!\left(\frac{3\gamma-1}{12}\right)\Gamma\!\left(\frac{3\gamma+7}{12}\right)\left(\frac{3\,e\,H\sin\chi}{2\pi m^3 c^5 \nu}\right)^{\frac{1}{2}(\gamma-1)} ,\\
&\int_0^\infty \mathcal{E}^{\gamma}\left(\frac{\nu}{\nu_c}\int_{\nu/\nu_c}^\infty K_{\frac{5}{3}}(\eta)\,d\eta\right) d\mathcal{E} = \tfrac{1}{4}\,\frac{\gamma+\frac{7}{3}}{\gamma+1}\,\Gamma\!\left(\frac{3\gamma-1}{12}\right)\Gamma\!\left(\frac{3\gamma+7}{12}\right)\left(\frac{3\,e\,H\sin\chi}{3\pi m^3 c^5 \nu}\right)^{\frac{1}{2}(\gamma-1)} ,
\end{aligned}\right\}$$

(5.42)

where $\Gamma(x)$ is the gamma function. We then get (see (5.35)) the following expression for the intensity of the radiation from a system of electrons with the energy spectrum (5.41) in a uniform magnetic field $\mathbf{H}$:

$$I_o(\mathbf{k}) = \frac{\sqrt{3}}{\gamma+1}\,\Gamma\left(\frac{3\gamma-1}{12}\right)\Gamma\left(\frac{3\gamma+19}{12}\right)\frac{e^3}{mc^2}\left(\frac{3e}{2\pi m^3c^5}\right)^{\frac{1}{2}(\gamma-1)}$$

$$\times K_e(\mathbf{k})(H\sin\chi)^{\frac{1}{2}(\gamma+1)}\,\nu^{-\frac{1}{2}(\gamma-1)}\,, \tag{5.43}$$

where $K_e(\mathbf{k})$ is the coefficient in (5.41).

We assume that the electron distribution can be taken to be uniform and isotropic, that is

$$N(\mathcal{E},\mathbf{R},\mathbf{k}) = \frac{1}{4\pi}N_e(\mathcal{E})\,,$$

where

$$N_e(\mathcal{E})\,d\mathcal{E} = K_e\,\mathcal{E}^{-\gamma}\,d\mathcal{E} \tag{5.44}$$

is the number of electron per unit volume with arbitrary directions of motion and energies in the range $\mathcal{E}$, $\mathcal{E}+d\mathcal{E}$. In that case

$$K_e(\mathbf{k}) = \frac{1}{4\pi}K_eL\,, \tag{5.45}$$

where K_e is the coefficient in (5.44) and L the extension of the emitting region along the line of sight. We note that in the general case $K_e(\mathbf{k})$ depends on the angle χ between the direction of the magnetic field and the line of sight.

In the case of a uniform field the degree of polarization depends solely on the index γ of the spectrum (5.41) and one can check from (5.30) and (5.42) that it is equal to

$$\Pi_0 = \frac{\gamma+1}{\gamma+\frac{7}{3}}\,, \tag{5.46}$$

which amounts to 75% for $\gamma = 3$ and to 69% for $\gamma = 2$.

Equations (5.43) and (5.46) are, in general, unsuitable for applications to the synchrotron radiation of cosmic electrons, as the radiation observed is gathered from a large region of space with the magnetic field being oriented differently in different parts. We rather assume that along the line of sight the magnetic field is randomly directed. In that case there is no polarization of the radiation and we can easily find the intensity by averaging (5.43) over all directions of the magnetic field. Since

$$\frac{1}{2}\int_0^{\pi}(\sin\chi)^{\frac{1}{2}(\gamma+1)}\sin\chi\,d\chi = \frac{\sqrt{\pi}}{2}\,\frac{\Gamma(\frac{1}{4}(\gamma+5))}{\Gamma(\frac{1}{4}(\gamma+7))}\,, \tag{5.47}$$

this averaging leads to the following expression for the intensity of the radiation for the case of a uniform and isotropic electron distribution with the energy spectrum (5.41) in a random magnetic field:

$$I_\nu \equiv I = a(\gamma)\,\frac{e^3}{mc^2}\left(\frac{3e}{4\pi m^3c^5}\right)^{\frac{1}{2}(\gamma-1)} H^{\frac{1}{2}(\gamma+1)}\, L K_e\, \nu^{-\frac{1}{2}(\gamma-1)}$$
$$= 1.35\times 10^{-22}\, a(\gamma)\, L K_e\, H^{\frac{1}{2}(\gamma+1)}\left(\frac{6.26\times 10^{18}}{\nu}\right)^{\frac{1}{2}(\gamma-1)} \frac{\text{erg}}{\text{cm}^2\,\text{sterad s Hz}}\,. \tag{5.48}$$

Here K_e is the coefficient in (5.44) corresponding to unit volume, by $H^{\frac{1}{2}(\gamma+1)}$ we must understand the average value of that quantity in the emitting region, and $a(\gamma)$ is a coefficient depending on the index γ of the energy spectrum:

$$a(\gamma) = \frac{2^{\frac{1}{2}(\gamma-1)}\,\sqrt{3}\,\Gamma\left(\frac{3\gamma-1}{12}\right)\Gamma\left(\frac{3\gamma+19}{12}\right)\Gamma\left(\frac{\gamma+5}{4}\right)}{8\sqrt{\pi}\,(\gamma+1)\,\Gamma\left(\frac{\gamma+7}{4}\right)} \tag{5.49}$$

The value of the coefficient $a(\gamma)$ as well as the values of other quantities introduced in what follows are given in Table 5.1.

TABLE 5.1

γ	1	1.5	2	2.5	3	4	5
$a(\gamma)$	0.283	0.147	0.103	0.0852	0.0742	0.0725	0.0922
$a'(\gamma)$	0.31	0.22	0.15	0.11	0.074	0.036	0.018
$y_1(\gamma)$	0.80	1.3	1.8	2.2	2.7	3.4	4.0
$y_2(\gamma)$	0.00045	0.011	0.032	0.10	0.18	0.38	0.65

It is clear from Eqs.(5.43) and (5.48) that a power-law energy spectrum of radiating particles with power index γ corresponds to a power-law frequency spectrum of the radiation

$$I_\nu \propto \nu^{-\alpha}\,, \quad \alpha = \tfrac{1}{2}(\gamma-1)\,. \tag{5.50}$$

In view of the important role played by Eq.(5.50) we shall also derive it by a simple approximate method. To do this we neglect the width of the spectrum of the radiation by a single electron, assuming that the whole emission occurs at a frequency $\nu = \nu_m$ corresponding to the maximum in the spectrum (see (5.40)). The energy of the electron can then be expressed in terms of the frequency ν: $\mathcal{E}^2 = (\nu/0.29)(4\pi m^3c^5/3eH_\perp)$. Moreover, the total power of the radiation by an ultra-relativistic electron is equal to the well known expression (see (5.24))

$$-\frac{d\mathcal{E}}{dt} = \mathcal{P} = \frac{2}{3} c \left(\frac{e^2}{mc^2}\right)^2 H_\perp^2 \left(\frac{\mathcal{E}}{mc^2}\right)^2 .$$

Under the above assumptions for the electron spectrum (5.44) the intensity of the radiation collected along the path length L is equal to

$$I_\nu \, d\nu = \frac{L}{4\pi} \mathcal{P} K_e \, \mathcal{E}^{-\gamma} \, d\mathcal{E}$$

$$= a'(\gamma) \frac{e^3}{mc^2} \left(\frac{3e}{4\pi m^3 c^5}\right)^{\frac{1}{2}(\gamma-1)} H^{\frac{1}{2}(\gamma+1)} L K_e \, \nu^{-\frac{1}{2}(\gamma-1)} d\nu , \qquad (5.51)$$

where $a'(\gamma) = 0.31(0.24)^{\frac{1}{2}(\gamma-1)}$ and where we have used the fact that for a random field $H_\perp^2 = \frac{2}{3} H^2$. Equation (5.51) differs from (5.48) only in that the coefficient $a(\gamma)$ is replaced by $a'(\gamma)$ and these factors differ for $1 < \gamma < 4$ less than a factor two (see Table 5.1).

Apart from the intensity I_ν one often uses the emissivity $\boldsymbol{\varepsilon}_\nu$ which is equal to the energy emitted per unit time from unit volume into unit solid angle. One sees easily that for an (on average) isotropic radiation collected along the path length L

$$\boldsymbol{\varepsilon}_\nu = I_\nu / L . \qquad (5.52)$$

Sometimes one also uses the emissivity of unit volume for all directions of emission. For isotropic radiation it equals $4\pi\boldsymbol{\varepsilon}_\nu$. In the case of synchrotron radiation by electrons with a power-law spectrum the intensity which appears here is given by Eq.(5.48). For mono-energetic electrons we have clearly

$$\boldsymbol{\varepsilon}_\nu = \frac{p(\nu)}{4\pi} N_e , \qquad (5.53)$$

where $p(\nu)$ is the power of the total emission (see (5.39)) and N_e the density of the radiating electrons (see also (5.37) with $N_e(\mathbf{k}) = N_e L/4\pi$).

The maximum emissivity, that is, the emission at the frequency ν_m (see (5.40)) is equal to

$$\boldsymbol{\varepsilon}_{\nu,m} = \frac{p_m}{4\pi} N_e \approx 0.13 \frac{e^3 H_\perp}{mc^2} N_e = 1.7 \times 10^{23} H_\perp N_e \quad \frac{\text{erg}}{\text{cm}^2 \text{ s sterad Hz}} . \qquad (5.54)$$

The maximum intensity obtained from isotropically distributed mono-energetic electrons is equal to

$$I_{\nu,m} = \int \boldsymbol{\varepsilon}_{\nu,m} \, dR = 1.7 \times 10^{-23} H_\perp \int N_e(\mathbf{R}) \, dR \quad \frac{\text{erg}}{\text{cm}^2 \text{ s sterad Hz}}$$

$$= 1.7 \times 10^{-26} H_\perp \int N_e(\mathbf{R}) \, dR \quad \frac{\text{W}}{\text{m}^2 \text{ sterad Hz}} . \qquad (5.55)$$

When one uses Eqs.(5.54) or (5.55) for an estimate of the density N_e, using the measured values of I_ν we get, of course, a minimum value for N_e.

We assumed above that the electron energy spectrum is a power-law spectrum (see (5.41) and (5.44)) in a sufficiently wide range of energies. We now give a quantitative estimate of that range. The errors introduced by replacing in (5.35) and (5.36) the limits of integration by 0 and ∞, respectively, do not exceed 10% for a given frequency ν for each of the limits, if the following conditions are satisfied:

$$\left.\begin{aligned} \mathcal{E}_1 &\leq mc^2\left(\frac{4\pi mc\nu}{3eHy_1(\gamma)}\right)^{\frac{1}{2}} \approx 2.5\times 10^2\left(\frac{\nu}{y_1(\gamma)H}\right)^{\frac{1}{2}} \text{ eV} , \\ \mathcal{E}_2 &\geq mc^2\left(\frac{4\pi mc\nu}{3eHy_2(\gamma)}\right)^{\frac{1}{2}} \approx 2.5\times 10^2\left(\frac{\nu}{y_2(\gamma)H}\right)^{\frac{1}{2}} \text{ eV} . \end{aligned}\right\} \quad (5.56)$$

The values of the factors $y_1(\gamma)$ and $y_2(\gamma)$ for different γ are given in Table 5.1. It is clear that the range of energies which gives the main contribution to the radiation at a given frequency depends strongly on the power index γ. For $\gamma \geq 1.5$ ($\alpha \geq 0.25$) more than 80% of the radiation at a given frequency comes from electrons with energies which differ by less than a factor 10. For $\gamma < 1.5$ this range of energies increases fast and as $\gamma \to \frac{1}{3}$ ($\alpha \to -\frac{1}{3}$) it becomes infinite. The fact is that in the range of frequencies ν less than ν_m the intensity of the radiation from a single particle $p_\nu \equiv p(\nu, \mathcal{E}) \propto (\nu/\nu_c)^{\frac{1}{3}} \propto \nu^{\frac{1}{3}}\ \mathcal{E}^{-\frac{2}{3}}$ and for the spectrum (5.41) the total intensity $I_\nu \propto \int p(\nu,\mathcal{E})\, N(\mathcal{E})\, d\mathcal{E} \propto \int d\mathcal{E}/\mathcal{E}^{\gamma-\frac{2}{3}}$ is unbounded, if the energy spectrum of the particles with an index $\gamma \leq \frac{1}{3}$ extends to arbitrarily large energies.

The value $\alpha = -\frac{1}{3}$ is, clearly, the minimum value for synchrotron radiation in vacuo, as already the radiation spectrum of a single particle does not contain parts with a faster increase of the intensity with frequency.

When one applies the theory in astrophysics one often encounters the problem of estimating the range of electron energies $(\mathcal{E}_1, \mathcal{E}_2)$ which produce radiation with the power-law spectrum (5.50) in the frequency range (ν_1, ν_2). If that range is sufficiently large $(\nu_2/\nu_1) \gg y_1(\gamma)/y_2(\gamma))$ we can conclude from the results given here that the electrons must have a power-law spectrum, at least in the energy range $\mathcal{E}_1 < \mathcal{E} < \mathcal{E}_2$, where

$$\left.\begin{aligned} \mathcal{E}_1 &= mc^2\left(\frac{4\pi mc\nu_1}{3eHy_1(\gamma)}\right)^{\frac{1}{2}} \approx 2.5\times 10^2\left(\frac{\nu_1}{y_1(\gamma)H}\right)^{\frac{1}{2}} \text{ eV} , \\ \mathcal{E}_2 &= mc^2\left(\frac{4\pi mc\nu_2}{3eHy_2(\gamma)}\right)^{\frac{1}{2}} \approx 2.5\times 10^2\left(\frac{\nu_2}{y_2(\gamma)H}\right)^{\frac{1}{2}} \text{ eV} . \end{aligned}\right\} \quad (5.57)$$

If, however, the range of frequencies is small or α is small (in practice $\alpha < 0.25$, that is $\gamma < 1.5$) one can only make a rough estimate of the electron energies, assuming that all of the emission by and electron with energy occurs at the frequency $\nu_m = 0.29\,\nu_c$ and putting $H_\perp = \sqrt{\frac{2}{3}}\,H$. We must then put in(5.57) $y_1(\gamma) = y_2(\gamma) = 0.24$.

We have given here expressions for the intensity of the synchrotron radiation in the two normally considered limiting cases: for a uniform and for a completely random field. The first of these is characterized by the maximum possible polarization and in the second one there is no polarization. The problems of whether a particular expression is applicable is solved primarily on the basis of polarization measurements. However, in those well known cases where one observes polarization of the cosmic synchrotron radiation, it turns out that it is, as a rule, much smaller than for the case of a uniform field (see (5.46)). This must, first of all, mean that the magnetic field in the emitting region is not uniform. The calculation of the degree of polarization in such an 'intermediate' case was given by Korchak and Syrovatskii (1962) for two magnetic field models (the result is also given by Ginzburg and Syrovatskii, 1964a, 1966a).

In most cases radio-astronomical observations reduce to a measurement of the intensity I. However, polarization measurements are beginning to play an ever larger role and it seems obvious to us that the general trend is towards measuring all Stokes parameters, both for radio-emission and in other frequency bands. Nonetheless, the intensity I remains the main characteristic of cosmic radiation. To be precise, in the case of cosmic synchrotron radio-emission, which can be distinguished from thermal radio-emission both through its spectrum and through a few other features (such as a very high intensity) measurements of the intensity are used to estimate the density and energy of the relativistic electron (the electron component of the cosmic rays) far from the Earth.

The angular sizes of galactic and extragalactic nebulae — discrete sources of non-thermal radio-emission — are, as a rule, small and the quantity measured is usually not the intensity I_ν but the spectral density of the radiation flux Φ_ν (see (5.33)). This quantity is defined as the radiative energy flux per unit frequency range, normally incident upon unit area :

$$\Phi_\nu = \int I_\nu \, d^2\Omega \quad , \tag{5.58}$$

where the integration is over the whole of the solid angle corresponding to the source. If the linear size L of the source is small compared to its distance from us R, and if we may assume that the absolute magnitude of the magnetic field and the density of the relativistic electrons are constant over volume of the source, we have from (5.48) and (5.58)

$$\Phi_\nu = a(\gamma)\frac{e^3}{mc^2}\left(\frac{3e}{4\pi m^3c^5}\right)^{\frac{1}{2}(\gamma-1)} \frac{K_V H^{\frac{1}{2}(\gamma+1)}}{R^2} \nu^{-\frac{1}{2}(\gamma-1)}$$

$$= 1.35\times10^{-22}\, a(\gamma)\frac{K_V H^{\frac{1}{2}(\gamma+1)}}{R^2}\left(\frac{6.26\times10^{18}}{\nu}\right)^{\frac{1}{2}(\gamma-1)} \frac{\text{erg}}{\text{cm}^2\text{ s Hz}} \tag{5.59}$$

where $K_V = K_e V$ is the coefficient in the electron energy spectrum for the whole volume of the (assumed spherical) source $V = \pi L^3/6$. We have assumed here that the electron energy spectrum has the form

$$N(\mathcal{E})\,d\mathcal{E} = K_V\,\mathcal{E}^{-\gamma}\,d\mathcal{E} \tag{5.60}$$

in the energy range

$$2.5\times10^2\left(\frac{\nu_1}{Hy_1(\gamma)}\right)^{\frac{1}{2}} \le \mathcal{E}(\text{eV}) \le 2.5\times10^2\left(\frac{\nu_1}{Hy_2(\gamma)}\right)^{\frac{1}{2}} \tag{5.61}$$

(cf. (5.57)) where ν_1 and ν_2 are the frequencies corresponding to the limits of the observed radio-band in which the spectral index $\alpha = \frac{1}{2}(\gamma-1)$ has a constant value.

If we express K_V in terms of the spectral density of the radiation flux Φ_ν at some frequency, we get

$$K_V = K_e V = \frac{7.4\times10^{21}R^2}{a(\gamma)H}\,\Phi_\nu\left(\frac{\nu}{6.26\times10^{18}}\right)^{\frac{1}{2}(\gamma-1)} . \tag{5.62}$$

From this we can determine the total number of relativistic electrons in the given energy range

$$N_e = \int_{\mathcal{E}_1}^{\mathcal{E}_2} K_V\,\mathcal{E}^{-\gamma}\,d\mathcal{E} = \frac{7.4\times10^{21}R^2\Phi_\nu}{(\gamma-1)\,a(\gamma)H}\left[\frac{y_1(\gamma)\nu}{\nu_1}\right]^{\frac{1}{2}(\gamma-1)}\left\{1-\left(\frac{y_2(\gamma)\nu_1}{y_1(\gamma)\nu_2}\right)^{\frac{1}{2}(\gamma-1)}\right\}. \tag{5.63}$$

This formula is, of course, approximative by nature as we have used the inequalities (5.61) when changing from $\mathcal{E}_1, \mathcal{E}_2$ to ν_1, ν_2 and each of these determines the limits only with an accuracy of 10%. As usually $\nu_1 \ll \nu_2$ and $y_1(\gamma) < y_2(\gamma)$ the number of electrons is practically determined for $\gamma > 1$ by the lower limit of the frequency range alone and is equal to

$$N_e(>\mathcal{E}_1) = \frac{7.4\times 10^{21} R^2 \Phi_\nu}{(\gamma-1)\,a(\gamma)H}\left[\frac{y_1(\gamma)\nu}{\nu_1}\right]^{\frac{1}{2}(\gamma-1)} . \qquad (5.64)$$

The values of the factors $a(\gamma)$ and $y_1(\gamma)$ are given in Table 5.1. Similarly we can write the total energy of the electrons in the source which are responsible for the emission in the observed frequency range $\nu_1 \leq \nu \leq \nu_2$ in the form

$$W_e = \int_{\mathcal{E}_2}^{\mathcal{E}_1} K_\nu\, \mathcal{E}^{-\gamma+1}\, d\mathcal{E} = A(\gamma,\nu)\,\frac{R^2\Phi_\nu}{H^{\frac{3}{2}}} , \qquad (5.65)$$

where

$$A(\gamma,\nu) = \begin{cases} \dfrac{2.96\times 10^{12}}{(\gamma-2)\,a(\gamma)}\,\nu^{\frac{1}{2}}\left[\dfrac{y_1(\gamma)\nu}{\nu_1}\right]^{\frac{1}{2}(\gamma-1)}\left\{1-\left[\dfrac{y_2(\gamma)\nu_1}{y_1(\gamma)\nu_2}\right]^{\frac{1}{2}(\gamma-2)}\right\} , & \gamma > 2 , \\ 1.44\times 10^{13}\,\nu^{\frac{1}{2}}\,\ln\left[\dfrac{y_2(\gamma)\nu_1}{y_1(\gamma)\nu_2}\right] , & \gamma = 2 , \\ \dfrac{2.96\times 10^{12}}{(2-\gamma)\,a(\gamma)}\,\nu^{\frac{1}{2}}\left[\dfrac{y_2(\gamma)\nu}{\nu_2}\right]^{\frac{1}{2}(\gamma-2)}\left\{1-\left[\dfrac{y_2(\gamma)\nu_1}{y_1(\gamma)\nu_2}\right]^{\frac{1}{2}(2-\gamma)}\right\} , & \frac{1}{3} < \gamma < 2 . \end{cases} \qquad (5.66)$$

When $\gamma < 1.5$ $(\alpha < 0.25)$ the formula which we have given for $A(\gamma,\nu)$ can in fact be used only for rough estimates and we must put on the right-hand side $y_2(\gamma) = y_1(\gamma) = 0.24$; this corresponds to the assumption that an electron with energy $\mathcal{E}$ emits only at the frequency $\nu = \nu_m = 0.29\,\nu_c$ (see (5.40a)).

Expression (5.65) enables us for a known distance R of the source and a known radiative flux Φ_ν to determine at some frequency the total energy of the relativistic electrons in the source provided we know the magnetic field strength H. Unfortunately, there are as yet no reliable independent methods for estimating the magnetic field strength in the sources (see, however, below) and for an evaluation of W_e we must thus make some additional assumptions.

As the primary assumption of this kind one usually assumes that the magnetic field energy in the source W_H and the energy of the relativistic particles (cosmic rays) $W_{c.r.}$ are of the same order of magnitude or, in first approximation, simply equal to one another. In fact, this assumption corresponds to a minimum of the total energy of the field + particles system† for a given

† The total energy of the particles and the magnetic field in the source as function of the field strength for a given radiative power is equal to $W = W_H + W_{c.r.} = C_1 H^2 + C_2 H^{-3/2}$, where C_1 and C_2 are coefficients which are independent of H (see (5.65) and (5.68)). If we determine the minimum of this expression with respect to H, we find that the total energy is a minimum when $W_H = \frac{3}{4} W_{c.r.}$

synchrotron radiation power. Moreover, a magnetic field with energy density appreciably smaller than the energy density of the relativistic particles could not contain the relativistic particles in the limited volume of the source and, as a result of their leaking out of the system, it would reach a state close to the state of energetic quasi-equilibrium between the magnetic field and the relativistic particles. We assume here, of course, that the system under the conditions which interest us will on the whole be in a quasi-stationary state. If we are, for instance, dealing with the ejection of a cloud of relativistic particles during the eruption of the nucleus of a galaxy there may exist also strongly non-equilibrium states in which the energy of the cosmic rays in the cloud during the period of time which interests us is appreciably larger than the magnetic field energy. One may surmise that the duration of the strongly non-equilibrium phase of the separation is nevertheless relatively short. In any case, there are grounds for assuming that in most cases

$$W_H = \kappa_H W_{c.r.} \quad , \quad \kappa_H \sim 1 \quad , \tag{5.67}$$

where κ_H is a numerical coefficient, $W_H = (H^2/8\pi)V$ is the total magnetic field energy, and $W_{c.r.}$ is the total energy of relativistic particles (cosmic rays and electrons) in the radio-emitting nebula.

The radio-astronomical data allow us to judge only how many electrons there are in the source and what their energy is; to determine the total energy of all relativistic particles $W_{c.r.}$ we must also establish a relation between it and the energy of the relativistic electrons W_e. There are at the moment no reliable methods for estimating the fraction W_e of the total energy $W_{c.r.}$ (see, however, below) and as a second essential assumption one usually takes the energy of all cosmic rays in the source to be simply proportional to the energy of the relativistic electrons :

$$W_{c.r.} = \kappa_e W_e \quad , \tag{5.68}$$

where κ_e is a numerical coefficient.

In Chapter 15 we shall dwell upon the astrophysics of cosmic rays, or, as one says more often, upon the problem of the origin of the cosmic rays. However, it is convenient now already to state that $\kappa_e \sim 10^2$ for the cosmic rays in our Galaxy ($\kappa_e \sim 10^2$ near the Earth and probably in most of the Galaxy, but not necessarily everywhere; it is thus well possible that near the centre of the Galaxy $\kappa_e \gg 10^2$; see Ginzburg 1973a). If the cosmic rays are generated on the Sun $\kappa_e \gg 1$. Theoretical considerations also lead to the conclusion that $\kappa_e \gg 1$. Indeed, when particles are accelerated due to their ejection in relativistic shock waves all particles acquire the same velocity and, hence, their energy is proportional to their mass. After that the energy of the electrons will be 'pulled' towards the energy of the protons and nuclei. However, on the other hand, electrons undergo synchrotron and Compton energy losses, which are practically absent for heavy particles. When the acceleration is statistical (Fermi mechanism) the energy of the electrons is less than that of a particle of mass M by a factor m/M. Finally, when they are accelerated in an electric field (and, in particular, accelerated when the 'freezing-in' of the magnetic field is broken) electrons and protons acquire, in general, the same momentum. However, even in that case the average energy of the electrons is, in general, in final reckoning less than that of heavy particles due to additional losses. Under cosmic conditions the inequality

$$\kappa_e \gg 1 \tag{5.69}$$

is thus the norm, although not necessarily always true.

If we make definite assumptions about the values of κ_H and κ_e we can use the observed radio-emission flux to determine both the magnetic field strength and the total energy of the cosmic rays and the electrons in the source, provided the spectrum, the angular size, and the distance of the source are known. It follows from (5.65), (5.67) and (5.68) that

$$W_H \equiv \frac{H^2}{8\pi} V = \kappa_H \kappa_e A(\gamma,\nu) \frac{R^2 \Phi_\nu}{H^{\frac{3}{2}}} , \tag{5.70}$$

whence

$$H = \left[48 \kappa_H \kappa_e A(\gamma,\nu) \frac{\Phi_\nu}{R\phi^3} \right]^{\frac{2}{7}} , \tag{5.71}$$

where $A(\gamma,\nu)$ is given by Eqs.(5.66), $V = \pi L^3/6$ is the volume (assumed to be spherical), and $\phi = L/R$ the angular size of the source. The total energy of the cosmic rays in the source then equals

$$W_{c.r.} = \kappa_e W_e = W_H/\kappa_H = 0.19\, \kappa_H^{-\frac{3}{7}} \left[\kappa_e A(\gamma,\nu)\, \Phi_\nu R^2 \right]^{\frac{4}{7}} (R\phi)^{\frac{9}{7}} . \tag{5.72}$$

Using the formulae given here and assuming that $\kappa_H \sim 1$ and $\kappa_e \sim 10^2$ one has obtained estimates for $W_{c.r.}$, W_e, and W_H in our Galaxy, in galactic sources of non-thermal radio-emission (first of all in supernova remnants), in other normal galaxies, in radio-galaxies, and in quasars. It is difficult to overestimate the value of all these results (see, in particular, Ginzburg and Syrovatskii, 1964a,1966a; Ginzburg, 1973a; and Chapters 15 to 17 later on). Moreover, for a further development of the astrophysics of cosmic rays it is extremely necessary to find means for determining independently all these quantities $W_{c.r.}$, W_e, and W_H or in some cases even only one of them without making assumptions about the values of the others (or, what is almost the same, without giving the coefficients κ_H and κ_e). There are in principle possibilities for this. For instance, one can determine the energy of the proton-nuclear components of the cosmic rays in distant sources using gamma-astronomy methods; to be precise, one uses the intensity of γ-rays which are formed in the decay of π^0-mesons which, in turn, are generated in the source as the result of collisions of cosmic rays (protons and nuclei) with nuclei in the interstellar gas. One might hope that such a method will in the near future bear fruit (Ginzburg, 1973a; for details see Chapter 17). A simultaneous determination of W_e and W_H (or H) is in principle also possible, for instance, by combining radio- and X-ray measurements. To be more precise, we are dealing with the (completely possible) case when the radio-emission of an object (say, a radio-galaxy) has a synchrotron character while its X-ray emission is caused by the inverse Compton scattering of relativistic electrons by the known field of the optical, infra-red, or radio-radiation. If the same relativistic electrons are responsible for the radio- and the X-ray emission (in other words, when the frequency ranges are appropriately chosen), we can in that case determine from a knowledge of the X-ray emission flux (and knowing the distance to the source, its size, and the density of the energy of the electrons which scatter the radiation, say, the infra-red radiation) the characteristics of the relativistic electrons in the source (see Chapter 16). Moreover, we can from the data about the flux and spectrum of the synchrotron radiation find also the field H in the source. Unfortunately so far it has not been possible to apply this approach to a single source due to the insufficient development of X-ray astronomy and to the difficulties connected with proving that the X-ray emission is caused by the Compton effect. However, one may hope also for success along this path in the future.

In conclusion we make a few remarks about the limits of applicability of the theory of the synchrotron emission given here. and also about the synchro-Compton emission.

We assumed above that only ultra-relativistic electrons were considered while the angle χ between **k** (or **v**, which is the same in the present case) and

$\mathbf{H} \equiv \mathbf{H}_0$ is sufficiently large (condition (5.1)) and that the classical theory of radiation is applicable, that is, we assumed that the condition

$$H \ll \frac{m^2c^3}{e\hbar}\frac{mc^2}{\mathcal{E}} = 4.4 \times 10^{13}\frac{mc^2}{\mathcal{E}} \text{ Oe} \tag{4.30}$$

is satisfied†.

In a number of cases it is also assumed that the radiative force is small compared to the Lorentz force, that is, we used the condition

$$\frac{\mathcal{E}}{mc^2} \ll \left[\frac{m^2c^4}{e^3 H \sin^2\chi}\right]^{\frac{1}{2}} \sim \frac{10^8}{\sqrt{(H\sin^2\chi)}} . \tag{4.44}$$

If the latter condition is not satisfied, the electron moves not along a circle, but along a rather sharply defined spiral curve with a decreasing radius (see, for instance, Shen, 1970; Suvarov and Chugunov, 1973; Lubart, 1974). This happens, however, only at spaces between successive pulses (see Fig.5.1; the same also refers to motion along a helical path). However, when condition (4.44) is violated, but condition (4.30) holds, the shape of each pulse is unchanged. The fact is, clearly, that the electron emits in a given direction only during a time $\Delta t' \sim mc/eH_\perp$. During that time the losses are small, provided

$$\mathfrak{R}\Delta t' \sim \frac{e^4H_\perp^2}{m^2c^3}\left(\frac{\mathcal{E}}{mc^2}\right)^2\frac{mc}{eH_\perp} \ll \mathcal{E},$$

which leads to the inequality

$$H_\perp = H\sin\chi \ll \frac{m^2c^4}{e^3}\frac{mc^2}{\mathcal{E}} , \tag{4.27a}$$

which is, in fact, the same as condition (4.27) which we derived without distinguishing between the fields H and $H_\perp$; the same can be said about the condition (4.30) (see last footnote). As a result, if inequality (4.30) is satisfied, inequality (4.27a) is certainly satisfied, as its right-hand side is larger by a factor $1/\alpha = \hbar c/e^2 \approx 137$. Thus, even if condition (4.44) is violated, but (4.30) holds, the shape of the continuous synchrotron spectrum (that is, the spectrum obtained by averaging over harmonics) is unchanged.

We have here completely neglected, moreover, possible effects of the medium (plasma) in which the radiating electrons move. In some cases the influence

---

† Condition (4.30) with the substitution, necessary in the general case, of H by $H_\perp = H\sin\chi$ is equivalent to the inequality $\hbar\omega_m \ll \mathcal{E}$, where $\omega_m \sim (eH_\perp/mc)(\mathcal{E}/mc^2)^2$ is the frequency corresponding to the maximum in the spectrum of the synchrotron radiation. Hence it is clear that for the 'tail' of the synchrotron radiation, that is, in the frequency region $\omega \gg \omega_m$, condition (4.30) must be replaced by a more rigorous one.

of the medium is very important and may completely change the picture (see Chapter 6).

When we considered the emission of a system of relativistic electrons we assumed that they radiated completely independently of one another and were in a given magnetic field **H**. However, if the density of radiating particles is large they can, first of all, change the external field (in the case of a magnetic field we are dealing here with a diamagnetic effect and with mutual induction- see §6 of Ginzburg, Sazonov, and Syrovatskii, 1968). Secondly, and this is usually more important, one must take into account reabsorption when the density of the radiating particles is sufficiently large, that is, in the case of the synchrotron mechanism the absorption of the radiation by the relativistic electrons themselves. We shall discuss this effect in Chapter 9.

The synchrotron theory given here can thus be applied only when a number of conditions are satisfied; this does not invalidate the theory as all these conditions are often fulfilled. However, the necessity to bear in mind the limitations and conditions for applicability both in the actual case discussed here as in physics in general is a very important fact. Many errors met with in the literature (not to mention unpublished errors of which there are, of course, incomparably more) are connected just with forgetting the limits of applicability of some of the formulae or expressions. It is not less important that as a result of rejecting one or more limitations often interesting possibilities may appear and one may find new mechanisms and effects.

Amongst the conditions for the validity of the synchrotron theory is the assumption that the emission occurs in a uniform and constant magnetic field. We did not mention this condition again as it is strictly the primary one and may be considered to be included in the definition: we call synchrotron radiation the emission by ultra-relativistic particles (charges) moving in a magnetic field **H** which is uniform in space and constant in time. The problem arises, of course, of the emission in a field $\mathbf{H}(\mathbf{r},t)$ or in an electromagnetic field $\mathbf{E}(\mathbf{r},t)$, $\mathbf{H}(\mathbf{r},t)$.

It is clear even from general considerations, and we emphasized this already at the beginning of Chapter 4, that many characteristic features of the emission by relativistic particles are not connected with the kind of external electromagnetic field in which the particles move. Apart from the motion in a constant magnetic field **H** and correspondingly the magneto-brems radiation, of particularly large interest is the radiation by a charge moving in the field of an electromagnetic wave with a frequency $\omega_0 = 2\pi\nu_0$. The radiation which occurs in that case with frequencies ω is usually called scattered radiation, as one can in this case talk about the scattering of a wave of frequency ω_0 by a moving particle (charge). At large energies when the relevant frequencies ω or ω_0 are comparable with the rest mass energy mc^2 of the particle, divided by $\hbar$, one usually speaks of Compton or inverse Compton scattering †. We shall consider that process in Chapter 16. Just now, however, we mention an interesting particular case which in the context of astrophysical applications attracted attention only recently and, in actual fact, after the discovery of pulsars. We are dealing with the motion and emission (scattering) of particles in the field of an electromagnetic wave with a very low frequency, say, a wave with a frequency $\Omega \equiv \omega_0$ which is emitted by a rotating magnetic neutron star (pulsar; for known pulsars $\Omega \leqslant 200\ s^{-1}$ and in most cases $\Omega = 2\pi/T_0 \sim 1$ to $10\ s^{-1}$; see Ginzburg, 1971; ter Haar, 1972). One can understand the features of this case, at least partially, by considering the motion of a charge in a variable magnetic field $\mathbf{H} = \mathbf{H}_0 \cos \Omega t$ (it is well known that under well defined conditions the induction electric field can be small compared to the magnetic field). It is then clear that the particle moves practically as in a constant field, as long as

$$\frac{\omega_H^*}{\Omega} = \frac{eH_0}{mc}\,\frac{mc^2}{\mathcal{E}}\,\frac{1}{\Omega} \gg 1 \quad . \tag{5.73}$$

The emission of the particle, on the other hand, will be approximately the same as in a constant field, provided

$$f = \frac{eH_{0,\perp}}{mc\Omega} \gg 1 \quad . \tag{5.74}$$

It is sufficient to note that under those conditions the characteristic time for the emission in the direction of the observer $\Delta t \sim mc/eH_{0,\perp}$ is small compared to the period of the wave $T_0 = 2\pi/\Omega$.

---

† Mostly one calls scattering of a photon by a particle at rest Compton scattering. The scattering of a soft photon by a fast moving particle, that is, a particle with a high energy, is, on the other hand, called inverse Compton scattering. It is completely obvious that in the two cases one is essentially dealing with the scattering process, but with different 'initial conditions' in a given frame of reference or with the same scattering process in different frames of reference which move with respect to one another with a velocity $v < c$.

In the wave zone of the emitter (say, the pulsar) the electric field in the wave $E = H$ and even under conditions (5.73) or (5.74) the motion and the emission differ from those in a purely magnetic field. All the same, when condition (5.74) holds, and even when $f \geqslant 1$, the nature of the radiation is in many respects close to that of the synchrotron radiation, and such radiation is sometimes called 'synchro-Compton radiation', (see Rees, 1971a,b; Gunn and Ostriker, 1971; Arons, 1972; Blandford, 1972). The most important difference between the synchro;Compton and the synchrotron radiation is that in the first case there occurs circular polarization, the degree of which is, in general, of order $1/f$ (the degree of circular polarization depends on the nature of the polarization of the low-frequency wave; for details see Rees, 1971a,b; Arons, 1972; Gunn and Ostriker, 1971; and Blandford, 1972); moreover, for the synchrotron radiation the circular polarization is generally characterized by the parameter $\xi = mc^2/\mathcal{E}$ (see earlier; we are dealing here with the radiation from a system of particles). Synchro-Compton radiation certainly deserves a detailed analysis, but here we restrict ourselves to the few remarks we have made and to referring to the literature.

Chapter VI

ELECTRODYNAMICS OF A CONTINUOUS MEDIUM

Hamiltonian method. Photons in a medium.
Emission by an oscillator in isotropic and anisotropic media.
Cherenkov radiation. Doppler effect. Ondulator in a medium.
Characteristic features of emission by particles moving in a medium.
Synchrotron emission in a plasma.

If an emitter such as a charge moves in a medium rather than in vacuo, the whole picture of the radiation may change radically. It is sufficient to say that for a given motion of a charge it may radiate in vacuo but not at all in a medium, or, on the other hand, it may radiate in a medium while there is no radiation in vacuo (in this last case we are primarily thinking of the emission by a uniformly moving charge). The theory of the emission in a medium (when we take the effect of the medium into account) should logically be based on the general electrodynamics of continuous media or, put differently, on macroscopic electrodynamics. However, in the plan of the present text it is much more natural to consider the theory of radiation in a medium at once after the theory of emission in vacuo. As far as the electrodynamics of continuous media is concerned, we assume that our readers know it, although we shall remind them of the basic formulae. Finally, we shall be interested here mainly in a few questions of principle and the physical essence will be our concern. We shall therefore not aim at maximum generality and, in particular, we shall usually neglect spatial dispersion, often assume the medium to be isotropic and transparent, and so on. We shall, however, in what follows (see Chapters 10 and 12, and also Agranovich and Ginzburg, 1966) elucidate some problems of the electrodynamics of continuous media at a more general level.

We write the field equations in a medium in the form

$$\left.\begin{aligned} \operatorname{curl} \mathbf{H} &= \frac{4\pi}{c}\mathbf{j} + \frac{1}{c}\frac{\partial \mathbf{D}}{\partial t}, \quad & \operatorname{div} \mathbf{D} &= 4\pi\rho, \\ \operatorname{curl} \mathbf{E} &= -\frac{1}{c}\frac{\partial \mathbf{H}}{\partial t}, \quad & \operatorname{div} \mathbf{H} &= 0. \end{aligned}\right\} \tag{6.1}$$

These equations differ from (1.1) by the substitution of the electric induction $\mathbf{D} = \mathbf{E} + 4\pi\mathbf{P}$ for the electric field $\mathbf{E}$ in the first two equations. We assume the medium to be non-magnetic (therefore, the magnetic induction $\mathbf{B} = \mathbf{H}$; one can usually also for a magnetic medium use this relation when spatial dispersion is taken into account: see Agranovich and Ginzburg, 1966 and Chapter 10). Moreover, as everywhere in the present book, we consider only a medium at rest (in the laboratory frame of reference, which we use, the velocity of the medium is identically equal to zero; see, however, Chapter 12). Finally, while in (1.1) the current density was written in the form $\rho\mathbf{v}$, in (6.1) we have introduced the current density $\mathbf{j}$. From the first two Eqs. (6.1) we find the charge conservation law

$$\mathrm{div}\,\mathbf{j} + \frac{\partial\rho}{\partial t} = 0\,. \tag{6.2}$$

It is well known that the set (6.1) becomes well defined only after we have expressed $\mathbf{D}$ in terms of $\mathbf{E}$ (or, in principle, in terms of $\mathbf{E}$ and $\mathbf{H}$). In an isotropic medium without spatial or temporal (frequency) dispersion

$$\mathbf{D}(\mathbf{r},t) = \boldsymbol{\varepsilon}(\mathbf{r},t)\,\mathbf{E}(\mathbf{r},t)\,. \tag{6.3}$$

If the medium is homogeneous in space and does not change with time, $\boldsymbol{\varepsilon}$ = constant. For optical or lower frequencies the spatial dispersion is usually small and we shall neglect it here. On the other hand, the temporal dispersion is, in general, always more or less important. This means that for an isotropic medium

$$\mathbf{D}(\mathbf{r},t) = \int_{-\infty}^{t} \hat{\boldsymbol{\varepsilon}}(\mathbf{r},t,t')\,\mathbf{E}(\mathbf{r},t')\,dt'\,, \tag{6.4}$$

where the limits of the integration over t reflect the requirement of causality; if the properties of the medium do not change with time, we have for the kernel $\hat{\boldsymbol{\varepsilon}}(\mathbf{r},t,t') = \hat{\boldsymbol{\varepsilon}}(\mathbf{r},t-t')$. Introducing the Fourier components

$$\mathrm{E}(\mathbf{r},\omega) = \frac{1}{2\pi}\int_{-\infty}^{+\infty} \mathbf{E}(\mathbf{r},t)\,e^{i\omega t}dt\,,\quad \mathbf{E}(\mathbf{r},t) = \int_{-\infty}^{+\infty} \mathbf{E}(\mathbf{r},\omega)\,e^{-i\omega t}\,d\omega\,, \tag{6.5}$$

and similarly for $\mathbf{D}$, we get

$$\mathbf{D}(\mathbf{r},\omega) = \boldsymbol{\varepsilon}(\mathbf{r},\omega)\,\mathbf{E}(\mathbf{r},\omega)\,,\quad \boldsymbol{\varepsilon}(\mathbf{r},\omega) = \int_0^{\infty} \hat{\boldsymbol{\varepsilon}}(\mathbf{r},\tau)\,e^{i\omega\tau}\,d\tau\,. \tag{6.6}$$

In a homogeneous medium $\boldsymbol{\varepsilon}(\mathbf{r},\omega) = \boldsymbol{\varepsilon}(\omega)$. In an anisotropic medium (without spatial dispersion and for the case where the properties of the medium are time-independent)

$$D_i(\mathbf{r},\omega) = \boldsymbol{\varepsilon}_{ij}(\mathbf{r},\omega)\,E_j(\mathbf{r},\omega)\,, \tag{6.7}$$

where $\boldsymbol{\varepsilon}_{ij}$ is a second rank tensor and where we assumed, as always, that one sums over repeated indices.

When we use the relations (6.6) or (6.7) we must, of course, in (6.1) also change to Fourier components. This is, however, not always convenient as, for instance, in the Hamiltonian method used below time-derivatives occur explicitly. Moreover, it turns out that when one applies the Hamiltonian method to begin with one can, in general, neglect completely the frequency dispersion (that is, the dependence of ε_{ij} or ε on ω) and afterwards in the final result replace the refractive index n (in the isotropic case $n=\sqrt{\varepsilon}$) by $n(\omega)$ and in that way completely take into account the frequency dispersion (Ginzburg, 1940b; Ryzhov, 1959; Ginzburg and Eidman, 1963). We shall come back to this, but in the present chapter we shall in the field equations put ε = constant (or ε_{ij} = constant) which refers directly only to a homogeneous dispersionless medium. Moreover, we assume the quantity ε to be real and positive (no absorption or total internal reflection; see also below).

Introducing potentials in the usual way,

$$\mathbf{E} = -\frac{1}{c}\frac{\partial \mathbf{A}}{\partial t} - \operatorname{grad}\phi , \quad \mathbf{H} = \operatorname{curl}\mathbf{A} , \tag{6.8}$$

we get from (6.1) for an isotropic medium

$$\begin{aligned} \nabla^2\mathbf{A} - \frac{\varepsilon}{c^2}\frac{\partial^2\mathbf{A}}{\partial t^2} - \operatorname{grad}\left(\frac{\varepsilon}{c}\frac{\partial\phi}{\partial t} + \operatorname{div}\mathbf{A}\right) &= -\frac{4\pi}{c}\mathbf{j} , \\ \nabla^2\phi + \frac{1}{c}\frac{\partial}{\partial t}\operatorname{div}\mathbf{A} &= -\frac{4\pi}{\varepsilon}\rho . \end{aligned} \tag{6.9}$$

If we choose a gauge in which

$$\operatorname{div}\mathbf{A} + \frac{\varepsilon}{c}\frac{\partial\phi}{\partial t} = 0 , \tag{6.10}$$

we have

$$\nabla^2\mathbf{A} - \frac{\varepsilon}{c^2}\frac{\partial^2\mathbf{A}}{\partial t^2} = -\frac{4\pi}{c}\mathbf{j} , \quad \nabla^2\phi - \frac{\varepsilon}{c^2}\frac{\partial^2\phi}{\partial t^2} = -\frac{4\pi}{\varepsilon}\rho . \tag{6.11}$$

In the gauge in which $\operatorname{div}\mathbf{A} = 0$ we get

$$\nabla^2\mathbf{A} - \frac{\varepsilon}{c^2}\frac{\partial^2\mathbf{A}}{\partial t^2} = -\frac{4\pi}{c}\mathbf{j} + \frac{\varepsilon}{c}\operatorname{grad}\frac{\partial\phi}{\partial t} , \quad \nabla^2\phi = -\frac{4\pi}{\varepsilon}\rho , \tag{6.12}$$

and

$$\mathbf{E} = \mathbf{E}_{tr} + \mathbf{E}_{\ell} , \quad \operatorname{div}\mathbf{E}_{tr} = 0 , \quad \mathbf{E}_{tr} = -\frac{1}{c}\frac{\partial\mathbf{A}}{\partial t} , \quad \mathbf{E}_{\ell} = -\operatorname{grad}\phi . \tag{6.13}$$

The Hamiltonian method is, in the electrodynamics of continuous media, developed completely analogously to what was done for a vacuum (see Chapter 1). We shall thus use Eqs.(6.12), (6.13) and the expansions (we assume that $\varepsilon \neq 0$)

$$\left.\begin{aligned} &\mathbf{A} = \sum_{\lambda, i=1,2} q_{\lambda i}\mathbf{A}_{\lambda i} , \quad \mathbf{A}_{\lambda 1} = \sqrt{8\pi}\,\frac{c}{n}\,\mathbf{e}_{\lambda}\cos(\mathbf{k}_{\lambda}\cdot\mathbf{r}) , \\ &\mathbf{A}_{\lambda 2} = \sqrt{8\pi}\,\frac{c}{n}\,\mathbf{e}_{\lambda}\sin(\mathbf{k}_{\lambda}\cdot\mathbf{r}) , \quad (\mathbf{e}_{\lambda}\cdot\mathbf{k}_{\lambda}) = 0 , \quad e_{\lambda} = 1 , \\ &n = \sqrt{\varepsilon} , \quad \int (\mathbf{A}_{\lambda i}\cdot\mathbf{A}_{\mu j})\, d^3\mathbf{r} = \frac{4\pi c^2}{\varepsilon}\,\delta_{\lambda\mu}\,\delta_{ij} , \end{aligned}\right\} \tag{6.14}$$

or

$$\left.\begin{aligned} &\mathbf{A} = \sum_\lambda (q_\lambda \mathbf{A}_\lambda + q_\lambda^* \mathbf{A}_\lambda^*) , \quad \mathbf{A}_\lambda = \sqrt{4\pi}\,\frac{c}{n}\,\mathbf{e}_\lambda \exp(i\mathbf{k}_\lambda\cdot\mathbf{r}) , \\ &(\mathbf{e}_\lambda\cdot\mathbf{k}_\lambda) = 0 , \quad n = \sqrt{\boldsymbol{\varepsilon}} , \quad \int (\mathbf{A}_\lambda\cdot\mathbf{A}_\mu^*)\, d^3\mathbf{r} = 4\pi\frac{c^2}{\boldsymbol{\varepsilon}}\,\delta_{\lambda\mu} \quad . \end{aligned}\right\} \tag{6.15}$$

Of course, the two expansions are equivalent; in different cases one or the other may be slightly more convenient to apply. A few provisos stated in Chapter 1, for instance, referring to the presence of two polarization vectors $\mathbf{e}_\lambda$, refer also to the expansions (6.14) and (6.15).

One can easily check that the energy of the transverse field is given by

$$\mathcal{H}_{tr} = \int \frac{\boldsymbol{\varepsilon} E_{tr}^2 + H^2}{8\pi}\, d^3\mathbf{r} = \tfrac{1}{2}\sum_{\lambda,i} (p_{\lambda i}^2 + \omega_\lambda^2 q_{\lambda i}^2) = \sum_\lambda (p_\lambda p_\lambda^* + \omega_\lambda^2 q_\lambda q_\lambda^*) , \tag{6.16}$$

where

$$p_{\lambda i} = \dot{q}_{\lambda i} , \quad p_\lambda = \dot{q}_\lambda , \quad \omega_\lambda^2 = \frac{c^2}{\boldsymbol{\varepsilon}}\, k_\lambda^2 \equiv \frac{c^2}{n^2}\, k_\lambda^2 . \tag{6.17}$$

We then get the equations of motion for the $q_{\lambda i}$ or the q_λ from (6.12) in the same way as for the vacuum case and they have the form

$$\ddot{q}_{\lambda i} + \omega_\lambda^2 q_{\lambda i} = \frac{1}{c}\int (\mathbf{j}\cdot\mathbf{A}_{\lambda i})\, d^3\mathbf{r} , \tag{6.18}$$

$$\ddot{q}_\lambda + \omega_\lambda^2 q_\lambda = \frac{1}{c}\int (\mathbf{j}\cdot\mathbf{A}_\lambda^*)\, d^3\mathbf{r} . \tag{6.19}$$

The most general case considered below corresponds to a particle of charge e, electric moment $\mathbf{p}(t)$, and magnetic moment $\boldsymbol{\mu}(t)$. If we assume the particle to be a point particle, which is usually admissable when calculating the emitted energy and, in general, the radiation field, we have

$$\mathbf{j} = \rho_e \mathbf{v} + c\,\mathrm{curl}\,\mathfrak{M} + \frac{\partial \mathbf{P}}{\partial t} =$$

$$= e\mathbf{v}\delta(\mathbf{r}-\mathbf{r}_i) + c\;\mathrm{curl}\{\boldsymbol{\mu}\delta(\mathbf{r}-\mathbf{r}_i)\} + \frac{\partial}{\partial t}\{\mathbf{p}\delta(\mathbf{r}-\mathbf{r}_i)\} , \tag{6.20}$$

where $\mathbf{r}_i$ is the position vector of the charge and $\mathbf{v} = \dot{\mathbf{r}}_i(t)$. For a charge (without moments, that is, putting in (6.20) $\mathbf{p} = 0$ and $\boldsymbol{\mu} = 0$) we get from (6.18) and (6.19)

$$\left.\begin{aligned} &\ddot{q}_{\lambda 1} + \omega_\lambda^2 q_{\lambda 1} = \sqrt{8\pi}\,\frac{e}{n}(\mathbf{e}_\lambda\cdot\mathbf{v})\cos(\mathbf{k}_\lambda\cdot\mathbf{r}_i) , \\ &\ddot{q}_{\lambda 2} + \omega_\lambda^2 q_{\lambda 2} = \sqrt{8\pi}\,\frac{e}{n}(\mathbf{e}_\lambda\cdot\mathbf{v})\sin(\mathbf{k}_\lambda\cdot\mathbf{r}_i) \; ; \end{aligned}\right\} \tag{6.21}$$

$$\ddot{q}_\lambda + \omega_\lambda^2 q_\lambda = \sqrt{4\pi}\,\frac{e}{n}(\mathbf{e}_\lambda\cdot\mathbf{v})\exp\{-i(\mathbf{k}_\lambda\cdot\mathbf{r}_i)\} . \tag{6.22}$$

As compared to the vacuum case the additional factor $1/n$ on the right-hand sides is not as important as the change in the relation between ω_λ and k_λ, that is, the appearance of the factor $\varepsilon^{-1} = n^{-2}$ in the relation $\omega_\lambda^2 = (c^2/n^2)\,k_\lambda^2$ (see (6.17)). The meaning of this change is clear: electromagnetic waves in a medium with a real permittivity $\varepsilon > 0$ propagate with a phase velocity

$$v_{ph} = \frac{c}{\sqrt{\varepsilon}} = \frac{c}{n} . \qquad (6.23)$$

This result is undoubtedly well known to the reader, but all the same we remind him that it is immediately clear from the homogeneous equations (6.11) or (6.12), that is, from the equations without charges or currents. For instance, (6.22) is in that case the equation for the free oscillations of an oscillator

$$\ddot{q}_\lambda + \omega_\lambda^2 q_\lambda = 0 , \quad \omega_\lambda = \frac{c}{n} k_\lambda ,$$

$$q_\lambda = c_1 \exp(i\omega_\lambda t) + c_2 \exp(-i\omega_\lambda t) ; \qquad (6.24)$$

from this it is clear that the expansion (6.15) is in the case when there are no charges or currents an expansion in plane waves of the kind

$$\exp\{\pm i[(\mathbf{k}_\lambda \cdot \mathbf{r}) - (c/n)k_\lambda t]\} ,$$

that is, waves with the phase velocity (6.23).

The Hamiltonian for a free radiation field in a medium has the form (6.16) and one quantizes as in the vacuum case (this also refers to the transverse field in a medium without charges or currents — the quantization corresponds to replacing $p_{\lambda i}$ and $q_{\lambda i}$ by operators which satisfy the commutation relations (1.45)). As a result we are led through the expansion of a free field in a medium in terms of waves of the type $\exp[\pm i(\mathbf{k}_\lambda \cdot \mathbf{r})]$ (see (6.15)) to the concept of 'photons in a medium' with an energy E_λ and a momentum $\mathbf{p}_\lambda$ given by

$$E_\lambda = \hbar\omega_\lambda , \quad \mathbf{p}_\lambda = \hbar\mathbf{k}_\lambda , \quad p_\lambda = \frac{\hbar\omega_\lambda n}{c} . \qquad (6.25)$$

As far as the energy is concerned this conclusion is obvious (see (1.43), (1.49), and (6.16)). In the case of the momentum one should find the eigenvalue of the momentum operator of the electromagnetic field, but the corresponding expression was for a long time written in different forms and there was a lot of discussion which of these was the correct one (the discussion was about the Minkowski and Abraham energy-momentum tensors). At the present time this problem has been relatively satisfactorily cleared up and we shall discuss it in Chapter 12. Fortunately, its solution (which, in any case, requires special considerations) is not needed for the solution of the problems which normally arise in the quantum theory of radiation in a medium (Ginzburg, 1940a; Jauch and Watson, 1948a,b; Watson and Jauch, 1949; Riazanov, 1957). The fact is that when evaluating the transition matrix elements which determine the probability for some radiation process, the vector potential operator has a factor of the kind $q_\lambda \exp[i(\mathbf{k}_\lambda \cdot \mathbf{r})]$; this factor occurs together with factors corresponding to the particle wavefunctions, that is, for free particles with factors $e^{i(\mathbf{p}\cdot\mathbf{r})\hbar}$. From this it is clear that in the momentum conservation laws radiation in a medium, as in vacuo, contributes an amount $\hbar\mathbf{k}_\lambda$. We have seen that in a medium $\hbar k_\lambda = \hbar\omega_\lambda n/c$, and this leads to (6.25). It is thus clear that, completely independent of the analysis of the problem of the form of the energy-momentum tensor in a medium, 'photons in a medium' which occur in general on the same basis as the photons in vacuo have an energy $\hbar\omega$ and a momentum $\hbar\omega n/c$ (for details see Ginzburg, 1973c; Walker and Lahoz, 1975; Ginzburg and Ugarov, 1976); we have just now discussed an isotropic medium, but the generalization to the case of an anisotropic medium is also obvious: in that case $n = n_\ell$ is the refractive index for the appropriate 'normal' wave; vide infra). The use of quantum representations or, to be more precise, of quantum language turns out to be rather convenient for the solution of a number of problems in the theory of radiation in a medium. We shall come back to this in Chapter 7.

We shall now use the Hamiltonian method (that is, in fact, the expansion in plane waves) to solve a number of problems from the theory of radiation in a medium.

We start with the emission by an oscillator which we considered in the vacuum case in Chapter 1. We put in (6.21)

$$\mathbf{r}_i \equiv \mathbf{r}(t) = \mathbf{a}_0 \sin\omega_0 t\,,\ \mathbf{v} = \dot{\mathbf{r}}(t) = \mathbf{v}_0 \cos\omega_0 t = \mathbf{a}_0\,\omega_0 \cos\omega_0 t\,,\ a_0 \ll \frac{1}{k} = \frac{\lambda}{2\pi} = \frac{c}{n\omega_0}\,,$$

where $\lambda = 2\pi c/n\omega_0$ is the wavelength of the radiation emitted by the oscillator; we then get the equation (see (1.80))

$$\ddot{q}_{\lambda 1} + \omega_\lambda^2 q_{\lambda 1} = \sqrt{8\pi}\,\frac{e}{n}(\mathbf{e}_\lambda \cdot \mathbf{v}_0)\cos\omega_0 t\,. \tag{6.26}$$

The rest of the calculations proceeds also completely analogously to the vacuum case, but the number of states is now equal to

$$\frac{k^2 dk\, d^2\Omega}{(2\pi)^3} = \frac{n^3\omega^2 d\omega\, d^2\Omega}{(2\pi c)^3} \tag{6.27}$$

and as a result we get †

---

† see footnote on next page

$$\frac{d\mathcal{H}_{tr}}{dt} = \frac{\mathcal{H}_{tr}}{t} \equiv \frac{dW_s}{dt} = \frac{e^2 a_0^2 \omega_0^4 n}{8\pi c^3} \sin^2\theta \, d^2\Omega \, . \tag{6.28}$$

The only difference with Eq.(1.85) for the vacuum case consists here in the appearance of the factor n (in (6.27) there occurs an extra factor n^3 as compared to the vacuum case but it is clear from (6.26) that the quantity q_λ^2 contains an extra factor $1/n^2$). If we take frequency dispersion into account we must put $n = n(\omega_0)$. This conclusion can, firstly, be based somewhat indirectly by comparing the results obtained by the Hamiltonian method with those obtained by other means. Secondly, if $\boldsymbol{\varepsilon} = \boldsymbol{\varepsilon}(\omega)$, we can write the field equations as before, say in the form (6.11) or (6.12), but we must consider $\boldsymbol{\varepsilon}$ to be an operator $\hat{\boldsymbol{\varepsilon}}$ such that $\hat{\boldsymbol{\varepsilon}}\, e^{-i\omega t} = \hat{\boldsymbol{\varepsilon}}(\omega)\, e^{-i\omega t}$. Furthermore, up to the integration in Eqs.(6.18) or (6.19) the wavevector $\mathbf{k}_\lambda$ occurs in them and we may assume that $n = n(k_\lambda)$, $\omega_\lambda = ck_\lambda/n(k_\lambda)$. Only in the radiation field is the frequency $\omega = \omega_\lambda = ck/n$, that is, are ω and k related through the usual dispersion relation. We only get to the radiation at the very end when we evaluate $\mathcal{H}_{tr}$ for large t (see Chapter 1 and (6.28)).

Therefore one may check (see Ginzburg, 1940b; Ryzhov, 1959; Ginzburg and Eidman, 1963; and Zheleznyakov, 1970, § 25) that taking the frequency dispersion into account must be done in the final expressions. However, the absence of an automatic procedure is undoubtedly a weak feature of the Hamiltonian method. In some sense the same can be said also about the evaluation not so much of the energy as of the fields themselves — to do this by the Hamiltonian method is somewhat more complicated than by other means (see Ryzhov, 1959 where absorption is also taken into account). This is, however, generally speaking true only for an isotropic medium where 'other means' are, indeed, well known. On the other hand, for an anisotropic medium the Hamiltonian method is hardly worse than any other one.

We shall discuss this below, but we shall first discuss the problem of the radiation in an isotropic medium by a vibrating electric and magnetic dipole (oscillator). In the first case the problem is the same as the one just considered, but we want now to use Eq.(6.20). If we assume in it that the charge

---

† As we emphasized already in Chapter 1, the summation in the expansions (6.14) and (6.15) is over a hemi-sphere of $\mathbf{k}_\lambda$-directions. Therefore if we change to an integration over angles we must introduce an extra factor $\frac{1}{2}$ (if we take, as usually, in the system of spherical polars $0 \le \phi \le 2\pi$ and $0 \le \theta \le \pi$, which corresponds to the whole sphere of directions; of course, this fact was taken in account in Chapter 1).

of the particle is zero and that it is at rest, when $\mathbf{r}_i = 0$, we get after substitution into (6.18) the equations ($\dot{\mathbf{p}} \equiv d\mathbf{p}/dt$)

$$\ddot{q}_{\lambda 1} + \omega_\lambda^2 q_{\lambda 1} = \frac{\sqrt{8\pi}}{n} (\mathbf{e}_\lambda \cdot \dot{\mathbf{p}}) \,, \tag{6.29}$$

$$\ddot{q}_{\lambda 2} + \omega_\lambda^2 q_{\lambda 2} = \frac{\sqrt{8\pi}\, c}{n} (\mathbf{e}_\lambda \cdot [\boldsymbol{\mu} \wedge \mathbf{k}_\lambda]) \,. \tag{6.30}$$

If we put $\mathbf{p} = e\mathbf{r} = e\mathbf{a}_0 \sin\omega_0 t$, Eq.(6.29) is the same as (6.26), as one should expect. For $n = 1$ Eq.(6.30) is the same as (2.19). We put $\boldsymbol{\mu} = \boldsymbol{\mu}_0 \sin\omega_0 t$ and completely analogous with what we did for the oscillator we find for the emitted power†

$$\frac{d\mathcal{H}_{tr}}{dt} \equiv \frac{dW_s}{dt} = \frac{\mu_0^2 \omega_0^2 n^3}{8\pi c^3} \sin^2\theta \, d^2\Omega \,, \quad \frac{dW}{dt} = \int \frac{dW_s}{dt} d^2\Omega = \frac{\mu_0^2 \omega_0^2 n^3}{3c^3} \,. \tag{6.31}$$

These expressions can be obtained at once for a vacuum from the well known formulae for (electric and magnetic) dipole radiation (see Landau and Lifshitz, 1975, §§ 67 and 71, and also (2.23)). The calculations for an isotropic medium can, of course, be carried out by the usual methods also.

We have given the result (6.31) in particular in order to emphasize the appearance of the factor n^3, as in the case of an electric dipole there occurs a factor n. The effect of these factors may be enormous. For instance, in a plasma in a magnetic field (this leads to anisotropy which is here unimportant) the refractive index $n_\ell(\omega)$ may for some of the characteristic waves (index ℓ) and frequencies ω reach values $n \sim 10^2$ to 10^3, and formally even larger values (for details see Ginzburg 1970b and Chapter 11). If we consider the magnetic dipole radiation of a pulsar in a medium, the above-mentioned fact and the appearance of the factor n^3 in (6.31), taking the effect of the medium into account appreciably changes the whole picture (see Ginzburg, 1971; we must bear in mind that the linear approximation discussed — the use of the linear relation (6.4) or (6.6) between $\mathbf{D}$ and $\mathbf{E}$ is, generally speaking, inappropriate near pulsars, but this is a special problem). Another, not less striking example of the effect of a medium is dipole radiation in an isotropic plasma when $n(\omega) = \sqrt{(1-\omega_p^2/\omega^2)}$, $\omega_p^2 = 4\pi e^2 N/m$ (N is the electron density in the non-relativistic plasma considered here). It is clear that in that case $n < 1$ and for an oscillator frequency $\omega_0 \sim \omega_p$ the refractive index n may be

---

† Clearly, $(\mathbf{e}_{\lambda 1} \cdot [\boldsymbol{\mu} \wedge \mathbf{k}_\lambda]) = -(\boldsymbol{\mu} \cdot [\mathbf{e}_{\lambda 1} \wedge \mathbf{k}_\lambda]) = -([\boldsymbol{\mu} \wedge \mathbf{e}_{\lambda 2}] \cdot \mathbf{k}_\lambda)$, where $\mathbf{e}_{\lambda 1,2}$ are the polarization vectors $((\mathbf{e}_{\lambda 1} \cdot \mathbf{e}_{\lambda 2}) = 0$, $(\mathbf{e}_{\lambda 1,2} \cdot \mathbf{k}_\lambda) = 0$; see Chapter 1).

close to zero; in that case the radiative power (6.28) or (6.31) is steeply decreased. Moreover, in an isotropic plasma the permittivity $\varepsilon = 1 - \omega_p^2/\omega^2$ turns out to be negative for an oscillator frequency $\omega_0 < \omega_p$. This means that waves with a frequency $\omega_0 < \omega_p$ can not propagate at all — they are damped in space according to the relation $E = E_0 \exp(-\omega\sqrt{|\varepsilon|}\, z/c)$ (for details see Ginzburg, 1970b and Chapter 11). Of course, the source does not radiate at all under such conditions.

We now consider the problem of the radiation in an anisotropic medium where we for the sake of simplicity assume the tensor ε_{ij} to be real and reduced to fixed principal axes. In that case $(x\to 1,\ y\to 2,\ z\to 3)$

$$D_1 = \varepsilon_1 E_1\ ,\ D_2 = \varepsilon_2 E_2\ ,\ D_3 = \varepsilon_3 E_3\ ; \tag{6.32}$$

we shall also use the notation $\mathbf{D} = \hat{\varepsilon}\mathbf{E}$, where $\hat{\varepsilon}$ is an operator, the meaning of which is clear from (6.32) or from the more general relation (6.7).

After introducing the potentials (6.8) we get the field equations for an anisotropic medium in the form

$$\left.\begin{aligned} \nabla^2\mathbf{A} - \frac{1}{c^2}\frac{\partial}{\partial t}\hat{\varepsilon}\frac{\partial \mathbf{A}}{\partial t} - \text{grad div}\,\mathbf{A} - \frac{1}{c}\frac{\partial}{\partial t}\hat{\varepsilon}\,\text{grad}\,\phi &= -\frac{4\pi}{c}\mathbf{j}\ , \\ \text{div}\left(\hat{\varepsilon}\,\text{grad}\,\phi + \frac{\hat{\varepsilon}\partial\mathbf{A}}{c\partial t}\right) &= -4\pi\rho\ . \end{aligned}\right\} \tag{6.33}$$

It is clear from (6.33) that neither a gauge such as (6.10), that is, the gauge $\text{div}\,\mathbf{A} + (1/c)(\partial\hat{\varepsilon}\phi/\partial t) = 0$, nor the gauge $\text{div}\,\mathbf{A} = 0$ leads to any simplifications and the well developed methods for integrating the d'Alembert equation (wave equation) are thus inapplicable. One should, however, add that a relativistically covariant form of writing the formulae (the tendency to retain this leads in the vacuum case to the very wide-spread use of the Lorentz gauge $\partial A^i/\partial x^i = 0$; see (1.7a)) is appreciably less important when a medium is present. The reason is that in the overwhelming majority of cases the medium is assumed to be (and, in fact, is) at rest in the laboratory frame of reference. The frame of reference fixed in the medium is then physically clearly distinguished and it is obvious to work in that frame.

We shall apply the gauge $\hat{\varepsilon}(\partial/\partial t)\,\text{div}\,\mathbf{A} = 0$ or, for the case (6.32) and neglecting dispersion, we put

$$\text{div}\,\mathbf{C} = 0\ ,\ \mathbf{C} = \hat{\varepsilon}\mathbf{A}\ ,\ C_1 = \varepsilon_1 A_1\ ,\ C_2 = \varepsilon_2 A_2\ ,\ C_3 = \varepsilon_3 A_3\ . \tag{6.34}$$

Equations (6.33) then become

$$\nabla^2\mathbf{A}-\frac{1}{c^2}\frac{\partial^2\mathbf{C}}{\partial t^2}-\operatorname{grad}\operatorname{div}\mathbf{A}-\frac{1}{c}\frac{\partial}{\partial t}\hat{\boldsymbol{\varepsilon}}\operatorname{grad}\phi=-\frac{4\pi}{c}\sum_i e_i\mathbf{v}_i\delta(\mathbf{r}-\mathbf{r}_i(t)), \tag{6.35}$$

$$\operatorname{div}(\hat{\boldsymbol{\varepsilon}}\operatorname{grad}\phi)=-4\pi\sum_i e_i\,\delta(\mathbf{r}-\mathbf{r}_i(t)), \tag{6.36}$$

where we are considering point charges e_i to fix the ideas (the radius vector of the charge is $\mathbf{r}_i(t)=\{x_i, y_i, z_i\}$, and $\dot{\mathbf{r}}_i=\mathbf{v}_i$).

Equation (6.36) has the same form as in electrostatics and we can write down its solution as follows:

$$\phi=\sum_i\frac{e_i}{\sqrt{\varepsilon_1\varepsilon_2\varepsilon_3}\,\sqrt{\{(x-x_i)^2/\varepsilon_1+(y-y_i)^2/\varepsilon_2+(z-z_i)^2/\varepsilon_3\}}}. \tag{6.37}$$

One can also show easily that the energy of the field in the whole of space equals (provided the field itself vanishes at infinity in the required manner)

$$\left.\begin{aligned}\mathcal{H}&=\frac{1}{8\pi}\int\{(\mathbf{E}\cdot\mathbf{D}+H^2)\}\,d^3\mathbf{r}=\mathcal{H}_{tr}+\mathcal{H}_\ell,\\ \mathcal{H}_{tr}&=\frac{1}{8\pi}\int\left\{\frac{\hat{\boldsymbol{\varepsilon}}}{c^2}\left(\frac{\partial\mathbf{A}}{\partial t}\cdot\frac{\partial\mathbf{A}}{\partial t}\right)+(\operatorname{curl}\mathbf{A})^2\right\}d^3\mathbf{r},\\ \mathcal{H}_\ell&=\frac{1}{8\pi}\int(\hat{\boldsymbol{\varepsilon}}\nabla\phi\cdot\nabla\phi)\,d^3\mathbf{r}.\end{aligned}\right\} \tag{6.38}$$

Here

$$\begin{aligned}\mathcal{H}_\ell&=\frac{1}{8\pi}\int\left(\varepsilon_1\frac{\partial^2\phi}{\partial x^2}+\varepsilon_2\frac{\partial^2\phi}{\partial y^2}+\varepsilon_3\frac{\partial^2\phi}{\partial z^2}\right)d^3\mathbf{r}\\ &=\frac{1}{2}\sum_{ij}\frac{e_ie_j}{\sqrt{\varepsilon_1\varepsilon_2\varepsilon_3}\,\sqrt{\{(x_i-x_j)^2/\varepsilon_1+(y_i-y_j)^2/\varepsilon_2+(z_i-z_j)^2/\varepsilon_3\}}}\end{aligned} \tag{6.39}$$

is the sum of the instantaneous Coulomb interactions energies of the charges (we neglect the self-energy of the charges). The energy $\mathcal{H}_{tr}$ is the analogue of the transverse field energy and can also be written in the form

$$\left.\begin{aligned}\mathcal{H}_{tr}&=-\frac{1}{8\pi c}\int\left(\mathbf{D}_{tr}\cdot\frac{\partial\mathbf{A}}{\partial t}\right)d^3\mathbf{r}+\frac{1}{8\pi}\int(\operatorname{curl}\mathbf{A})^2\,d^3\mathbf{r},\\ \mathbf{D}_{tr}&=-\frac{\hat{\boldsymbol{\varepsilon}}}{c}\frac{\partial\mathbf{A}}{\partial t}\equiv-\frac{\partial\mathbf{C}}{\partial t},\ \operatorname{div}\mathbf{D}_{tr}=0.\end{aligned}\right\} \tag{6.40}$$

We now expand in series:

$$\left.\begin{aligned}\mathbf{A}&=\sum_{\lambda,\ell}\{q_{\lambda\ell}(t)\,\mathbf{A}_{\lambda\ell}(\mathbf{r})+q^*_{\lambda\ell}(t)\,\mathbf{A}^*_{\lambda\ell}(\mathbf{r})\},\\ \mathbf{C}&=\hat{\boldsymbol{\varepsilon}}\mathbf{A}=\sum_{\lambda,\ell}\{q_{\lambda\ell}(t)\,\mathbf{C}_{\lambda\ell}(\mathbf{r})+q^*_{\lambda\ell}(t)\,\mathbf{C}^*_{\lambda\ell}(\mathbf{r})\},\end{aligned}\right\} \tag{6.41}$$

$$\mathbf{A}_{\lambda\ell} = \sqrt{4\pi}\, c\, \mathbf{a}_{\lambda\ell} \exp\{i(\mathbf{k}_\lambda \cdot \mathbf{r})\} ,$$
$$\mathbf{C}_{\lambda\ell} = \sqrt{4\pi}\, c\, \mathbf{b}_{\lambda\ell} \exp\{i(\mathbf{k}_\lambda \cdot \mathbf{r})\} ,$$

where the index ℓ (= 1,2) corresponds to the two possible polarizations of the normal waves in an anisotropic medium without spatial dispersion (if (6.32) holds, this polarization is linear). By virtue of the first of Eqs.(6.34) we have $(\mathbf{b}_{\lambda\ell} \cdot \mathbf{k}_\lambda) = 0$, and we may put $(\mathbf{b}_{\lambda 1} \cdot \mathbf{b}_{\lambda 2}) = 0$; we then get from (6.34) and (6.41)

$$(\mathbf{k}_\lambda \cdot \hat{\boldsymbol{\varepsilon}} \mathbf{a}_{\lambda\ell}) \equiv k_{\lambda 1}\, \varepsilon_1\, a_{\lambda\ell,1} + k_{\lambda 2}\, \varepsilon_2\, a_{\lambda\ell,2} + k_{\lambda 3}\, \varepsilon_3\, a_{\lambda\ell,3} = 0 ,$$
$$(\hat{\boldsymbol{\varepsilon}} \mathbf{a}_{\lambda 1} \cdot \hat{\boldsymbol{\varepsilon}} \mathbf{a}_{\lambda 2}) = 0 . \tag{6.42}$$

We impose two more conditions upon the $\mathbf{a}_{\lambda\ell}$ (note that there is no summation in the first of these conditions over the repeated index ℓ):

$$(\hat{\boldsymbol{\varepsilon}} \mathbf{a}_{\lambda\ell} \cdot \mathbf{a}_{\lambda\ell}) = 1 \quad , \quad (\hat{\boldsymbol{\varepsilon}} \mathbf{a}_{\lambda 1} \cdot \mathbf{a}_{\lambda 2}) = 0 . \tag{6.43}$$

The first of these is a normalization condition, and the second one corresponds to a choice of polarizations which corresponds to the normal waves; it is clear from (6.42) and (6.43) that the vectors $\mathbf{k}_\lambda$, $\mathbf{a}_{\lambda\ell}$, and $\mathbf{b}_{\lambda\ell}$ are coplanar (both for $\ell = 1$ and for $\ell = 2$).

We now substitute (6.41) into (6.35); after multiplying by $\mathbf{A}^*_{\mu\ell}$ and using Eqs.(6.42), (6.43), and the normalization condition

$$\int (\mathbf{A}_{\lambda\ell} \cdot \mathbf{A}^*_{\mu m})\, d^3\mathbf{r} = 4\pi c^2 (\mathbf{a}_{\lambda\ell} \cdot \mathbf{a}_{\mu m})\, \delta_{\lambda\mu} , \tag{6.44}$$

we get (for details see Ginzburg, 1940b)

$$\ddot{q}_{\lambda\ell} + \omega^2_{\lambda\ell}\, q_{\lambda\ell} = \sqrt{4\pi} \sum_i e_i (\mathbf{v}_i \cdot \mathbf{a}_{\lambda\ell}) \exp\{-i(\mathbf{k}_\lambda \cdot \mathbf{r}_i)\} , \tag{6.45}$$

$$\omega^2_{\lambda\ell} = \frac{k^2_\lambda c^2}{n^2_{\lambda\ell}} = [\mathbf{k}_\lambda \wedge \mathbf{a}_{\lambda\ell}]^2 c^2 = \{k^2_\lambda\, a^2_{\lambda\ell} - (\mathbf{k}_\lambda \cdot \mathbf{a}_{\lambda\ell})^2\}\, c^2 . \tag{6.46}$$

Equation (6.45) for the field oscillators is thus in a form which is the same as for the vacuum case and the whole difference is included in Eq.(6.46) for the frequencies $\omega_{\lambda\ell}$ and Eqs.(6.42) and (6.43) for the polarization vectors $\mathbf{a}_{\lambda\ell}$.

We have already mentioned that we are dealing with expansions in normal waves and the relations given here for $\omega_{\lambda\ell}$ and $\mathbf{a}_{\lambda\ell}$ are just the equations which determine the relations between ω and $\mathbf{k}$ and the polarization for normal waves. To be more precise, when we are considering the homogeneous field equations, it is clear from (6.45) that, if there are no charges, $\omega^2 = \omega^2_{\lambda\ell}$ and (6.46) is then the dispersion (Fresnel) equation which connects $\omega_\ell = kc/n_\ell$ with $\mathbf{k}$ in the normal waves (clearly, $n_\ell = ck/\omega_\ell$ is the refractive index; for details see Agranovich and Ginzburg, 1966 and Chapter 10.

One can easily show (see (6.40), (6.41), (6.44), and so on) that in terms of the variables $q_{\lambda\ell}$ we have

$$\mathcal{H}_{tr} = \sum_{\lambda,\ell} (p_{\lambda\ell} p^*_{\lambda\ell} + \omega^2_{\lambda\ell} q_{\lambda\ell} q^*_{\lambda\ell}) \ , \ p_{\lambda\ell} = \dot{q}_{\lambda\ell} \ . \tag{6.47}$$

The rest of the calculations in the solution of the various problems about the emission by charges proceeds in the same way as in the vacuum case or the case of an isotropic medium. For instance, we get for an oscillator in the dipole approximation instead of (6.26)

$$\ddot{q}_{\lambda\ell} + \omega^2_{\lambda\ell} q_{\lambda\ell} = \sqrt{4\pi}\, e\, (\mathbf{a}_{\lambda\ell} \cdot \mathbf{v}_0)\cos\omega_0 t \ . \tag{6.48}$$

The remaining calculations proceed as for an isotropic medium, but

$$\frac{k^2 dk\, d^2\Omega}{(2\pi)^3} = \frac{n^3_\ell \omega^2 d\omega\, d^2\Omega}{(2\pi c)^3} \ . \tag{6.49}$$

As a result we get

$$\frac{d\mathcal{H}_{tr,\ell}}{dt} \equiv \frac{dW_{s,\ell}}{dt} = \frac{e^2\omega_0^2(\mathbf{a}_\ell \cdot \mathbf{a}_0)^2 n^3_\ell}{8\pi c^3} d^2\Omega \ , \ \ell = 1,2, \tag{6.50}$$

where, as everywhere above, $n_\ell = n_\ell(\omega, \theta, \phi)$, that is, n_ℓ depends not only on the frequency, but also on the direction — in fact, on the orientation of the wavevector $\mathbf{k}$ relative to the symmetry axes of the medium (in the case (6.32) these axes are the x-, y-, and z-axes if $\boldsymbol{\varepsilon}_1$, $\boldsymbol{\varepsilon}_2$, and $\boldsymbol{\varepsilon}_3$ are different). The polarization vectors $\mathbf{a}_\ell = \mathbf{a}_\ell(\omega, \theta, \phi)$ depends also on the direction and the frequency, and, of course, on the wave type (index $\ell = 1,2$). In an isotropic medium $\mathbf{a}_{\lambda\ell} = \mathbf{e}_{\lambda\ell}/\sqrt{\boldsymbol{\varepsilon}} = \mathbf{e}_{\lambda\ell}/n$ and we take one of the $\mathbf{e}_{\lambda\ell}$ vectors to be at right angles to $\mathbf{a}_0$; in that case (6.50) goes over into (6.28). The result (6.50) is very obvious — the oscillator 'pumps' the emission into each of the normal waves and the pumping intensity is proportional to $(\mathbf{a}_{\lambda\ell} \cdot \mathbf{a}_0)^2$, that is, the square of the component of the amplitude $\mathbf{a}_0$ along the polarization vector (electric vector) of the corresponding wave.

We note that for the free field of radiation in an anisotropic medium, one can easily bring the energy of the transverse field (which in this case is the same as the total energy of the system) into canonical form (see the transition from (1.54) to (1.58)) and therefore one can quantize it by standard methods. As a result one can introduce the concept of 'photons in an anisotropic medium', for which

$$E = \hbar\omega \ , \ \mathbf{p}_\ell = \hbar\mathbf{k}_\ell \ , \ p_\ell = \frac{\hbar\omega n_\ell(\omega,\theta,\phi)}{c} \ , \tag{6.51}$$

where θ and ϕ are angles characterizing the direction of the vector $\mathbf{k}_\ell$ with respect to the symmetry axes of the medium (for a given frequency ω the dispersion equation determines the values of $k_\ell = (\omega/c)\, n_\ell\,(\omega,\theta,\phi)$ for the normal waves which can propagate in the medium in a given direction; when we neglect the spatial dispersion and strictly longitudinal waves, there are two such waves, that is, $\ell = 1,2$). This result generalizes (6.25) and was mentioned earlier.

A very important and characteristic feature of the electrodynamics in a medium is the possibility that emission may occur even for a uniformly moving charge. We must mention here two effects: the Vavilov-Cherenkov effect (Cherenkov radiation; see Tamm and Frank, 1937; Ginzburg 1940a, 1960; Jelley, 1958; Bolotovskii, 1960, 1962; Landau and Lifshitz, 1960, Zrelov, 1968 and Ter-Mikaelyan, 1972; and transition radiation (Frank and Ginzburg, 1945; Ginzburg and Frank, 1946; Garibyan, 1960, 1970; Bass and Yakovenko, 1965; Tamoykin, 1972; Ginzburg and Tsytovich, 1974a, 1978; Ginzburg, 1975b).† Transition radiation is discussed in Chapter 7. Now, however, we shall discuss Cherenkov radiation which occurs in a uniform medium when the condition

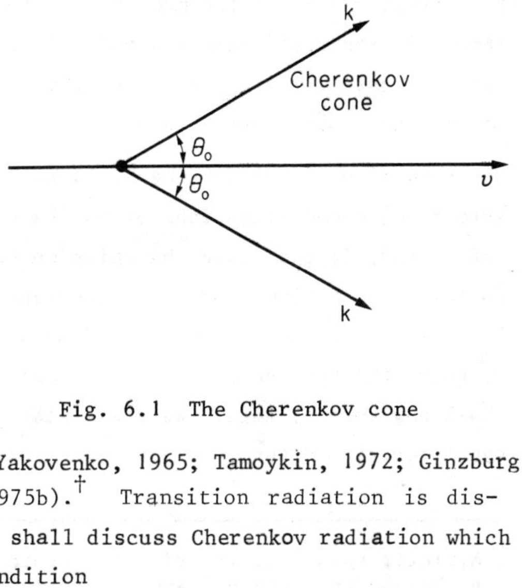

Fig. 6.1 The Cherenkov cone

$$v > v_{ph} = \frac{c}{n(\omega)} \tag{6.52}$$

is satisfied, that is, provided the particle velocity v = constant is larger than the phase velocity of waves in the medium, $v_{ph} = c/n(\omega)$. In that case

† The literature dealing with Cherenkov and transition radiation is huge (see the reviews by Jelley (1958); Ginzburg (1960); Bolotovskii (1960, 1962); Bass and Yakovenko (1965); Zrelov (1968); Tamoykin (1972); and Ter-Mikaelyan (1972); where one can find both original results and bibliographies.

a wave of frequency ω is emitted at an angle θ_0 to the velocity **v**, where (Fig.6.1)

$$\cos\theta_0 = \frac{v_{ph}}{v} = \frac{c}{v\,n(\omega)}\,. \tag{6.53}$$

Of course, condition (6.52) is a consequence of (6.53) since $\cos\theta_0$ must be ≤ 1 (the value $\cos\theta_0 = 1$ corresponds to the threshold for emission when under realistic conditions the intensity of the radiation vanishes; Ginzburg, 1960). The result (6.53) has, so to speak, a kinematic nature (condition for interference; for details see, for instance, Bolotovskii, 1960, 1962; Frank, 1972) and refers to any kind of wave. In that sense condition (6.53) had been known for a long time in applications to sound waves (Mach cone). Clearly we must in the case of electromagnetic waves in an anisotropic medium understand by $v_{ph} = c/n(\omega)$ the phase velocity corresponding to a normal wave, and in that case $n = n_\ell(\omega,\theta_0,\theta,\phi)$, where θ_0 is the angle between **k** and **v** while the angles θ and ϕ determine the orientation of the vector **k** with respect to the symmetry axes of the medium (in the case of a moving medium which is isotropic in its rest frame the velocity of the medium **u** plays the role of the symmetry axis). We restrict ourselves for the sake of simplicity in what follows to an isotropic medium.

The condition (6.53) for Cherenkov emission can be obtained not only from interference considerations or as the result of evaluating the radiation field (of course, in that case the emission condition follows automatically from the formulae)† but also from the energy and momentum conservation laws (see Chapter 7) or as a resonance condition (Ginzburg, 1939a,b). This last one occurs when we solve the Cherenkov radiation problem by the Hamiltonian method, and we shall now turn to this. We shall start from the equations in the form (6.21), substituting in them for the case of a uniformly moving charge the radius

---

† Strictly speaking, in order to obtain the condition (6.53) one need not go beyond the original expression (4.13) for the Liénard-Wiechert potentials, generalized to the case where there is a medium by replacing c by $v_{ph} = c/n$ (of course, this should be done only where c plays the role of the wave velocity and not where it is a coefficient of the current density, for instance, on the right-hand side of Eqs. (1.8) or (6.11)). As the result from such a substitution we get, for instance, $\mathbf{A} = e\mathbf{v}/\{cr[1-(vn/c)\cos\theta']\}$, and it is clear that in the wave zone, where $\theta' = \theta$, the potential has a singularity just at the angle $\theta = \theta_0$, where $\cos\theta_0 = c/nv$. (One should not confuse the angle θ between **k** and **v** with the angle θ which together with the angle ϕ determines the orientation of the vector **k** relative to the symmetry axes of the medium; this latter angle is hardly used in the calculations.)

vector $\mathbf{r}_i = \mathbf{v}t$, $\mathbf{v} = \text{const}$. We then get

$$\begin{aligned} \ddot{q}_{\lambda 1} + \omega_\lambda^2 q_{\lambda 1} &= \sqrt{8\pi}\,\frac{e}{n}\,(\mathbf{v}\cdot\mathbf{e}_\lambda)\cos(\mathbf{k}_\lambda\cdot\mathbf{v})t\,, \\ \ddot{q}_{\lambda 2} + \omega_\lambda^2 q_{\lambda 2} &= \sqrt{8\pi}\,\frac{e}{n}\,(\mathbf{v}\cdot\mathbf{e}_\lambda)\sin(\mathbf{k}_\lambda\cdot\mathbf{v})\,t\,. \end{aligned} \tag{6.54}$$

The resonance condition and, hence, the existence of a solution for the energy $\mathcal{H}_{tr}$ which increases with time, has the form

$$\omega_\lambda \equiv \frac{c}{n}\,k_\lambda = (\mathbf{k}_\lambda\cdot\mathbf{v}) = k_\lambda v\cos\theta_0\,, \tag{6.55}$$

which is the same as (6.53).

The solution of Eqs.(6.54) with initial conditions $q_{\lambda 1,2} = \dot{q}_{\lambda 1,2} = 0$ is the following:†

$$\left.\begin{aligned} q_{\lambda 1} &= \frac{\sqrt{8\pi}\,e\,(\mathbf{e}_\lambda\cdot\mathbf{v})}{n\,\omega_\lambda^2\,(1-(v^2/c^2)\,n^2\cos^2\theta_\lambda)}\,\times \\ &\quad\times\left\{\cos\left(\frac{nv\omega_\lambda t}{c}\cos\theta_\lambda\right) - \cos\omega_\lambda t\right\}, \\ q_{\lambda 2} &= \frac{\sqrt{8\pi}\,e\,(\mathbf{e}_\lambda\cdot\mathbf{v})}{n\,\omega_\lambda^2\,(1-(v^2/c^2)\,n^2\cos^2\theta_\lambda)}\,\times \\ &\quad\times\left\{\cos\left(\frac{nv\omega_\lambda t}{c}\cos\theta_\lambda\right) - \frac{nv}{c}\cos\theta_\lambda\sin\omega_\lambda t\right\}, \end{aligned}\right\} \tag{6.56}$$

where θ_λ is the angle between $\mathbf{k}_\lambda$ and $\mathbf{v}$.

Substituting (6.56) into (6.16) and changing from a summation to an integration (see (6.27)) we get

$$\begin{aligned} \mathcal{H}_{tr} &= \frac{8\pi e^2 v^2}{(2^2 c)^3} \\ &\times\int_0^{2\pi} d\phi \int_0^{\pi/2} d\theta \int_0^{\omega_{max}} \left\{\frac{n\sin^2\theta(1+(v^2/c^2)\,n^2\cos^2\theta)\left[1-\cos\{1-(v/c)\,n\cos\theta\}\,\omega t\right]}{(1+(v/c)\,n\cos\theta)^2\,(1-(v/c)\,n\cos\theta)^2}\right. \\ &\qquad\left. + \frac{n\sin^2\theta\sin\omega t\sin\{[(v/c)\,n\cos\theta]\,\omega t\}}{(1+(v/c)\,n\cos\theta)^2}\right\}\sin\theta\,d\omega\,. \end{aligned} \tag{6.57}$$

We have used here the fact that $(\mathbf{v}\cdot\mathbf{e}_{\lambda 1})^2 + (\mathbf{v}\cdot\mathbf{e}_{\lambda 2})^2 = v^2\sin^2\theta_\lambda \equiv v^2\sin^2\theta$ or, equivalently, that we can choose the polarization vector $\mathbf{e}_{\lambda 1}$ to lie in the plane determined by the vectors $\mathbf{v}$ and $\mathbf{k}_\lambda$. It is clear that the radiation

---

† Of course, when we want to obtain the Cherenkov emission for large t, the exact initial conditions are immaterial, and the ones chosen are only simpler than others.

is polarized in that plane (that is, that the electric vector of the wave lies in that plane). The second term in (6.57) does not lead to solutions which increase with time, but the first term shows δ-function behaviour for large t (see (1.84)). Integrating over θ and using Eq.(1.84) we can thus easily obtain the Tamm-Frank (1937) formula for the Cherenkov radiation power (we have also performed the trivial integration over ϕ):

$$\frac{d\mathcal{H}_{tr}}{dt} \equiv \frac{dW}{dt} = \frac{e^2 v}{c^2}\int\left(1 - \frac{c^2}{v^2 n^2(\omega)}\right)\omega\, d\omega = \frac{e^2 v}{c^2}\int \sin^2\theta_0\, \omega\, d\omega \tag{6.58}$$

(the integration is here over the region $vn(\omega)/c \geq 1$).

If we are interested in the emission over a path length L, we must, clearly, in (6.58) replace v (the path length per unit time) by L.

Dispersion is taken into account in (6.58) due to the fact that we put in (6.57) $n = n(\omega)$; we discussed the basis for this procedure earlier. Even in the case of an isotropic medium the calculation given here is very simple and it can in that sense compete with other methods (Tamm and Frank, 1937; Landau and Lifshitz, 1960). In an anisotropic medium, on the other hand, the Hamiltonian method (expansion in normal waves) is for us the simplest. We must bear in mind that the indexes $n_\ell(\omega,\theta,\phi)$ in that case play the role of the refractive index $n(\omega)$ and that this must be taken into account, of course, when we integrate over angles in expressions such as (6.57) (this was forgotten in an earlier paper (Ginzburg, 1940c) and this led to errors; Bolotovskii (1960, 1962) gave the correct expressions for the Cherenkov radiation power in crystals).

As the next example of a very important illustration of the role played by the medium we discuss the Doppler effect when a source moves with velocity $\mathbf{v}$ in a medium with refractive index $n(\omega)$. We give at once the result — a generalization of Eqs.(4.5) and (4.11):

$$\omega(\theta) = \frac{\omega_{00}\sqrt{(1 - v^2/c^2)}}{|1 - (v/c)\, n(\omega)\cos\theta|} = \frac{\omega_0}{|1 - (v/c)\, n(\omega)\cos\theta|}, \tag{6.59}$$

where ω_{00} is the frequency in the 'eigen' frame of reference of the emitter (in that frame the velocity of the centre of mass of the emitter $v = 0$), ω_0 the frequency of the emitter in the laboratory frame (say, the frequency of the external electric field in the ondulator) and, as usual, θ the angle between $\mathbf{k}$ and $\mathbf{v}$.

We obtain (6.59), first of all, by means of the general rule for changing from the theory of radiation in vacuo to the theory of radiation in a medium — by making in (4.5) and (4.11) the substitution

$$c \to v_{ph} = \frac{c}{n(\omega)} . \tag{6.60}$$

Of course, in an anisotropic medium $n_\ell(\omega,\theta,\phi)$ plays the role of $n(\omega)$. Nevertheless, one cannot perform the substitution (6.60) completely automatically — it is sufficient to note that in the case of (4.5) it must be performed in the denominator, but not in the numerator; in the numerator of (6.5) there appears, as before, $\sqrt{(1-v^2/c^2)}$. This is, of course, completely natural as the change from ω_{00} to ω_0 in (6.59) is connected with the relativistic time contraction, while the value of the denominator is determined by the motion of the emitter and we can get it completely consistently from the expression for the radiation field. Whatever method we use (see, for instance, (6.22)) we find in the expressions for the potentials (and thus also for the fields) phase factors of the form $\exp\{i[(\mathbf{k}\cdot\mathbf{r}) - (\mathbf{k}\cdot\mathbf{r}_i) - \omega t]\}$; in the dipole approximation we have for the emitter $\mathbf{r}_i = \mathbf{v}t$ and hence there occurs the frequency $\omega = \omega_0 + (\mathbf{k}\cdot\mathbf{v}) = \omega_0 + (\omega/c)\, vn(\omega) \cos\theta$, whence follows (6.59). We give in Chapter 7 another derivation of this formula which is based upon the conservation laws.

We note an important feature of Eq.(6.59) — the occurrence of the absolute signs in the denominator (Frank, 1942, 1959) which is important when the particle moves with a 'super-light' velocity $v > v_{ph} = c/n$ (see condition (6.52)). From the necessity to find a positive value for the frequency ω it is immediately clear that we must take the absolute value of the denominator in (6.59) when

$$\frac{v}{c} n(\omega) \cos\theta > 1 . \tag{6.61}$$

The range of angles satisfying condition (6.61) is called the region of the anomalous or 'super-light' Doppler effect. Of course, that region exists only for the super-light velocities (6.52). If, on the other hand

$$\frac{v}{c} n(\omega) \cos\theta < 1 , \tag{6.62}$$

we have the normal Doppler effect. The regions of the normal and the anomalous Doppler effect are separated by the Cherenkov cone (Fig.6.2). Under real conditions, the Doppler effect in a medium is rather complicated (Frank, 1942, 1959) as dispersion must be taken into account (the ω-dependence of n). We restrict ourselves here to the remark that if one neglects dispersion, the frequency $\omega(\theta = \theta_0)$ would become infinite on the Cherenkov cone itself where $(v/c)\, n \cos\theta = 1$ (see (6.53)). In fact, as $\omega \to \infty$ the refractive index $n(\omega) \to 1$

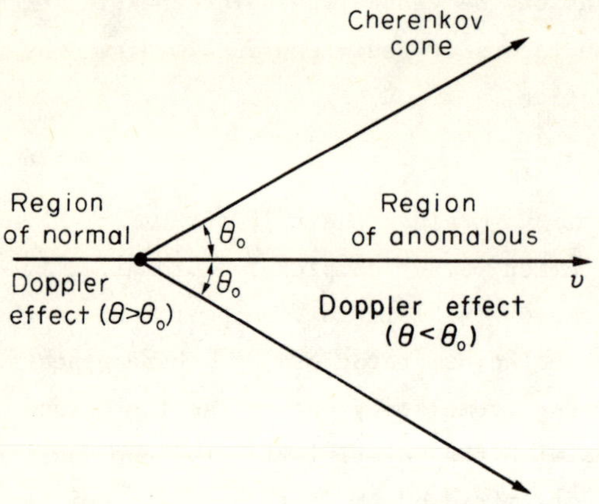

Fig.6.2 Regions of the normal and the anomalous Doppler effect

and strictly speaking there is no emission whatever on the Cherenkov cone (we assume that $v < c$ and we do not just now consider transition radiation). However, the tendency of the frequency to increase as the angle θ approaches

$$\theta_0 = \arccos(c/n(\omega)v)$$

is, of course, retained also when dispersion is taken into account.

The role played by the Cherenkov cone in radiation in a medium becomes even clearer when we consider not only the change in frequency, but also the intensity distribution. We shall do this by considering the example of an ondulator in a medium, that is, we generalize what we said about an ondulator in Chapter 4 to the case when the charge moves in a transparent medium with refractive index $n(\omega)$. We shall assume the motion of the charge to be the same as in vacuo, as we may assume the energy losses (which are in first instance ionization losses) to be small for a particle with a sufficiently large energy (we do not yet consider the possibility to have a gap or channel in the medium; see Ginzburg 1947, 1972b,c; Alferov, Bashmakov and Bessonov, 1975 and Chapter 7). The introduction of the medium is thus simply realized by replacing c by c/n (see (6.60)) in Eq.(4.14). As a result we get instead of (4.15)

$$\left.\begin{aligned} \frac{dU_s}{d\Omega} &= \frac{e^2\omega_0^4 a_0^2 nL\{(1-(v/c)n\cos\theta)^2 - (1-(v^2/c^2)n^2)\sin^2\theta\cos^2\phi\}}{8\pi c^3 v|1-(v/c)n\cos\theta|^5}, \\ a_0^2 &= \frac{e^2E_0^2}{m^2\omega_0^4}\left(\frac{mc^2}{\mathcal{E}}\right)^2, \quad \omega(\theta) = \frac{\omega_0}{|1-(v/c)n(\omega)\cos\theta|}. \end{aligned}\right\} \tag{6.63}$$

Hence it is clear that the power of the radiation is greatest near the Cherenkov cone (we assume here that $n > 1$ and that the Cherenkov condition (6.53) can be satisfied for a rather wide range of frequencies). We assume, for instance, that

$$\left.\begin{aligned} n(\omega) &= n = \text{const}, \quad \omega \le \omega_m, \\ n(\omega) &= 1, \quad \omega > \omega_m. \end{aligned}\right\} \tag{6.64}$$

Instead of requiring that $n(\omega > \omega_m) = 1$ we can in what follows just as well assume that the radiation is completely absent when $\omega > \omega_m$, due to strong absorption. In the case (6.64) and as $v/c \to 1$ we may take the second term which is proportional to $1 - (v^2/c^2)n^2 \approx 1 - n^2$ to be the main one. If we now integrate over the angles we get (bearing in mind that $\sin^2\theta_0 = 1 - c^2/n^2(\omega)v^2$)

$$U(n>1) = \int \frac{dU_s}{d\Omega} d^2\Omega = \frac{e^2 \omega_0^4 a_0^2 L(n^2-1)\sin^2\theta}{16 c^4} \left(\frac{\omega_m}{\omega_0}\right)^2$$

$$= \frac{(n^2-1)^2}{16 n^2} \left(\frac{e^2}{mc^2}\right)^2 \left(\frac{mc^2}{\mathcal{E}}\right)^2 \left(\frac{\omega_m}{\omega_0}\right)^4 E_0^2 L . \quad (6.65)$$

Comparing this expression with the vacuum expression (4.16) we see how large the difference between the two expressions is under the given conditions

$$\frac{U(n>1)}{U(n=1)} = \frac{3(n^2-1)^2}{16 n^2} \left(\frac{\omega_m}{\omega_0}\right)^4 \left(\frac{mc^2}{\mathcal{E}}\right)^4 . \quad (6.66)$$

Incidentally, the energy of the Cherenkov radiation over a path length L is in the case (6.64) equal to (see (6.58))†

$$U = \frac{dW}{dt}\frac{L}{v} = \frac{e^2(1 - c^2/v^2n^2)\,\omega_m^2 L}{2c^2} . \quad (6.67)$$

Expression (6.63), or the more general one obtained from (4.14) through the substitution (6.60), rather completely reflects the characteristic features of the radiation by a particle moving in a medium. If $n > 1$, the Cherenkov angle $\theta_0 = \arccos(c/vn)$ plays the role of the angle $\theta = 0$ around which (in a range of angles $\theta \sim mc^2/\mathcal{E}$) the emission from an ultra-relativistic particle in vacuo is concentrated (Fig. 6.3). Near the Cherenkov cone both the frequency of the emitted waves, and their intensity are largest. We need only remember the role played by dispersion, due to which there are many Cherenkov cones (indeed, $\theta_0(\omega) = \arccos[c/vn(\omega)]$). For a

---

† We must bear in mind that for a charged particle (in which case both Cherenkov radiation and the radiation described by equations such as (6.63) and (6.65) occurs during motion in an ondulator) the 'ondulator' radiation appreciably 'draws' from the Cherenkov radiation in the sense that the amplification of the radiation (6.63),(6.65) is accompanied by a diminishing of the power of the Cherenkov radiation (see Gailitis, 1964; Musakhanyan and Nikishov, 1974). This fact makes it difficult to separate the ondulator emission from the Cherenkov background (we did not take this into account in an earlier paper (Ginzburg 1972c) and this led to an incorrect estimate for the possibility of the production of an 'ondulator counter in a medium' which would be suitable to measure the energy of relativistic particles).

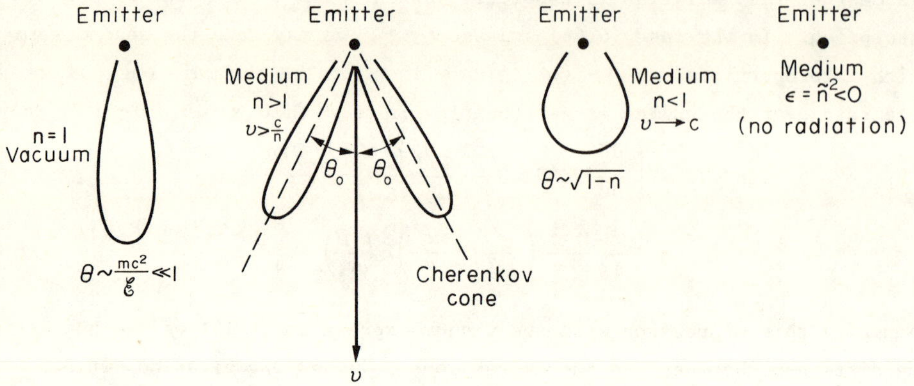

Fig.6.3 Sketch of the polar diagrams for an emitter in vacuo and in a medium

non-transparent medium† there is no radiation independently of the nature of the emitter and its motion (of course, we are talking about frequencies for which the medium is non-transparent).

Finally, we must especially look at the case

$$0 < n(\omega) < 1 , \tag{6.68}$$

which is very important, as in an isotropic non-relativistic plasma

$$\varepsilon(\omega) = n^2(\omega) = 1 - \frac{\omega_p^2}{\omega^2} , \quad \omega_p = \left(\frac{4\pi e^2 N}{m}\right)^{\frac{1}{2}} = 5.64 \times 10^4 \sqrt{N} , \tag{6.69}$$

where we have neglected absorption (collisions) while N is the electron density.

Under cosmic conditions the region where Eq.(6.69) is applicable is very wide (see Ginzburg, 1970b; Ginzburg and Syrovatskii, 1966a, 1969; Ginzburg, Sazonov and Syrovatskii, 1968, and Chapter 11) and one must thus very often take the effect of the medium into account, using just Eq.(6.69) according to which condition (6.68) is just satisfied in the transparency region.

In the region (6.68) even an ultra-relativistic particle can radiate, like a non-relativistic one. This is formally clear from Eqs.(6.59) and (6.63) for the frequency and the intensity and it is essentially a direct consequence of the fact that the ratio of the velocity v to the phase velocity of light

† A non-transparent medium can be non-absorbing, for instance, in the case of a real dielectric permittivity $\varepsilon(\omega) = \tilde{n}^2(\omega) < 0$ (we denote by $\tilde{n}$ the complex refractive index; see Chapters 10 and 11).

$v_{ph} = c/n$ plays a deciding role for emission in a medium. Therefore, unless we talk about somewhat exotic radiation sources which move with a velocity $v > c$ (see below, Chapter 8), for $n < 1$ we have always $v < v_{ph}$ and, if the difference $v_{ph} - v$ is sufficiently large, all specific properties of emitters moving in vacuo with relativistic velocities disappear. To be precise, the maximum frequency in a medium with $n < 1$ (see (6.59); we assume for the sake of simplicity that $n(\omega) =$ constant or, at any rate, that the dispersion is unimportant) is equal to

$$\omega(0) = \frac{\omega_0}{1 - (v/c)n} \approx \frac{\omega_0}{1 - n}, \tag{6.70}$$

where, clearly, the last expression refers to the case $v \to c$; in that case, if $1 - n \ll 1$, the radiation is concentrated in the range of angles $\theta \sim \surd(1 - n)$ (see Fig.6.3).

We shall find the condition that a medium with $n < 1$ will not affect the emission by relativistic particles. To do that it is sufficient to write out in detail the denominator in Eq.(6.59) for the Doppler effect as $v/c \to 1, \theta \to 0$:

$$1 - \frac{v}{c} n \cos\theta = 1 - \frac{v}{c}\cos\theta + (1 - n)\frac{v}{c}\cos\theta \approx$$

$$\approx 1 - \frac{v}{c} + (1 - n) \approx \frac{1}{2}\left(\frac{mc^2}{\mathcal{E}}\right)^2 + (1 - n) .$$

Hence it is clear that the role of the medium is unimportant, if

$$2\{1 - n(\omega)\}\left(\frac{\mathcal{E}}{mc^2}\right)^2 \ll 1 . \tag{6.71}$$

In the case of synchrotron radiation in a medium we must replace Eq.(5.39), for instance, by the following one, as can easily be checked (Ginzburg and Syrovatskii, 1966a, 1969; Ginzburg, Sazonov and Syrovatskii, 1968):

$$\left.\begin{aligned} p(\nu) &= \surd 3\,\frac{e^3 H \sin\chi}{mc^2}\left[1 - (1 - n^2)\left(\frac{\mathcal{E}}{mc^2}\right)^2\right]^{-\frac{1}{2}} \frac{\nu}{\nu_{c,n}} \int_{\nu/\nu_{c,n}}^{\infty} K_{\frac{5}{3}}(\eta)\, d\eta , \\ \nu_{c,n} &= \nu_c\left[1 + (1 - n^2)\left(\frac{\mathcal{E}}{mc^2}\right)^2\right]^{-\frac{3}{2}}, \quad \nu_c = \frac{3eH\sin\chi}{4\pi mc}\left(\frac{\mathcal{E}}{mc^2}\right)^2 . \end{aligned}\right\} \tag{6.72}$$

The whole approximation used here (which is clear from what was said in Chapter 5) is suitable only when $1 - n \ll 1$, in which case we can put $1 - n^2 = 2(1 - n)$. It is clear that therefore the condition that the effect of the medium be small, which can be deduced from (6.72), is exactly the same as (6.71). Let us apply this last condition to a plasma for which

$$1 - n^2 = 2(1 - n) = \frac{\omega_p^2}{\omega^2} = \frac{4\pi e^2 N}{m\omega^2}. \qquad (6.73)$$

The role of the medium (plasma) is then small provided

$$\omega^2 = (2\pi\nu)^2 \gg \omega_p^2 \left(\frac{\mathcal{E}}{mc^2}\right)^2 = \frac{16\pi^2 e c N}{3H_\perp} \nu_c , \qquad (6.74)$$

or for the main part of the synchrotron emission (frequencies $\nu \sim \nu_c$), provided

$$\nu \sim \nu_c \gg \frac{4\,ecN}{3H_\perp} \approx 20\,\frac{N(\mathrm{cm}^{-3})}{H_\perp(\mathrm{Oe})}\ \mathrm{Hz}\,. \qquad (6.75)$$

In the interstellar medium, where $N \approx 1\ \mathrm{cm}^{-3}$ and $H_\perp \sim 10^{-6}$ to 10^{-5} Oe condition (6.75) is well satisfied for most frequency ranges used in radio-astronomy. In denser objects (some galaxies, shells around supernovae, and so on) condition (6.75) is more rigid. At sufficiently long wavelengths one must always reckon with the possibility that this condition is violated, provided other factors such as, for instance, reabsorption do not play an even greater role (see Chapter 9).

Chapter VII

CHERENKOV EFFECT, DOPPLER EFFECT, TRANSITION RADIATION

Cherenkov and Doppler effect from the quantal point of view. Radiation reaction in a medium. Cherenkov emission and absorption in an isotropic and magneto-active plasma. Cherenkov emission by dipoles. Emission in channels and gaps. Application of the reciprocity theorem. Transition radiation.†

When one analyzes various problems connected with the emission, absorption, and amplification of electromagnetic waves when charges or other 'systems' (such as atoms, bunches, or antennae) move in a medium, elementary quantal pictures have turned out to be very fruitful. It is important that this is the case also when the problem is essentially classical so that the final results within the approximations used will be independent of the quantum constant $\hbar$.

As the starting point for our use of quantum-mechanical considerations we take the concept of quanta, or photons in a medium, with an energy $\hbar\omega$ and a momentum

$$\hbar\mathbf{k} = \frac{\hbar\omega n_\ell(\omega,\mathbf{s})}{c}\,\mathbf{s}\,,$$

where $\mathbf{k} = k\mathbf{s}$ is the wavevector and $n_\ell(\omega,\mathbf{s})$ the refractive index for a normal wave of the given kind (ℓ) which propagates in the medium considered — which, in general, may be anisotropic and gyrotropic. In Chapter 6 (for details see Ginzburg, 1940a; Jauch and Watson, 1948a,b; Watson and Jauch, 1949; Riazanov, 1957) we outlined a scheme for the quantization of the electromagnetic field in a medium. Of course, such an approach is correct only where a phenomenological theory is applicable. We must also bear in mind that the momentum of a photon in a medium introduced here is its total momentum, including both the field momentum and the momentum imparted to the medium in the emission of the wave (vide infra; for more precise details see Chapter 12).

---

† Except for the part dealing with transition radiation we follow in the present chapter an earlier review paper (Ginzburg, 1960). In the Russian literature the Cherenkov effect is called the Vavilov-Cherenkov effect.

From a quantal point of view the kinematics of the emission, that is, the conditions imposed upon the frequency and the direction of the emission, are determined by the energy and momentum conservation laws (this is also true for the conditions for absorption). If, for instance, before the emission the 'system' (electron, atom, antenna) had an energy E_0, while its energy after the emission was E_1 with the corresponding momenta being equal to $\mathbf{p}_0$ and $\mathbf{p}_1$, the following conservation laws must be satisfied in the emission of the photon:

$$E_0 - E_1 = \hbar\omega , \tag{7.1}$$

$$\mathbf{p}_0 - \mathbf{p}_1 = \hbar\mathbf{k} = \frac{\hbar\omega n}{c}\frac{\mathbf{k}}{k} \equiv \frac{\hbar\omega n}{c}\mathbf{s} , \tag{7.2}$$

where, for the sake of simplicity, we have assumed the medium to be isotropic so that we could drop the index ℓ of n.

For a 'system' which moves uniformly in vacuo (that is, when $n = 1$) emission without a change in its internal state is impossible (for instance, an electron, moving uniformly in vacuo, cannot radiate; see Chapter 1). This well known fact follows, in particular, also from Eqs.(7.1) and (7.2) as in the case when $n = 1$ they have only the solution $\omega = 0$ for a particle without internal degrees of freedom. If, however, $n \neq 1$, we can substitute the expressions

$$E_{0,1} \equiv \mathcal{E}_{0,1} = \sqrt{(m^2c^4 + c^2p^2_{0,1})} \; ; \; \mathbf{p}_{0,1} = m\mathbf{v}_{0,1}/\sqrt{(1 - v^2_{0,1}/c^2)}$$

into (7.1) and (7.2) and we then get as the condition for emission without a change in the internal state (Ginzburg, 1940a)

$$\cos\theta_0 = \frac{c}{n(\omega)v_0}\left(1 + \frac{\hbar\omega(n^2-1)}{2mc^2}\sqrt{1 - \frac{v_0^2}{c^2}}\right) ,$$

$$\hbar\omega = \frac{2(mc/n)(v_0\cos\theta_0 - c/n)}{(1 - v_0^2/c^2)^{\frac{1}{2}}\,(1 - 1/n^2)} , \tag{7.3}$$

where θ_0 is the angle between $\mathbf{v}_0$ and $\mathbf{k}$.

When $\hbar\omega/mc^2 \ll 1$ this condition changes to the classical condition (6.53) for emission, as would be expected (when $\hbar\omega/mc^2 \ll 1$, the 'recoil' connected with the emission of a quantum is relatively small).† It is clear from (7.3), of course, that emission is possible (that is, $\cos\theta_0 < 1$ and $\omega > 0$) only when the

† It is clear from (7.3) that if we want to be more precise the condition for classical behaviour must be written in somewhat different form, namely,

$$\{\hbar\omega(n^2-1)/2mc^2\}\sqrt{[1 - v_0^2/c^2]} \ll 1.$$

motion is with a velocity exceeding the speed of light, that is, when the inequality $v_0 n/c > 1$ holds (see (6.52)).

When the result does not contain $\hbar$ a quantum-mechanical calculation has only a methodological value, but it often turns out to be more convenient. Essentially this reduces to using the conservation laws which have a wider range of applicability in the sense that they can be used also without invoking quantum mechanics. Let us, in fact, assume that it follows from the classical theory of an electromagnetic field in a medium that the connection between the energy $\mathcal{H} = \mathcal{H}_{em}$ and the total momentum $\mathbf{G}$ of the radiation in the medium is found to be $\mathbf{G} = (\mathcal{H}n/c)\mathbf{s}$.†

Moreover, if the changes in energy and momentum are sufficiently small we have for free motion of the charge $\Delta\mathcal{E} = (\mathbf{v}\cdot\Delta\mathbf{p})$, since

$$\frac{d\mathcal{E}}{d\mathbf{p}} = \frac{d}{d\mathbf{p}}\left(\sqrt{\left[m^2c^4 + c^2p^2\right]}\right) = \frac{c^2\mathbf{p}}{\mathcal{E}} = \mathbf{v}\ . \tag{7.4}$$

Because we have assumed the change $\Delta\mathcal{E}$ to be small we need not distinguish between v_0 and v_1 and we denote the velocity of the source $v_0 \approx v_1$ by v.

From (7.4) and the conservation laws (7.1) and (7.2) we get, replacing $\hbar\omega$ by $\mathcal{H}$,

$$\Delta\mathcal{E} = \mathcal{H} = (\mathbf{v}\cdot\Delta\mathbf{p}) = \frac{\mathcal{H}n}{c}(\mathbf{s}\cdot\mathbf{v})\ ,$$

or $\cos\theta_0 = c/nv$, that is, we get the Cherenkov condition (6.53). However, it is simpler to introduce at once the quanta $\hbar\omega$ and this is a completely natural thing to do not only in the quantal, but also in the classical case. We shall proceed in this way.

If we are considering the motion of a 'system' rather than a 'structureless' particle so that its internal energy can change, we have

$$\mathcal{E}_0 = \sqrt{\{(m+m_0)^2c^4 + c^2p_0^2\}}\ ,\ \mathcal{E}_1 = \sqrt{\{(m+m_1)^2c^4 + c^2p_1^2\}}\ ,$$

where $(m+m_0)c^2 = mc^2 + w_0$ is the total energy in the lower state and $(m+m_1)c^2 = mc^2 + w_1$ the total energy in the upper state. Clearly, $w_1 - w_0 = \hbar\omega_i > 0$ is the energy difference of the two levels of the system (atom, ...) considered here.

---

† It will be shown in Chapter 12 that the momentum of the field equals $G_{em} = \mathcal{H}/nc$, while the momentum of the force imparted to the dielectric in the radiation, $G^{(m)} = (n^2-1)G_{em} = \mathcal{H}(n^2-1)/nc$. The total momentum lost to the emitter is thus $G = G_{em} + G^{(m)} = \mathcal{H}n/c$.

If we now apply the conservation laws (7.1) and (7.2) with $\hbar\omega/mc^2 \ll 1$ we get exactly the Doppler Eq.(6.59) with $\omega_{00} \equiv \omega_i$. If, however, we do not neglect terms of order $\hbar\omega/mc^2$, we find, as in the case of the Cherenkov emission (see (7.3)), a somewhat more complicated expression (Ginzburg and Frank, 1947a). In practice, however, we can restrict ourselves to the usual form (6.59) of the formula for the Doppler effect. In a quantal calculation, moreover, another important fact is clarified which completely escapes attention in a classical derivation of Eq.(6.59); namely, in the region of the normal Doppler effect, that is, when (see (6.62))

$$\frac{v}{c}\, n(\omega) \cos\theta < 1 , \tag{7.5}$$

the emission corresponds to a transition of the system from an upper level with energy w_1 to a lower level with energy w_0 (the direction of the transition is determined by the condition that the energy of the quantum emitted must be positive, that is, formally from the requirement $\omega > 0$). If, however, the quantum is emitted inside the Cherenkov cone, that is, if we are dealing with the anomalous Doppler effect and (see (6.61))

$$\frac{v}{c}\, n(\omega) \cos\theta > 1 , \tag{7.6}$$

the emission of the quantum is accompanied by a transition of the system from the lower level w_0 to the upper level w_1. The energy of the quantum, and also the energy exciting the radiating system, is then derived from the kinetic energy of its translational motion.

It is clear from this example that in quantum theory, in contrast to classical theory, when one finds the conditions for radiation themselves one determines at the same time the direction of the process, that is, whether the transition is up or down. It is just this fact which together with the possibility of such a simple calculation of the induced emission (vide infra) makes quantal calculations so valuable for obtaining conditions for radiation, the condition for amplification (instability) of waves in beams, and so on.

If the system has only two discrete levels 0 and 1, we see that when $(v/c)n < 1$ (subluminal motion) the stationary state of the radiator corresponds to its being in the lower level 0 (we assume that, say, the system moves in a channel in the medium and that there are no extraneous sources of excitation). In other words, if the level 1 is excited, the system will be de-excited after some time and go over into state 0. If, however, $(v/c)n > 1$ (superluminal motion) there is also in stationary states a non-vanishing probability to find the system in level 1 and it radiates all the time both normal and anomalous Doppler waves. The population of the levels 0 and 1, and also the intensity of the emission of normal and anomalous waves, are clearly determined by the ratio of the total probabilities for the emission of these waves. For a system with many levels (Ginzburg and Fain, 1959) the emission of anomalous Doppler waves while the system makes a transition upwards leads to the possibility of the amplification of 'transverse oscillations' and, for instance, to the ionization of an atom.

To be more exact, there are here two cases (Ginzburg and Eidman, 1959b). In the first case the average energy of the transverse oscillations of the system decreases while it moves. This means that for a wavepacket consisting of wave-functions with different, but approximately equal, energies (we think, for instance, of an electron moving along a magnetic field) the centre of mass of the packet in the energy scale diminishes. The difference between subluminal and superluminal motions lies in this case in the rate at which the average energy changes and also in the nature of the spreading of the packet. Thus, when the system moves subluminally states with energies larger than those represented in the initial spectrum of the packet never become occupied. If the system moves superluminally, however, there is a finite probability to find the system (we assume, of course, that we have an ensemble of systems) in any, however highly excited, level which can be reached with condition (7.6) being valid, notwithstanding the fact that the average energy decreases.

In the second of the above mentioned cases the system is unstable even 'on average', that is, its average energy (and we are talking here about the energy of the oscillations, that is, the energy of the excitations) increases with time and, moreover, the nature of the spreading of the packet is changed.

Elucidation of with which of the possibilities we are dealing requires actual calculations of the transition probabilities. In that respect a quantal calculation is, generally speaking, superfluous for a classical system, whatever its advantages, and it is natural to use the classical theory of radiation, as we shall do below.

We now note that quantum considerations like those given are nevertheless useful also for an analysis of the problems, already mentioned, of the absorption and amplification of waves in particle beams (in the case of wave amplification the beam, in fact, becomes unstable). In this way one easily obtains criteria for the instability of a particle beam moving in an isotropic plasma (vide infra). It is, moreover, clear that when beams of 'systems' with two or more levels move superluminally, as a rule, amplification (negative absorption) must occur rather than absorption (re-absorption) of anomalous Doppler waves (Zheleznyakov, 1959). This is connected with the fact that when a quantum corresponding to the region of anomalous Doppler waves (that is, flying at an angle $\theta < \theta_0$ to the velocity of the system) is absorbed, the system will make a transition not upwards, as in the normal effect, but downwards.† On the other hand, the upwards transition of the system corresponds now to induced emission which in the region of the normal effect corresponds to a downwards transition of the system. Therefore, if all systems (atoms, electrons in a magnetic field) in a beam moving superluminally are, for instance, in the lower level, normal Doppler waves, emitted by one of the systems, will be absorbed in this beam, and the anomalous Doppler waves will be amplified: they will along their path transfer other systems upwards through induced emission, that is with the emission of yet another anomalous Doppler quantum.

We shall in Chapter 9 dwell on the use of the Einstein coefficients for the spontaneous emission and absorption probabilities. However, it is already now convenient to use these coefficients in order to make what has been said above somewhat more quantitative.

If both the upper and lower levels 1 and 0 are populated, the coefficient for absorption in the beam of normal Doppler waves is equal to (see Ginzburg and Zheleznyakov, 1959a, 1965; Ginzburg, Zheleznyakov and Eidman, 1962; Zheleznyakov, 1959, 1970; Ginzburg, 1970b)

$$\mu_n = -\frac{dI_\omega}{I_\omega dz} = A_1^0 \frac{8\pi^3 c^2 N_1 (N_0/N_1 - 1)}{\omega^2 n^2}, \quad I_\omega = I_\omega(0)\, e^{-\mu z}, \quad \omega = 2\pi\nu, \tag{7.7}$$

where $A_1^0(\theta)$ is the probability per unit solid angle for a $1 \rightarrow 0$ spontaneous transition with emission of a quantum at an angle θ to the velocity, N_1 and N_0 are the particle densities in the beam in levels 1 and 0, respectively, and n is the refractive index of the medium for the frequency ω considered when the wave propagates at an angle θ (we assume for the sake of simplicity that the dipole moment for the transition $1 \rightleftarrows 0$ is for all particles parallel to the velocity). In order that normal Doppler waves are amplified, the number of particles in the upper level 1 must exceed their number in the lower level 0 (in that case $N_0/N_1 < 1$ and $\mu_n < 0$). Such a distribution over the levels

---

† Absorption is the process which is the inverse of emission and what we say here follows therefore at once from the calculations performed for emission. The terms 'upwards' and 'downwards' are used here everywhere as referring to the energy scale of the excited levels of the emitting system.

does not occur in thermal equilibrium and its production is, generally speaking, connected with well defined difficulties. The position is changed in the case of anomalous Doppler waves, when emission of waves (of a photon in the medium) takes place in the transition $0 \to 1$, and absorption in the transition $1 \to 0$. In that case

$$\mu_{an} = A_1^0 \frac{8\pi^3 c^2 N_0 (N_1/N_0 - 1)}{\omega^2 n^2} , \tag{7.8}$$

and $\mu_{an} < 0$ when $N_1/N_0 < 1$. From this it is, of course, also clear that when the anomalous Doppler effect occurs (that is, if $(v/c)n > 1$) a beam of particles which are solely in the lower level 0, shows negative absorption and waves emitted by separate particles are amplified. This fact is possibly very favourable for using beams of particles moving in a dielectric gap or a retarding system for generating and amplifying micro-waves (Ginzburg 1947, 1972b,c).

The role of the system (particle) which emits anomalous Doppler waves may, as we mentioned already, be played by electrons oscillating as a result from being acted upon by an applied field or moving along a helical path along a magnetic field which is parallel to the axis of the beam. For small amplitudes such electrons (if we forget about Cherenkov radiation) emit just like the corresponding oscillators which move with a velocity v which is equal to the component $v_{\parallel}$ of the electron velocity along the beam axis.

In an electron beam the transverse velocities $v_{\perp}$ are usually distributed in such a way that the distribution function $f(v_{\perp})$ decreases with increasing $v_{\perp}$ (this occurs, for instance, for the distribution $f(v_{\perp}) = \text{const.} \times \exp\{-mv_{\perp}^2/k_B T\}$. In similar circumstances the normal Doppler waves will be damped due to reabsorption in the beam; however, the anomalous Doppler waves will, on the contrary, be amplified. The amplification of waves in an electron beam means that the amplitude of the oscillations increases and the beam becomes unstable. The electrons will then, in general, start to bunch and coherent radiation emerges. The quantum condition for the instability of the beam (the condition $(v_{\parallel}/c)\, n(\omega) > 1$) is the same as the condition which one can obtain by solving the classical problem of the instability of an electron beam in a magnetic field (Zheleznyakov, 1959). The instability of electron beams noted here is, in particular, possible in a magneto-active plasma and is of interest for the theory of sporadic radio-emission of the Sun (Ginzburg and Zheleznyakov, 1958b, 1959b; Zheleznyakov, 1970).

The condition (6.53) for Cherenkov emission has in the classical case an interference character, and it is therefore universal for any kind of wave — of course, replacing the phase velocity of light, $c/n(\omega)$, by the phase velocity v_{ph} of the waves considered which may be sound waves, capillary waves, ... The same is also true for the results given here which are obtained by using the energy and momentum conservation laws employing either a quantal or a classical approach to them. In that case the quantal approach, that is, the introduction of quanta, is appreciably simpler, not only for light, but also for (longitudinal) plasma waves[†] and sound. In the latter case the energy of a sound quantum (phonon) is equal to $E = \hbar\omega$ and its momentum to $\mathbf{p} = \hbar\mathbf{k} = (E/u)\,\mathbf{s}$, where u is the sound velocity; for sound the dispersion is usually unimportant and one need not distinguish between the phase and group velocities. Of course, as in electrodynamics, in the case of supersonic motion the emitting acoustic system will make an 'upwards' transition — that is, will be excited — in the region of the anomalous Doppler effect and thus to a certain extent 'be amplified' (Tamm, 1959).

Let us now consider an interesting feature connected with the fact that the directions of the phase and the group velocities are not the same, as may be the case in an anisotropic medium or when one takes into account spatial dispersion (see Agranovich and Ginzburg, 1966). If the projection of the group velocity $d\omega/d\mathbf{k}$ on the direction at right angles to the velocity of the particle, that is, the quantity $d\omega/dk_r$, where k_r is the component of $\mathbf{k}$ at right angles to $\mathbf{v}$, is negative, no energy, apparently, leaves the emitter, but energy is absorbed by it. However, under similar circumstances one must use the advanced rather than retarded potentials (Mandel'shtam, 1947; Pafamov, 1957, 1959). If we choose the vector $\mathbf{k}$ always to be directed along the phase velocity, this vector will in the case when $d\omega/dk_r < 0$ in the Cherenkov and Doppler waves be directed towards the particle trajectory, while the energy — as should be the case — will go away from the trajectory. In the case of Cherenkov radiation the difference between the cases $d\omega/dk_r > 0$ and $d\omega/dk_r < 0$ is clear from Fig.7.1. The angle θ_0 is in the case $d\omega/dk_r < 0$ determined as before by the Cherenkov condition (6.53) as is clear from the choice made for the direction of $\mathbf{k}$ from interference considerations, and also from the conservation laws (7.1) and (7.2). The latter statement follows from the fact

---

† The quanta of plasma waves are often called plasmons. If we understand by photons in a medium the quanta of any electromagnetic field — where we are, strictly speaking, dealing with a free field, that is, a field when there are no charges or currents — plasmons are a particular case of photons in a medium.

that we have used plane waves of the form

$$\exp\{i(\mathbf{k}\cdot\mathbf{r}) - i\omega t\}$$

for which the momentum of the corresponding quantum is equal to $(\hbar\omega n/c)(\mathbf{k}/k)$; when we use this form of waves there is no difference whatever in the directions of **k** in Figs.7.1a and 7.1b, as in terms of the plane waves the disposition of the wavefronts is the same in both cases — we consider fronts with the vector **k** lying on the Cherenkov cone. Equation (6.59) for the Doppler effect for the case $d\omega/dk_r < 0$ also remains valid. However, the physical difference between the two cases is, of course, very important and is connected with the different directions of the group velocity. In fact, in an isotropic medium in the normal case (see Fig.7.1a) the group velocity is parallel to **k**. However, in the case depicted in Fig. 7.1b the group velocity $d\omega/d\mathbf{k}$ is directed at an obtuse angle $\theta_1 = \pi - \theta_0$ to the particle velocity. Under such circumstances the Cherenkov radiation will, when particles pass through a plate of finite thickness, emerge from the back surface of the plate, and will also be refracted in an unusual way at that surface (this is clear from Mandel'shtam's work (1947)).

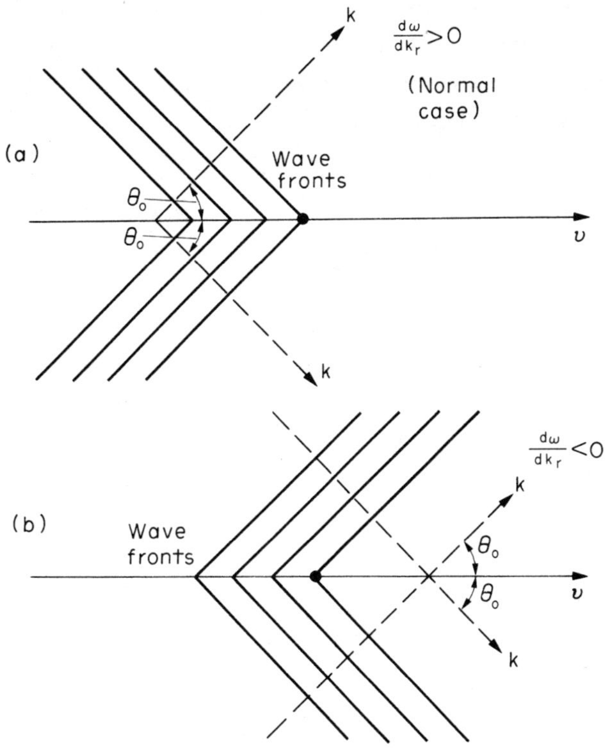

Fig. 7.1 Cherenkov radiation when $d\omega/dk_r > 0$ and when $d\omega/dk_r < 0$. k_r is the component of the wave-vector at right angles to the particle velocity **v**

For electrons moving with superluminal velocities in a plasma or a retarding medium in the presence of a magnetic field and also in similar cases of oscillatory motion of electrons, usually only the classical region is of interest — the quantum numbers corresponding to the transverse motion are large. Under such circumstances the problem of the emission of waves and of the damping or amplification of the transverse oscillations of the electrons may and in practice must be solved by means of classical calculations. The corresponding calculations reduce essentially to the evaluation of the radiation reaction force when the charge moves in the medium. Let us consider that problem in a somewhat wider context.

As the presence of the medium may radically change the character of the electromagnetic waves radiated by the moving particle (see Chapter 6) it is clear that the radiation reaction force in the medium is also changed, and sometimes very considerably. For instance, an oscillator with a frequency ω in an isotropic plasma with index of refraction $n = \{1 - 4\pi Ne^2/m\omega^2\}^{\frac{1}{2}}$ does not radiate at all if $\omega_p^2 = 4\pi Ne^2/m > \omega^2$, when $\varepsilon = \tilde{n}^2 < 0$; in a magneto-active plasma there is in the non-relativistic approximation no radiation by an electron revolving in a magnetic field $\mathbf{H}_0$ with a frequency $\omega_H = eH_0/mc$ (see Ginzburg and Zheleznyakov, 1958a; Ginzburg 1970b). In both these cases the radiation force, of course, vanishes, while in vacuo it equals $\mathbf{f} = (2e^2/3c^3)\dddot{\mathbf{r}}$. On the other hand, in the case of uniform motion in a medium there is, if for some frequencies the velocity $v > c/n(\omega)$, the Cherenkov radiation force $\mathbf{f}_{Ch}$ which produces per unit time an amount of work equal to $(\mathbf{f}_{Ch} \cdot \mathbf{v}) = -dW/dt$. It is thus clear from (6.58) that

$$\mathbf{f}_{Ch} = -\frac{e^2\mathbf{v}}{c^2v}\int_{c/n(\omega)\le v}\left[1 - \frac{c^2}{v^2c^2(\omega)}\right]\omega\, d\omega\,. \tag{7.9}$$

In the light what we have said so far the problem naturally arises of how to evaluate the radiation reaction force for arbitrary motion of a charge in an arbitrary medium. In the past this problem has not particularly attracted attention. The point is, apparently, that the radiation force when a charge moves in a medium is usually appreciably smaller than the braking force connected with ionization losses. For example, losses due to Cherenkov emission which might be called radiative, are even in a transparent, but dense medium only a small fraction of the total losses. The position is, generally speaking, not changed for the case of non-uniform motion of a charge. There are, however, interesting and practically important cases when taking the radiation forces into account when a charge moves in a medium is very important — motion in a magneto-active plasma, motion in channels, gaps, and close to the surface of a medium.

We give here a scheme to evaluate the radiation force in a medium (Ginzburg and Eidman, 1959b) where, as elsewhere, we shall not be afraid to repeat ourselves. For a point charge with charge density $\rho = e\delta(\mathbf{r} - \mathbf{R})$ the equations for the field and the equations of motion have the form

$$\left.\begin{aligned} &\mathrm{curl}\,\mathbf{H} = \frac{4\pi}{c} e\mathbf{v}\,\delta(\mathbf{r}-\mathbf{R}) + \frac{1}{c}\frac{\partial \mathbf{D}}{\partial t}\,, \quad \mathrm{div}\,\mathbf{D} = 4\pi e\delta(\mathbf{r}-\mathbf{R})\,,\\ &\mathrm{curl}\,\mathbf{E} = -\frac{1}{c}\frac{\partial \mathbf{H}}{\partial t}\,, \quad \mathrm{div}\,\mathbf{H} = 0\,, \end{aligned}\right\} \tag{7.10}$$

$$\frac{d}{dt}\frac{m\mathbf{v}}{\sqrt{1-v^2/c^2}} = e\left\{\mathbf{E}_0 + \frac{1}{c}\left[\mathbf{v}\wedge\mathbf{H}_0\right]\right\} + $$
$$+ e\int\left\{\mathbf{E}(\mathbf{r}) + \frac{1}{c}\left[\mathbf{v}\wedge\mathbf{H}(\mathbf{r})\right]\right\}\delta(\mathbf{r}-\mathbf{R})\,d^3\mathbf{r}\,. \tag{7.11}$$

Here $\mathbf{R}(t)$ is the position of the charge ($\mathbf{v} \equiv \dot{\mathbf{R}} \equiv d\mathbf{R}/dt$), $\mathbf{E}_0$ and $\mathbf{H}_0$ are the external fields, and $\mathbf{E}$ and $\mathbf{H}$ the fields produced by the charge itself (for the sake of simplicity we assume the medium to be non-magnetic).

The only effective method of solving the problem for an arbitrary medium is by expanding the fields in terms of normal plane waves, that is, using the method which we have called the Hamiltonian one. As a result we have

$$\left.\begin{aligned} &\tilde{D}_\alpha(\omega) = \varepsilon_{\alpha\beta}(\omega)\,\tilde{E}_\beta(\omega)\,, \quad \alpha,\beta = 1,2,3\,;\\ &\mathbf{E} = -\frac{1}{c}\frac{\partial \mathbf{A}}{\partial t} - \nabla\phi\,; \quad \mathbf{H} = \mathrm{curl}\,\mathbf{A}\,,\\ &\tilde{\mathbf{A}} = \sqrt{4\pi}\,c\sum_{\lambda,j=1,2}\frac{q_{\lambda j}(t)\,\mathbf{a}_{\lambda j}}{n_{\lambda j}}\exp\left[i(\mathbf{k}_\lambda\cdot\mathbf{r})\right]\,,\\ &n^2_{\lambda j} = \varepsilon_{\alpha\beta}(\omega)(a_{\lambda j})_\beta\,(a^*_{\lambda j})_\alpha\,, \end{aligned}\right] \tag{7.12}$$

$$\varepsilon_{\alpha\beta}\frac{\partial \tilde{A}_\alpha}{\partial x_\beta} + \mathrm{c.c.} = 0 \tag{7.13}$$

where condition (7.13) is chosen for the sake of convenience, c.c. indicates the complex conjugate quantity, summation is understood to be performed over indexes which occur twice, and the argument ω shows that we are talking about Fourier components; the real fields are equal to $\mathbf{D} = \tilde{\mathbf{D}} + \tilde{\mathbf{D}}^* = \tilde{\mathbf{D}} + \mathrm{c.c.}$, $\mathbf{E} = \tilde{\mathbf{E}} + \tilde{\mathbf{E}}^*$, and so on. In Eqs.(7.12) and (7.13) $n_{\lambda j}$ is the refractive index and the $\mathbf{a}_{\lambda j}$ are the complex polarization vectors corresponding to the jth normal wave.

The equations for the potentials, obtained from (7.10), (7.12), and (7.13) have the form

$$\left.\begin{aligned} &\nabla^2\tilde{\mathbf{A}} - \mathrm{grad\,div}\,\tilde{\mathbf{A}} - \frac{1}{c^2}\varepsilon_{\alpha\beta}\frac{\partial^2\tilde{\mathbf{A}}_\beta}{\partial t^2}\mathbf{e}_\alpha - \frac{1}{c}\varepsilon_{\alpha\beta}\frac{\partial^2\tilde{\phi}}{\partial t\partial x_\beta}\mathbf{e}_\alpha + \mathrm{c.c.}\\ &\quad = -\frac{4\pi}{c}\mathbf{j}_e = -\frac{4\pi}{c}e\mathbf{v}\,\delta(\mathbf{r}-\mathbf{R})\,,\\ &\varepsilon_{\alpha\beta}\frac{\partial^2\tilde{\phi}}{\partial x_\alpha\partial x_\beta} + \mathrm{c.c.} = -4\pi e\delta(\mathbf{r}-\mathbf{R})\,, \end{aligned}\right\} \tag{7.14}$$

where $\mathbf{e}_\alpha$ is the unit vector along the α-axis and $\mathbf{j}_e = e\mathbf{v}\,\delta(\mathbf{r}-\mathbf{R})$ is the current density corresponding to the particle considered; the somewhat different notation and definition of various quantities as compared to those used in Chapter 6 is because of convenience and in order to refer the reader to the papers by Eidman and the author (Ginzburg and Eidman, 1959b; Eidman, 1960) where the calculations are given in detail using the same notation.

Substitution of the expansion (7.12) into (7.14) produces a set of oscillator equations for the field amplitudes $q_{\lambda j}$ which can be integrated in an elementary way. It is necessary to substitute the fields obtained this way into the equation of motion (7.11). As a result we get

$$\frac{d}{dt}\frac{m\mathbf{v}}{\sqrt{1-v^2/c^2}} = \mathbf{F}_0 - \frac{e^2}{2\pi^2}\sum_{j=1,2}\int_0^t\int_0^{k_{max}}\left\{-\frac{\mathbf{a}_j(\mathbf{v}'\cdot\mathbf{a}_j^*)}{n_j^2}\cos\left[\omega_j(t-t')\right]\right.$$
$$\left. - i\left[\mathbf{v}\wedge\left[\mathbf{k}\wedge\mathbf{a}_j\right]\right]\frac{(\mathbf{v}'\cdot\mathbf{a}_j^*)}{n_j^2\omega_j}\sin\left[\omega_j(t-t')\right]\right\} e^{i(\mathbf{k}\cdot\{\mathbf{R}-\mathbf{R}'\})}\,dt'\,d^3\mathbf{k} + \text{c.c.}$$
$$= \mathbf{F}_0 + \mathbf{f}_{rad}\,, \tag{7.15}$$

where $\mathbf{R} = \mathbf{R}(t')$, $\mathbf{v}' = \mathbf{v}(t')$, and $\mathbf{F}_0 = e\left\{\mathbf{E}_0 + \frac{1}{c}\left[\mathbf{v}\wedge\mathbf{H}_0\right]\right\}$.

The method applied here for calculating the radiation reaction force is convenient in a number of cases even for an isotropic medium or a vacuum. This was checked in Chapter 2 for the latter case. Moreover, for the case of a particle moving uniformly in an isotropic medium with a refractive index $n > c/v$ we get from (7.15) a formula for the braking force (7.9) due to Cherenkov radiation.

Ginzburg and Eidman (1959b) considered the superluminal motion of an oscillator. In an isotropic medium we have for the case of an oscillator vibrating parallel to the translational velocity $\mathbf{v}_0$

$$\mathbf{R} = \{0,0,v_0t + R_0\sin\Omega t\}\;;\; \mathbf{v} = \{0,0,v_0 + v_\sim\cos\Omega t\}\,,\; v_\sim = R_0\Omega\,;$$
$$\mathbf{a}_1 = \{1,0,0\}\,,\mathbf{a}_2 = \{0,\cos\theta\,,-\sin\theta\}\,,\; \mathbf{k} = \{0\,,k\sin\theta\,,k\cos\theta\}\,, \tag{7.16}$$

and in what follow we shall be dealing only with the dipole approximation, that is, the case when

$$kR_0 = \frac{\omega}{c}\,n(\omega)\,R_0 \ll 1\,. \tag{7.17}$$

Under those circumstances we get the following expression for the work done by the radiation field on the particle:

$$A = \int_0^T (\mathbf{v}\cdot\mathbf{f}_{rad})\,dt = v_0\int_0^T f_{rad,z}\,dt + v_\sim \int_0^T \cos\Omega t\, f_{rad,z}\,dt = A_0 + A_\sim\,, \tag{7.18}$$

$$A = -\frac{e^2R_0^2T}{4c^3\beta_0}\left\{\int_{\beta_0 n(\omega)\cos\theta<1} \omega^3\left[1-\frac{1}{\beta_0^2n^2(\omega)}\left(1-\frac{\Omega}{\omega}\right)^2\right]d\omega\right.$$
$$\left.+\int_{\beta_0 n(\omega)\cos\theta>1} \omega^3\left[1-\frac{1}{\beta_0^2n^2(\omega)}\left(1+\frac{\Omega}{\omega}\right)^2\right]d\omega\right\}, \tag{7.19}$$

where

$$\omega = \frac{\Omega}{|1-\beta_0 n(\omega)\cos\theta|}\,, \quad \beta_0 = \frac{v_0}{c}\,. \tag{7.20}$$

If the dispersion law has a 'step-function' character, that is,

$$n(\omega) = n = \text{constant, for } \omega \le \omega_m\,; \quad n(\omega) = 1\,, \text{ for } \omega > \omega_m\,, \tag{7.21}$$

one can write the result (7.19) in the form

$$A = -\frac{e^2\Omega^4R_0^2nT}{4c^3}\int\frac{\sin^3\theta\,d\theta}{|1-\beta_0 n\cos\theta|^5}\,, \tag{7.22}$$

where for the anomalous Doppler effect

$$0 \le \theta \le \arccos\frac{1+\Omega/\omega_m}{\beta_0 n}$$

and for the normal Doppler effect

$$\arccos\frac{1-\Omega/\omega_m}{\beta_0 n} \le \theta \le \pi\,.$$

The quantity $U \int_0^T (dW/dt)\,dt = -A > 0$ equals the energy emitted during a period T by the particle. Expression (7.22) is the analogue of formula (6.63) which refers to an oscillator vibrating at right angles to the translational velocity.

The work of the radiation field expended while amplifying or reducing the energy of the particle oscillations is, in agreement with (7.19), equal to

$$A_\sim = A - A_0 = \frac{e^2\Omega R_0^2T}{4c^3\beta_0}\left\{\int_{\beta_0 n(\omega)\cos\theta>1} \omega^2\left[1-\frac{1}{\beta_0^2n^2(\omega)}\left(1+\frac{\Omega}{\omega}\right)^2\right]d\omega\right.$$
$$\left.-\int_{\beta_0 n(\omega)\cos\theta<1} \omega^2\left[1-\frac{1}{\beta_0^2n^2(\omega)}\left(1-\frac{\Omega}{\omega}\right)^2\right]d\omega\right\}. \tag{7.23}$$

In the case (7.21)

$$A_{\sim} = \frac{e^2R_0^2\Omega^4 nT}{4c^3}\left\{\int_0^{\mathfrak{a}} \frac{\sin^3\theta\, d\theta}{(1-\beta_0 n\cos\theta)^4} - \int_{\mathfrak{b}}^{\pi} \frac{\sin^3\theta\, d\theta}{(1-\beta_0 n\cos\theta)^4}\right\},$$

$$\mathfrak{a} \equiv \text{arc cos}\frac{1}{\beta_0 n}\left(1+\frac{\Omega}{\omega_m}\right), \quad \mathfrak{b} \equiv \text{arc cos}\frac{1}{\beta_0 n}\left(1-\frac{\Omega}{\omega_m}\right). \tag{7.24}$$

The radiation propagating outside the Cherenkov cone, which corresponds to the second integral in (7.23) or (7.24), leads thus to a damping of the oscillations while the radiation inside this cone (the anomalous Doppler effect), corresponding to the first integral in (7.23) or (7.24), amplifies the oscillations.† This result is in complete agreement with the quantummechanical considerations (vide supra). One sees easily that the second integral in (7.23) is larger than the first one, and that the same is true in (7.24). Hence it follows that the vibrations of the oscillator are always damped in an isotropic medium and $A_{\sim} \to 0$ only if $\beta_0 n(\omega) \to \infty$ in an appreciable region of integration.

Ginzburg and Eidman (1959b; Eidman, 1960) considered also the motion of an oscillator vibrating at right angles to its translational velocity $\mathbf{v}_0$ and the helical motion of a charge in a magnetic field. They showed that as in the preceding case in an isotropic medium the vibrations are always damped (this result does not necessarily follow for other radiating systems such as, for instance, a sufficiently long antenna).

It is convenient for the calculation of some of the features of the superluminal motion of charges in anisotropic media to consider the motion of an oscillator along the optic axis of a uni-axial non-gyrotropic crystal with the electron assumed to be oscillating in the same direction. In that case

$$\left.\begin{aligned} &\mathbf{R} = \{0,0,v_0t + R_0\sin\Omega t\}, \quad \mathbf{k} = \{0, k\sin\theta, k\cos\theta\}, \\ &\mathbf{a}_1 = \{0, \cos\theta + K_1\sin\theta, -\sin\theta + K_1\cos\theta\}, \quad \mathbf{a}_2 = \{1,0,0\}, \\ &K_1 = \frac{(n_1^2 - \boldsymbol{\varepsilon}_\perp)\cos\theta}{\boldsymbol{\varepsilon}_\perp \sin\theta}, \quad \frac{1}{n_1^2} = \frac{\sin^2\theta}{\boldsymbol{\varepsilon}_\parallel} + \frac{\cos^2\theta}{\boldsymbol{\varepsilon}_\perp}, \quad kR_0 \ll 1, \end{aligned}\right\} \tag{7.25}$$

where n_1 is the refractive index for the extra-ordinary wave which in the

---

† If the work $A_{\sim}$, or part of it, is positive, it corresponds to an amplification of the oscillations, as $A_{\sim}$ is the work done by the radiation force on the particle.

given case is just the one which is radiated. The quantity K_1 is the ratio of the components of the electric field strength vector in the extra-ordinary wave which are parallel to and at right angles to the vector **k**; the electric vector itself is parallel to the polarization vector $\mathbf{a}_1$ (see Chapter 6).

We can now obtain the expressions which correspond to Eqns.(7.19) and (7.23)

$$A = -\frac{e^2R_0^2T}{4c^3\beta_0}\int_{L_1+L_2}\omega^3\,\frac{\boldsymbol{\varepsilon}_\perp^2(\omega)\,\sin^2\theta\,|1-(\cot\theta/n_1)(\partial n_1/\partial\theta)|^{-1}}{[\boldsymbol{\varepsilon}_\perp(\omega)\sin^2\theta+\boldsymbol{\varepsilon}_\parallel\cos^2\theta]^2}\,d\omega\,,$$

$$A_\sim = \frac{e^2R_0^2\Omega T}{4c^3\beta_0}\Big\{-\int_{L_1}\omega^2\,\frac{\boldsymbol{\varepsilon}_\perp^2(\omega)\,\sin^2\theta\,|1-(\cot\theta/n_1)(\partial n_1/\partial\theta)|^{-1}}{[\boldsymbol{\varepsilon}_\perp(\omega)\sin^2\theta+\boldsymbol{\varepsilon}_\parallel(\omega)\cos^2\theta]^2}\,d\omega$$

$$+\int_{L_2}\omega^2\,\frac{\boldsymbol{\varepsilon}_\perp^2(\omega)\sin^2\theta\,|1-(\cot\theta/n_1)(\partial n_1/\partial\theta)|^{-1}}{[\boldsymbol{\varepsilon}_\perp(\omega)\sin^2\theta+\boldsymbol{\varepsilon}_\parallel(\omega)\cos^2\theta]^2}\,d\omega\Big\}\,. \qquad (7.26)$$

The integration domains L_1 and L_2 are here determined by the Doppler relations

$$1-\beta_0 n(\omega,\theta)\cos\theta = \Omega/\omega \qquad (7.27)$$

for normal Doppler frequencies (domain L_1) and

$$\beta_0 n(\omega,\theta)\cos\theta - 1 = \Omega/\omega \qquad (7.28)$$

for the anomalous Doppler frequencies (domain L_2). One sees easily that both integrals in Eqn.(7.26) for $A_\sim$ are always positive. This means that the radiation at the normal Doppler frequencies (first integral in (7.26)) corresponds to a damping of the oscillations, while the radiation at anomalous Doppler frequencies corresponds to an amplification of the oscillations.

We must note that such a subdivision is somewhat arbitrary and that, of course, only the force which is the difference of the two integrals has a physical meaning.

In contrast to the isotropic case, in the present problem the oscillations may not only be damped, but they can also be amplified — we are, of course, speaking about the sign of the whole of the work $A_\sim$ and not just about its parts. For instance, let $\boldsymbol{\varepsilon}_\parallel$ and $\boldsymbol{\varepsilon}_\perp$ be frequency-independent and let $\boldsymbol{\varepsilon}_\parallel < 0$, $\boldsymbol{\varepsilon}_\perp > 0$; in that case $n_1^2(\theta_\infty) \to \infty$ at an angle determined by the condition (see Eqn.(7.25) for n_1^2)

$$\boldsymbol{\varepsilon}_\perp \sin^2\theta_\infty + \boldsymbol{\varepsilon}_\parallel \cos^2\theta_\infty = 0 \qquad (7.29)$$

In such a medium the extra-ordinary waves can propagate at an angle $|\theta| < |\theta_\infty|$, but for angles $\frac{1}{2}\pi > \theta > \theta_\infty$ we have already $\tilde{n}_1^2 < 0$, and the waves cannot propagate. Moreover, n_1^2 is a minimum and equal to $\boldsymbol{\varepsilon}_\perp$ when $\theta = 0$. If now $\beta_0 \boldsymbol{\varepsilon}_\perp > 1$, we can always choose $\boldsymbol{\varepsilon}_\parallel$ such that the Cherenkov angle θ_0 is larger than θ_∞, where $\beta_0 n \cos\theta_0 = 1$. Clearly, under such conditions Cherenkov radiation is totally absent — the angle θ_0 corresponds to a value $\tilde{n}_1^2 < 0$ — and only anomalous Doppler waves are emitted in the forward direction — for $\theta < \frac{1}{2}\pi$, and in fact for $\theta < \theta_\infty$. In the backward direction — for $\pi < \theta < \theta_\infty$ — normal Doppler waves are emitted, but now $1 - \beta_0 n_1 \cos\theta = 1 + \beta_0 n_1 |\cos\theta|$ and the total work $A_\sim$ is positive. One can check this by using (7.27) and (7.28) to change in (7.26) to an integration over θ, as a result of which we get for the case considered

$$A_\sim = \frac{e^2\Omega^4 R_0^2 T}{4c^3\beta_0}\left\{\int_0^{\mathfrak{a}} \frac{n_1^5(\theta)\sin^3\theta\, d\theta}{\boldsymbol{\varepsilon}_\parallel^2 [\beta_0 n_1(\theta)\cos\theta - 1]^4} - \int_0^{\mathfrak{a}} \frac{n_1^5(\theta')\sin^3\theta'\, d\theta'}{\boldsymbol{\varepsilon}_\parallel^2 [\beta_0 n_1(\theta')\cos\theta' + 1]^4}\right\},$$

$$\mathfrak{a} \equiv \arc\tan\sqrt{|\boldsymbol{\varepsilon}_\parallel|/\boldsymbol{\varepsilon}_\perp}\,, \tag{7.30}$$

where $\theta' = \pi - \theta$. Here $A_\sim > 0$ by virtue of the fact that the first integral in (7.30) is always larger than the second one. In the case considered the oscillations are thus amplified.

Eidman (1960) has studied the problem of the motion of charges in a magnetized plasma and showed that in that case under well-defined conditions the oscillations are amplified or, to be more precise, that spiral lines along which the particle moves in the magnetic field are 'unwound'. Amplification thus occurs, for instance, when

$$\frac{\omega_p^2}{\omega_H^2} = \beta_0 \equiv \frac{v_0}{c} \ll 1\,, \quad \omega_H = \frac{eH_0}{mc}\,, \quad \omega_p^2 = \frac{4\pi N e^2}{m}\,,$$

where H_0 is the uniform magnetic field in which the plasma is situated and N the electron density in the plasma. Amplification also occurs for the following values of the parameters — which are found through numerical integration: $\beta_0 = 0.01$, $\omega_p^2/\omega_H^2 = 10$ and $\beta_0 = 0.99$, $\omega_p^2/\omega_H^2 = 10$. If on the other hand, we have, for instance, $\beta_0 = 0.99$ and $\omega_p^2/\omega_H^2 = 0.01$ the transverse motion of the particle is damped.

When the oscillations are amplified the energy of the translational motion (in this case the motion along the field) changes into energy in the transverse motion. As a result the translational velocity v_0 diminishes and amplification necessarily ceases when the velocity v_0 reaches the minimum light velocity c/n_{max} in the given medium.

The difference in the sign of the force acting upon the oscillatory motion of the particle in the case of the normal and anomalous Doppler radiation clearly agrees completely with the conclusion reached by using the conservation laws (vide supra). We have already stressed that in the isotropic case that difference leads to a weakening of the 'friction' or even its virtual vanishing but it cannot cause amplification of the vibrations of the oscillator; the quantal amplification of the vibrations which is connected with the spreading of the packet in 'energy space' occurs, of course, in the case of superluminal emission, also in an isotropic medium. Amplification of the oscillations is possible in an anisotropic medium and, in particular, in a magnetized plasma.

It is totally obvious that the instability of 'superluminal' particle beams which occurs in the classical approximation even in the case of an isotropic medium is closely connected with the radiation reaction of a single particle considered here.

We note also that we have here considered a medium at equilibrium or, at any rate, such a medium that the normal waves in it are absorbed when damping (conductivity) is taken into account, that is, that their amplitude diminishes as they propagate through the medium.

In media with a negative conductivity — which are sometimes called inverted media — the normal waves are amplified (maser effect) and the problem of the radiation reaction needs special consideration (Ginzburg and Eidman, 1963; Gavrilov and Kolomenskii, 1971, 1972). In that case the oscillations can be amplified even for subluminal motion ($v < c/n$), for instance in the case of the non-relativistic motion of an oscillator.

As plasma physics is at present in the centre of attention we shall briefly dwell here upon some aspects connected with the theory of radiation for superluminal velocities ($v > c/n$).

In an isotropic plasma, that is when there is no external magnetic field $\mathbf{H}_0$, we have for transverse waves

$$n_{1,2}^2 = 1 - \frac{4\pi N e^2}{m(\omega^2 + \nu_{eff}^2)} < 1$$

(the phase velocity of the waves $v_{ph} = c/n > c$), and, hence, Cherenkov radiation is impossible (see, however, Chapter 8; we assumed above that $v < c$). However, when we take into account thermal motion in an isotropic plasma, longitudinal plasma waves can propagate[†] which have a refractive index equal to (see Ginzburg, 1970b and Chapter 11)

$$n_3^2 = \frac{c^2k^2}{\omega^2} = \frac{1-\omega_p^2/\omega^2}{3\,\beta_T^2}\,,\quad \beta_T^2 = \frac{k_B T}{mc^2}\,,\quad \omega_p^2 = \frac{4\pi N e^2}{m}\,. \tag{7.31}$$

Here e and m are the electron charge and mass, N is the electron density, k_B Boltzmann's constant, and T the absolute temperature. Equation (7.31) is equivalent to the dispersion equation

$$\omega^2 = \omega_p^2 + 3\,\frac{k_B T}{m}\,k^2$$

and leads to the following expressions for the phase and group velocities:

$$v_{ph} = \frac{\omega}{k} = \frac{c}{n_3} = \left(\frac{3\,k_B T/m}{1-\omega_p^2/\omega^2}\right)^{\frac{1}{2}},$$

$$v_{gr} = \frac{d\omega}{dk} = \frac{3k_B T}{m\omega}\,k = \left[\frac{3k_B T}{m}\left(1-\frac{\omega_p^2}{\omega^2}\right)\right]^{\frac{1}{2}}. \tag{7.32}$$

The plasma waves form one of the three normal wave branches in a plasma which appear on equal footing. The phase velocity of the plasma waves can be less than the speed of light in vacuo c and hence the Cherenkov effect can occur for those waves for 'normal' motion of the source (particle) with a velocity $v < c$. Such radiation, indeed, occurs when charged particles move through a plasma — the energy which they lose as a result of 'distant' collisions just goes into the 'Cherenkov' emission of plasma waves. As a result of this radiation a particle of charge e and speed v which is appreciably above the thermal velocity $v_T = \sqrt{k_B T/m}$ loses per unit time an amount of energy (Pines and Bohm, 1952):

$$\frac{d\mathcal{E}}{dt} = -\,\frac{e^2\omega_p^2}{2v}\,\ln\left(1+\frac{2v^2}{v_T^2}\right). \tag{7.33}$$

One does not usually call the emission of plasma waves by a moving particle Cherenkov radiation. The question of terminology is, of course, not very important and, moreover, is a question of taste or habit. It seems to us, nonetheless, that in the case of the emission of plasma waves — in contrast, say, to the emission of sound — it would be very appropriate to speak just of the Cherenkov effect. Firstly, the high-frequency longitudinal (plasma) waves

---

† We neglect the motion of the ions and thus leave the quasi-acoustic (low-frequency) longitudinal waves out of our considerations (see Ginzburg, 1970b and Chapter 11). We also neglect the absorption due to collisions by putting the collision frequency equal to zero, $\nu_{eff} = 0$.

occur, as we have already noted, on equal footing with the electromagnetic (transverse) waves. Secondly, and this is more important, in a magnetized plasma — that is, when there is an external magnetic field $\mathbf{H}_0$ present — there occur in the general case three normal waves which are neither longitudinal nor transverse. The distinguishing of plasma waves in those circumstances becomes completely arbitrary (Ginzburg, 1970b); it is usually conventional also to split the waves emitted when a charge moves in a magnetized plasma into Cherenkov electromagnetic waves and plasma-type waves. Moreover, when one lets the external magnetic field $\mathbf{H}_0$ tend to zero (transition to the isotropic case) the emission of the waves which are called the Cherenkov waves does not vanish, but it goes continuously over into the above-mentioned emission of plasma waves (Ginzburg and Zheleznyakov, 1958a).

We note that what we have said refers not only to a gaseous plasma but also to other media in which we can in a well defined approximation talk about the propagation of plasma waves. In this respect optically anisotropic media (crystals) are the analogue of a magnetized plasma.

In solids and liquids the plasma frequencies $\omega_p = (4\pi N e^2/m)^{\frac{1}{2}}$ are very large — they lie in the ultra-violet range of the spectrum. It therefore has turned out to be important to take quantization into account and the above-mentioned concepts of the quanta of plasma waves — the plasmons — are introduced (Pines, 1963) which have an energy $\hbar\omega \approx \hbar\omega_p$ (we assume here that the medium is isotropic). It is clear that the difference between plasmons and electromagnetic field quanta in the medium — the 'photons in the medium' — corresponds merely to the difference between transverse and longitudinal waves (vide supra). In an anisotropic medium that difference does, in general, not exist and one is fully justified to consider the so-called discrete energy losses when electrons pass through thin layers (Pines, 1963) as the result of the Cherenkov effect (Ginzburg, 1958). When studying the discrete energy losses it turns out to be important to take also the momentum of the photons or plasmons into account (Pines, 1963).

In the case of a gaseous (rarefied) plasma when the frequencies ω_p are relatively low — we think here of the condition $\hbar\omega_p \ll \frac{1}{2}Mv^2$, $\hbar\omega_p \ll k_B T$, where M is the mass and v the velocity of the radiating particle, and T the plasma temperature — it is not necessary to use quantum language. However, also under those conditions, it may turn out to be very convenient and effective, as in the case of electromagnetic waves, to apply the quantum theory of emission and absorption of plasma waves and the plasmon concept itself. As an example we mention the calculation of the reabsorption of plasma waves and the establishment of a criterion of the stability of a particle beam moving in the plasma (Ginzburg and Zheleznyakov, 1959a, 1965; Ginzburg, Zheleznyakov and Eidman, 1962).

The instability of a beam arises if the perturbations (waves) which are formed in it grow. From the quantal point of view this means that the absorption coefficient for waves (quanta) in the wave must be negative (that is, $\mu < 0$) as occurs if, firstly, the particles in the beam in general can emit waves and, secondly, the velocity distribution function of the particles in the beam guarantees the prevalence of induced emission over absorption. A sufficiently fast particle ($v \gg \sqrt{k_B T/m}$) moving in an isotropic plasma emits, as we have mentioned, Cherenkov plasma waves. The absorption coefficient μ is negative, that is, there are more induced emission processes than absorption processes, if there are more particles in the beam at the 'upper' levels than at the 'lower' levels (see (7.7) and (7.8) and for details see Chapter 15). For particles without inner degrees of freedom, that is, for free electrons, protons, ... , or when we can neglect changes in the inner state the upper level corresponds simply to a larger velocity. From this it follows at once that beams are unstable in which in some range of velocities there are more fast than slow particles, that is, when the velocity distribution function of the particles in the beam $f_s(v)$ has a positive derivative. The same condition for instability $df_s/dv > 0$ is obtained also classically (Bohm and Gross, 1949), but as the result of a special investigation. The quantum method for obtaining a stability criterion is not less but in a certain sense much more effective — because of the greater complexity of the problem — for the above-mentioned case of the motion of a beam of charged particles in a magnetized plasma; in this case we must take into account the change in the velocity components of the particle at right angles to the magnetic field, or in quantum terms, take into account transitions between energy levels for the motion at right angles to the field, which is quantized (Zheleznyakov, 1959, 1970).

To illustrate what we have said we shall consider for the sake of simplicity the emission in the direction of the velocity $\mathbf{v}$ by a one-dimensional beam of particles; in the general case the function $f_s(v_k)$, where $v_k = v \cos\theta$ is the velocity component of the particles in the beam along the wavevector $\mathbf{k}$ of the emitted waves, plays the role of $f_s(v)$. For a distribution function of the kind

$$f_s(v_k) = \text{const} \times \exp\left\{-\frac{M}{2k_BT_s}(v_k - v_0\cos\theta)^2\right\},$$

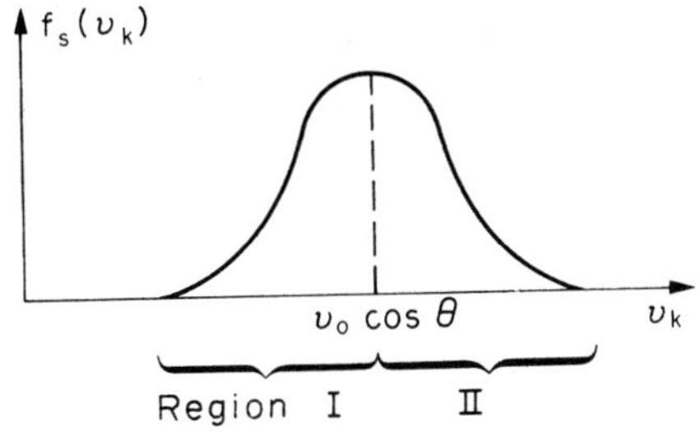

Fig.7.2 The distribution function $f_s(v_k)$ for particles in a beam

shown in Fig.7.2, the coefficient $\mu > 0$ in region II, where $df_s(v_k)/dv_k < 0$, while $\mu < 0$ in region I where $df_s(v_k)/dv_k > 0$. By virtue of the Cherenkov condition $v\cos\theta_0 = v_k = c/n_3(\omega)$ and waves emitted by particles with different values of v_k (in particular, values corresponding to the regions I and II in Fig.7.2) have different frequencies and hence cannot cancel one another and guarantee stability, even if $\mu < 0$ for only a small part of v_k. We note that for any isotropic three-dimensional electron velocity distribution $f = f(v^2)$ the function $f(v_k) = \int f(v^2)\, dv_\perp$, where $v_\perp$ is the velocity component along the direction perpendicular to **k** , does nowhere have a positive derivative and that the distribution is stable.

The possibility of the emission of Cherenkov waves leads, of course, also to the possibility that the particles can absorb the same kind of waves from whatever origin. Hence it is clear that apart from absorption of waves connected with collisions† in a plasma Cherenkov type absorption must occur. Thus, plasma waves must be absorbed also when there are no collisions. Of course, there is no such absorption for transverse waves in an isotropic plasma, as there is also no Cherenkov emission.‡

The necessary occurrence of absorption of plasma waves was already explained quite some time ago by completely different means (Landau, 1946) without including considerations of Cherenkov radiation.

In fact, let us consider the linearized kinetic equation for the plasma electrons (see, for instance, Ginzburg, 1970b and Chapter 11)

---

† Bremsstrahlung occurs when particles collide. The inverse process consists in the absorption of waves as a result of collisions.

‡ When we employ the usual Maxwellian velocity distribution we can formally conclude that there is a non-vanishing, albeit very weak absorption. Such a conclusion is, however, erroneous and is connected with the fact that the non-relativistic Maxwell distribution does not guarantee the complete absence of particles with velocities $v > c$.

$$\frac{\partial f_1}{\partial t} + (\mathbf{v}\cdot\nabla_{\mathbf{r}})f_1 + \frac{e}{m}(\mathbf{E}\cdot\nabla_{\mathbf{v}})f_0 = 0 \ , \ f = f_0 + f_1 \ , \ |f_0| \gg |f_1| \tag{7.34}$$

(we have neglected here collisions, and $f_0(\mathbf{v})$ is the zeroth-order distribution function, that is, in the case of equilibrium a Maxwellian distribution). Fourier transforms, that is, substituting $f_1(\mathbf{v},\mathbf{r},t) = g(\mathbf{v})\exp\{i(\mathbf{k}\cdot\mathbf{r}) - i\omega t\}$, then lead to the equation

$$i\{\omega - (\mathbf{k}\cdot\mathbf{v})\} f_1 = \frac{e}{m}(\mathbf{E}\cdot\nabla_{\mathbf{v}}) f_0 \ . \tag{6.35}$$

If $\omega \neq (\mathbf{k}\cdot\mathbf{v})$, we can divide by $\{\omega - (\mathbf{k}\cdot\mathbf{v})\}$ and obtain a well defined expression for f_1; if we then substitute f_1 into the field equation

$$\text{curl curl}\ \mathbf{E} + \frac{1}{c^2}\frac{\partial^2\mathbf{E}}{\partial t^2} = -\frac{4\pi}{c^2}\frac{\partial \mathbf{j}_t}{\partial t} \ , \ \mathbf{j}_t = e\int \mathbf{v} f_1\, d^3\mathbf{v} \ ,$$

we get a dispersion relation connecting ω with $\mathbf{k}$; we can write that equation in the form $c^2k^2/\omega^2 = n^2_{1,2,3}$, where $n_{1,2,3}$ are the refractive indexes used above for waves of the appropriate type — transverse waves ($n_{1,2}$) and longitudinal waves (n_3). If, however, $\omega = (\mathbf{k}\cdot\mathbf{v})$, we cannot divide Eqn.(7.35) by $\{\omega - (\mathbf{k}\cdot\mathbf{v})\}$ and one can show that a longitudinal wave which propagates in the plasma is damped (Landau, 1946). However, the condition

$$\omega = (\mathbf{k}\cdot\mathbf{v}) = \frac{\omega n v}{c}\cos\theta \ , \ k^2 = \frac{\omega^2 n^2}{c^2} \tag{7.36}$$

is just the Cherenkov condition (6.53), (6.55). As we said, in an isotropic plasma it can be satisfied only for plasma waves for which the absorption in the given case is precisely the inverse Cherenkov effect — the wave is then weakened and the plasma electrons with velocities satisfying condition (7.36) obtain an extra energy.†

When there is an external magnetic field present the emission of waves — in a magnetized plasma — occurs as a result of collisions (bremsstrahlung), by virtue of the Cherenkov effect, and thanks to the acceleration of the particle in the magnetic field (magneto-bremsstrahlung). Corresponding to these there are three absorption mechanisms. One should, however, state that the division of emission and absorption into Cherenkov and magneto-brems processes is somewhat arbitrary. We know that a particle (electron) in a magnetic field moves along a spiral, revolving with a frequency $\omega_H^* = \omega_H mc^2/\mathcal{E} = (eH_0/mc)(mc^2/\mathcal{E})$,

---

† The physical interpretation of the damping of the plasma waves which is not connected with collisions was essentially given already by Bohm and Gross (1949), but they did not refer explicitly to Cherenkov radiation.

where $\mathcal{E}$ is the total energy. In vacuo such motion leads to radiation with frequencies $s\omega_H^*$ ($s = 1, 2, \ldots$; we have neglected here the Doppler shift of the frequency for the sake of simplicity). However, when there is a plasma, the nature of the radiation — its intensity, directionality, and polarization — is changed and apart from the frequencies $s\omega_H^*$ radiation may occur with a continuous spectrum which is clearly Cherenkov radiation (when the particle moves strictly along the field magneto-brems radiation is completely missing). Moreover, when the particle moves, for instance, along a circle in the plane at right angles to the field $\mathbf{H}_0$ only discrete frequencies $s\omega_H^*$ are emitted, that is, we are only dealing with magneto-bremsstrahlung, to use the terminology employed here. However, it is clear physically that also in that case the radiation spectrum is practically continuous provided the radius of the circle is sufficiently large and $\mathcal{E}/mc^2 \gg 1$, and its nature is in the appropriate frequency range close to the spectrum of the Cherenkov radiation. In view of what we have said the only consistent method is in the general case to consider magneto-brems and Cherenkov radiation and absorption in a unified way (Eidman, 1958, 1962).

Let us dwell upon the defermination of the frequencies emitted (and absorbed) in a magnetized plasma in somewhat more detail. To do this we write down the equation of the field amplitudes $q_{\lambda j}$ which we introduce earlier (see (7.12)),

$$\ddot{q}_{\lambda j} + \omega_{\lambda j}^2 q_{\lambda j} = \sqrt{4\pi}\,\frac{e}{n_{\lambda j}}\,(\mathbf{v}\cdot\mathbf{a}_{\lambda j}^*)\exp\{-i(\mathbf{k}_\lambda\cdot\mathbf{R})\} \equiv f(t)\,, \tag{7.37}$$

where $\omega_{\lambda j}^2 = c^2k_\lambda^2/n_{\lambda j}^2$, while $\mathbf{R}(t)$ and $\mathbf{v} = d\mathbf{R}/dt$ are the radius vector and velocity of the radiating particle.

Equation (7.37) is obtained by substituting the expansion (7.12) into Eqn. (7.14) for the vector potential, multiplying that equation by $\mathbf{a}_{\lambda j}^*\exp\{-i(\mathbf{k}_\lambda\cdot\mathbf{r})\}$ and integrating over space. If we disregard a constant factor the form of the 'force' $f(t)$ in (7.37) is at once clear, as

$$\int(\mathbf{j}_e\cdot\mathbf{a}_{\lambda j}^*)\exp\{-i(\mathbf{k}_\lambda\cdot\mathbf{r})\}\,d^3\mathbf{r} = e(\mathbf{v}\cdot\mathbf{a}_{\lambda j}^*)\exp\{-i(\mathbf{k}_\lambda\cdot\mathbf{R})\}$$

when $\mathbf{j}_e = e\mathbf{v}\,\delta(\mathbf{r}-\mathbf{R})$ (see (7.14)).

Equation (7.37) has a solution for $q_{\lambda j}$ which grows with time which corresponds to radiation at only the frequencies $\omega_{\lambda j}$ which are represented in the spectrum of the 'force' $f(t)$. If, for instance, the electron moves uniformly, we have $\mathbf{R} = \mathbf{v}t$ and only the frequency $\omega = (\mathbf{k}\cdot\mathbf{v})$ occurs in the spectrum of the 'force' f. The condition for radiation thus takes the form $\omega_{\lambda j} = \omega = (\mathbf{k}\cdot\mathbf{v})$, that is, we get at once the Cherenkov condition (7.36) as we mentioned already in Chapter 6.

For an electron in a magnetic field $\mathbf{H}_0$ along the z-axis we have

$$\left.\begin{aligned} \mathbf{R} &= \left\{R_0 \cos \omega_H^* t\,,\, R_0 \sin \omega_H^* t\,,\, v_z t\right\}, \\ \mathbf{v} &= \left\{-v_\perp \sin \omega_H^* t\,,\, v_\perp \cos \omega_H^* t\,,\, v_z\right\}, \quad v_\perp = R_0 \omega_H^*, \\ f(t) &= \text{const} \times \left(-a_x^* v_\perp \sin \omega_H^* t + a_y^* v_\perp \cos \omega_H^* t + a_z^* v_z\right) \times \\ &\quad \times \exp\left\{-i\left(kR_0 \sin\alpha \sin \omega_H^* t + k v_z t \cos\alpha\right)\right\}, \end{aligned}\right\} \tag{7.38}$$

where, for the sake of simplicity, we have put $k_x = 0$, and α is the angle between $\mathbf{k}$ and $\mathbf{H}_0$ (the z-axis). Using the expansion of a plane wave in Bessel functions,

$$\exp\left\{-i\, k_\lambda R_0 \sin\alpha \sin \omega_H^* t\right\} = \sum_{s=-\infty}^{+\infty} J_s(k_\lambda R_0 \sin\alpha) \exp(-is\omega_H^* t)$$

we easily get the resonance condition

$$\omega = \left|s\,\omega_H^* + k v_z \cos\alpha\right| \;;\; s = 0, \pm 1, \pm 2, \pm 3, \ldots . \tag{7.39}$$

For $s = 0$ this condition is exactly the Cherenkov condition (7.36) with $v = v_z$; at the same time all terms with $s \neq 0$ are absent only when the motion is strictly along the field, when $R_0 = 0$. When $s \neq 0$ we can write instead of (7.39) $(n = ck/\omega)$

$$\omega = \frac{s\,\omega_H^*}{1 - (v_z n/c)\cos\alpha}\,,\; s > 0 \;;\; \omega = \frac{s\,\omega_H^*}{(v_z n/c)\cos\alpha - 1}\,,\; s < 0\,, \tag{7.40}$$

where, as everywhere before, the frequency is positive.

If the velocity $v_\perp \ll v_z = v\cos\theta$, the electron in the magnetic field radiates like two appropriately chosen dipoles, moving along the field with a velocity $v_z \approx v$; to this case correspond the values $s = \pm 1$ — to be more precise, the intensity of the higher overtones is small, provided

$$kR_0 \sin\alpha = (\omega n/c)(v_\perp/\omega_H^*) \sin\alpha \ll 1.$$

Equations (7.40) are for $s = \pm 1$ essentially the same as Eqn.(6.59) for the Doppler effect in a medium; clearly for the motion considered in a magnetic field

$$\omega_{00}(1 - v^2/c^2)^{\frac{1}{2}} = \omega_{00} mc^2/\mathcal{E} = \omega_H mc^2/\mathcal{E} = \omega_H^*,$$

as ω_H is the frequency in the system in which the centre of mass of the radiator is at rest.

Changing from emission to absorption we see that waves with the frequencies (7.39) must be absorbed in a magnetized plasma; these frequencies correspond to magneto-brems and Cherenkov radiation (including the Doppler effect). We note that one can arrive at the same result (Gertsenstein, 1954) by considering for the motion of an electron in a magnetic field the frequency spectrum of the force acting upon that electron in the field of the wave; the frequency of the force is not equal to the frequency of the field **E**, as the electron moves around and is at different times in a field with a different field strength.

Above we considered only the conditions for emission and absorption. The calculation of the radiative intensity and of the absorption coefficient is yet an independent and sometimes very complicated problem. It is solved either by the kinetic equation method, or by other methods. A summary of the corresponding results is given elsewhere (Ginzburg, 1970b; see also Chapter 11). It is, perhaps, not superfluous to note here that the absorption of waves in a magnetic field which is not connected with collisions plays an important role and not only at very high temperatures — in apparatus connected with thermonuclear reactions — but also, for instance, in the solar corona where the temperature $T \sim 10^6$ °K (see Zheleznyakov, 1970; Ginzburg, 1970b).

One usually considers only Cherenkov radiation of point charges or charged bunches (packets). At the same time it is, of course, fully obvious that Cherenkov radiation can be emitted by any source moving with a velocity v larger than the phase velocity of light in the medium $v_{ph} = c/n$. In other words, the condition (6.53) for emission is retained also for any multipole, in particular, for an electrical or magnetic dipole (for references to the literature see Jelley, 1958; Ginzburg, 1960, Bolotovskii, 1960, 1962; Zrelov, 1968); however, the intensity of the radiation is appreciably changed and already for a dipole — not to mention the higher-order multipoles — it is usually considerably lower than for a charge. For instance, for $v \sim c$ and $n \sim 1$ the intensity of the radiation of an electrical dipole with moment $p = ed$ is, as to order of magnitude, smaller than the intensity of the radiation of a charge e by a factor $p^2\omega^2/e^2c^2 \sim (d/\lambda)^2$; in the case of a magnetic dipole μ this ratio is of the order $\mu^2\omega^2/e^2c^2$ (the appearance of the factor $(d/\lambda)^2$ can be particularly easily understood by considering the dipole to be two charges $+e$ and $-e$ at a distance d; see Frank, 1952).

For elementary particles, such as electrons, neutrons, ..., or for atomic nuclei, magnetic dipole Cherenkov radiation is very weak and of no interest.

The position is changed when we consider bunches of particles which under well defined circumstances radiate like point particles with a charge and multipole moments corresponding to the whole bunch. Such a situation can just occur when bunches or current rings move in a magnetized plasma or along the axis of a channel or a gap, or also close to a system, which retards the electromagnetic waves, such as a waveguide, and so on. Moreover, the calculation of the intensity of the Cherenkov radiation of a dipole is a well known methodical process which is, in particular, used to reach some conclusions about the magnetic moment of particles with different spins (see Ginzburg, 1960 for references). Even in the problem of the Cherenkov radiation of a magnetic dipole there were for a long time some points of lack of clarity. Finally, it is rather interesting to elucidate how the Cherenkov radiation of a dipole is altered when it moves in a channel or a gap. For these reasons we shall dwell here for a moment upon the Cherenkov radiation of dipoles and we shall follow earlier expositions (Ginzburg and Eidman, 1959a; Ginzburg, 1960).

Let us consider a point particle of charge e, electrical dipole moment $\mathbf{p}$ and magnetic moment $\boldsymbol{\mu}$ moving with a velocity $\mathbf{v} = \text{const}$. The current density connected with the particle is then equal to (ρ_e is the charge density, $\mathfrak{M}$ the magnetization, and $\mathbf{P}$ the polarization; see also (6.20))

$$\mathbf{j} = \rho_e \mathbf{v} + c\,\text{curl}\,\mathfrak{M} + \frac{\partial \mathbf{P}}{\partial t} = e\mathbf{v}\delta(\mathbf{r}-\mathbf{v}t) + c\,\text{curl}\left\{\boldsymbol{\mu}\delta(\mathbf{r}-\mathbf{v}t)\right\} + \frac{\partial}{\partial t}\left\{\mathbf{p}\delta(\mathbf{r}-\mathbf{v}t)\right\}. \quad (7.41)$$

For the sake of simplicity we shall assume that the medium is isotropic and non-magnetic (the magnetic permeability $\mu = 1$; do not confuse the magnetic moment $\boldsymbol{\mu}$ and the permeability μ!). Assuming that the vector potential $\mathbf{A}$ satisfies the condition $\text{div}\,\mathbf{A} = 0$ we then get the equations (compare, for instance, (7.14) with $\boldsymbol{\varepsilon}_{\alpha\beta} = \boldsymbol{\varepsilon}\delta_{\alpha\beta}$ and Chapter 6).

$$\nabla^2\mathbf{A} - \frac{\boldsymbol{\varepsilon}}{c^2}\frac{\partial^2\mathbf{A}}{\partial t^2} = -\frac{4\pi}{c}\mathbf{j} + \frac{\boldsymbol{\varepsilon}}{c}\frac{\partial}{\partial t}\,\text{grad}\,\phi\ , \quad \nabla^2\phi = -\frac{4\pi\rho}{\boldsymbol{\varepsilon}}, \quad (7.42)$$

$$\left.\begin{aligned} &\mathbf{A} = \sum_{\lambda,j}\left(q_{\lambda j}\mathbf{A}_{\lambda j} + q^*_{\lambda j}\mathbf{A}^*_{\lambda j}\right), \quad \mathbf{A}_{\lambda j} = c\left(\frac{4\pi}{\boldsymbol{\varepsilon}}\right)^{\frac{1}{2}}\mathbf{a}_{\lambda j}\exp\left\{i\left(\mathbf{k}_\lambda\cdot\mathbf{r}\right)\right\}, \\ &(\mathbf{a}_{\lambda i}\cdot\mathbf{a}_{\lambda j}) = \delta_{ij}\ , \quad (\mathbf{k}_\lambda\cdot\mathbf{a}_{\lambda j}) = 0\ , \quad i, j = 1, 2\ , \\ &\mathcal{H}_{tr} = \int\frac{\boldsymbol{\varepsilon}E^2_{tr} + H^2}{8\pi}\,d^3\mathbf{r} = \sum_{\lambda,j}\left(p_{\lambda j}p^*_{\lambda j} + \omega^2_{\lambda j}q_{\lambda j}q^*_{\lambda j}\right), \end{aligned}\right\} \quad (7.43)$$

$$\left.\begin{aligned} \omega_{\lambda j}^2 &= \omega_\lambda^2 = \frac{e^2 k_\lambda^2}{\boldsymbol{\varepsilon}}, \quad \mathbf{E}_{tr} = -\frac{1}{c}\frac{\partial \mathbf{A}}{\partial t}, \\ \mathbf{H} &= \text{curl}\,\mathbf{A}, \quad p_{\lambda j} = \frac{dq_{\lambda j}}{dt} \equiv \dot{q}_{\lambda j}. \end{aligned}\right\} \tag{7.43}$$

Clearly, ϕ is here the scalar potential, $\mathcal{H}_{tr}$ the energy of the transverse field, and the $\mathbf{a}_{\lambda j}$ are the polarization vectors. Substituting expression (7.41) into (7.42) and integrating over space after multiplying by $\mathbf{A}^*_{\lambda j}$ we find

$$\begin{aligned} \ddot{q}_{\lambda j} + \omega_\lambda^2 q_{\lambda j} &= \frac{1}{c}\int (\mathbf{j}\cdot\mathbf{A}^*_{\lambda j})\, d^3\mathbf{r} \\ &= \left(\frac{4\pi}{\boldsymbol{\varepsilon}}\right)^{\frac{1}{2}} \Big\{ e\,(\mathbf{a}_{\lambda j}\cdot\mathbf{v}) - ic(\mu\cdot[\mathbf{k}_\lambda \wedge \mathbf{a}_{\lambda j}]) \\ &\quad - i(\mathbf{a}_{\lambda j}\cdot\mathbf{p})(\mathbf{k}_\lambda\cdot\mathbf{v})\Big\} \exp\left\{-i(\mathbf{k}_\lambda\cdot\mathbf{v})\,t\right\}. \end{aligned} \tag{7.44}$$

We can find the energy $\mathcal{H}_{tr}$ by integrating Eqn.(7.44), for instance, with the initial conditions $q_{\lambda j}(0) = \dot{q}_{\lambda j}(0) = 0$. This energy contains a part which increases with time and is connected with the appearance of resonance at the Cherenkov condition $\omega_\lambda = (\mathbf{k}\cdot\mathbf{v})$. The part of $\mathcal{H}_{tr}$ which increases with time, and that is the only one to be discussed in what follows, is independent of the initial conditions and can easily be evaluated by introducing the density of states

$$dZ_i(\omega) = \frac{\boldsymbol{\varepsilon}^{\frac{3}{2}}\omega^2\, d\omega\, d^2\Omega}{(2\pi c)^3}$$

and integrating over the angle θ between $\mathbf{k}$ and $\mathbf{v}$; in that case $d^2\Omega = \sin\theta\, d\theta\, d\phi$. All these operations were already performed in Chapters 1 and 6.

It is clear from (7.44) that the radiation of the charge e and that of the dipoles with moments $\mathbf{p}$ and $\boldsymbol{\mu}$ are shifted in phase by $\frac{1}{2}\pi$, and there is therefore no interference between the radiations from the charge and from the dipoles. In other words, the energy emitted per unit time is thus equal to the sum of expression (6.58) for a charge and the expression for the energy of the Cherenkov emission by dipoles

$$\frac{d\mathcal{H}_{tr}}{dt} \equiv \frac{dW}{dt} = \frac{1}{2\pi v c^2} \sum_{j=1,2} \int d\omega \int_0^{2\pi} n^2\omega^3 \left\{\left(\boldsymbol{\mu}\cdot[\mathbf{s}\wedge\mathbf{a}_j]\right) + \frac{1}{n}(\mathbf{a}_j\cdot\mathbf{p})\right\}^2 d\phi, \tag{7.45}$$

where $n^2(\omega) = \boldsymbol{\varepsilon}(\omega)$ is the dielectric permittivity of the medium,

$$\cos\theta = \cos\theta_0 = c/n(\omega)v, \quad \mathbf{s} = \mathbf{k}/k,$$

and θ and ϕ are the polar and azimuthal angles in the system of coordinates with the z-axis along the direction of the velocity $\mathbf{v}$. The integration over the frequency in (7.45) is over the region where $vn(\omega)/c \geq 1$. We emphasized in Chapter 6 that this calculation takes into account the dispersion, that is, the ω-dependence of n, although this is not immediately obvious.

For a magnetic dipole of moment $\boldsymbol{\mu}$ directed along the velocity we get from (7.45) (Ginzburg, 1940a)

$$\frac{dW}{dt} = \frac{\mu^2}{vc^2} \int n^2 \left(1 - \frac{c^2}{v^2 n^2}\right) \omega^3 \, d\omega \, . \tag{7.46}$$

We also get at once from (7.45) for an electrical dipole the already well known expressions (Frank, 1942, 1952, 1959). The above mentioned lack of clarity arose when one considers a magnetic dipole moving at right angles to the axis of the dipole. If in the rest frame of the particle this moment is equal to $\boldsymbol{\mu}_0$ and $\mathbf{p}_0 = 0$, in the laboratory frame we have the well known relations $\boldsymbol{\mu} = \boldsymbol{\mu}_0$ and $\mathbf{p} = \frac{1}{c}[\mathbf{v} \wedge \boldsymbol{\mu}]$. For that case we find, when $\mathbf{v} \perp \boldsymbol{\mu}$

$$\frac{dW}{dt} = \frac{\mu^2}{2vc^2} \int n^2 \omega^3 \left\{2\left(1 - \frac{1}{n^2}\right)^2 - \left(1 - \frac{v^2}{n^2 c^2}\right)\left(1 - \frac{c^2}{n^2 v^2}\right)\right\} d\omega \, . \tag{7.47}$$

This expression is the same as the one obtained by Frank (1942), but differs from the result of other calculations. Frank (1952), for instance, found instead of Eqn. (7.47) the relation

$$\frac{dW}{dt} = \frac{\mu^2 v}{2c^4} \int n^4 \omega^3 \left(1 - \frac{c^2}{n^2 v^2}\right)^2 d\omega \, . \tag{7.48}$$

The cause of this disagreement lies in the fact that Frank (1952) and some other authors used 'true' magnetic dipoles formed from magnetic poles; the calculation was first performed for magnetic poles from which the dipole was then formed. However, moving 'true' magnetic dipoles are equivalent to a current moment only in vacuo. Indeed, when one uses magnetic poles with a density $\rho_m(\mathbf{r})$ the field equations have the form (see, e.g., Vainshstein, 1957)

$$\left.\begin{aligned} \operatorname{curl} \mathbf{H} &= \frac{1}{c}\frac{\partial \varepsilon \mathbf{E}}{\partial t}, & \operatorname{div} \varepsilon \mathbf{E} &= 0, \\ \operatorname{curl} \mathbf{E} &= -\frac{1}{c}\frac{\partial \mu \mathbf{H}}{\partial t} - \frac{4\pi}{c}\rho_m \mathbf{v}, & \operatorname{div} \mu \mathbf{H} &= 4\pi\rho_m, \end{aligned}\right\} \tag{7.49}$$

where we have assumed that $\rho = 0$, $\mathbf{j} = 0$, and $\mathbf{B} = \mu \mathbf{H}$ (we remind ourselves once more that one must distinguish the same symbol μ for the magnetic permeability and for the magnetic moment).

Hence we have

$$\operatorname{curl}\operatorname{curl}\mathbf{H} + \frac{\varepsilon\mu}{c^2}\frac{\partial^2\mathbf{H}}{\partial t^2} = -\frac{4\pi}{c}\,\varepsilon\,\frac{\partial(\rho_m\mathbf{v})}{\partial t},$$

$$\operatorname{curl}\operatorname{curl}\mathbf{E} + \frac{\varepsilon\mu}{c^2}\frac{\partial^2\mathbf{E}}{\partial t^2} = -\frac{4\pi}{c}\operatorname{curl}(\rho_m\mathbf{v}).$$

When electric charges and current are present, but $\rho_m = 0$, we have on the other hand

$$\operatorname{curl}\operatorname{curl}\mathbf{H} + \frac{\varepsilon\mu}{c^2}\frac{\partial^2\mathbf{H}}{\partial t^2} = \frac{4\pi}{c}\operatorname{curl}(\rho\mathbf{v}),$$

$$\operatorname{curl}\operatorname{curl}\mathbf{E} + \frac{\varepsilon\mu}{c^2}\frac{\partial^2\mathbf{E}}{\partial t^2} = -\frac{4\pi}{c}\,\mu\,\frac{\partial(\rho\mathbf{v})}{\partial t}.$$

The equations for magnetic poles are thus obtained from those for charges through the substitutions

$$\mathbf{E}\to\mathbf{H},\quad \mathbf{H}\to-\mathbf{E},\quad \rho\to\rho_m,\quad \mu\to\varepsilon.$$

The current moment is thus for $\mu = 1$, indeed, equivalent to the 'true' magnetic moment only in vacuo, when $\varepsilon = 1$. However, in a medium with $\varepsilon \neq 1$ a moving 'true' magnetic moment possesses an electric moment equal to $\frac{\varepsilon}{c}[\mathbf{v}\wedge\boldsymbol{\mu}]$, and not to $[\mathbf{v}\wedge\boldsymbol{\mu}]/c$. Such a substitution is equivalent to taking into account the electrical polarization of the medium produced by the dipole itself (Ginzburg, 1952b). In other words, a 'true' magnetic dipole is equivalent to a current moment 'manufactured' from material with a permittivity ε and therefore polarized. It is interesting that such a situation may exist for bunches; for this it is necessary, as we have remarked, that for the frequency considered the dielectric permittivity ε in the bunch itself be equal to the permittivity of the surrounding medium (for instance, a plasma in a magnetic field).

If we use the Pauli or Dirac equations (and also the equations for particles with spin 1, $\frac{3}{2}$, and 2) for a quantummechanical calculation (see Ginzburg, 1940a, and Ginzburg 1960 for references), we find, in particular, equations such as (7.46) and (7.47). If then the spin is parallel or antiparallel to the velocity $\mathbf{v}$, an expression such as (7.47) is obtained only for transitions with spin-flip, as only for such a flip the components of the spin operator at right angles to $\mathbf{v}$ are important. In fact, the Cherenkov radiation of a magnetic dipole does therefore not possess any specifically quantal features.

A characteristic feature of Eqn.(7.47) as compared to (6.58) and (7.46) consists in the fact that the integrand in it does not vanish at the threshold, when $\cos\theta_0 = c/nv = 1$, when

$$\frac{dW}{dt} = \frac{\mu^2}{vc^2}\int n^2\omega^3\left(1-\frac{1}{n^2}\right)d\omega\,. \qquad (7.47a)$$

One cannot, however, consider this result to be paradoxical as the power W itself at the threshold vanishes and beyond that increases smoothly. Indeed, when the velocity increases the radiation starts at a frequency corresponding to the maximum value of $n(\omega)$, when dispersion is taken into account. Moreover, when recoil is taken into account, which is done automatically in a quantal calculation, one obtains Eqn. (7.3) in which in the case of a bunch the mass of the whole bunch clearly plays the role of the mass m. By virtue of (7.3) even when $n = \text{const}$ the radiation starts with increasing velocity $\mathbf{v}$ at a single frequency, in this case at $\omega = 0$; the region of integration and the radiation power (7.47) itself therefore increase gradually with increasing velocity v.

We note that in a quantal calculation one can also obtain an expression such as (7.48); to do that one must add to the Dirac equation for a charged particle an additional term proportional to $\gamma_i\gamma_k G_{ik}$, and for a particle with a non-kinematic magnetic moment replace $\gamma_i\gamma_k F_{ik}$ by $\gamma_i\gamma_k H_{ik}$, where $F_{ik} = \{\mathbf{H}, i\mathbf{E}\}$, $H_{ik} = \{\mathbf{H}, i\mathbf{D}\}$, $G_{ik} = F_{ik} - H_{ik}$, and the γ_i are the well known Dirac matrices. However, there is no basis for introducing these changes for applications to separate particles and it would have no sense to apply such a quantal calculation to the case of bunches.

To conclude the exposition of the theory of the Cherenkov radiation we consider the case of the motion of a source (charge, dipole) in channels or gaps; for the sake of simplicity we assume that in the channel or gap $\varepsilon = 1$ and $\mu = 1$. This problem is important, firstly, for exploring the possibility to reduce ionization losses which, roughly speaking, are concentrated in the immediate vicinity of particle trajectories. At the same time, Cherenkov radiation is produced in a region of dimensions of the order of the wavelength $\lambda = 2\pi c/\omega n = \lambda_0/n$. Secondly, the consideration of radiation in channels and gaps is of some methodological interest.

It follows from the calculations (see Ginzburg and Frank, 1947b, and below) that in the case of Cherenkov radiation the intensity of the radiation for $a/\lambda \ll 1$ is the same as in a continuous medium (a is the radius of the channel or the width of the gap). One obtains this result from the above-mentioned intuitive idea that a channel or gap with dimensions $a \ll \lambda$ should not affect the radiation produced in a region with dimensions of the order of λ. In fact, however, such a conclusion is for dipoles and other multipoles true only in particular cases.

To evaluate the effect of thin channels or gaps on Cherenkov radiation it is convenient to use the reciprocity theorem

$$\int_{(1)} \left(\mathbf{j}_\omega^{(1)} \cdot \mathbf{E}_\omega^{(2)}\right) d^3\mathbf{r} = \int_{(2)} \left(\mathbf{j}_\omega^{(2)} \cdot \mathbf{E}_\omega^{(1)}\right) d^3\mathbf{r}\,, \tag{7.50}$$

where $\mathbf{j}_\omega^{(1,2)} \equiv \mathbf{j}^{(1,2)}(\omega)$ are the Fourier components of the external current density in the regions 1 and 2; the field $\mathbf{E}_\omega^{(2)}$ is produced by the current 2 in the region 1 and the field $\mathbf{E}_\omega^{(1)}$ by the current 1 in the region 2 (see, for instance, Vainshtein, 1957; Ginzburg, 1970b).†

If we write the current in the form $\mathbf{j} = \rho_e \mathbf{v} + c\,\mathrm{curl}\,\mathfrak{m} + \partial\mathbf{P}/\partial t$, we get

$$\int_{(1)} \left[\left(\{\rho_e \mathbf{v}\}_\omega^{(1)} \cdot \mathbf{E}_\omega^{(2)}\right) - i\omega\left\{\left(\mathbf{P}_\omega^{(1)} \cdot \mathbf{E}_\omega^{(2)}\right) - \mu_1\left(\mathfrak{m}_\omega^{(1)} \cdot \mathbf{H}_\omega^{(2)}\right)\right\}\right] d^3\mathbf{r}$$

$$= \int_{(2)} \left[\left(\{\rho_e \mathbf{v}\}_\omega^{(2)} \cdot \mathbf{E}_\omega^{(1)}\right) - i\omega\left\{\left(\mathbf{P}_\omega^{(2)} \cdot \mathbf{E}_\omega^{(1)}\right) - \mu_2\left(\mathfrak{m}_\omega^{(2)} \cdot \mathbf{H}_\omega^{(1)}\right)\right\}\right] d^3\mathbf{r}\,, \tag{7.51}$$

where $\mu_{1,2}$ is the magnetic permeability of the medium in the points 1,2. In the case of Cherenkov radiation by a point charge moving along the z-axis

$$(\rho_e \mathbf{v})_\omega^{(1)} = \frac{e}{2\pi}\mathbf{v}\exp(i\omega z/v)\,\delta(x)\,\delta(y)\,, \tag{7.52}$$

and if we put an electric dipole at some point 2 well away from the trajectory, with moment $\mathbf{p}^{(2)} = \int \mathbf{P}^{(2)}\, d^3\mathbf{r}$ we have

$$\frac{e}{2\pi}\int \left(\mathbf{v} \cdot \mathbf{E}^{(2)}(0,0,z)\right)\exp\left(i\frac{\omega}{v}z\right) dz = -\,i\omega\left(\mathbf{p}^{(2)} \cdot \mathbf{E}(2)\right), \tag{7.53}$$

where $\mathbf{E}(2) \equiv \mathbf{E}^{(1)}(2)$ is the radiation field at the point 2 in which we are interested (we have dropped the index ω).

The quantity $(\mathbf{v} \cdot \mathbf{E}^{(2)}(0,0,z))$ remains the same as for a continuous medium when a charge moves in a thin channel or a narrow gap, that is, when $a/\lambda \ll 1$, as the tangential components of the field $\mathbf{E}^{(2)}$ are continuous. Therefore the radiation field also remains the same as in the case of a continuous medium, as is clear from (7.53).

† In the given form the reciprocity theorem is valid for any linear non-moving medium, provided there is no external magnetic field present. When there is a magnetic field present, so that the tensors ε_{ij} and μ_{ij} are asymmetric, only a generalized reciprocity theorem is valid (see Ginzburg, 1970b, § 29).

For a radiating electric dipole $\mathbf{P}^{(1)} = \mathbf{p}\,\delta(x)\,\delta(y)\,\delta(z-vt)$ we have

$$\frac{1}{2\pi}\int\left(\mathbf{p}\cdot\mathbf{E}^{(2)}(0,0,z)\right)\exp\left(i\frac{\omega}{v}z\right)dz = \left(\mathbf{p}^{(2)}\cdot\mathbf{E}(2)\right). \tag{7.54}$$

If the dipole with moment $\mathbf{p}=\mathbf{p}^{(1)}$ is parallel to the axis of the channel or lies in the plane of the gap, the radiation field again remains the same as for a continuous medium when $a/\lambda \ll 1$. As the component of $\mathbf{D} = \boldsymbol{\varepsilon}\mathbf{E}$ normal to the dividing boundary is continuous we have for a dipole at right angles to the plane of the gap

$$\left(\mathbf{p}\cdot\mathbf{E}^{(2)}(0,0,z)\right) = \varepsilon(\omega)\left(\mathbf{p}\cdot\mathbf{E}_0^{(2)}(0,0,z)\right), \tag{7.55}$$

where $\mathbf{E}_0^{(2)}$ is the field produced by dipole 2 in the continuous medium (Fig. 7.3). If the field of the Cherenkov radiation of dipole 1, with moment $\mathbf{p}^{(1)}=\mathbf{p}$, in the continuous medium is denoted by $\mathbf{E}_0$, we have from the reciprocity theorem

$$\frac{1}{2\pi}\int\left(\mathbf{p}\cdot\mathbf{E}_0^{(2)}\right)\exp\left(\frac{i\omega z}{v}\right)dz = \left(\mathbf{p}^{(2)}\cdot\mathbf{E}_0(2)\right). \tag{7.56}$$

When there is a gap present, we have by virtue of (7.55) and the reciprocity theorem

$$\frac{1}{2\pi}\int\left(\mathbf{p}\cdot\mathbf{E}_0^{(2)}\right)\exp\left(\frac{i\omega z}{v}\right)dz =$$
$$= \frac{1}{2\pi}\varepsilon\int\left(\mathbf{p}\cdot\mathbf{E}_0^{(2)}\right)\exp\left(\frac{i\omega z}{v}\right)dz = \left(\mathbf{p}^{(2)}\cdot\mathbf{E}(2)\right). \tag{7.57}$$

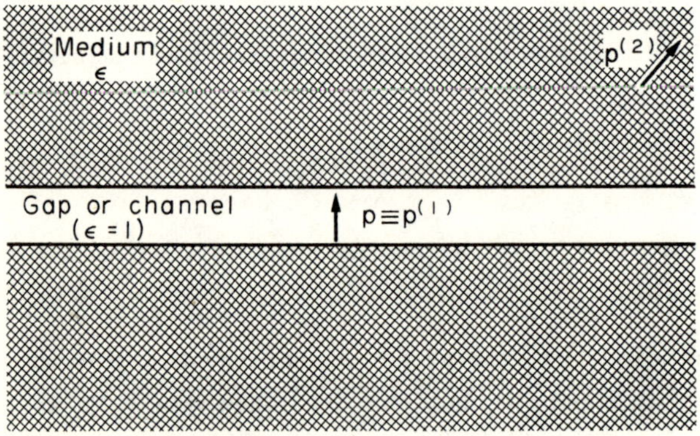

Fig.7.3 Radiation of a dipole in a channel or a gap

From the last two relations it follows that the field of the Cherenkov radiation $\mathbf{E}=\varepsilon\mathbf{E}_0$, that is, it is larger by a factor $\boldsymbol{\varepsilon}$ than for a dipole moving in a continuous medium. For a dipole directed at right angles to the axis of a thin channel which has the shape of a circular cylinder, $\mathbf{E} = (2\varepsilon/(\varepsilon+1))\,\mathbf{E}_0$. As the magnetic field in the wave zone is proportional to the electric field, the radiated energy in the cases considered of a gap and of a channel increases, respectively, by factors ε^2 and $(2\varepsilon/(\varepsilon+1))^2$.† An arbitrarily oriented dipole may be assumed to consist of dipoles parallel and perpendicular to the axis of the channel (gap) and, thus, by virtue of the superposition principle that problem reduces to the previous ones. We see from (7.51) that for a magnetic dipole of moment $\boldsymbol{\mu}$ when the magnetic permeability $\mu = 1$ the presence of a narrow channel does not affect the radiation. If there are at the same time magnetic and electric dipoles, the fields radiated by them (but not, of course, the energies) add up, that is, the problem can again easily be solved.

A moving current moment and a 'true' magnetic moment placed in an empty cavity must, of course, produce the same radiation. This conclusion was also verified by a direct calculation (Bogdankevich, 1960) of the radiation of various dipoles moving in a circular channel; in the particular case of a thin channel one found, as should be the case, the result given above, that is, an amplification of the field of an electric dipole by a factor $2\varepsilon/(\varepsilon+1)$.

In connection with the fact that the Cherenkov radiation of a moving electric dipole, and also of a magnetic dipole, provided the magnetic permeability $\mu \neq 1$, depends on the shape of an arbitrarily narrow cavity there arises the problem of the validity of Eqns.(7.45), (7.46), and (7.47) for the motion of dipoles in a continuous medium. It is clear from the reciprocity theorem that we are here dealing with the possibility of considering the field $\mathbf{E}_{eff}$ which acts upon the dipole to be the average macroscopic field $\mathbf{E}$. This is, in general, not the case for fixed dipoles which are introduced into the medium (that is, $\mathbf{E}_{eff} \neq \mathbf{E}$). However, for a particle with a charge or with dipole moments which moves along a given trajectory the average field is at once also the macroscopic field. We also reach the same conclusion that the starting Eqns.(7.42) to (7.44) are valid for the motion of a particle in a continuous medium when we obtain these equations by averaging the equations of microscopic electrodynamics. Therefore, in our opinion there is no reason to doubt the validity of Eqns.(7.45) to (7.47) for the Cherenkov radiation of point dipoles in a continuous medium.

We could see above the great efficiency of the method based on the use of the reciprocity theorem when we calculated the Cherenkov radiation in narrow channels. This method has also been applied for considering transition radiation

---

† When using Eqn.(7.45) to evaluate the radiated energy we must replace $\mathbf{p}$ when there is a channel or a gap present by the appropriate expression determined by using (7.54), that is, for instance, for a dipole at right angles to the axis of a circular channel by $(2\varepsilon/(\varepsilon+1))\mathbf{p}$.

(Frank and Ginzburg, 1945; Ginzburg and Frank, 1946), and it is useful for solving a whole range of other problems in the theory of Cherenkov and transition radiation when there are boundaries present, even apart from studies of many other electrodynamic problems.[†]

In concluding this chapter we dwell upon transition radiation arising in the case when at the position of the source the properties of the medium are changed — to fix the idea, say, the refractive index $n(\omega)$ is changed. Such a formulation is a very general one — but, apparently, not generally accepted — and it must be sorted out.

When a source moves in vacuo it emits electromagnetic waves either when its velocity $v > c$ or, when it is accelerated, that is, formally when the parameter v/c changes with time. It is just this last possibility which is usually considered in the theory of radiation, since a separate particle, apart from the particularly hypothetical tachyons, only can have a velocity $v < c$ (we shall consider in the next chapter sources moving with a velocity $v > c$). For motion in a medium the parameter $v/v_{ph} = vn(\omega)/c$ plays the role of the parameter v/c (we are now considering a transparent medium; in the general case the electromagnetic properties of a linear medium are characterized by the tensor $\varepsilon_{ij}(\omega, \mathbf{k})$). Now, the superluminal regime, when $vn/c > 1$, is, firstly, realized without any particular difficulty, and radiation occurs even when $vn/c = \text{const}$ (Cherenkov effect). Secondly, it is clear by analogy with the vacuum case that independently of the value of the parameter vn/c radiation occurs when it changes with time. But this is possible not only when the particle is accelerated, when $d\mathbf{v}/dt \neq 0$, but also in the case when n changes in space and/or time. Indeed, the radiation is determined by the change in n at the position of the particle (radiator) or close to its position in the zone where the wave is formed. In other words, the value $n(t, \mathbf{r}_i(t))$ is important, where $\mathbf{r}_i(t)$ is the position vector of the charge; for the sake of simplicity we neglect now dispersion and the possibility that n changes not on the trajectory $\mathbf{r}_i(t)$ itself, but near it. Such radiation, for which the dependence of n on t and $\mathbf{r}$ is 'responsible' is called transition radiation. Of course, in its 'pure form' transition radiation occurs only when $\mathbf{v} = \text{const}$, $v < c/n$. If, however, $v > c/n$ and/or $dv/dt \neq 0$, transition radiation occurs in combination with Cherenkov radiation and/or bremsstrahlung. Later on, in order not

---

[†] Bolotovskii (1960, 1962) and Zrelov (1968) have reviewed the results of solving problems in the theory of Cherenkov radiation when boundaries are present.

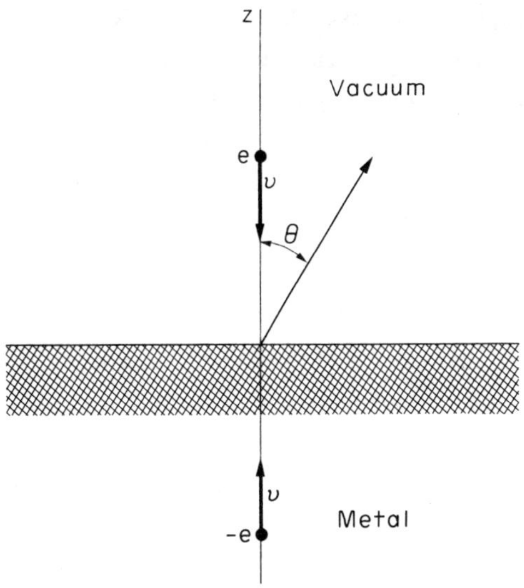

Fig.7.4 The transition radiation when a charge e crosses a vacuum-metal boundary corresponds to the 'annihilation' of the charge and its mirror image

to introduce unnecessary complications we assume that

$$\mathbf{v} = \text{const} \ , \ v < \frac{c}{n(\omega)} \qquad (7.58)$$

The simplest and most realistic situation in which transition radiation occurs arises when a charge crosses the dividing boundary of two media (Frank and Ginzburg, 1945). Apart from the general considerations which can be given, the fact that in that case radiation occurs is particularly clear, one might say obvious, for the example when a charge goes from a vacuum into a metal. In the low-frequency region, up to and including optical frequencies, one can consider a metal to a good approximation to be perfectly reflecting (a perfect conductor). Therefore the field of a charge in vacuo, as long as it moves in the vacuum, is the sum of the fields of a point charge e and of its mirror image (a charge −e which lies in the metal; Fig.7.4). When the charge crosses the boundary it and its mirror image 'annihilate one another' as the field of the charge when it is inside the metal is totally screened in the approximation of perfect (infinite) conductivity. Thus, exactly the same radiation arises as for an actual disappearance of the charges e and −e, for instance, an electron and a positron.

The evaluation of the field and the radiative energy in transition radiation for different cases can be found in various publications by Frank and Ginzburg (1945), Ginzburg and Frank (1946), Bass and Yakovenko (1965), Garibyan (1960, 1970), Ter-Mikaelyan (1972), Tamoykin (1972), Ginzburg (1975b) and Ginzburg and Tsytovich (1974a), as well as in the literature quoted in those papers. Here we give merely a few separate remarks about this topic.†

† Among the problems which are not going to be discussed we may mention the macroscopic mass renormalization in transition radiation and the motion in channels and gaps (Tsytovich, 1962a, 1964; Ginzburg and Tsytovich, 1974a; Ginzburg, 1975b). Ginzburg and Tsytovich (1978) have elucidated a whole set of problems in the theory of transition radiation and transition scattering (see Chapter 14).

To find the emitted energy when the boundary between a vacuum and a metal (perfect conductor) is crossed we can most simply use the formulae for the bremsstrahlung of charges when they are suddenly annihilated (see Landau and Lifshitz, 1975, § 69, where the quantity dU_s is denoted by $d\mathcal{E}_{n\omega}$)

$$\frac{d^2U_s}{d^2\Omega d\omega} = W_\omega = \frac{1}{4\pi^2c^3}\left\{\sum_i e_i\left(\frac{[\mathbf{v}_{i2}\wedge\mathbf{s}]}{1-(\mathbf{s}\cdot\mathbf{v}_{i2})/c} - \frac{[\mathbf{v}_{i1}\wedge\mathbf{s}]}{1-(\mathbf{s}\cdot\mathbf{v}_{i1})/c}\right)\right\}^2 \tag{7.59}$$

where e_i is the charge of the ith particle, the velocity of which suddenly changes from $\mathbf{v}_{i1}$ to $\mathbf{v}_{i2}$, while $\mathbf{s}=\mathbf{k}/k$ is the direction of the wavevector; the suddenness of the change in $\mathbf{v}$ means that its change occurs during a time $\tau \ll T = 2\pi/\omega$, where ω is the frequency considered.

In the case of interest to us the charge e, moving with a velocity $\mathbf{v}$ is suddenly stopped together with the charge $-e$ moving with the velocity $-\mathbf{v}$. In the non-relativistic approximation we find (the axis of the polar system of coordinates θ,ϕ is the direction along $-\mathbf{v}$)

$$\left.\begin{aligned} W_\omega(\theta,\phi) &= \frac{e^2[\mathbf{v}\wedge\mathbf{s}]^2}{\pi^2c^3} = \frac{e^2v^2\sin^2\theta}{\pi^2c^3}\,, \\ W_\omega &= \int_0^{2\pi} d\phi \int_0^{\frac{1}{2}\pi} W_\omega(\theta,\phi)\sin\theta\, d\theta = \frac{4e^2v^2}{3\pi c^3}\,. \end{aligned}\right\} \tag{7.60}$$

These values are four times larger than the energy emitted into the hemisphere of directions when a single charge is stopped (in the non-relativistic case the field of a charge and its mirror image simply add up, that is, are duplicated). In the general case (for any velocity v)

$$\left.\begin{aligned} W_\omega(\theta,\phi) &= \frac{e^2v^2}{\pi^2c^3}\,\frac{\sin^2\theta}{[1-(v^2c^2)\cos^2\theta]^2} \\ W_\omega &= \frac{4e^2v^2}{3\pi c^3}\left\{\frac{3(v^2/c^2+1)}{8(v/c)^3}\ln\left(\frac{1+v/c}{1-v/c}\right) - \frac{3}{4v^2/c^2}\right\} \\ &= \frac{e^2}{\pi c}\left\{\frac{1+v^2/c^2}{2v/c}\ln\left(\frac{1+v/c}{1-v/c}\right) - 1\right\} \end{aligned}\right\} \tag{7.61}$$

Exactly the same radiation occurs when a charge leaves a metal to go into a vacuum (in that case the z-axis is directed along $\mathbf{v}$). For a comparison we note that when a single charge is suddenly (instantaneously) stopped in vacuo, we get easily from (7.59) (for a detailed calculations see Landau and Lifshitz, 1975, § 69)

$$W_\omega = \frac{e^2}{\pi c}\left\{\frac{c}{v}\ln\frac{1+v/c}{1-v/c} - 2\right\}. \tag{7.62}$$

As $v/c \to 1$ (ultrarelativistic case) Eqns.(7.61) and (7.62) become the same. This can be explained by the fact that the radiation is mainly directed along the velocity of the charge, that is, along $\mathbf{v}$, or for the mirror image, along $-\mathbf{v}$. However, the radiation of a charge 'departing' into the metal when the charge moves from the vacuum into the metal, or the radiation of the mirror image 'departing' into the metal when the particle moves into the vacuum are not observed. In other words, in the vacuum the radiation turns out to be the same as for a sudden stopping (acceleration) of a single charge. In the non-relativistic approximation, by the way, the quantity W_ω given by (7.62) is smaller by a factor two than according to (7.60), as it is a single charge which radiates in the case (7.62) (leading to a decrease by a factor four), but in the whole of space (giving an enhancement by a factor two).

In the case where a charge crosses the boundary between a vacuum and a medium medium with a complex permittivity $\varepsilon(\omega) = \boldsymbol{\varepsilon} + i\varepsilon$ the calculation is rather complicated and we shall merely give here the result (Frank and Ginzburg, 1945; Garibyan, 1960, 1970; Bass and Yakovenko, 1965; Ter-Mikaelyan, 1972)

$$W_\omega(\theta,\phi) = \frac{e^2v^2\sin^2\theta}{\pi^2c^3}\,F\,,$$

$$F = \frac{\cos^2\theta}{[1-(v^2/c^2)\cos^2\theta]^2}\left|\frac{(\varepsilon-1)[1-v^2c^2+(v/c)\{\boldsymbol{\varepsilon}-\sin^2\theta\}^{\frac{1}{2}}]}{[\boldsymbol{\varepsilon}\cos\theta+\{\boldsymbol{\varepsilon}-\sin^2\theta\}^{\frac{1}{2}}][1+(v/c)\{\boldsymbol{\varepsilon}-\sin^2\theta\}^{\frac{1}{2}}]}\right|^2 . \qquad (7.63)$$

One accomplishes the transition to the case of a perfect conductor by letting $|\boldsymbol{\varepsilon}| \to \infty$ which leads to (7.61) where $F = [1-(v^2/c^2)\cos^2\theta]^{-2}$. As $v/c \to 1$ practically the whole of the backwards radiation is concentrated within angles $\theta \sim [1-v^2/c^2]^{\frac{1}{2}} \ll 1$, and therefore

$$F = \left|\frac{\sqrt{\boldsymbol{\varepsilon}}-1}{\sqrt{\boldsymbol{\varepsilon}}+1}\right|^2 \frac{1}{[1-(v^2/c^2)\cos^2\theta]^2}\,, \quad \boldsymbol{\varepsilon} = \boldsymbol{\varepsilon}(\omega)\,,$$

$$W_\omega = \int W_\omega(\theta,\phi)\,d^2\Omega = \frac{e^2}{\pi c}\left|\frac{\sqrt{\boldsymbol{\varepsilon}}-1}{\sqrt{\boldsymbol{\varepsilon}}+2}\right|^2\left\{\ln\frac{2}{1-v/c}-1\right\}. \qquad (7.64)$$

This formula generalizes Eqn.(7.61) as $v/c \to 1$ and, of course, changes into it for large values of $|\boldsymbol{\varepsilon}| \gg 1$. Moreover, it is clear from (7.64) that the radiative energy decreases as $\sqrt{\boldsymbol{\varepsilon}(\omega)}$ approaches 1 and therefore the radiation itself is concentrated in the optical range. The situation changes for transition radiation occurring when the charge goes from a medium into a vacuum. The general expression for $W_\omega(\theta,\phi)$ is in that case obtained from (7.63) by changing v to −v; moreover, it is now more convenient to take for the angle θ the angle between **v** and **k** rather than between **−v** and **k**. The appearance of the factor $[1-(v/c)\{\varepsilon-\sin^2\theta\}^{\frac{1}{2}}]^2$ in the denominator leads in that case to an increase of the role played by the high frequencies as $v/c\to 1$; for those frequencies the permittivity is close to unity as in the X-ray region

$$\varepsilon \approx 1-\frac{\omega_p^2}{\omega^2}\ , \quad \omega_p^2=\frac{4\pi Ne^2}{m}\ , \tag{7.65}$$

where N is the total electron density or, for lower frequencies, their density in the part of the atomic shells. As a result (Garibyan, 1960, 1970) we find under the conditions for X-ray transition radiation in the forward direction (see (7.63) as $v/c\to 1$ with v replaced by −v, and (7.65))

$$\left.\begin{aligned} W_\omega(\theta,\phi) &= \frac{e^2\theta^2}{\pi^2 c}\left\{\frac{1}{1-v^2/c^2+\theta^2}-\frac{1}{1-v^2/c^2+\omega_p^2/\omega^2+\theta^2}\right\}^2 , \\ W_\omega &= \frac{2e^2}{\pi c}\left\{\left[\frac{1}{2}+\frac{\omega^2(1-v^2/c^2)}{\omega_p^2}\right]\ln\left[1+\frac{\omega_p^2}{(1-v^2/c^2)\omega^2}\right]-1\right\} \end{aligned}\right\} \tag{7.66}$$

or approximately

$$\left.\begin{aligned} W_\omega &= \frac{2e^2}{\pi c}\ln\frac{\omega_c}{\omega}\ , \ \omega\ll\omega_c\ ; \quad W_\omega=\frac{e^2}{6\pi c}\frac{\omega_c^4}{\omega^4}\ , \ \omega\gg\omega_c\ ; \\ \omega_c &= \frac{\omega_p}{[1-v^2/c^2]^{\frac{1}{2}}}=\omega_p\frac{\mathcal{E}}{Mc^2}\ , \end{aligned}\right\} \tag{7.67}$$

$$U=\int W_\omega\, d\omega=\frac{e^2\omega_p}{3c}\ \frac{\mathcal{E}}{Mc^2}\ . \tag{7.68}$$

The transition radiation in the forward direction lies thus for ultra-relativistic particles mainly in the X-ray range (for dense media $\omega_p\sim 10^{15}$ to $10^{16}\,s^{-1}$) and the characteristic frequency $\omega_c\sim(10^{15}$ to $10^{16})\,\mathcal{E}/Mc^2\,s^{-1}$ and the total emitted energy U increases linearly with increasing $\mathcal{E}/Mc^2$. It is clear from (7.67) and (7.68) that in the main in the case which is now under consideration photons with energies $\hbar\omega\sim\hbar\omega_c=\hbar\omega_p\,(\mathcal{E}/Mc^2$ are emitted, while on average their number when the particle crosses the medium-vacuum boundary is equal to $U/\hbar\omega\sim e^2/\hbar c\approx 1/137$. The application of transition radiation for registering high-energy particles is therefore only realistic when one uses many boundaries, for instance, in the case of a set of plates (see Garibyan, 1960, 1970; Bass and Yakovenko, 1965; Arutyunyan et al., 1971, 1972; Wang et al., 1972; Ter-Mikaelyan, 1972; Alikhanyan et al., 1972; Cherry et al., 1972). For plates, and in general for an inhomogeneous medium, the ratio of the

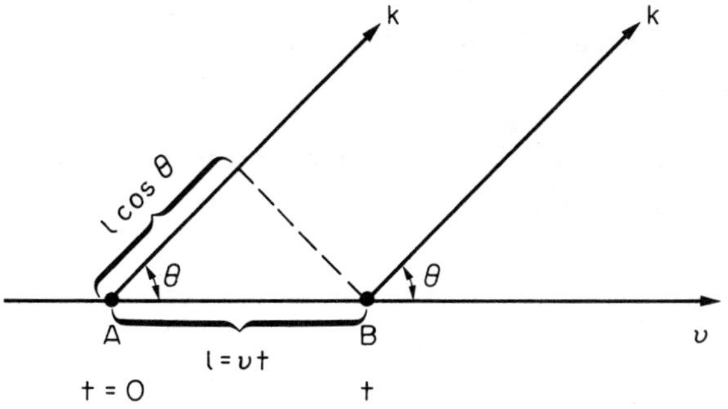

Fig.7.5 The zone ℓ where the radiation is formed

thickness of the plate — or another length which characterizes the inhomogeneity of the medium — to the zone where the radiation is formed plays an essential role. For a radiator at rest and when it moves with a velocity $v \ll c/n$ the radiation is formed in a region of size $\ell \sim c/n\omega = \lambda/2\pi$ — and the phase of the wave $\phi = k\ell = (2\pi/\lambda)\ell$ changes just over such a distance by an amount of the order of unity. In the case of Cherenkov radiation when a change moves in a channel or a gap with a characteristic size a we also verified above that when $a/\lambda \ll 1$ the radiation remains the same as in a continuous medium, that is, it is formed in a region with dimensions of the order of λ. However, for fast moving sources the zone where the radiation is formed is, generally speaking, not at all equal to the length $\lambda/2\pi$; for the case of Cherenkov radiation the zone ℓ of the formation is of the order of $\lambda/2\pi$ only in the direction at right angles to the velocity which is just so important in the case of channels and gaps. To find the zone ℓ where the radiation is formed for the case of a source moving with a velocity $\mathbf{v}$ and emitting waves at an angle θ to $\mathbf{v}$, we consider Fig.7.5. Let at time $t = 0$ the source be at the point A and the phase of the wave emitted by it in the direction of $\mathbf{k}$ be equal to ϕ_A. The zone of formation ℓ is defined as that distance along the trajectory of the source (distance between the points A and B) for which the phase ϕ_B of the waves emitted at the point B in the same direction $\mathbf{k}$ differs by π from the phase ϕ_A of the waves emitted in the point A. We then have

$$|\phi_A - \phi_B| = |k\ell\cos\theta - \omega t| = \left|\frac{\omega n}{c}\ell\cos\theta - \omega\frac{\ell}{v}\right| = \pi ,$$

whence

$$\ell = \frac{\pi v}{\omega|1 - (v/c)\,n(\omega)\,\cos\theta|} = \frac{(vn/c)\lambda}{2|1 - (v/c)\,n(\omega)\,\cos\theta|} , \quad \lambda = \frac{2\pi c}{n\omega} . \qquad (7.69)$$

There is, of course, a certain amount of arbitrariness in the definition of the zone of formation ℓ. Moreover, according to (7.69) $\ell \to 0$ as $v \to 0$ (that is, for a charge at rest). At the same time the concept of a non-vanishing zone of formation has a well-defined meaning also for a radiator at rest — in that case (in the simplest situations) $\ell \sim \frac{2\pi c}{\omega} = \lambda$.

Of course, for the Cherenkov angle $\theta_0 = \arccos(c/nv)$ the waves emitted along the trajectory are in phase and the emitted energy is therefore proportional to the length L of the trajectory — and is formally infinite as $L \to \infty$, as would be assumed when one neglects the radiation at the end and the beginning of the path. If we apply this to transition radiation the formation zone plays the role of the size of those regions in vacuo and in the medium which are responsible for the radiation (for details and more precise calculations see Garibyan, 1960, 1970 and Ter-Mikaelyan, 1972). To be precise, when an ultrarelativistic particle crosses from a medium into a vacuum and when the radiation is in the forward direction, which is important at high frequencies (vide supra) the formation zone in vacuo (for $\theta = 0$) equals

$$\ell = \frac{\lambda}{2(1 - v/c)} = \lambda \left(\frac{\mathcal{E}}{Mc^2}\right)^2 \sim \lambda_p \frac{\mathcal{E}}{Mc^2}, \tag{7.70}$$

where when going over to the last of Eqns.(7.70) we have taken for the wavelength λ a wavelength $\lambda_c = 2\pi c/\omega_c = \lambda_p (Mc^2/\mathcal{E})$; here $\lambda_p = 2\pi c/\omega_p$ (see (7.67)). When $\lambda_p \sim 10^{-5}$ cm and $\mathcal{E}/Mc^2 \sim 10^4$ (for instance, for electrons with energies of 5 GeV) we have $\ell \sim 0.1$ cm. In that connection, the use of transition radiation from a laminated medium is effective only within well defined limits — the intensity of the radiation from the boundaries adds up (provided we neglect the oscillations due to interference) only when the distance between the boundaries is larger than the formation zone. If we are dealing with a single plate both its boundaries also 'work' effectively only as long as the thickness of the plate $d \gtrsim \ell$. If, however, the plate is thin ($d \ll \ell$) the transition radiation from it is appreciably weaker than from a single boundary. In the limit of an arbitrarily thin plate ($d \to 0$) there is no transition radiation, as becomes clear already from general considerations.† An application of the transition radiation theory for the case of a single separating boundary to thin plates, small pellets, and so on can, of course, lead to gross mistakes — this occurred in some attempts to evaluate the X-ray transition radiation occurring in interstellar space for the case when fast electrons passed through

---

† We have here in mind a plate of some real material, and not a perfect conductor for which the limit as $d \to 0$ has no physical meaning, if we use that concept literally.

cosmic dust grains; Yodh et al. (1973) and Durand (1973) have given a correct discussion of this problem.

Apart from the transition radiation from separate boundaries or stacks of plates, the transition radiation in a statistically inhomogeneous medium when a charge flies past different bodies is also of interest (see Bass and Yakovenko, 1965; Ter-Mikaelyan, 1972; Tamoykin, 1972). We note in particular the case of non-stationary media when the properties of the medium change not only spatially, but also temporally. In particular transition radiation must occur also in a spatially uniform medium, provided its properties or, more precisely, its refractive index n changes with time. We shall now occupy ourselves just with such a problem, when suddenly at time $t = 0$, n changes from a value n_1 $(t < 0)$ to a value n_2 $(t > 0)$ (Ginzburg and Tsytovich, 1974a; Ginzburg, 1975b).

Therefore, let a charge with velocity $v = \text{const}$, $v < c/n_{1,2}$ move in a homogeneous medium, the properties of which discontinuously change at $t = 0$. Such a change in n can, for instance, be realized by a sudden increase in pressure — and thus in the density of the medium — as the result of a change in the strength of the external field in which the medium is placed (we have in mind, for instance, a medium in a capacitor). By a sudden change in the properties of the medium we understand here a change over a time $\tau \ll 2\pi/\omega$, where ω is the frequency of the radiation in which we are interested.

Let us turn to the original equations (6.1) for the field in the medium. For a discontinuity (sudden change) in the properties of the medium, generally speaking, only the time-derivatives will be large. We therefore integrate the first and third of Eqns.(6.1) over t over an interval Δt which includes the discontinuity. We then get as $\Delta t \to 0$ the boundary conditions

$$\mathbf{D}_1 = \mathbf{D}_2 \ , \quad \mathbf{H}_1 = \mathbf{H}_2 \ , \quad t = 0 \ , \tag{7.71}$$

where the indexes 1 and 2 refer to the medium before $(t < 0)$ and after $(t > 0)$ the discontinuity. We shall solve the problem by the Hamiltonian method and for the sake of convenience we repeat a few formulae. We write

$$\left.\begin{aligned}
&\mathbf{E}_{tr} = -\frac{1}{c}\frac{\partial \mathbf{A}}{\partial t} \ , \ \mathbf{H} = \operatorname{curl} \mathbf{A} \ , \ \operatorname{div} \mathbf{A} = 0 \ , \ \mathbf{D} = \varepsilon \mathbf{E} \ , \\
&\mathbf{A} = \sum_\lambda (q_\lambda \mathbf{A}_\lambda + q_\lambda^* \mathbf{A}_\lambda^*) \ , \\
&\mathbf{A}_\lambda = \sqrt{4\pi}\,\frac{c}{n}\,\mathbf{e}_\lambda \exp\left[i(\mathbf{k}_\lambda \cdot \mathbf{r})\right] \ , \ e_\lambda^2 = 1 \ , \ (\mathbf{e}_\lambda \cdot \mathbf{k}_\lambda) = 0 \ .
\end{aligned}\right\} \tag{7.72}$$

The longitudinal part of the field $\mathbf{E}_\ell = \mathbf{E} - \mathbf{E}_{tr}$ will not be of interest to us and one can check that if one takes it into account the expressions used in what follows are unchanged. For a uniformly moving charge $\mathbf{j} = e\mathbf{v}\delta(\mathbf{r} - \mathbf{v}t)$, $\mathbf{v} = \text{const}$ and in the usual way we get from (6.1) and (7.72) (we assume that $n = \sqrt{\varepsilon} = \text{const}$; taking dispersion into account does not change the result as we saw in Chapter 6)

$$\ddot{q}_\lambda + \omega_\lambda^2 q_\lambda = \frac{e}{c}(\mathbf{v} \cdot \mathbf{A}_\lambda^*(\mathbf{v}t)) = \frac{\sqrt{4\pi}\, e}{n}(\mathbf{e}_\lambda \cdot \mathbf{v}) \exp[-i(\mathbf{k}_\lambda \cdot \mathbf{v}t)] ; \quad \omega_\lambda^2 = \frac{c^2}{n^2} k_\lambda^2 . \tag{7.73}$$

As the functions $\exp[i(\mathbf{k}_\lambda \cdot \mathbf{r})]$ and $\exp[-i(\mathbf{k}_\mu \cdot \mathbf{r})]$ for $\mu \neq \lambda$ as well as the functions $\mathbf{e}_\lambda \exp[\pm i(\mathbf{k}_\lambda \cdot \mathbf{r})]$ and the longitudinal field $\mathbf{E}_\ell$ are orthogonal onto each other (and this fact has, of course, already been used in deriving Eqns.(7.73)) the boundary conditions (7.71) take the form

$$n_1 \dot{q}_{\lambda,1} = n_2 \dot{q}_{\lambda,2} , \quad q_{\lambda,1}/n_1 = q_{\lambda,2}/n_2 , \quad t = 0 . \tag{7.74}$$

We shall assume that for $t < 0$ there was no radiation field, that is, that the charge was surrounded merely by the field it carries along with itself — we remind ourselves that we have assumed for the sake of simplicity that $v < c/n_{1,2}$ so that neither before nor after the change in the refractive index n there is any Cherenkov radiation. We must therefore for $t < 0$ take only the induced solution of Eqns.(7.73), that is,

$$q_{\lambda,1} = \frac{\sqrt{4\pi}\, e(\mathbf{e}_\lambda \cdot \mathbf{v}) \exp[-i(\mathbf{k}_\lambda \cdot \mathbf{v}t)]}{n_1\{k_\lambda^2 c^2/n_1^2 - (\mathbf{k}_\lambda \cdot \mathbf{v})^2\}} , \quad t < 0 . \tag{7.75}$$

For $t > 0$ there is both the field which is carried along as well as the radiation field — the solution of the homogeneous Eqns.(7.73) —

$$q_{\lambda,2} = \frac{\sqrt{4\pi}\, e(\mathbf{e}_\lambda \cdot \mathbf{v}) \exp[-i(\mathbf{k}_\lambda \cdot \mathbf{v}t)]}{n_2\{k_\lambda^2 c^2/n_2^2 - (\mathbf{k}_\lambda \cdot \mathbf{v})^2\}} + C_+ \exp\left(ik_\lambda \frac{c}{n_2} t\right) + C_- \exp\left(-ik_\lambda \frac{c}{n_2} t\right) . \tag{7.76}$$

Using the conditions (7.74) for $t = 0$ we then get from these equations

$$C_\pm = \frac{\sqrt{4\pi}\, e(\mathbf{e}_\lambda \cdot \mathbf{v})}{2c^2 k_\lambda^2} \left\{ n_2 \left[\frac{1}{1 - (\mathbf{s}_\lambda \cdot \mathbf{v})^2 n_1^2/c^2} - \frac{1}{1 - (\mathbf{s}_\lambda \cdot \mathbf{v})^2 n_2^2/c^2} \right] \mp \frac{(\mathbf{s}_\lambda \cdot \mathbf{v})}{c} \left[\frac{n_1^2}{1 - (\mathbf{s}_\lambda \cdot \mathbf{v})^2 n_1^2/c^2} - \frac{n_2^2}{1 - (\mathbf{s}_\lambda \cdot \mathbf{v})^2 n_2^2/c^2} \right] \right\} . \tag{7.77}$$

where $\mathbf{s}_\lambda = \mathbf{k}_\lambda / k_\lambda$; we shall in what follows put $(\mathbf{s}_\lambda \cdot \mathbf{v})^2 = v^2\cos^2\theta$ and $(\mathbf{e}_\lambda \cdot \mathbf{v})^2 = v^2 \sin^2\theta$, as the second possible direction of the polarization vector $\mathbf{e}_\lambda$ can be chosen to be at right angles to the velocity $\mathbf{v}$. The energy of the transverse field is

$$\mathcal{H}_{tr} = \int \frac{\varepsilon E_{tr}^2 + H^2}{8\pi} d^3\mathbf{r} = \sum_\lambda (p_\lambda p_\lambda^* + \omega_\lambda^2 q_\lambda q_\lambda^*) . \tag{7.78}$$

We are interested in the total energy of the radiation field which appears as a result of the change in n. This means that we must substitute into (7.78) the solution

$$q'_{\lambda 2} = C_+ \exp\left(ik_\lambda \frac{c}{n_2} t\right) + C_- \exp\left(-ik_\lambda \frac{c}{n_2} t\right) ,$$

where we must restrict ourselves to that part of $\mathcal{H}_{tr}$ which is independent of t. One sees easily that that part is equal to $(d^2\Omega = \sin\theta d\theta d\phi)$

$$\mathcal{H}_{tr} = 2\sum_\lambda \omega_\lambda^2 (C_+^2 + C_-^2) = \int W_\omega(\theta,\phi)\, d\omega\, d^2\Omega ,$$

$$W_\omega(\theta,\phi) = \frac{(C_+^2 + C_-^2)\,\omega^2 k^2}{(2\pi)^3} \frac{dk}{d\omega} , \quad \omega^2 = \frac{c^2k^2}{n_2^2} ; \tag{7.79}$$

we have used here the fact that in (7.72) and (7.78) the summation is over a hemisphere of directions of $\mathbf{k}_\lambda$, while in (7.79) all directions are taken into consideration.† We finally get

$$W_\omega(\theta,\phi) = \frac{e^2v^2\sin^2\theta}{\pi^2c^3} F , \tag{7.80}$$

$$F = \frac{1}{4n_2}\left\{ n_2^2 \left[\frac{1}{1-(v^2/c^2)\,n_1^2\cos^2\theta} - \frac{1}{1-(v^2/c^2)\,n_2^2\cos^2\theta}\right]^2 + \right.$$

$$\left. + \frac{v^2\cos^2\theta}{c^2}\left[\frac{n_1^2}{1-(v^2/c^2)\,n_1^2\cos^2\theta} - \frac{n_2^2}{1-(v^2/c^2)\,n_2^2\cos^2\theta}\right]^2 \right\} . \tag{7.81}$$

If

$$\frac{v^2n_{1,2}^2}{c^2}\cos^2\theta \ll 1 , \tag{7.82}$$

we have

$$F \approx \frac{(n_1^2 - n_2^2)^2\, v^2\cos^2\theta}{4\,n_2 c^2} . \tag{7.83}$$

---

† In this method we do not distinguish between waves emitted at angles θ and $\pi - \theta$. The expression (7.81) therefore gives half the sum of the intensities for the angles θ and $\pi - \theta$. This is not important for the non-relativistic case (7.83); for the relativistic case the formulae for the intensity at all θ can be found in the paper by Ginzburg and Tsytovich (1974a).

One can, of course, obtain Eqns.(7.80) and (7.83) also directly by performing the appropriate simplifications which are connected with the condition (7.82) in Eqns.(7.75) and (7.76).

For a comparison we remind ourselves that when a non-relativistic electron crosses the boundary between vacuum and a medium with a refractive index n_2 one gets, as can be seen from (7.63), for the energy of the transition radiation expression (7.80) with

$$F = \left(\frac{n_2^2 - 1}{n_2^2 \cos\theta + \{n_2^2 - \sin^2\theta\}^{\frac{1}{2}}}\right)^2 \cos^2\theta \quad ;$$

if, however, medium 2 is a perfect conductor, we have $F = 1$; this can also be obtained directly from the previous formulae taking $n_2^2 \to \infty$. The given kind of transition radiation in a non-stationary medium and the transition radiation when a particle crosses the dividing boundary between two media are thus, indeed, related to one another. Moreover, we note that, as should be the case, when $n_1 = n_2$, the factor $F = 0$ (see (7.81) and (7.83)). For a given difference $(n_1^2 - n_2^2)^2$ the quantity W_ω can increase when we take into account the decrease in the denominators $1 - (v^2 n_{1,2}^2/c^2)\cos^2\theta$ in (7.81). One might hope to use that fact if one tries to apply the transition radiation discussed here as an indicator for the change in the refractive index n with time. Of course, these changes can be noted also, for instance, through the change in phase of electromagnetic waves passing through the medium. However, such a clear situation is not always realizable, particularly not under astronomical conditions. The possibility to use fast particles and the transition radiation produced by them as a 'probe', although it is not tempting, must not be forgotten. Under real conditions the characteristics of the radiation will, probably, be determined at once by several processes and circumstances, in particular, the inhomogeneity of the medium, the changes in time of its properties, bremsstrahlung, and so on. Only the actual analysis can in each case sort out which of these radiation mechanisms dominates.

Each radiation mechanism is connected with the corresponding absorption mechanism for waves; this fact follows in the classical case from the invariance of the equations under time reversal and in the quantal case the connection between the direct and the inverse processes is clear from the fact that the absolute magnitudes of the transition matrix elements for these processes are equal. Transition radiation — or, to be more precise, the corresponding transitions in the system consisting of the charges and the radiation field — can manifest itself also in more complex processes — in induced radiation processes or in spontaneous or stimulated (induced) Raman scattering, and so on. Finally, transition radiation, like any other radiation, leads to a change in energy of the emitting particle. It is true, that when we solve the problem of the transition radiation — and also that of Cherenkov radiation — we may assume that the velocity of the source is given and constant. Such a way of stating the problem is the more legitimate when the corresponding energy losses can be compensated by work done by external sources (forces). However, when such a compensation is not present the particle loses, of course, as a result of the radiation some part of its energy — or sometimes gains some energy, in particular, when we are dealing with media with negative absorption or, as one sometimes says, with an inverted dielectric (Ginzburg and Eidman, 1963; Gavrilov and Kolomenskii, 1971, 1972). We note, moreover, that when a charge moves in an inhomogeneous medium it is, in general, not at all possible to equate the work done by the radiative force and the energy emitted. The fact is that when the properties of the medium change there appears not only radiation, but the energy of the field carried along with the particles also changes (Tsytovich, 1962a, 1964; Ginzburg and Tsytovich, 1974a, 1978). Therefore for a given velocity the work done by the radiative force on the particles equals the sum (with the opposite sign) of the emitted energy and the change in the energy of the field carried along by the particles — in other words, the external forces guaranteeing the uniformity of the particle motion in the inhomogeneous medium lose energy to it which is equal to the sum of the radiated energy and the energy connected with the change in the field which is carried along.

Transition radiation is a rather elementary and universal effect and hundreds of papers have already been devoted to it. It is in that connection interesting that after the publication of the theory of transition radiation (Frank and Ginzburg, 1945; Ginzburg and Frank, 1946) 13 years elapsed before the first experimental investigation occurred (Goldsmith and Jelley, 1959). This was, apparently, not due to any special experimental difficulties, but to 'whims of fashion' which one meets with quite often, even in physics.

Chapter VIII

ON SUPERLUMINAL RADIATION SOURCES

Apparent and real superluminal velocities of radiation sources. Cherenkov effect and Doppler effect for the motion of sources with a velocity larger than the velocity of light in vacuo.

The velocity of light in vacuo $c = 3\times 10^{10}$ cm/s is the limiting, largest velocity encountered in Nature. In this way one can formulate 'in zeroth approximation' the conclusion following from the theory of relativity and confirmed experimentally. It has, however, been known for a very long time that already 'in the first approximation' this statement is incorrect or, at any rate, that it needs to be made more precise. The simplest example is the phase velocity of light $v_{ph} = c/n$ which can be arbitrarily large as $n \to 0$ (of course, there are real media with $n < 1$; we just remind ourselves of a plasma where for certain conditions $n = (1 - \omega_p^2/\omega^2)^{\frac{1}{2}}$, $\omega_p^2 = 4\pi Ne^2/m$). In connection with these and other examples one states more precisely that the velocity of signals, perturbations, particles, radiation sources, and so on, must be less than the velocity of light, but that this is not necessary for, say, the velocity with which a constant phase 'moves', that is, a phase velocity. However, even this statement needs further refinements and if it is incorrectly taken it leads to paradoxes and contradictions. For instance, the velocity of a signal is usually assumed to be equal to the group velocity

$$v_{gr} = \frac{d\omega}{dk} = \frac{c}{d(n\omega)/d\omega}$$

(we restrict ourselves here for the sake of simplicity exclusively to the case of an isotropic medium when $k = (\omega/c)n(\omega)$ and the velocity $\mathbf{v}_{gr}$ is in the direction of $\mathbf{k}$). This velocity v_{gr} may well turn out to be larger than c, for instance, in an anomalous dispersion region, where $dn/d\omega < 0$. The solution of such an apparent contradiction was obtained more than sixty years ago (Sommerfeld, 1914) by analyzing the propagation of a signal through a dispersive medium. The fact is that the concept of a group velocity $v_{gr} = d\omega/dk$ has, in general, an exact meaning only when we neglect the spreading of the signal and its absorption (for details see Agranovich and Ginzburg, 1966; Ginzburg, 1970b; Vainshtein, 1957, 1976). The velocity of the main part of the signal is thus different from $d\omega/dk$, in particular in the region where the anomalous dispersion is most strongly expressed. One can, moreover, show that the velocity of the leading front of the signal is strictly equal to c. If we forget about evaluating the field behind the front (the 'precursor' region), the conclusion just stated about the velocity of the front is clear also without calculations. Indeed, the expansion of the field of the signal (the wave train) in a Fourier integral will always contain also very high frequencies. However as $\omega \to \infty$ (and practically already in the X-ray region) the refractive index $n \to 1$ as the particles in the medium can not react on the field of the wave. Such high frequencies just form the 'precursor' of the signal which moves with the velocity c. Another fact is that for signals with a relatively low carrier frequency ω_0 the energy included in the 'precursor' is negligibly small. The main part, or 'body', of the signal moves, when we neglect absorption, usually just with the group velocity $v_{gr} = d\omega/dk$, but in those cases when $d\omega/dk > c$ (although not only then) the signal is strongly deformed and the idea of the group velocity becomes inapplicable and, in any case, transfer of energy with a velocity larger than c does not occur. We want to underline especially that what we have said has been well known for a long time. It is, however, interesting that some hypothetical effect of a statement about the impossibility to exceed the velocity of light in vacuo c continues to be effective even in our time. As an example we may mention the widely held view that it is impossible to observe the Cherenkov effect or the anomalous Doppler effect in vacuo or in media with $n < 1$ — in particular for transverse waves in an isotropic plasma. Another example is connected with the continuing discussions about quasar distances. It is, indeed, not easy to determine these distances as the only known direct method in those cases is connected with measuring the red shift of spectral lines. Assuming the shift to be cosmological, that is, connected with the expansion of the Universe, we get the appropriate distances, but opponents of such an interpretation base themselves on the absence of a proof of the cosmological nature of the red shift for quasars. In our opinion it is difficult to doubt the cosmological nature of the red shift for quasars, but that is not the point of the present discussion. It is interesting that as an argument against using the cosmological distances for quasars data were mentioned in the literature about changes in the quasar (radio-source) structure with a velocity $u > c$. Schematizing we have here a radiation source, say, a radio-source, with an angular size which increases in time with an angular velocity

$\Omega = d\phi/dt$. The angle ϕ under which we see the source at the Earth and the angular velocity Ω are the observed quantities. If the distance from the source is R, the velocity with which the boundary of the source changes on the celestial sphere is $u_\perp = \Omega R$. It is just this velocity which is perpendicular to the line of sight which sometimes turns out to be larger than the velocity c if we use the cosmological distance R. Starting from the assumption that necessarily $u < c$ one concluded from this that the quasars must be relatively close by; the red shift in their spectra might then be caused by a gravitational effect or by the quasar having a large velocity relative to the galaxies in its neighbourhood.

In fact, however, the observed 'apparent' velocity $u_\perp = \Omega R$ can be arbitrarily large and from it there follows no restriction whatever on R. Indeed, basing ourselves completely upon the theory of relativity and keeping away from any hypothetical (and, most probably, inadmissible) possibilities of objects moving with a velocity $v > c$, such as tachyons, we have, at the same time, no reasons for identifying the velocity v of the object with the above mentioned velocity $u_\perp$. Let us, for instance, consider a screen, which may be a supernova shell or a quasar shell, which is illuminated by a radiation source and is visible (observable) in scattered radiation or luminescent radiation. It is then completely clear that the screen can 'flare up' simultaneously at all points and in that case the velocity $u \to \infty$. Illuminating the screen in a definite way, starting at some point, one can obtain also any other value for the 'velocity' of expansion of the luminescent screen. A less trivial example of obtaining 'apparent' superluminal velocities u is the observed expansion velocity of some shells (Rees, 1967; Rees and Simon, 1968; Ginzburg and Syrovatskii, 1969; Cavaliere et al., 1971).

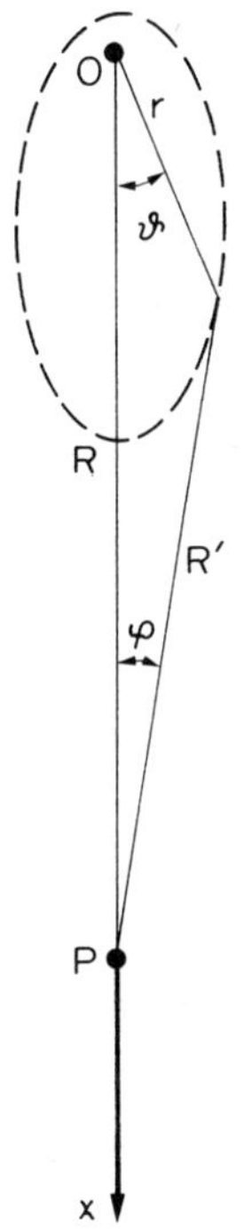

Fig.8.1 Evaluation of the 'apparent' (observed) velocity of an expanding shell

In fact, let us consider the observation of the outside surface of a spherical shell, say, the product of an explosion, which is moving with a velocity v.

Let the explosion have taken place at the point O (Fig.8.1) at time t_0', and let the signal be received at the point of observation P at time $t = 0$. Clearly, $t_0' = -R/c$, where R is the distance between the points O and P and where we have assumed that it is possible to neglect the influence of the medium on the propagation of the signal (light, radiowaves). Let us now find the geometric locus of the points ('visible' shell) from which radiation reaches the observer at some time $t > 0$. We characterize the points on this 'visible' shell by their distance r from the point O and the angle ϑ between **r** and the line OP (see Fig.8.1). The time of emission t', corresponding to the point (r, ϑ) and the time of observation t are connected by the relation

$$t' = t - \frac{R'}{c} \approx t - \frac{R}{c} + \frac{r}{c}\cos\vartheta ,$$

where $R' \approx R - r\cos\vartheta$, as we have assumed that $R \gg r$. On the other hand, $t' - t_0' = t + R/c = r/v$, as the trajectory of a point r in the shell proceeds with the velocity v. Combining the two relations for t which we have written down we get

$$r = \frac{vt}{1 - (v/c)\cos\vartheta} \tag{8.1}$$

The factor $[1 - (v/c)\cos\vartheta]^{-1}$ is here the same as in the formula for the Doppler effect and has the same nature — it is connected with the fact that the propagation velocity of light c is finite. Due to the fact that the velocity of light is finite, in the case discussed here light (radio-waves) arrive at the moment of observation t at the point of observation from points of the shell which correspond to different times $t' - t_0'$ after the time of the explosion $t_0' = -R/c$. The situation is here similar to the one which occurs when one observes (photographs) a fast moving object when one also must distinguish the form of the object at the time when it is observed (when the light arrives) at a given point and the shape of the object for a given time of emission corresponding, say, to simultaneous events in the reference frame (laboratory frame) considered (see, for instance, Weisskopf, 1960; McGill, 1968; Smorodinskii and Ugarov, 1972). Turning now to the expanding shell we find the 'apparent' (visible) velocity of its expansion in the direction at right angles to the line of sight. Clearly (see (8.1))

$$u_\perp = \frac{dr}{dt}\sin\vartheta = \frac{v\sin\vartheta}{1 - (v/c)\cos\vartheta}, \quad \Omega = \frac{d\phi}{dt} = \frac{u_\perp}{R} . \tag{8.2}$$

The velocity $u_\perp$ is a maximum when $du_\perp/d\vartheta = 0$ for an angle $\vartheta_{max} = \arccos(v/c)$ and

$$u_{\perp,max} = \frac{v}{[1-v^2/c^2]^{\frac{1}{2}}}, \quad \Omega_{max} = \frac{v}{R[1-v^2/c^2]^{\frac{1}{2}}} \tag{8.3}$$

The velocity

$$u = \frac{dr}{dt} = \frac{v}{1-(v/c)\cos\vartheta}$$

is itself a maximum when $\vartheta = 0$ and $u_{max} = v/(1-v/c)$. It is clear that the apparent velocity $u_{\perp,max}$ may be larger than c although the velocity of the shell $v < c$. It is true, this happens only if the velocity v is sufficiently large, that is, it is a relativistic effect.

Might it be possible that not only the apparent velocity (in the sense given here), but also the real velocity of a radiation source could exceed the velocity of light c? One must also give a positive answer to this problem, and we shall follow here the paper by Bolotovskii and Ginzburg (1972). It is, strictly speaking, sufficient for a proof to give the example of a 'spot' from a rotating source which moves on a far-away screen. The velocity of the spot is equal to

$$v = \Omega R, \tag{8.4}$$

where Ω is the angular velocity of the source ('lighthouse') and R the distance from the source to the screen. The lighthouse model is now the generally accepted one for pulsars (see, for instance, Ginzburg, 1971; ter Haar, 1972) and in that case the velocity of the lighthouse spot at the Earth exceeds for all known pulsars the velocity of light c. For the best known pulsar PSR 05 32 in the Crab Nebula $\Omega \approx 200\ s^{-1}$ and $R \approx 6\times 10^{21}$ cm (1500 pc), whence $v = \Omega R \approx 1.2\times 10^{24}$ cm/s (!) If we rotate the beam of a laser with an angular velocity $\Omega = 10^5\ s^{-1}$, we have $v = \Omega R > c$ already for distances $R > 3$ km.

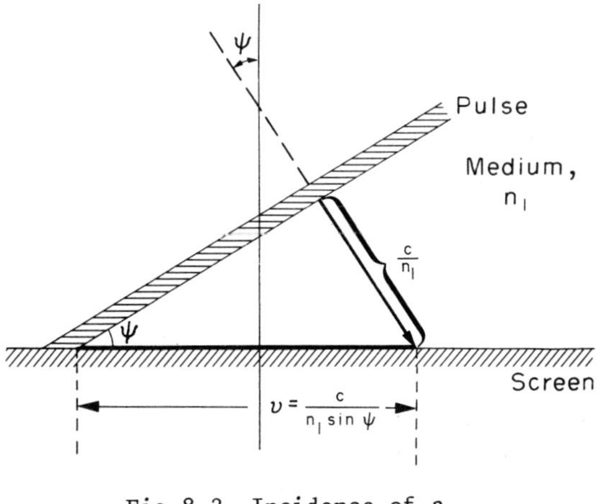

Fig.8.2 Incidence of a pulse on a plane screen

As the simplest model or example of motion with a superluminal velocity we can mention a light pulse of plane waves which is obliquely incident upon some plane dividing boundary (screen) (Frank, 1942). If we denote the angle of incidence of the wave on the screen by Ψ (clearly, Ψ is the angle between the wavevector **k** and the normal to the screen; Fig.8.2) the intersection of the pulse with the screen. that is, the spot of light on the screen, moves along the screen with a velocity

$$v = \frac{c}{n_1 \sin\Psi}, \tag{8.5}$$

where $n_1 > 1$ is the index of refraction of the medium above the screen, which for the sake of simplicity we assume to be non-dispersive — in fact, it is important for us only that the velocity of the light pulse is taken to equal c/n_1. Clearly, the velocity of the light spot, or to be more precise, the streak of light, can always be made larger than c by changing the angle of incidence Ψ, and in vacuo it exceeds c in general for all angles Ψ as in that case

$$v = \frac{c}{\sin\Psi}. \tag{8.6}$$

A beam of electrons moving along the normal to the front of the beam with a velocity $u < c$ can, of course, play the role of the light pulse; in that case

$$v = \frac{u}{\sin\Psi} \tag{8.7}$$

and a superluminal velocity of the spot is also always admissible. Moreover, the velocity v in all cases (8.5) to (8.7) can be made arbitrarily large — when normal incidence is approached (as $\Psi \to 0$) the velocity $v \to \infty$. This is completely understandable as for normal incidence the pulse intersects with the screen simultaneously over the whole of its surface. Scissors can be used as the mechanical analogue of a pulse which is incident on a screen — the point of intersection of the two blades which form the scissors play in that case the role of the spot.

In the case of a rotating source, mentioned above, and also in the case of a pulse intersecting with a screen a large velocity of the spot is reached through diminishing the angle between the constant phase surface (the wave-front) and the screen. Indeed, if we consider for the sake of simplicity a cylindrical source in vacuo which rotates with an angular velocity Ω we can write the field in the wave zone in the form †

---

† This formula gives the solution of a scalar problem. The function E satisfies a wave equation for $r > r_0$ and the boundary condition $E = f(\phi - \Omega t)$ at the surface of the cylinder $r = r_0$. In the system of coordinates which rotates around the z-axis with velocity Ω the field is thus a static one.

$$E = \sum_{s=1}^{\infty} \frac{\exp[is\{(\Omega/c)r + \phi - \Omega t\}]}{\sqrt{r}} \tag{8.8}$$

The constant phase surface is determined by the equation

$$\frac{\Omega}{c} r + \phi - \Omega t = \text{const} \tag{8.9}$$

or

$$r = \text{const} \div c \left(t - \frac{\phi}{\Omega}\right). \tag{8.10}$$

Equation (8.10) is the equation of a spiral. For a cylindrical screen of radius R at a large distance the equal phase surface intersects the screen along the generatrix of the cylinder for which

$$R = \text{const} + c\left(t - \frac{\phi_0}{\Omega}\right), \tag{8.11}$$

where the angle ϕ_0 which determines the generatrix under consideration changes with time according to $d\phi_0/dt = \Omega$. In other words, the intersection (spot) moves along the screen with a velocity

$$v = R\frac{d\phi_0}{dt} = \Omega R\,. \tag{8.12}$$

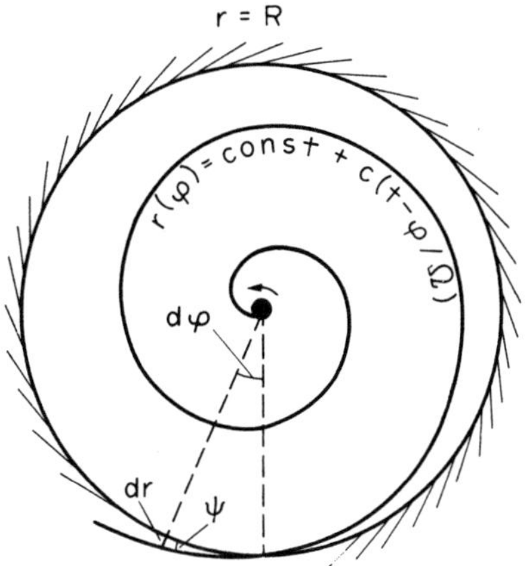

Fig.8.3 Spot from a rotating source (lighthouse) on a spherical or cylindrical screen

We have thus in a more formal way obtained the obvious — or, at any rate well known — result (8.4). It is important that the angle Ψ between the equal phase surface and the screen is given by the condition (Fig.8.3)

$$\tan\Psi = -\frac{dr}{R\,d\phi} = \frac{c}{\Omega R} = \frac{c}{v}\,. \tag{8.13}$$

For small angles Ψ we have, of course, $\tan\Psi \approx \sin\Psi \approx \Psi$ and $v \approx c/\sin\Psi$, in agreement with (8.6). In other words, the large velocity is, as we noted earlier, caused (for instance, when $v \gg c$) by the angle Ψ between the wavefront and the screen being small.

We have here in fact not made any assumptions about the nature of the field and only — and this for the sake of simplicity — assumed that its speed of propagation was equal to c. Hence it is clear that spots with velocities $v > c$ can be obtained not only in the case of electromagnetic waves, but also in that of gravitational waves. Using the ray treatment we are led to the possibility that we can have spots moving with any velocities both for neutrinos (velocity c) and for any other kind of particle (velocity $u < c$).† The fact that the appearance of a velocity $v > c$ for spots does not contradict the theory of relativity cannot cause any shadows of doubt. It is sufficient to say that this result is obtained for completely realistic examples, for instance, when a pulse of light or of electrons impinges on a screen (see Fig. 8.2). Nevertheless, as a supplement we note that the application of the velocity of light for synchronizing watches, which is normally used in an exposition of the theory of relativity is, firstly, not a unique method, but only one of a number of possible methods. Secondly, such a method is, indeed, in most cases the most convenient and appropriate one, not because the velocity of light is the maximum possible speed, but because it is universal – that is, it is the same for all inertial systems of reference (of course, provided one chooses identical rulers and clocks in all systems). Finally, when one nevertheless speaks of the velocity of light in vacuo c as the maximum possible speed, one has in mind the velocity with which perturbations, interactions, or 'signals' can be transferred. Such a statement is, indeed, valid – at least in the framework of the theory of relativity and in the whole of physics known to us. Light and other spots which we discussed do not violate that statement, although they can move with velocities $v > c$, that is, they cannot be used to transfer a signal with a velocity $v > c$. Indeed, let us consider a pulse (of light or of electrons) the intersection of which with a screen (spot) moves along the screen along the x-axis with a velocity $v > c$ and reaches the points 1 and 2 with coordinates

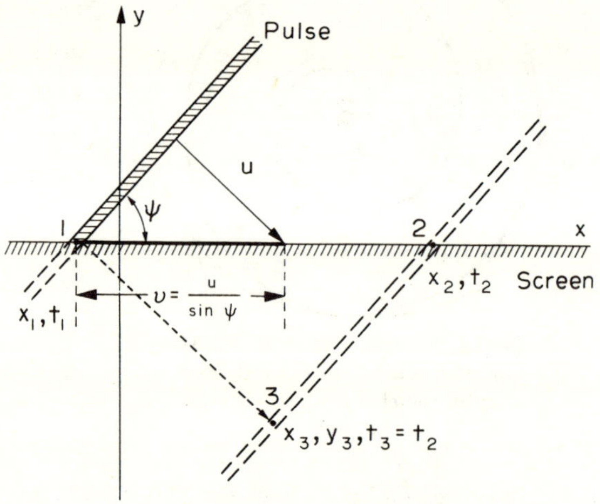

Fig.8.4 Intersection of a pulse and a plane screen

† When a rotating source emits particles with a velocity u the trajectories of such particles are: $r = r_0 + u(t - t_0)$, $\phi = \Omega t_0$, whence $r = r_0 + u(t - \phi/\Omega)$, where t_0 is the time of emission.

x_1 and x_2 at time t_1 and t_2 (Fig.8.4). Clearly, $x_2 = x_1 + v(t_2 - t_1)$ and if $v = u/\sin\Psi > c$, the events 1 and 2 are separated by a spatial interval that is, $(x_2 - x_1)^2 > c\,(t_2 - t_1)^2$. The perturbation ('cut') which is 'plotted' in the point 1 on the moving pulse at time t_1 turns out to be at time t_2 at the point 3 with coordinates $x_3 = x_1 + u\sin\Psi\,(t_2 - t_1)$, $y_3 = u\,(t_2 - t_1)\cos\Psi$ and $(x_3 - x_1)^2 + y_3^2 = u^2(t_2 - t_1)^2 \le c^2(t_2 - t_1)^2$. This perturbation does, however, not hit the point 2. Nonetheless, the superluminal velocity of the spot is, of course, in no sense an apparent one; it is just as real as any other velocity of a macroscopic structure or body. Therefore we emphasize that the superluminal velocities of spots are of a different nature from the apparent velocities such as $u_{\perp,max}$ (see (8.3)); the velocity $u_{\perp,max}$ can exceed c because we are dealing with the observation at a given time t of signals which are emitted at different times (vide supra). The retardation caused by the fact that the speed of propagation of light is finite is then important.

Taking retardation into account also affects appreciably the behaviour of spot when they are observed at some point. We shall restrict ourselves here to the simplest example of a light spot which moves with a constant velocity v along a plane screen and is observed in the point O' (Fig.8.5). By observation we understand here the reception of the light emitted by the spot when it hits the screen, that is, as the result of scattering, or due to luminescence of the screen when it is illuminated. If $v \le c$, the spot will be observed in the 'normal' way, as a spot moving on the screen from top to bottom. Let us now assume that $v \to \infty$, that is, that the whole track of the spot is outlined instantaneously. In that case the spot will first of all be noted in the point O which is closest to O' (the straight line OO' is perpendicular to the screen). After that the observer sees, clearly, two spots which move away from the point O in opposite directions. When $c < v < \infty$ sometimes one can also observe two spots.

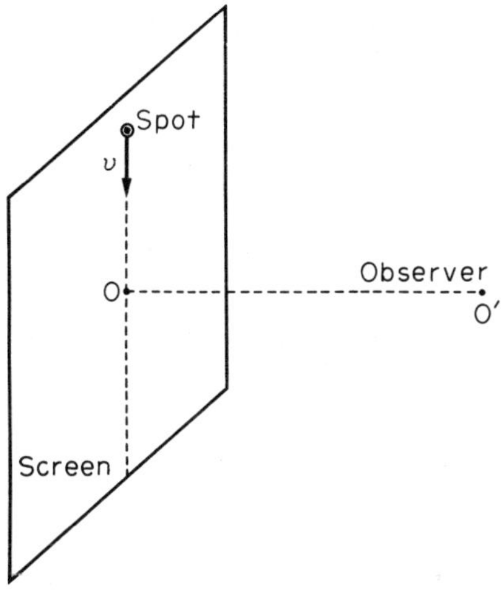

Fig. 8.5 Observation of a spot on a screen

The existence of superluminal velocities and of the above mentioned kind of superluminal sources, as we shall simply call in what follows sources moving with a velocity $v > c$, has been well known for a long time.† In the dark remained for a long time only the fact that such sources are 'in no sense worse' than subluminal sources in the framework of a macroscopic theory or in a general macroscopic approach. Macroscopic we understand here in the sense that a superluminal source is not a single, arbitrarily small particle, but always must be connected with a collection of such (microscopic) particles.‡ Moreover, in any real statement of the problem the number of particles corresponding to the motion of a superluminal source (spot) turns out to be very large. The usual field theory and in particular Eqns. (6.1) with a current density $\mathbf{j} = \rho\mathbf{v}$ which in principle can change and shift with any frequency and velocity forms an adequate theoretical basis for describing the radiation of superluminal sources.

---

† In general, we call sources moving with a velocity $v > v_{ph} = c/n$ superluminal sources. Such a terminology is reasonable, but by calling in the present chapter only those sources for which the velocity $v > c$ we shall hardly cause confusion, especially as we have stated clearly what we are doing.

‡ The concept 'macroscopic' with which we are dealing here is rather relative and appreciably 'weaker' than the conditions connected with the transition from the equations of microscopic electrodynamics — or, in the old terminology, from the equations from electron theory — to macroscopic electrodynamics. Indeed, from the equations of electrodynamics only the equation of continuity follows and for the rest the motion of the charges may be given 'externally' — whether this motion is compatible with the equations of motion for the particles is another question. Hence it is clear that already in the framework of the electron theory one can without contradictions assume the current density $\mathbf{j} = \rho\mathbf{v}$ to be within wide limits arbitrary and, in particular, one can put $v > c$. In that sense the calculations by Sommerfeld (1904, 1905, 1964) which were made as long ago as in 1904 where he considered the radiation by a charge moving with a velocity $v > c$ were completely correct. It is true that Sommerfeld has in mind the motion of a single charge when, indeed, $v < c$ — this conclusion follows from the special theory of relativity created by Einstein in 1905, provided we forget about the tachyons. In his papers in 1904 and 1905 Sommerfeld essentially anticipated the theory of the Cherenkov effect. It is very interesting that for more than 30 years, up to the work by Tamm and Frank (1937) nobody suspected that it would be interesting either to replace in the problem of the radiation of a uniformly moving source the velocity c by the phase velocity c/n or to consider a spot-like source moving with a velocity $v > c$.

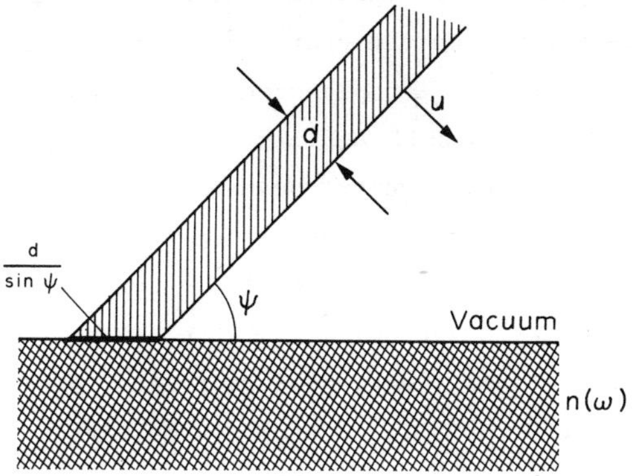

Fig.8.6 A charged filament incident upon a screen

Let us consider a charged filament which is incident with velocity u at an angle Ψ upon the boundary of a transparent medium with refractive index $n(\omega)$. In other words, we have a situation which is schematically depicted in Fig.8.6 and which is similar to the one shown in Fig.8.2. Up to its intersection with the boundary of the medium the charges, say, electrons or protons, which constitute the filament move uniformly. However, after the intersection with the boundary the charges are braked and as a result there appears a (polarization) current which moves with a velocity $v = u/\sin\Psi$ corresponding to the velocity with which the intersection with the boundary of the medium moves. Such a current occurs also when we neglect the braking of the charges by virtue of the transition effect — the change in the parameters of the medium along the path of the charge — leading to the emission of transition radiation. Clearly, one can describe the situation by saying that on reaching the medium the charges are stopped and after that they are, say, neutralized by currents in the medium. As a result there moves along the surface of the medium a charge q with a velocity v. For the sake of simplicity we shall assume that the filament has a square cross-section (with edgelength d) and consists of charges e and density N. The area of the intersection of the filament with the boundary of the medium, that is, the area of the spot, equals then $S = d^2/\sin\Psi$ and on that area we have a charge $q = Ned^3 \cot\Psi$ (the boundary of the medium is per unit time crossed by a charge $Ned^2 v \cos\Psi$ and to unit length along the velocity there corresponds a charge $Ned^2 \cos\Psi$ and, hence, the length $d/\sin\Psi$ of the spot corresponds just to this charge q). The solution of the problem of the radiation of a charge moving at the vacuum-medium boundary is known (Bolotovskii, 1960, 1962). The result for the emitted radiation can be written in the form

$$\frac{dW}{dt} = \frac{q^2 v}{c^2} \int \left(1 - \frac{c^2}{n^2(\omega)v^2}\right) F\,\omega\,d\omega \,. \tag{8.14}$$

It is clear that this formula changes into Eqn.(6.58) for a homogeneous medium when $F = 1$. The factor $F(\omega, \ldots)$ takes into account the effect of the boundary, the size of the source, and so on. One may think from general considerations that the same formula would also be applicable for a superluminal source with $v > c$, where $F = F(\omega, \Psi, d, \ldots)$ and also depends on the charge distribution in vacuo.† We can give the actual form of the factor F only when we have made an exact calculation and also used a completely well defined model of the source. This will be done in what follows. Now, however, we note that in any case the integration in (8.14) is over the range of frequencies satisfying the Cherenkov condition (6.53). We must then, of course, put $n = 1$ in the vacuum (we have assumed that the medium is bounded by a vacuum). If $v > c$, there occurs therefore in the vacuum (above the medium) always some radiation, provided only that $F \neq 0$. However, the factor F must in fact always be very small for waves with wavelengths $\lambda = 2\pi c/\omega$ which are less than the projection of the size of the spot on the direction of the wavevector **k**. In the

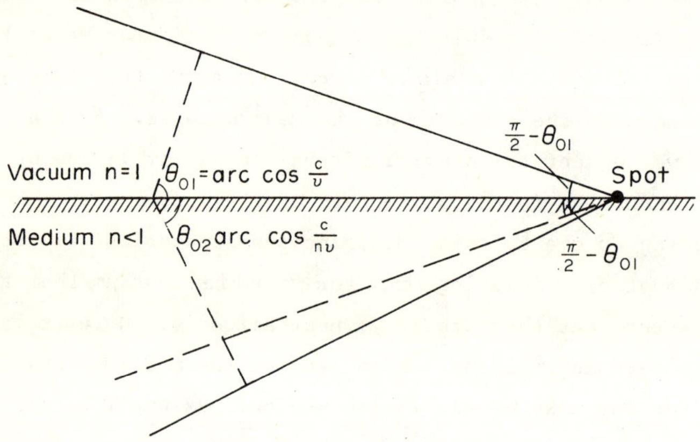

Fig.8.7 Cherenkov radiation of a spot

medium the situation is the same when $v > c$ and $n(\omega) > 1$, but when $n(\omega) < 1$ the condition $v > v_{ph} = c/n$ must also play the role of a cut-off factor — radiation in the medium is possible only when it is satisfied. One can also state in the general case that the radiation is characterized by the angle $\theta_{01} = \arccos(c/v)$ in vacuo and the angle $\theta_{02} = \arccos(c/n(\omega))$ in the medium, where θ is the angle between **k** and **v** (Fig.8.7). As the velocity of the

---

† It would be more precise to write the right-hand side of Eqn.(8.14) as a sum of two terms

$$\int \left(1 - \frac{c^2}{v^2}\right) F_1 \, \omega \, d\omega + \int \left(1 - \frac{c^2}{n^2(\omega) v^2}\right) F_2 \, \omega \, d\omega \,,$$

where the first term corresponds to the radiative power in vacuo and the second term to the radiative power in the medium. However, as long as we have not given the actual form of the factor F Eqn.(8.14) has a purely symbolic character and can thus be retained.

leading front of the electromagnetic waves in any medium is equal to c, when one takes dispersion into account, the radiation of a superluminal source in a medium is characterized not only by the angle θ_{02}, but also by the angle $\theta_{01} = \arccos(c/v)$ which in this case determines the opening of the cone corresponding to the leading front of the wave. Therefore, if $\theta > \theta_{01}$ the field in the medium vanishes. If we talk about the main part of the radiation, and not about the leading front, an analogous situation also occurs for the Cherenkov effect in a dispersive medium where the group velocity

$$v_{gr} = d\omega/dk = c(d(\omega n)/d\omega)^{-1}$$

is less than the phase velocity $v_{ph} = c/n$. There is here no special reason to discuss that side of the problem (see Tamm, 1939; Motz and Schiff, 1953).

Let us now consider the exact solution of the problem when a filament is incident upon a perfectly conducting plane. The geometry of the problem is the same as in Fig.8.6, but the medium with the refractive index $n(\omega)$ is replaced by a perfect conductor. A charge incident upon the conductor (crossing its boundary) vanishes for an external observer, that is, if we are concerned with the radiation mechanism, we are dealing here with transition radiation; we are, however, interested in the result of the interference of such radiation from a moving filament while it is known a priori that the resulting radiation will be directed at an angle $\theta_{01} = \arccos(c/v)$. The field of the filament in vacuo is the sum of the fields of the filament iteslf and its mirror image, that is, it is produced by a current with density

$$\mathbf{j} = Q\delta(z)\,\mathbf{u}_1\delta((\mathbf{b}_1\cdot\mathbf{r}) - ut)\ ,\quad y > 0\ ;$$
$$\mathbf{j} = -Q\delta(z)\,\mathbf{u}_2\delta((\mathbf{b}_2\cdot\mathbf{r}) - ut)\ ,\quad y < 0\ . \tag{8.15}$$

Here Q is the charge per unit length of the filament, $\mathbf{u}_1 = u\mathbf{b}_1$ and $\mathbf{u}_2 = u\mathbf{b}_2$ are the velocities of the filament and its mirror image ($b_1 = b_2 = 1$, $b_{1x} = b_{2x}$, $b_{1y} = -b_{2y}$, $b_{1z} = b_{2z} = 0$; the filament lies in the xy-plane and, for the sake of simplicity, it is assumed to be infinitesimally thin). The Fourier components of the current density are equal to

$$\mathbf{j}_\omega = \frac{1}{2\pi}\int \mathbf{j}\, e^{i\omega t}\, dt = \frac{Q\delta(z)}{2\pi}\left\{\mathbf{b}_1 \exp\left[\frac{i\omega}{u}(\mathbf{b}_1\cdot\mathbf{r})\right] - \mathbf{b}_2 \exp\left[\frac{i\omega}{u}(\mathbf{b}_2\cdot\mathbf{r})\right]\right\}\ .$$

At large distances from the screen we have for the Fourier components of the vector potential

$$\mathbf{A}_\omega = \frac{e^{ikR}}{cR}\int \mathbf{j}_\omega(\mathbf{r}')\exp[-i(\mathbf{k}\cdot\mathbf{r}')]\,d^3\mathbf{r}'$$

$$= i\,\frac{Qe^{ikR}}{cR}\left\{\frac{\mathbf{b}_1}{(\omega/u)b_{1y}-k_y} - \frac{\mathbf{b}_2}{(\omega/u)b_{2y}-k_y}\right\}\delta\left(\frac{\omega}{u}b_{1x}-k_x\right), \qquad (8.16)$$

where $\mathbf{k} = (\omega/c)\,\mathbf{s} = k\,\mathbf{s}$ is the wavevector of the emitted wave; clearly $s^2 = 1$, $k = \omega/c$. Moreover, one can easily find the magnetic field $\mathbf{H}_\omega = i\,[\mathbf{k}\wedge\mathbf{A}_\omega]$ and then the integral

$$\frac{c}{4\pi}\int_{-\infty}^{+\infty} H^2\,dt = \frac{c}{4\pi}\int_{-\infty}^{+\infty} dt\int_{-\infty}^{+\infty} d\omega\int_{-\infty}^{+\infty} d\omega'\,(\mathbf{H}_\omega\cdot\mathbf{H}_{\omega'})\,e^{i(\omega+\omega')t}$$

$$= \tfrac{1}{2}c\int_{-\infty}^{+\infty} d\omega\int_{-\infty}^{+\infty} d\omega'\,(\mathbf{H}_\omega\cdot\mathbf{H}_{\omega'})\,\delta(\omega+\omega')$$

$$= \tfrac{1}{2}c\int_{-\infty}^{+\infty} |\mathbf{H}_\omega|^2\,d\omega = c\int_0^{\infty} |\mathbf{H}_\omega|^2\,d\omega$$

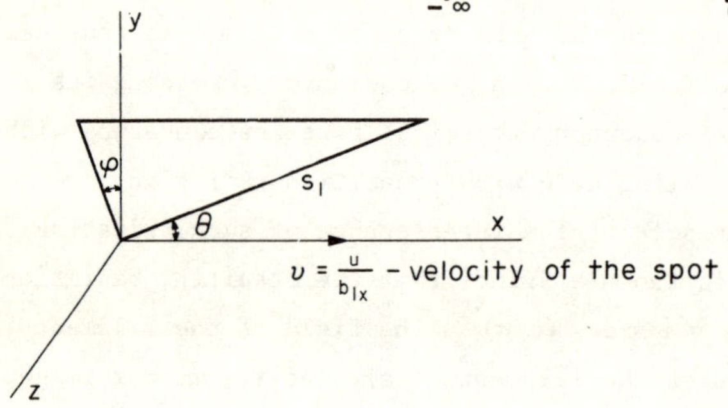

Fig.8.8 Evaluation of the Cherenkov radiation of a spot

We shall take the x-axis, along which the spot moves, as the polar axis and let the wavevector of the radiation $\mathbf{k} = (\omega/c)\,\mathbf{s}$ make an angle θ with the polar axis; we denote the azimuthal angle by ϕ (Fig.8.8) and in vacuo $-\frac{1}{2}\pi \le \phi \le \frac{1}{2}\pi$.

It is clear from Eqn.(8.16) that $\mathbf{A}_\omega$ is proportional to the delta-function of the argument $(\omega/u)b_{1x} - k_x$. It is clear that the magnetic field $\mathbf{H}_\omega$ will also be proportional to the delta-function and the energy of the radiation to the square of the delta-function. The integral of the square of a delta-function diverges, indicating that the energy of the radiation is infinite. This infinity is physically easily explainable — we assume that the filament intersects with the screen during an infinite period of time. To find a finite result we can consider the motion of the filament during a large, but finite period T. It is clear that the energy of the radiation will be proportional to T. The following formal procedure leads to the same result. We write

$$\delta^2\left(\frac{\omega}{u}b_{1x}-k_x\right)=\frac{u}{b_{1x}}\,\delta\left(\omega-\frac{k_x u}{b_{1x}}\right)\delta\left(\frac{\omega}{u}b_{1x}-k_x\right).$$

We now expand the first factor in a Fourier integral:

$$\delta^2\left(\frac{\omega}{u}b_{1x}-k_x\right)=\frac{u}{2\pi b_{1x}}\,\delta\left(\frac{\omega}{u}b_{1x}-k_x\right)\int_{-\infty}^{+\infty}\exp\left[i\left(\omega-\frac{k_x u}{b_{1x}}\right)t\right]dt\,.$$

Because of the presence of a delta-function in this product we can put the index of the exponent in the integrand equal to zero and we than find

$$\delta^2\left(\frac{\omega}{u}b_{1x}-k_x\right)=\frac{vT}{2\pi}\,\delta\left(\frac{\omega}{u}b_{1x}-k_x\right),$$

where T is the total time during which the filament moves and $v=u/b_{1x}$ is the velocity of the source (spot). Acting thus, we obtain the following expression for the energy emitted per unit time into a solid angle $d^2\Omega=\sin\theta\,d\theta\,d\phi$ in the frequency range $d\omega$:

$$\begin{aligned}\frac{dW_\omega(\theta,\phi)}{dt}&=\frac{1}{T}c\,|\mathbf{H}_\omega|^2\,R^2\sin\theta\,d\theta\,d\phi\,d\omega\\&=\frac{Q^2v}{2\pi\omega}\left\{\frac{[\mathbf{s}\wedge\mathbf{b}_1]}{(c/u)b_{1y}-s_y}-\frac{[\mathbf{s}\wedge\mathbf{b}_2]}{(c/u)b_{1y}+s_y}\right\}^2\delta\left(\frac{c}{v}-s_x\right)\sin\theta\,d\theta\,d\phi\,d\omega\,.\end{aligned}$$

As a result of the presence of the corresponding delta-function it is clear from this that waves are emitted only with wavevectors $\mathbf{k}$ which satisfy the condition $s_x=\cos\theta=c/v=\cos\theta_{01}$, as should be the case. After integrating over θ we find

$$\left.\begin{aligned}&\frac{dW_\omega(\phi)}{dt}=\frac{Q^2v}{2\pi\omega}\left\{\frac{[\mathbf{s}\wedge\mathbf{b}_1]}{(c/u)b_{1y}-s_y}-\frac{[\mathbf{s}\wedge\mathbf{b}_2]}{(c/u)b_{1y}+s_y}\right\}^2 d\phi\,d\omega\,,\\&\mathbf{b}_1=\{\sin\Psi,-\cos\Psi,0\}\,,\quad\mathbf{b}_2=\{\sin\Psi,\cos\Psi,0\}\,,\\&\mathbf{s}=\{\cos\theta_{01},\sin\theta_{01}\cos\phi,\sin\theta_{01}\sin\phi\}\,,\\&\cos\theta_{01}=\frac{c}{v}\,,\quad v=\frac{u}{\sin\Psi}\,,\end{aligned}\right\}\qquad(8.17)$$

where Ψ is the angle between the particle velocity $\mathbf{u}$ and the x-axis.

Finally we get

$$\frac{dW}{dt}=\frac{2Q^2v}{\pi}\frac{c^2}{u^2}\int_0^\infty\frac{d\omega}{\omega}\int_{-\pi/2}^{\pi/2}d\phi\,\times\frac{(1-u^2/v^2)^2-(1-c^2/v^2)(1-u^4/c^2v^2)\cos^2\phi}{[(c^2/u^2)\cos^2\Psi-(1-c^2/v^2)\cos^2\phi]^2}\,.\qquad(8.18)$$

It is clear from (6.58) that a charge q moving in a homogeneous medium would radiate 'in the range' $d\omega\, d\phi$ with a power

$$\frac{dW_\omega(\phi)}{dt} = \frac{q^2 v}{2\pi c^2}\left(1 - \frac{c^2}{v^2}\right)\omega \; d\phi \; d\omega \quad ,$$

where we have put $n = 1$. Comparing this expression with (8.17) we see that the filament is equivalent to a charge

$$q = Q\left|\frac{[\mathbf{s} \wedge \mathbf{b}_1]}{(c/u)b_{1y} - s_y} - \frac{[\mathbf{s} \wedge \mathbf{b}_2]}{(c/u)b_{1y} + s_y}\right| \frac{c}{\omega}\frac{1}{[1 - c^2/v^2]^{\frac{1}{2}}} . \tag{8.19}$$

As Q is the charge per unit length of the filament, the factor Q in (8.19) is the effective length of the filament responsible for the emission in the direction of **k**. This length is nothing but the length over which the transition radiation in the direction of **k** is formed. The integrals of (8.17) and (8.18) diverge as $\omega \to 0$; this is simply connected with the assumption that the filament is infinitely extended. The radiative power decreases with increasing ω, clearly, because with increasing ω the length over which the transition radiation is formed decreases. The frequency dependence may be different in other problems of a similar nature (vide infra).

We have already mentioned that one can consider the radiation mechanism when separate particles or a filament as a whole crosses the boundary of a conductor to be transition radiation. However, one can just as well (and with the same final result) assume that one gets bremsstrahlung as the result of the instantaneous stopping of the charges and their mirror images at the boundary (in the case of a perfect conductor these two possibilities are indistinguishable when one calculates the field in the vacuum; see Chapter 7). In general, the mechanism of the 'elementary radiation process' which ultimately leads to the Cherenkov effect is in a well known sense unimportant — the nature of the Cherenkov radiation (and in first instance one is concerned with the condition $\cos\theta_0 = c/vn(\omega)$) is determined by the interference of the waves emitted along the path of the source. What we have said is, of course, in complete agreement with the Huygens principle. The radiation by a charged filament, incident upon a screen, which we have considered is thus just the Cherenkov effect for $v > c$ and at the same time takes place in vacuo (it is true that the presence of some boundary with a medium is here necessary). The intensity of the radiation and its angular dependence on ϕ will change depending on the properties of the two media (of course, in order that one can observe the Cherenkov radiation at least one of the media must be transparent; above we assumed that medium 1 was a vacuum). For an anisotropic medium we must in condition (6.53) take the refractive index $n(\omega)$ for each normal wave separately and the value of n depends also on the angles to the axes of symmetry (crystal axes, direction of an external magnetic field, and so on). We mention especially radiation of waves in waveguides. In general there arise here a multitude of problems similar to the ones one meets with in the theory of Cherenkov radiation for $v < c$ (see Jelley, 1958; Ginzburg, 1960; Bolotovskii, 1960, 1962; Zrelov, 1968). It is also clear that the sources (spots) considered radiate also in the subluminal regime, that is, when $c/n < v < c$. Such sources are of interest also when one considers, for instance, the excitation of different kinds of surface waves as the result of the Cherenkov effect or of transition radiation on an inhomogeneous surface; in the last case the condition $v > c/n$ does, of course, not occur. What we have said is valid also in the case of waves which are not electromagnetic in kind; as an example we mention the possibility of exciting second sound waves in helium II by a moving source, say, by a laser beam moving along the surface of helium.

The radiation of a superluminal source to no extent reduces to the Cherenkov effect. For instance, even for uniform motion, but with 'modulations' of the source with a frequency ω_0 one will observe radiation with the Doppler frequency $\omega = \omega_0 |1 - (v/c)\, n \cos\theta|^{-1}$ (see (6.59)). One can realize the modulation in different ways - by an additional oscillation of the beam, a change in its density (along the beam), application of a 'lattice' (periodic inhomogeneities) on the screen, and so on. Of course, the peculiar features of the superluminal radiation with $v > c$, as in the case $c/n < v < c$ appear also when the source moves non-uniformly. As an example we may mention synchrotron (or, rather, quasi-synchrotron) radiation which occurs when a source moves along a circle. Such a case is realized when particles or photons which are emitted by a rotating source are incident upon a spherical or cylindrical screen. A more concrete model of this experiment is (Eidman, 1974a,b): a rotating source, say, a pulsar, emits a directed beam of γ-rays which is incident upon a more or less dense medium (a plasma) which is at a distance R from the source. The γ-rays, incident upon the screen are scattered by electrons which due to recoil acquire momentum and therefore produce a radial polarization which 'moves' along the screen with a velocity $v = \Omega R$. As a result there flows along the screen a current with density,

$$\mathbf{j} = \frac{\partial}{\partial t}\left[\mathbf{p}(t)\,\delta(\mathbf{r} - \mathbf{R}(t))\right], \quad \mathbf{p}(t) = p\{\cos\Omega t, \sin\Omega t, 0\},$$
$$\mathbf{R}(t) = R\{\cos\Omega t, \sin\Omega t, 0\}, \qquad (8.20)$$

where **p** is the electric dipole moment corresponding to the polarization produced which is assumed to be a point dipole; this is possible if we consider radiation with a wavelength λ which is appreciably longer than the size ℓ of the source.

The radiation occurring is when $v = \Omega R > c$ in its nature analogous to synchrotron radiation in a medium under conditions when $v > c/n$ (see Chapter 6); the total emitted power is equal to

$$\frac{dW}{dt} \approx \frac{p^2(1 + v^2/c^2)}{2v^3} \int_{\Omega \ll \omega \ll c/\ell} \omega^3 \, d\omega \,. \tag{8.21}$$

The integral has a cut-off at high frequencies because of the finite size of the dipole — this was not taken into account in (8.20) and (8.21); incidentally, Eidman (1974a,b) assumed that the dipole **p** in (8.20) was parallel to the z-axis rather than to the radius, that is, he put $\mathbf{p} = p\{0, 0, 1\}$, but this will probably only lead to a different numerical coefficient in Eqn.(8.21). In pulsar models a perturbation moving with a velocity $v > c$ in the plasma can be produced also by magnetic dipole radiation or by particle beams emerging from the pulsar.

In connection with the development of laser techniques there is special interest in the possibility to use light to produce a superluminal source. The use of a rotating beam is not really so easy, even if we apply a laser, when we require that the field strength in the spot would be sufficiently large when $v = \Omega R > c$. It is therefore simpler to achieve the incidence of a pulse on a screen — a dividing boundary (see Fig.8.2 and Eqns.(8.5) and (8.6) above). If the screen is a perfectly plane dividing boundary between two media and if we can consider the problem in the linear (weak field) approximation, we are dealing with the usual problem of the reflection and refraction of light. It is thus immediately clear — and, of course, follows from the field equations — that a pulse incident at an angle Ψ_1 is reflected also at an angle $\Psi_1' = \Psi_1$, while the angle of refraction Ψ_2 is determined by the law of refraction (Fig.8.9)

$$\frac{\sin \Psi_2}{\sin \Psi_1} = \frac{n_1}{n_2} \,, \quad \Psi_1' = \Psi_1 \,. \tag{8.22}$$

It is interesting that, as was already noted long ago by Frank (1942), the conditions (8.22) are the same as the conditions for the occurrence of the Cherenkov effect for the pulse considered, the intersection of which with the screen moves with a velocity $v = (c/n_1) \sin \Psi_1$ (see (8.5)). Indeed, the

Cherenkov angle in medium 1 is determined by the condition $\cos\theta_{01} = c/n_1 v = \sin\Psi_1$, whence we get $\Psi_1 = \Psi_1' = \frac{1}{2}\pi - \theta_{01}$, as should be the case (see Fig. 8.9). For medium 2 we have $\cos\theta_{02} = c/n_2 v = (n_1/n_2)\sin\Psi_1$, which is the same as (8.22) since $\cos\theta_{02} = \sin\Psi_2$. One can say literally that 'we did not know that we spoke prose' — the superluminal (for $n_1 > 1$ the more general) Cherenkov condition had been known already for several centuries. What we have said about the correspondence between the reflection and refraction laws and the Cherenkov condition is, however, natural as all these relations are obtained in the same way from the Huygens principle. To obtain any new results we must consider a problem where non-linearity for various media is taken into account — in particular for piezo-electrics.

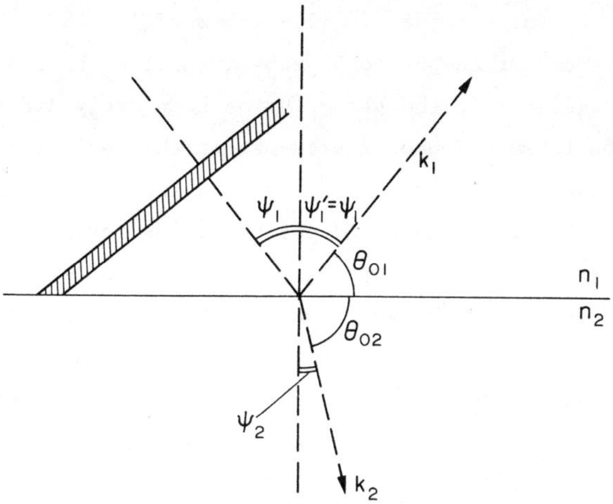

Fig.8.9 Reflection and refraction of a light pulse at a dividing boundary between two media

There is yet another remark which we want to make here; it concerns light spots in the case of rough or luminescent screens. In the last case the radiation emerging from the spot is, in general, incoherent. The same is practically also the case for rough screens as we are in that case usually dealing with rather large light spots with dimensions which are appreciably larger than the wavelength of the light. If the radiation is incoherent, interference is impossible and such specific features as the sharp directionality of the Cherenkov radiation disappear.

The historical fate of the studies of the radiation of sources moving with a velocity exceeding the phase velocity of light is very unusual. We are dealing with classical effects which are qualitatively clear already in the framework of the simplest optical representations, such as the Huygens principle or interference, and which are quantitatively described by the Maxwell equations. We have seen that the elementary reflection and refraction laws of light at a plane dividing boundary between two media in fact are the same as the conditions for Cherenkov radiation from a source which moves along the boundary.

The Cherenkov condition for a charge — a superluminal source (with velocity $v > c$) was obtained in 1904 (Sommerfeld, 1904). Nonetheless, the Cherenkov effect turned out to be experimentally observed only in 1934, and that by accident — in the sense that a completely different problem was studied — and the formulation of the theory of this effect (Tamm and Frank, 1937) required large and rather lengthy efforts (Frank, 1972). It is also interesting that in the first stage the possibilities for applying the Cherenkov effect in physics, both for measuring purposes and for understanding various phenomena, turned out to be very modest. However, the Cherenkov effect and related phenomena are now, in fact, widely used, and one can say that their study is a whole chapter of physics to which a vast number of papers and a number of surveys have been devoted (Jelley, 1958; Ginzburg, 1960; Bolotovskii, 1960, 1962; Zrelov, 1968). It would appear that the problem, if not exhausted, at any rate has already been studied fully and from all sides. However, this turned out to be not completely true, as witness the paper by Bolotovskii and Ginzburg 1972) and the present chapter. Indeed, the opinion was widely propagated (and the present author himself subscribed to it himself) that the Cherenkov effect and the anomalous Doppler effect could be observed only for waves which correspond to a refractive index $n(\omega) > 1$ (and to the condition $c/n < v < c$). In accordance with this it was thought that the corresponding effects were impossible in vacuo. Meanwhile there existed superluminal sources moving with a velocity $v > c$. These sources could be considered to a large extent on the same basis as 'ordinary' sources with velocities $v < c$. To be precise, superluminal sources are able to generate Cherenkov radiation in any medium, including a vacuum or a medium with $n(\omega) < 1$. General kinds of superluminal sources have on the whole the same peculiar features as sources moving with a velocity satisfying $c/n < v < c$, such as an anomalous Doppler effect. From the point of view of radiation theory the important difference between superluminal $(v > c)$ and subluminal $(v < c)$ sources consists in that a superluminal source can not be a separate 'elementary' particle and therefore is always extended. It is just the size of a superluminal source which in the first instance determines, especially for radiation in vacuo, the shortwavelength limit of the emitted spectrum. It is in that connection difficult to rely upon the use of superluminal sources, for instance, for the production of X-rays (such a possibility might turn out to be tempting as the fact that the refractive index $n(\omega)$ tends to unity at high frequencies, which is an obstacle for the use of the Cherenkov effect in the X-ray range for sources with $v < c$, does not play a critical role when $v > c$). We should not be amazed, however, if in the future some interesting applications would be found also for superluminal sources. Moreover, one might meet with superluminal sources in astronomy. Independent of this, the radiation by such sources (with $v > c$) of electromagnetic, gravitational — and, possible, neutrino — waves and a whole host of problems which then arise is, in our opinion, of undoubted physical interest.

Chapter IX

REABSORPTION AND RADIATIVE TRANSFER

Reabsoption and maser effect (wave amplification)
Radiative transfer equations.
Einstein coeffecient method and its application to the case of polarized radiation.
Reabsorption and amplification of synchrotron radiation in vacuo and when there is a 'cold' plasma present.

When we considered the synchrotron radiation of a system of particles (see Chapter 5) we assumed that different relativistic electrons radiated independently of one another — or, as one says, radiated incoherently. This referred not only to radiation in vacuo, but also to the radiation when the effect of a non-relativistic, 'cold' plasma was taken into account (see Chapter 6). Moreover, when along the line of sight there are a sufficiently large number of radiating particles, absorption and stimulated (induced) emission by the radiating particles themselves begins to play a role. One usually calls this process reabsorption. Reabsorption can, in principle, appreciably change the intensity and polarization of the radiation. Furthermore, negative reabsorption, that is, amplification of the radiation, is possible under some conditions. Such an amplification, or negative reabsorption, is called a maser effect. Of course, the nature of the reabsorption is closely connected with the nature of the radiation considered, that is, in the case of reabsorption of synchrotron radiation by the radiators one has in mind charged relativistic particles moving in a magnetic field. Not only such radiation, but also its reabsorption can be changed considerably if in the radiating volume we have apart from the relativistic electrons also a cold plasma. For instance, in the case of radiation in vacuo, reabsorption in any system of relativistic electrons which has an isotropic distribution of its velocity directions is positive, that is, under those conditions we have absorption. However, where there is a cold plasma present reabsorption of synchrotron radiation can become negative (vide infra). This means that the corresponding system, such as a layer or a cloud, of relativistic electrons with an isotropic velocity distribution will act like a maser.

† This chapter is mainly based upon section 3 of the review by Ginzburg and Syrovatskii (1969).

Although we shall discuss in what follows, as we have said, only the reabsorption of synchrotron radiation, much of what will be said below also refers to any other kind of radiation.

For a study of reabsorption one often uses expressions for the intensity of the radiation from separate particles averaged over all directions. The conditions for the applicability and even the nature itself of such an approach are unclear *a priori* and for a determination of the changes in the polarization it is totally unsuitable. It is sufficient to say that the synchrotron radiation has all the same a non-zero angular distribution while its polarization properties depend strongly on the angle $\psi = \chi - \alpha$ between the direction of the velocity and the direction of the radiation (see (5.21) to (5.23)). In a study of reabsorption, and especially, negative reabsorption, including polarization of the radiation it is thus necessary to give a more rigorous analysis of the angular and polarization properties of the synchrotron radiation. We must also add that a cold plasma in a magnetic field is anisotropic (magnetoactive) and far from always, even in a weak field, can it be taken to be isotropic to a good approximation, with a refractive index $n \approx 1 - \omega_p^2/2\omega^2$. The polarization characteristics are particularly sensitive in this respect as the rotation of the polarization plane (the Faraday effect) is an integral effect — it increases with the length of the path traversed by the wave (see, for instance, Ginzburg, 1970b and Chapter 11).

The general problem which is usually studied in separate particular cases is the following. The distribution function of relativistic electrons, $N_e(\mathbf{p}, \mathbf{R})$, the cold plasma density, $N(\mathbf{R})$, and the magnetic field strength, $\mathbf{H}(\mathbf{R})$, are given in some region (the 'source'). One must determine the radiation field both in the region considered (the source) and particularly far from it. One is usually dealing with the radiation from the source itself, but one may encounter the necessity to determine the effect of such a 'source' on the radiation from another source passing through it, where the other source may be situated further from the position of the receiver (this is just the reason why the term source acquires an arbitrary character).

We have already assumed above that the source is stationary so that the time t nowhere occurs. We must give up such a restriction in the case of moving or expanding sources. In practice, however, other simplifications, apart from the assumption of stationarity, are also possible. For instance, under cosmic conditions the anisotropy of the electron velocity distribution vanishes rather fast or, at least, is strongly diminished as the result of the existence of a whole number of instabilities (see Chapter 15). One can therefore most often assume that the distribution function of the relativistic electrons depends solely on their energy, that is, we can use the quantity $N_e(\mathcal{E}, \mathbf{R})$. Moreover, the position dependence of N_e, N, and $\mathbf{H}$ is under cosmic conditions always very slow, when we are dealing with distances of the order of the wavelength of the radiation. Hence, in general one can apply the geometric optics approximation, and often can simply assume that all quantities are constant along the line of sight in a region of some dimension L. Another possibility is to assume that the densities N_e and N are constant along a length L, but that the field $\mathbf{H}$ is on average random with a field strength H.

To describe the radiation in the general case we must use the Stokes parameters I, Q, U, and V, which can be combined into a tensor $I_{\alpha\beta}$ $(\alpha,\beta = 1,2)$

$$I_{\alpha\beta} = \frac{1}{2}\begin{pmatrix} I+Q & U+iV \\ U-iV & I-Q \end{pmatrix}; \quad \left.\begin{array}{ll} I = I_{11}+I_{22}, & V = i(I_{21}-I_{12}), \\ Q = I_{11}-I_{22}, & U = I_{21}+I_{12} \end{array}\right\} \tag{9.1}$$

The indexes $1,2$ correspond here to x- and y-axes at right angles to the line of sight. The connection between the Stokes parameters and the radiation intensity I, the degree of polarization Π, the ratio of the axes of the polarization ellipse p, and the angle $\tilde{\chi}$ which determines the orientation of that ellipse, is as follows (for details see, for instance, Gardner and Whiteoak, 1966; Ginzburg and Syrovatskii, 1966a; and also Zheleznyakov, 1970, § 6)

$$I = I, \quad \Pi = \frac{\{Q^2+U^2+V^2\}^{\frac{1}{2}}}{I}, \quad \beta = \arctan p,$$

$$\sin 2\beta = \frac{V}{\{Q^2+U^2+V^2\}^{\frac{1}{2}}}, \quad \tan 2\tilde{\chi} = \frac{U}{Q}. \tag{9.2}$$

From this it is, of course, obvious that the Stokes parameter I is just the intensity.

The Stokes parameters used — and any other parameters which can be expressed in terms of them — refer to radiation in some frequency range $\Delta\nu \ll \nu$ and correspond to expressions which are quadratic in the fields averaged over a time $\Delta t \gg 1/\Delta\nu$. In an anisotropic and, in particular, in a magneto-active medium the electric field $\mathbf{E}$ is, in general, not at right angles to $\mathbf{k}$, whereas the induction $\mathbf{D}$ is always orthogonal to the wavevector $\mathbf{k}$. It is therefore more convenient to define the tensor $I_{\alpha\beta}$ as $I_{\alpha\beta} = D_\alpha D_\beta^*$ $(\alpha, \beta = 1, 2;$ Zheleznyakov, 1968; 1974). The Stokes parameters and the quantities (9.2) will now also refer to the vector $\mathbf{D}$, and not to $\mathbf{E}$. We must bear in mind that the intensity $I = \mathrm{Tr}\, I_{\alpha\beta} \equiv D_{11}D_{11}^{*} + D_{22}D_{22}^{*}$ is in the general case no longer proportional to the energy flux. When we receive the radiation far from the source (in vacuo or, to be more precise, outside the anisotropic medium) this fact is clearly unimportant.

One uses the transfer equation to determine the tensor $I_{\alpha\beta}$; this equation has in recent years been studied and discussed in a number of papers (in great detail by Zheleznyakov (1968, 1974)). The transfer equation has in a uniform medium for the stationary case, where $I_{\alpha\beta}$ is independent of t, the form

$$\frac{dI_{\alpha\beta}}{dz} = \boldsymbol{\varepsilon}_{\alpha\beta} + (\mathcal{R}_{\alpha\beta\gamma\delta} - \mathcal{K}_{\alpha\beta\gamma\delta})\, I_{\gamma\delta}\,. \tag{9.3}$$

Here

$$\boldsymbol{\varepsilon}_{\alpha\beta} = \int R^2\, p_{\alpha\beta} N_e\, (\mathcal{E}, \mathbf{R}, \boldsymbol{\tau})\; d\mathcal{E}\, d^2\Omega_\tau \tag{9.4}$$

is the emissivity per unit volume, that is, the power of the spontaneous emission of unit volume into unit solid angle per unit frequency range; in the present chapter we do not use the notation $\boldsymbol{\varepsilon}_{ij}$ for the permittivity tensor and the remainder of the notation is the same as in Chapter 5. To fix the ideas we shall restrict ourselves to the case of synchrotron radiation when the quantities $p_{\alpha\beta}$ are given by Eqns. (5.21) to (5.23), (5.26) and (5.27); in this case it may turn out to be necessary to add to Eqn. (5.23) for p_{12} terms of higher order in $\xi = mc^2/\mathcal{E}$ (see Sazonov and Tsytovich, 1968). When there is a plasma present one must also in those formulae replace

$$\xi = mc^2/\mathcal{E} \quad \text{by} \quad \eta = \{\xi^2 + (\omega_p^2/\omega^2)\}^{\frac{1}{2}} \approx \{1 - n^2v^2/c^2\}^{\frac{1}{2}}$$

(see Ginzburg, Sazonov and Syrovatskii, 1968; Ginzburg and Syrovatskii, 1969). The tensors $\mathcal{R}_{\alpha\beta\gamma\delta}$ and $\mathcal{K}_{\alpha\beta\gamma\delta}$ in (9.3) characterize the changes in $I_{\alpha\beta}$ due, respectively, to Faraday roation and absorption of the radiation. The tensors $\mathcal{R}_{\alpha\beta\gamma\delta}$ and $\mathcal{K}_{\alpha\beta\gamma\delta}$ can be expressed in terms of parameters which characterize the 'normal' waves which can propagate in the medium considered.

In an anisotropic medium two 'normal' waves can propagate when we neglect spatial dispersion; in the case of uni-axial crystals or of a magnetized plasma they are called the ordinary (o or index 2) and extra-ordinary (e or index 1) waves. For normal waves in a uniform medium all quantities (the fields $\mathbf{E}, \mathbf{D}, \mathbf{H}$) depend exponentially on t and $\mathbf{R}$, for instance,

$$\mathbf{D}_{o,e} = A_{o,e}\, \boldsymbol{\Upsilon}_{o,e} \exp(-\kappa_{o,e} z)\, \exp[-i(\omega t - \delta_{o,e} - k_{o,e} z)]\,. \tag{9.5}$$

As in (9.3) the waves are here assumed to propagate along the z-axis, $\kappa_{o,e}$ is the amplitude absorption coefficient (the power absorption coefficient $\mu_{o,e}$ equals $2\kappa_{o,e}$; one often denotes by κ the absorption index $c\mu/2\omega$), $\omega = 2\pi\nu$, and $\mathbf{k}_{o,e}$ is the wavevector ($k_{o,e} = (\omega/c)\, n_{o,e}$, where $n_{o,e}$ is the refractive index). The complex vectors $\boldsymbol{\gamma}_{o,e}$ characterize the polarization of the normal waves; $A_{o,e}$ and $\delta_{o,e}$ are the arbitrary amplitudes and phases of these waves. In a magnetized plasma we can, if we neglect absorption (that is, in practice, for sufficiently weak absorption), put

$$\gamma_{\alpha o}\gamma^*_{\alpha o} = \gamma_{\alpha e}\gamma^*_{\alpha e} = 1\ ,\ \gamma_{\alpha o}\gamma^*_{\alpha e} = \gamma^*_{\alpha o}\gamma_{\alpha e} = 0\ , \tag{9.6}$$

where we have summed over $\alpha = 1,2$, as always over repeated Greek indexes (in other words, $\gamma_{\alpha o}\gamma_{\alpha o} = \mathrm{Tr}\ \gamma_{\alpha o}\gamma_{\beta o}$, and so on; for a detailed discussion of normal waves in a magnetized plasma see, for instance, Zhelezhnyakov, 1970; Ginzburg, 1970b, and Chapter 11). The components of the induction of an arbitrary radiation field in the frequency range $\Delta\omega$ have the form

$$D_\alpha(z,t) = \int_{\Delta\omega} A_e\gamma_{\alpha e}\exp\{-\kappa_e z - i(\omega t - \delta_e - k_e z)\}\,d\omega\ +$$
$$+ \int_{\Delta\omega} A_o\gamma_{\alpha o}\exp\{-\kappa_o z - i(\omega t - \delta_o - k_o z)\}\,d\omega\ .$$

If we now form the tensor $D_\alpha D^*_\beta$ from these components and also evaluate the derivative $\frac{d}{dz}(D_\alpha D^*_\beta)$ and average over time for a sufficiently narrow frequency range $\Delta\omega$ we can at once go over to Eqn.(9.3) (Zheleznyakov, 1968; 1974) where (do not confuse the index γ with the polarization vector $\boldsymbol{\gamma}$!)

$$\mathcal{R}_{\alpha\beta\gamma\delta} = -i(k_e - k_o)(\gamma_{\alpha e}\gamma^*_{\beta o}\gamma^*_{\gamma e}\gamma_{\delta o} - \gamma_{\alpha o}\gamma^*_{\beta e}\gamma^*_{\gamma o}\gamma_{\delta e})\ , \tag{9.7}$$

$$\mathcal{K}_{\alpha\beta\gamma\delta} = (\kappa_e + \kappa_0)(\gamma_{\alpha e}\gamma^*_{\beta o}\gamma^*_{\gamma e}\gamma_{\delta o} + \gamma_{\alpha o}\gamma^*_{\beta e}\gamma^*_{\gamma o}\gamma_{\delta e}) +$$
$$+ 2\kappa_e\gamma_{\alpha e}\gamma^*_{\beta e}\gamma^*_{\gamma e}\gamma_{\delta e} + 2\kappa_o\gamma_{\alpha o}\gamma^*_{\beta o}\gamma^*_{\gamma o}\gamma_{\delta o}\ . \tag{9.8}$$

If the absorption is sufficiently important, one can no longer assume the normal waves to be orthogonal (see (9.6)) and Eqns.(9.7) and (9.8) are no longer correct. This occurs, in particular, under conditions when relativistic particles (a hot plasma) and a cold plasma make comparable contributions to the real and/or imaginary parts of the dielectric permittivity tensor. Sazonov and Tsytovich (1968 ; Sazonov 1969) have considered the transfer Eqn. (9.3) without the assumption (9.6), but only in conditions when the effect of the plasma is not too strong.

If there is only one kind of waves (ordinary or extra-ordinary) in the medium, that is, if only fields of kind e or of kind o enter into the tensor $I_{\alpha\beta}$, we have $\mathfrak{R}_{\alpha\beta\gamma\delta} I_{\gamma\delta} = 0$. One can easily obtain this result formally, but it is straightaway obvious, as the polarization of the normal waves by definition does not change in a uniform medium. It is also clear that for a single normal wave $\mathfrak{K}_{\alpha\beta\gamma\delta} I_{\gamma\delta} = 2\kappa_{e,o} I_{\alpha\beta}$ and, if there are no sources, the transfer Eqn. (9.3) becomes

$$\frac{dI_{\alpha\beta}^{(e,o)}}{dz} = -2\kappa_{e,o} I_{\alpha\beta}^{(e,o)} = -\mu_{e,o}(\mathbf{k}) I_{\alpha\beta}^{(e,o)} . \tag{9.9}$$

Equation (9.9) is immediately obvious as it reflects the fact that in the normal waves the field vectors — in particular, the vector $\mathbf{D}$ — change due to absorption as $\exp(-\kappa_{e,o} z)$ (see (9.5)). The quantities $2\kappa_{e,o} = \mu_{e,o}(\mathbf{k})$ are the power (intensity) absorption coefficients along the wavevector $\mathbf{k}$. If the directions of the phase and group velocities, that is, the directions of the vectors $\mathbf{k}$ and $\mathbf{v}_{gr} = d\omega/d\mathbf{k}$, are the same, the quantities $2\kappa_{e,o}$ are, of course, the same as the absorption coefficients along the ray $\mu_{e,o}$. In the general case $\mu_{e,o} = 2\kappa_{e,o} \cos\phi_{e,o}$, where the $\phi_{e,o}$ are the angles between $\mathbf{k}_{e,o}$ and $\mathbf{v}_{gr,e,o}$. Under conditions when one can use (9.9) only the intensity of the radiation $I = I_{xx} + I_{yy}$, for which $dI^{(e,o)}/dz = -2\kappa_{e,o} I^{(e,o)}$, varies along $\mathbf{k}$ (that is, along the z-axis). As to the quantities Π, p (or β), and $\tilde{\chi}$, they remain unchanged for normal waves, as we have already mentioned. Formally, the same follows from (9.2) and (9.9) and is connected with the fact that the quantities Π, p, and $\tilde{\chi}$ depend only on ratios of Stokes parameters. It is also obvious that Π, p, and $\tilde{\chi}$ remain constant also when there are radiation sources of only one kind present in the medium. In that case

$$\frac{dI^{(e,o)}}{dz} = \boldsymbol{\varepsilon}_{e,o} - 2\kappa_{e,o} I^{(e,o)} \equiv \boldsymbol{\varepsilon}_{e,o} - \mu_{e,o}(\mathbf{k}) I^{(e,o)} . \tag{9.10}$$

One can generalize this equation for the case of a non-uniform medium, provided the geometric optics approximation is valid and, hence, provided it is admissible to use the ray concept — the possibility to use a ray treatment is restricted also by the condition that the absorption be weak (Ginzburg, 1970b). The corresponding transfer equation for the intensity $I^{(e,o)}$ of one kind of waves has the form

$$\frac{1}{v_{gr}} \frac{\partial I}{\partial t} + \frac{k^2}{|\cos\phi|} \frac{\partial}{\partial \ell} \left(\frac{I|\cos\phi|}{k^2} \right) = \boldsymbol{\varepsilon} - \mu I. \tag{9.11}$$

We have here taken into account the possibility that the problem is not stationary, $I = I^{(e,o)}$, and all the other expressions also refer to waves of type e or o with frequency ν; moreover, v_{gr} is the magnitude of the group velocity, ϕ is the angle between $\mathbf{k}$ and $\mathbf{v}_{gr}$, $k = (\omega/c)\, n$ is the length of the wavevector, and $\mu = 2\kappa \cos\phi$ is the absorption coefficient along the ray (a length element of the ray is $d\ell$). We note that if, in general, in (9.10) the quantity $I^{(e,o)}$ in a magnetized plasma, is not proportional to the energy flux (vide supra), we are in (9.11) dealing with the intensity in the real sense of the word, that is, with an energy flux per unit solid angle.

As far as we know, Eqn.(9.11) has not yet been generalized to the case where there are simultaneously two kinds of radiation present. For a uniform and stationary medium such a generalized equation reduces, of course, to (9.3). This equation is, undoubtedly, also valid when the functions $\boldsymbol{\varepsilon}_{\alpha\beta}$, $\mathfrak{R}_{\alpha\beta\gamma\delta}$, and $\mathcal{K}_{\alpha\beta\gamma\delta}$ change sufficiently slowly with changes in the coordinates. However, it is clear from comparing (9.11) and (9.3) that Eqn.(9.3) can only be valid in a non-uniform medium when one can neglect refraction (ray bending) and the derivatives dn/dz as compared to $dI_{\alpha\beta}/dz$. Moreover, the usual geometric optics approximation must, of course, be valid, that is, all quantities must change little along a wavelength in the medium, $\lambda = 2\pi c/n\omega$. For instance, the condition

$$\lambda \frac{d\boldsymbol{\varepsilon}_{\alpha\beta}}{dz} \ll \boldsymbol{\varepsilon}_{\alpha\beta}$$

must be satisfied. However, in general, the more rigid condition

$$\lambda \left| \frac{dn_{e,o}}{dz} \right| \ll |n_e - n_o|$$

must also be satisfied. Such an inequality is, like the condition for the validity of the geometric optics approximation, typical for a weakly anisotropic medium when calculating the polarization (see Zheleznyakov, 1970; and Ginzburg, 1970b, § 26).

In the above we have attempted to elucidate the problem of the transfer of radiation from a rather general point of view. It became then explicitly clear that one may encounter very complicated or, at least, cumbersome and difficult to visualize solutions for $I_{\alpha\beta}$ or the Stokes parameters. The situation becomes even more complicated when the cold plasma is rather dense and when the magnetic fields are rather strong. It is under such conditions no longer possible to take the effect of the plasma into account by replacing the quantity $\xi = mc^2/\mathcal{E}$ by $\eta = \{\xi^2 + \omega_p^2/\omega^2\}^{\frac{1}{2}}$ (for a discussion of this problem see Eidman, 1958, 1962; Ramaty, 1968; Melrose, 1968). A peculiar situation also occurs when the velocity distribution of the relativistic electrons turns out to be anisotropic (Sazonov and Tsytovich, 1968; Zheleznyakov and Suvorov, 1968, 1972; Sazonov, 1970, 1973). Moreover, even when the velocity distribution of the electrons is isotropic one must analyze in particular the case when the function $N_e(\mathcal{E})$ changes fast with energy. However, the function $N_e(\mathcal{E})$ may be assumed to be sufficiently smooth and one can use the expressions for the reabsorption coefficient given below provided $N_e(\mathcal{E})$ changes little over an energy interval $\Delta\mathcal{E}$ corresponding to radiation of neighbouring frequency overtones

$$\Omega_H = \frac{\omega_H^*}{\sin^2\chi} = \frac{eH}{mc} \frac{mc^2}{\mathcal{E}\sin^2\chi} .$$

The emitted frequency is $\omega = n\Omega_H$ $(n = 1, 2, 3, \dots)$ and, hence,

$$|\Delta\omega| = n\frac{eH}{mc} \frac{mc^2}{\mathcal{E}\sin^2\chi} \frac{\Delta\mathcal{E}}{\mathcal{E}} \sim \Omega_H$$

$$\Delta\mathcal{E} \sim \frac{\mathcal{E}}{n} = \frac{\Omega_H\mathcal{E}}{\omega} = \frac{eH}{mc\omega\sin^2\chi} mc^2 = \frac{eH\lambda}{2\pi\sin^2\chi} ,$$

where $\lambda = 2\pi c/\omega$ is the wavelength (we consider radiation in vacuo). The above-mentioned condition for the smoothness of the changes in the function $N_e(\mathcal{E})$ thus has the form

$$\frac{dN_e}{d\mathcal{E}}\Delta\mathcal{E} \sim \frac{dN_e}{d\mathcal{E}} \frac{eH}{mc\omega\sin^2\chi} mc^2 = \frac{dN_e}{d\mathcal{E}} \frac{eH\lambda}{2\pi\sin^2\chi} \ll N_e . \quad (9.12)$$

This condition is necessary for $\chi \approx \frac{1}{2}\pi$ and for $\chi < \frac{1}{2}\pi$ it is sufficient, but not necessary because of the χ-dependence of Ω_H (for details see Ginzburg and Syrovatskii, 1969).

Condition (9.12) can hardly be violated in the majority of cases encountered in astrophysics: the energy interval $\Delta\mathcal{E} = eH\lambda_0/2\pi$ is even in the metre band less than or of the order of 10^5 H eV and can be sufficiently important only in regions with a strong field, $H \gg 1$ Oe.

The discussion of the whole range of problems which we have touched upon here would require at least a special survey. Moreover, a whole set of problems referred to here have not yet been considered. In what follows we shall therefore only discuss two much more restricted problems, namely, the reabsorption of synchrotron radiation in vacuo and in a plasma for the case of quasi-longitudinal propagation. These cases are, however, probably the most important ones for applications to radio-astronomy. Before we dwell upon the corresponding calculations it is advisable to make a few remarks referring to the Einstein coefficients method and its application to polarized radiation.

When studying the transfer Eqn.(9.3) and other, similar equations for the intensity of normal waves or from Stokes parameters it is necessary to evaluate the coefficients $\boldsymbol{\varepsilon}_{\alpha\beta}$, $\mathfrak{R}_{\alpha\beta\gamma\delta}$, and $\mathcal{K}_{\alpha\beta\gamma\delta}$ in (9.3), the coefficients $\boldsymbol{\varepsilon}_{e,o}$ and $\mu_{e,o} = 2\kappa_{e,o}$ in (9.10), and so on. To find the emissivity $\boldsymbol{\varepsilon}_{\alpha\beta}$ one must use Eqn.(9.4). One can evaluate the other quantities in the general case by the kinetic equation method (Sazonov and Tsytovich, 1968; Zheleznyakov and Suvorov, 1968, 1972) where one must use the classical relativistic kinetic equation, when one is dealing with the classical region (condition $\hbar\omega \ll \mathcal{E}$). The corresponding calculations are rather cumbersome. Because of this and also due to a natural tendency to obtain results by the simplest and most translucent ways, the Einstein coefficients method plays a large role in the analysis of reabsorption. This method is in general well known, but its application to the case of a medium and especially an anisotropic medium, and also when the polarization of the radiation is taken into account has some peculiar features (see Ginzburg and Zheleznyakov, 1959a, 1965; Ginzburg, Zheleznyakov and Eidman, 1962; Zheleznyakov, 1970, § 27; Ginzburg, 1970b, § 12).

We explained in Chapter 6 (see also Chapter 12) that one may assume in a weakly absorbing (formally, a transparent) medium that the quanta in normal waves have an energy $\hbar\omega$ and a momentum $\hbar\mathbf{k}_\ell = (\hbar\omega/c)\, n_\ell(\omega, \mathbf{s})\,\mathbf{s}$, where $\mathbf{k}_\ell = \mathbf{k}_\ell \mathbf{s}$, $|\mathbf{s}| = 1$, and the subscript ℓ corresponds to a given wave (in a magnetized plasma we are dealing with ordinary, extra-ordinary, and plasma waves). In the classical region the results of the calculations are independent of the quantum constant $\hbar = h/2\pi$, but one can bring no arguments to bear against using quantal considerations, if they are the convenient ones. The energy flux and energy density in waves of type ℓ are equal to $I_\ell\, d\omega\, d^2\Omega$ and $\rho_\ell\, d\omega\, d^2\Omega$, where $d^2\Omega$ is an element of solid angle and where we have temporarily used spectral densities referring to the range $d\omega = 2\pi\, d\nu$. We have also the relation

$$I_\ell = \rho_\ell v_{gr,\ell} = \rho_\ell \left|\frac{d\omega}{d\mathbf{k}_\ell}\right| = \rho_\ell \frac{c}{|\cos\phi_\ell| \left|\dfrac{\partial(\omega n_\ell)}{\partial\omega}\right|} \tag{9.13}$$

($\mathbf{v}_{gr,\ell} = d\omega/d\mathbf{k}_\ell$ is the group velocity of waves of kind ℓ).

We introduce the Einstein coefficients A_m^n, B_m^n, and B_n^m where $A_m^n\, d\omega\, d^2\Omega$ is the probability for spontaneous emission per unit time in a transition between the states $m \to n$ with emission of a quantum of the given normal wave in the ranges $d\omega$ and $d^2\Omega$, $B^n_m \rho\, d\omega\, d^2\Omega$ the probability for the same induced transition, and $B^m_n \rho\, d\omega\, d^2\Omega$ the probability for the absorption of a quantum in the transition $n \to m$. The coefficients A^n_m, B^n_m, and B^m_n are connected through the relations

$$B^n_m = B^m_n \; ; \; B^n_m = \frac{(2\pi c)^3}{n^2_\ell \hbar \omega^3 \left| \frac{\partial (\omega n_\ell)}{\partial \omega} \right|} A^n_m \, . \tag{9.14}$$

Hence we get for the vacuum the usual relations (see, for instance, Heitler, 1947)

$$B^n_m = \frac{(2\pi c)^3}{\hbar \omega^3} A^n_m = \frac{2\pi c^3}{h \nu^3} A^n_m \, . \tag{9.15}$$

We can here understand by n and m any two states in momentum space for which the energy difference $\mathcal{E}_m - \mathcal{E}_n = \hbar\omega = h\nu$. If we were dealing with a transition between energy levels, we should take into account the statistical weights of these levels. By definition the relation (9.15) refers to waves with a single polarization. If we define the probability for an induced transition in vacuo as $\tilde{B}^n_m I_\nu d\nu\, d^2\Omega$ — as was done, for instance, by Ginzburg and Syrovatskii (1966a) — we have $\tilde{B}^n_m = (c^2/h\nu^3)\tilde{A}^n_m$, where $\tilde{A}^n_m d\nu\, d^2\Omega = 2\pi A^n_m d\nu\, d^2\Omega$ is the probability for spontaneous emission in the ranges $d\nu$ and $d^2\Omega$. Finally, if we understand by $\tilde{A}^n_m d\nu\, d^2\Omega$ the probability for the emission of waves with both polarizations, we can use the relation $\tilde{B}^n_m = (c^2/2h\nu^3)\tilde{A}^n_m$, which was applied by Ginzburg and Syrovatskii (1966a). However, we have not given here a sufficiently complete and well defined definition of the various relations. First of all, there is no basis for the transition made here to unpolarized radiation, although one might expect that in the way given one would obtain the average value of μ for the two possible polarizations. Secondly, in vacuo or in an isotropic medium one has polarization degeneracy (the possibility of choosing normal waves with arbitrary polarizations), due to which one can only get the polarization relations from an extra analysis.

Let us denote the electron densities in the states n and m with energies $\mathcal{E}_n$ and $\mathcal{E}_m$ by N_n and N_m; here $\mathcal{E}_m - \mathcal{E}_n = \hbar\omega \equiv h\nu$. By virtue of (9.14) the absorption coefficient μ_ℓ along the ray for a wave of type ℓ is then equal to

$$\mu_\ell = -\frac{\Delta I_\ell}{I_\ell} = \frac{\hbar\omega \sum (N_n B^m_n \rho_\ell - N_m B^n_m \rho_\ell)}{\rho_\ell} \left| \frac{d\omega}{d\mathbf{k}_\ell} \right|^{-1}$$

$$= \frac{8\pi^3 c^2}{\omega^2 n^2_\ell} |\cos\phi_\ell| \sum_{m \gtrless n} A^n_m (N_n - N_m) . \tag{9.16}$$

For the sake of simplicity we shall at once assume that $|n_\ell - 1| \ll 1$ and $|\cos\phi_\ell| \approx 1$. Moreover, if we consider the ultra-relativistic case ('needle' radiation, that is, radiation solely in the direction of the particle velocity) and assume the distribution function to be isotropic, we can put

$$N_n - N_m = N_e(\mathbf{p} - \hbar\mathbf{k}) - N_e(\mathbf{p}) = N_e\left(p - \frac{h\nu}{c}\right) - N_e(p) \approx -\frac{h\nu}{c}\frac{dN_e}{dp}.$$

We have used here the fact that in the classical case considered $h\nu \ll cp \approx \mathcal{E}$. Finally, the emissivity in the range $d\nu$ is equal to $\boldsymbol{\varepsilon}_\ell = \sum \tilde{A}^n_m N_m h\nu = \sum 2\pi A^n_m N_m h\nu$, and it is, for instance from a comparison with (9.4), clear that we must in (9.16) replace $A^n_m = \tilde{A}^n_m/2\pi$ by $(R^2/2\pi h\nu)\, p_\ell(\nu, \mathcal{E})$, where $p_\ell(\nu,\mathcal{E})$ is a function like the $p_{\alpha\beta}(\nu)$ in (5.26) and (5.27), but referring to a wave of the kind ℓ. We shall in what follows make clear what this means. We now give the final expression for μ_ℓ under the assumptions made:

$$\mu_\ell = -\frac{c}{\nu^2}\int \frac{dN_e(p)}{dp}\, q_\ell(\nu,\mathcal{E})\, p^2\, dp =$$

$$= -\frac{c^2}{4\pi\nu^2}\int \mathcal{E}^2 \frac{d}{d\mathcal{E}}\left(\frac{N_e(\mathcal{E})}{\mathcal{E}^2}\right) q_\ell(\nu,\mathcal{E})\, d\mathcal{E}, \qquad (9.17)$$

$$q_\ell(\nu,\mathcal{E}) = \int R^2\, p_\ell(\nu,\mathcal{E})\, d^2\Omega, \qquad (9.18)$$

where we have used the relations $\mathcal{E} = cp$ and $N_e(p)\, 4\pi p^2\, dp = N_e(\mathcal{E})\, d\mathcal{E}$; moreover, we must make clear that when replacing in (9.16) the summation by an integration the element of phase space equals $p^2\, dp\, d^2\Omega$, where $d^2\Omega$ is an element of solid angle into which the spontaneous emission takes place (by assumption the angle between $\mathbf{p}$ and $\mathbf{k}$ is small). According to (9.17) and (9.18) the problem of evaluating μ_ℓ consists of making clear the meaning of the quantities $p_\ell(\nu,\mathcal{E})$ or $q_\ell(\nu,\mathcal{E})$. In an anisotropic medium such a procedure is completely clear as q_ℓ is the spectral density of the power emitted by an electron in the form of a wave of kind ℓ. However, in vacuo or in an isotropic medium where we have polarization degeneracy, it is necessary to make clear which waves must be considered to be the normal ones when the reabsorption coefficient μ_ℓ is evaluated.

It is true that at first sight it may seem that the result of the calculations should not depend on the choice of the polarization of the normal waves as it is just this independence which leads to the polarization degeneracy. Of course, this is just the case when one performs the calculations consistently using the kinetic equation method: a well defined choice of polarization of the normal waves in the vacuum case and, in principle, the use itself of normal waves in any medium is not obligatory. However, in the Einstein coefficients method one works only with probabilities (intensities), and not with probability amplitudes (fields). That is why the coherence of the different normal waves which occurs, in general, in the case of degeneracy can not be taken into account in the Einstein coefficients method. In other words, by the very nature of the method its use is, generally speaking, connected with fixing the type of wave for which the absorption coefficient is evaluated.

For a 'pure' vacuum it is, of course, impossible to state which waves are normal waves. However, in that case there is no problem of reabsorption. If, nevertheless, one speaks of 'reabsorption in vacuo' one has merely in mind the possibility to neglect the effect of the cold plasma on the emission and reabsorption. However, a relativistic plasma in the source shows by the very nature of the reabsorption problem an effect on the absorption of the waves. It must also show some effect on the refractive index when the medium is anisotropic. This is obvious as we are dealing with relativistic particles (the plasma) in a magnetic field and hence there is in the system a physically preferred direction — that of the magnetic field. We showed in Chapter 5 that if the distribution function of the ultra-relativistic particles is not steeply anisotropic their radiation is linearly polarized with the electric vector in the waves a maximum in the direction at right angles to the projection $\mathbf{H}_\perp$ of the vector $\mathbf{H}$ onto the plane of the figure; in what follows we shall for the sake of simplicity call such waves polarized perpendicular to the field, and waves with the $\mathbf{E}$ vector parallel to $\mathbf{H}_\perp$ we call waves polarized along the field. It is under such conditions natural to expect that the normal waves will also be polarized along and perpendicular to the field (we remind ourselves that we restrict ourselves to angles $\chi \gg mc^2/\mathcal{E}$, that is, that we do not consider radiation by particles with velocities in a direction which makes a small angle $\chi \lesssim mc^2/\mathcal{E}$ with the field direction; in the case considered linear polarization also occurs only provided $\chi \gg mc^2/\mathcal{E}$). Sazonov (1969) has confirmed this statement by a quantitative analysis. When we apply Eqns. (9.17) and (9.18) to evaluate the reabsorption coefficients of synchrotron radiation by ultra-relativistic particles in vacuo we must thus calculate the coefficients $\mu_\perp$ and $\mu_\parallel$ for polarizations perpendicular to and along the field (in other words, the index ℓ is replaced by $\perp$ or $\parallel$). It is clear from all that has been said that we must take for $p_\perp(\nu, \mathcal{E})$ and $p_\parallel(\nu, \mathcal{E})$ in (9.18) expression (5.34) multiplied by $2\pi \sin\chi$. Hence

$$
\left.\begin{aligned}
q_\perp(\nu,\mathcal{E}) &= \frac{\sqrt{3}\,e^2\omega_H^*\sin\chi}{2c\eta}\,\frac{\nu}{\nu_c}\left[\int_{\nu/\nu_c}^{\infty}K_{\frac{5}{3}}(z)\,dz + K_{\frac{2}{3}}\left(\frac{\nu}{\nu_c}\right)\right],\\
q_{\parallel}(\nu,\mathcal{E}) &= \frac{\sqrt{3}\,e^2\omega_H^*\sin\chi}{2c\eta}\,\frac{\nu}{\nu_c}\left[\int_{\nu/\nu_c}^{\infty}K_{\frac{5}{3}}(z)\,dz + K_{\frac{2}{3}}\left(\frac{\nu}{\nu_c}\right)\right],
\end{aligned}\right\} \quad (9.19)
$$

$$
\left.\begin{aligned}
\nu_c &= \frac{3\sin\chi}{4\pi}\,\frac{\omega_H^*}{\eta^3}\,,\quad \omega_H^* = \frac{eH}{mc}\,\frac{mc^2}{\mathcal{E}}\,,\\
\eta &= \left\{\left(\frac{mc^2}{\mathcal{E}}\right)^2 + \frac{\omega_p^2}{\omega^2}\right\}^{\frac{1}{2}}\,,\quad \omega_p^2 = \frac{4\pi e^2 N}{m}
\end{aligned}\right\} \quad (9.20)
$$

For the sake of convenience we have given here the expressions which are also valid for an isotropic plasma $n = 1 - \omega_p^2/2\omega^2$, $|1-n| \ll 1$, although we shall in what follows in this chapter usually put $\eta = \xi = mc^2/\mathcal{E}$. In the ultra-relativistic case considered with an isotropic (or weakly anisotropic velocity distribution of radiating particles no waves are emitted with elliptical polarization (if we neglect terms of order $\eta = \{(mc^2/\mathcal{E})^2 + (\omega_p^2/\omega^2)\}^{\frac{1}{2}}$). Because of this we can when analyzing the particular radiation of a source limit ourselves to the Stokes parameters I and Q or the intensities $I_\perp = \frac{1}{2}(I+Q)$ and $I_{\parallel} = \frac{1}{2}(I-Q)$. The spectral density of the power of the total radiation is equal to

$$
q(\nu,\mathcal{E}) = q_\perp + q_{\parallel} = p(\nu,\mathcal{E}) = \frac{\sqrt{3}\,e^3\omega_H^*\sin\chi}{c\eta}\,\frac{\nu}{\nu_c}\int_{\nu/\nu_c}^{\infty}K_{\frac{5}{3}}(z)\,dz\,. \quad (9.21)
$$

In vacuo

$$
p(\nu,\mathcal{E}) = \frac{\sqrt{3}\,e^3 H_\perp}{mc^2}\,\frac{\nu}{\nu_c}\int_{\nu/\nu_c}^{\infty}K_{\frac{5}{3}}(z)\,dz\,, \quad (9.22)
$$

which is, of course, the same as (5.39).

We introduce the notation

$$
\mu_\perp(\chi) = \mu(\chi) + \lambda(\chi)\,,\quad \mu_{\parallel} = \mu(\chi) - \lambda(\chi)\,. \quad (9.23)
$$

The expression $\mu(\chi) = \frac{1}{2}(\mu_\perp + \mu_{\parallel})$ is exactly the same as the expression one would obtain (see, for instance, Ginzburg and Syrovatskii, 1966a) if we used the above-mentioned relation $\tilde{B}_m^n = (c^2/2h\nu^3)\,\dot{A}_m^n$ at once for waves with both polarizations. This result is obvious as $\mu(\chi)$ is the arithmetic average of $\mu_\perp$ and $\mu_{\parallel}$. For a power-law spectrum $N_e(\mathcal{E}) = K_e\mathcal{E}^{-\gamma}$ we have (Ginzburg, Sazonov and Syrovatskii, 1968)

$$
\mu(\chi) = \frac{\gamma + \frac{10}{3}}{\gamma + 2}\,\lambda(\chi) = g(\gamma)\,\frac{e^3}{2\pi m}\left(\frac{3e}{2\pi m^3 c^5}\right)^{\frac{1}{2}\gamma} K_e H_\perp^{\frac{1}{2}(\gamma+2)}\nu^{-\frac{1}{2}(\gamma+4)}\,. \quad (9.24)
$$

Table 9.1

γ	1	2	3	4	5
$g(\gamma)$	0.96	0.70	0.65	0.69	0.83
$\overline{g(\gamma)}$	0.69	0.47	0.40	0.44	0.46

In Table 9.1 we have given only the numerical values of $g(\gamma)$; the formula for $g(\gamma)$ is given by Ginzburg and Syrovatskii (1966a).

For the case of a power-law spectrum of the electrons we get for the degree of polarization of the synchrotron radiation in vacuo, neglecting reabsorption (compare (5.46))

$$\Pi_0 = \frac{I_\perp^{(o)} - I_\parallel^{(o)}}{I_\perp^{(o)} + I_\parallel^{(o)}} = \frac{\gamma+1}{\gamma+\frac{7}{3}}, \quad \frac{I_\perp^{(o)}}{I_\parallel^{(o)}} = \frac{1+\Pi_0}{1-\Pi_0} = \frac{3\gamma+5}{2} . \tag{9.25}$$

Moreover, according to (9.23) and (9.24) we have

$$\mu_\perp = \mu + \lambda = \frac{6\gamma+16}{3\gamma+10}\mu, \quad \mu_\parallel = \mu - \lambda = \frac{4}{3\gamma+10}\mu, \quad \frac{\mu_\parallel}{\mu_\perp} = \frac{\mu-\lambda}{\mu+\lambda} = \frac{2}{3\gamma+8} . \tag{9.26}$$

The transfer equation of the type (9.10) has in the case considered clearly the form

$$\frac{dI_{\perp,\parallel}}{dz} = \boldsymbol{\varepsilon}_{\perp,\parallel} - \mu_{\perp,\parallel} I_{\perp,\parallel} , \tag{9.27}$$

where

$$\boldsymbol{\varepsilon}_{\perp,\parallel} = \int q_{\perp,\parallel}(\nu, \mathcal{E}) N_e(\mathcal{E})\, d\mathcal{E} . \tag{9.28}$$

One can easily calculate the emissivity (9.28) for a power-law spectrum, using Eqns.(9.19) and (5.42). We restrict ourselves here to the remark that when there is no reabsorption it is clear from (9.25) that we get for the self-radiation of a uniform source of size L

$$I_{\perp,\parallel}^{(o)} = \boldsymbol{\varepsilon}_{\perp,\parallel} L , \quad \frac{I_\perp^{(o)}}{I_\parallel^{(o)}} = \frac{\boldsymbol{\varepsilon}_\perp}{\boldsymbol{\varepsilon}_\parallel} = \frac{3\gamma+5}{2} \tag{9.29}$$

When we take reabsorption into account we get by integrating Eqn.(9.27) with the condition that at the start of the layer (where $z=0$) $I_{\perp,\parallel} = 0$

$$I_\perp = \frac{\boldsymbol{\varepsilon}_\perp}{\mu_\perp}[1 - \exp(-\mu_\perp z)] , \quad I_\parallel = \frac{\boldsymbol{\varepsilon}_\parallel}{\mu_\parallel}[1 - \exp(-\mu_\parallel z)] . \tag{9.30}$$

For a thin layer (source of size L) $\mu_{\perp,\parallel} L \ll 1$ and

$$\frac{I_\perp}{I_\parallel} = \frac{I_\perp^{(o)}}{I_\parallel^{(o)}} = \frac{\boldsymbol{\varepsilon}_\perp}{\boldsymbol{\varepsilon}_\parallel} = \frac{3\gamma+5}{2} , \quad \Pi = \frac{I_\perp - I_\parallel}{I_\perp + I_\parallel} = \Pi_0 = \frac{\gamma+1}{\gamma+\frac{7}{3}} \tag{9.31}$$

For a thick layer $\mu_{\perp,\parallel} L \gg 1$ and

$$\frac{I_\perp}{I_\parallel} = \frac{\varepsilon_\perp \mu_\parallel}{\varepsilon_\parallel \mu_\perp} = \frac{3\gamma+5}{3\gamma+8} < 1 \quad , \quad \Pi = \left| \frac{I_\perp - I_\parallel}{I_\perp + I_\parallel} \right| = \frac{3}{6\gamma+13} \ . \tag{9.32}$$

Of course, those parts of expressions (9.31) and (9.32) which do not contain the index γ have a general value and refer not solely to a power-law spectrum. We remind ourselves that when we use a power-law spectrum in our calculations we assume that $\gamma > \frac{1}{3}$ (see Chapter 5).

Assume that the magnetic field along the line of sight is on average random in direction. We further assume that when waves propagate in such a field their polarization does not change when the direction of the field changes (this occurs when the geometric optics approximation is inapplicable to describe the polarization of the normal waves due to the fact that conditions such as $\lambda |dn_{e,o}/dz| \ll |n_e - n_o|$, which we mentioned earlier, are not satisfied; for details see Zheleznyakov, 1970, § 24). In such cases the anisotropy of the absorption vanishes for the propagation of waves in a random field and waves with any polarization will be absorbed identically with some absorption coefficient μ. For a given angle χ the average absorption coefficient is equal to

$$\tfrac{1}{2}(\mu_\perp + \mu_\parallel) = \mu(\chi)$$

To obtain $\bar{\mu}$, that is, the average of $\mu(\chi)$ over the angles χ between the field **H** and the line of sight (the velocity of the emitting electrons) it is natural to evaluate the expression

$$\left.\begin{aligned} \bar{\mu} &= \tfrac{1}{2}\int_0^\pi \mu(\chi)\sin\chi\, d\chi = \\ &= \overline{g(\gamma)}\, \frac{e^3}{2\pi m}\left(\frac{3e}{2\pi m^3 c^5}\right)^{\frac{1}{2}\gamma} K_e H^{\frac{1}{2}(\gamma+2)}\, \nu^{-\frac{1}{2}(\gamma+4)} \ , \\ \overline{g(\gamma)} &= \frac{\sqrt{3\pi}}{8}\,\frac{\Gamma(\frac{1}{4}(\gamma+6))}{\Gamma(\frac{1}{4}(\gamma+8))}\,\Gamma\left(\frac{3\gamma+2}{12}\right)\Gamma\left(\frac{3\gamma+22}{12}\right) . \end{aligned}\right\} \tag{9.33}$$

Ginzburg, Sazonov and Syrovatskii (1968) have shown that Eqn.(9.33) which we did not derive rigorously is, indeed, the reabsorption coefficient for a random field. We have given the numerical values of the function $\overline{g(\gamma)}$ in Table 9.1. For the sake of convenience we give also the following expression:

$$\bar{\mu} = \overline{g(\gamma)} \times 0.019\,(3.5\times 10^9)^\gamma\, K_e H^{\frac{1}{2}(\gamma+2)}\, \nu^{-\frac{1}{2}(\gamma+4)}\ \text{cm}^{-1} \ . \tag{9.34}$$

Ginzburg and Ozernoi (1966) have discussed reabsorption in a non-uniform field. We give below (see (9.47)) the formula for μ for the case of a mono-energetic electron spectrum.

The problem of the region of applicability of these formulae arises now, of course, that is, the question of whether one can neglect the effect of the cold plasma (the electron density N). To be able to do this it is, first of all, necessary that the cold plasma does not affect the emission of the relativistic electrons. Hence we arrive at the condition (see (6.75) and below in the present chapter)

$$\nu \gg \frac{4ecN}{3H_\perp} = \frac{4\omega_p^2}{3\pi\,\omega_H \sin\chi} \approx 20\,\frac{N}{H_\perp} = 20\,\frac{N(\mathrm{cm}^{-3})}{H(\mathrm{Oe})\sin\chi}\ \mathrm{s}^{-1}\,. \qquad (9.35)$$

Secondly, the rotation of the plane of polarization by the cold plasma must be small, whence we get the condition (see, for instance, Ginzburg and Syrovatskii, 1966a)†

$$\nu \gg 10^2\sqrt{NHL\cos\chi}\,. \qquad (9.36)$$

This condition is, of course, not necessary, if the polarization of the normal waves is determined by the relativistic particles (this is the case when the inequality, which is the opposite of the inequality (9.41) given below, holds). Thirdly, the normal waves are linearly polarized only when the same inequality, which is the opposite of inequality (9.41), holds. All three conditions, written down here, are together sufficient for completely neglecting the effect of the plasma. However, this neglecting is under certain circumstances also possible under less stringent requirements.

When in the radiating region there is a cold plasma present, we must take its influence into account, first of all, on the radiation process and, secondly on the wave propagation. For calculating the radiation one can, in general, assume the plasma to be isotropic provided

$$\left.\begin{aligned} \omega &\gg \omega_H = \frac{eH}{mc} = 1.76\times10^7\ H(\mathrm{Oe})\ \mathrm{s}^{-1}\,,\\ \omega &\gg \omega_p = \left\{\frac{4\pi Ne^2}{m}\right\}^{\frac{1}{2}} = 5.64\times10^4\,N(\mathrm{cm}^{-3})^{\frac{1}{2}}\ \mathrm{s}^{-1} \end{aligned}\right\} \qquad (9.37)$$

in which case

$$n_e = n_o = n = 1 - \frac{\omega_p^2}{2\omega^2}\,, \quad |1-n| \ll 1\,. \qquad (9.38)$$

In that case the effect of the plasma on the radiation is already reflected, for instance, in Eqns.(9.19) to (9.21).

As to the wave propagation, conditions (9.37) are, of course, insufficient for neglecting anisotropy. However, the considerable simplification when these

---

† Here and below we equate the angle α between **H** and **k** to the angle χ between **H** and the electron velocity **v**, as we are dealing with synchrotron radiation.

conditions are satisfied is first of all connected with the possibility to assume in most cases that the wave propagation is quasi-longitudinal, in which case[†]

$$n_e = 1 - \frac{\omega_p^2}{2\omega(\omega - \omega_L)} , \quad n_o = 1 - \frac{\omega_p^2}{2\omega(\omega + \omega_L)} ,$$
$$n_e - n_o = \frac{\omega^2 \omega_L}{\omega^3} , \quad \omega_L = \omega_H \cos\chi . \tag{9.39}$$

Here we have already assumed that $|n_{e,o} - 1| \ll 1$. The e and o waves are both circularly polarized with opposite directions of rotation of the field vectors, and in the extra-ordinary wave these vectors rotate in the same direction as an electron in the magnetic field. The condition for the applicability of the quasi-longitudinal approximation (9.39) are for the conditions which are of interest to us

$$\frac{\omega_H^2 \sin^4\chi}{4\omega^2\cos^2\chi} \ll 1 , \quad \frac{\omega_H^2}{2\omega^2}\sin^2\chi \ll 1 . \tag{9.40}$$

One sees easily that Eqns.(9.39) are practically always applicable in radio-astronomy, provided the effect of the relativistic particles on the refractive index is small compared to the effect of the cold plasma which is taken into account in (9.39).

As a result of the effect of the relativistic particles (Sazonov, 1969, 1970, 1973)

$$|n - 1| \sim \frac{c}{2\omega}\,\mu(\chi) = \frac{\lambda}{4\pi}\,\mu(\chi) ,$$

where $\mu(\chi)$ is the reabsorption coefficient given by (9.24) or by (9.33) and (9.34). Hence, we can neglect the role of the relativistic particles when evaluating n provided $(n_o - n_e) \gg c\mu/2\omega$, which leads to

$$N \gg mc^2 \left(\frac{3e}{2\pi m^3 c^5}\right)^{\frac{1}{2}\gamma} \frac{(\sin\chi)^{\frac{1}{2}}(\gamma + 2)}{\cos\chi} K_e H^{\frac{1}{2}\gamma} \nu^{-\frac{1}{2}\gamma}$$
$$\sim 10^{-6}\,(3.5\times10^9)^{\gamma}\,\frac{(\sin\chi)^{\frac{1}{2}}(\gamma + 2)}{\cos\chi} K_e H^{\frac{1}{2}\gamma}\ \mathrm{cm}^{-3} . \tag{9.41}$$

When Eqns.(9.39) are applicable, the problem of radiation transfer is appreciably simplified. The tensors $\mathfrak{R}_{\alpha\beta\gamma\delta}$ and $\mathcal{K}_{\alpha\beta\gamma\delta}$ become then very simple so that one can write Eqn.(9.3) in the following form

---

† We have given elsewhere (Ginzburg, 1970b) details about all the conditions given here and the formulae for n_e and n_o.

$$\left.\begin{aligned}\frac{dI}{dz} &= \boldsymbol{\varepsilon}_I - \tfrac{1}{2}(\mu_e+\mu_o)\,I + \tfrac{1}{2}(\mu_e-\mu_o)\,V\,,\\ \frac{dV}{dz} &= \boldsymbol{\varepsilon}_V - \tfrac{1}{2}(\mu_e+\mu_o)\,V + \tfrac{1}{2}(\mu_e-\mu_o)\,I\,,\\ \frac{dQ}{dz} &= \boldsymbol{\varepsilon}_Q - \tfrac{1}{2}(\mu_e+\mu_o)\,Q + (k_e-k_o)\,U\,,\\ \frac{dU}{dz} &= \boldsymbol{\varepsilon}_U - \tfrac{1}{2}(\mu_e+\mu_o)\,U - (k_e-k_o)\,Q\,,\end{aligned}\right\} \quad (9.42)$$

where we have changed to the Stokes parameters (Zheleznyakov, 1968, 1974). Here $k_{e,o} = (\omega/c)\,n_{e,o}$ and $\boldsymbol{\varepsilon}_{I,V,Q,U}$ are combinations of the $\boldsymbol{\varepsilon}_{\alpha\beta}$ corresponding to the transition from the tensor $I_{\alpha\beta}$ to the Stokes parameters (see (9.1); for instance, $\boldsymbol{\varepsilon}_I = \boldsymbol{\varepsilon}_{11} + \boldsymbol{\varepsilon}_{22}$). The Faraday effect is determined by the difference $n_e - n_o = (c/\omega)(k_e - k_o)$ and does not affect the equations for the intensity I and the degree of circular polarization $\Pi_c = V/I$, but it affects the degree of linear polarization $\Pi_\ell = \{Q^2 + U^2\}^{\frac{1}{2}}/I$ and the orientation of the ellipse $\tilde{\chi}$ (we remind ourselves that $\tan 2\tilde{\chi} = U/Q$). It is convenient to introduce the intensities of the extra-ordinary and the ordinary radiation

$$I_e = \tfrac{1}{2}(I - V)\ , \quad I_o = \tfrac{1}{2}(I + V)\ . \qquad (9.43)$$

Using (9.42) and (9.43) we can write

$$\frac{dI_{e,o}}{dz} = \boldsymbol{\varepsilon}_{e,o} - \mu_{e,o}\,I_{e,o}\ , \quad \boldsymbol{\varepsilon}_e = \tfrac{1}{2}(\boldsymbol{\varepsilon}_I - \boldsymbol{\varepsilon}_V)\ , \quad \boldsymbol{\varepsilon}_o = \tfrac{1}{2}(\boldsymbol{\varepsilon}_I + \boldsymbol{\varepsilon}_V)\ . \qquad (9.44)$$

This result is rather obvious from the start: in the linear medium which we consider the intensity (energy flux) of each of the normal waves is independent of the intensity of the other wave. This conclusion refers to any normal wave, but if their polarization is arbitrary (elliptical) the intensities I_e and I_o are complicated expressions in terms of the Stokes parameters and it is not clear that it is expedient to use the Stokes parameters. However, also in the case of quasi-longitudinal propagation we must use all four Stokes parameters to characterize fully the radiation (see Zheleznyakov, 1968, 1974 for the solution of Eqns. (9.42)).

Nonetheless we shall restrict ourselves in what follows merely to discussing the problem of the changes in the intensity of the e and o waves, that is, we shall use Eqns. (9.44). When waves of only one kind are present, the polarization is given and Eqn. (9.44) completely describes the radiation. Such a situation occurs, in particular, when there is negative reabsorption and the layer is sufficiently thick. Indeed, in the case of negative reabsorption the intensity of the waves grows exponentially when they pass through the layer when its thickness increases. Therefore, when leaving a thick layer the radiation consisting of those normal waves for which the absolute magnitude of the reabsorption coefficient μ is largest will dominate.

We have already mentioned that, if (9.37) holds the effect of the plasma on the radiation is taken into account by Eqns.(9.19) to (9.21). Up to terms of order $mc^2/\mathcal{E}$ half of the total radiation power $q(\nu,\mathcal{E}) \equiv p(\nu,\mathcal{E})$, given by Eqn. (9.21), 'goes over' into each normal, circularly polarized wave. Thus we have $q_{e,o} = \frac{1}{2} p(\nu,\mathcal{E})$ and, from (9.17),

$$\left.\begin{aligned}
\mu_e = \mu_o &= -\frac{c^2}{8\pi\nu^2}\int_0^\infty \mathcal{E}^2 \frac{d}{d\mathcal{E}}\left(\frac{N_e(\mathcal{E})}{\mathcal{E}^2}\right) p(\nu,\mathcal{E})\, d\mathcal{E} = \\
&= \frac{c^2}{8\pi\nu^2}\int_0^\infty \frac{N_e(\mathcal{E})}{\mathcal{E}^2}\frac{d}{d\mathcal{E}}\{\mathcal{E}^2 p(\nu,\mathcal{E})\}\, d\mathcal{E}\ , \\
p(\nu,\mathcal{E}) &= \sqrt{3}\,\frac{e^3 H_\perp}{mc^2}\left[1+\frac{\nu_p^2}{\nu^2}\left(\frac{\mathcal{E}}{mc^2}\right)^2\right]^{-\frac{1}{2}} \frac{\nu}{\nu_c}\int_{\nu/\nu_c}^\infty K_{\frac{5}{3}}(z)\, dz\ , \\
\nu_c &= \frac{3\, eH_\perp}{4\pi mc}\left(\frac{\mathcal{E}}{mc^2}\right)^2\left[1+\frac{\nu_p^2}{\nu^2}\left(\frac{\mathcal{E}}{mc^2}\right)^2\right]^{-\frac{3}{2}}, \\
\frac{\nu_p^2}{\nu^2} &= 1-n^2\ ,\quad \nu_p^2 = \frac{\omega_p^2}{4\pi^2} = \frac{Ne^2}{\pi m}\ .
\end{aligned}\right\} \qquad (9.45)$$

For a better understanding of these formulae and for a comparison with other expressions we note that

$$1+\frac{\nu_p^2}{\nu^2}\left(\frac{\mathcal{E}}{mc^2}\right)^2 = \left(\frac{\mathcal{E}}{mc^2}\right)^2\left[\left(\frac{mc^2}{\mathcal{E}}\right)^2+\frac{\omega_p^2}{\omega^2}\right] =$$

$$= \left(\frac{\mathcal{E}}{mc^2}\right)^2 \eta^2 \approx \left(\frac{\mathcal{E}}{mc^2}\right)^2\left(1-n^2\frac{v^2}{c^2}\right).$$

It is clear from (9.45) that the effect of the plasma on synchrotron radiation and its reabsorption is unimportant when

$$\frac{\nu_p^2}{\nu^2}\left(\frac{\mathcal{E}}{mc^2}\right)^2 \ll 1\ . \qquad (9.46a)$$

This condition leads to the earlier inequality (9.35) (see (6.74)). In the region (9.46a) the integrand in Eqn.(9.45) for $\mu_{e,o}$ is always positive and from this it follows that in that case we have always $\mu_{e,o} > 0$. As condition (9.46a) is always satisfied for a vacuum, μ is always positive in vacuo.† If however

$$\frac{\nu_p^2}{\nu^2}\left(\frac{\mathcal{E}}{mc^2}\right)^2 \gg 1\ , \qquad (9.46b)$$

† see footnote on next page.

the effect of the plasma is pronounced and for a suitable choice of the electron spectrum $N_e(\mathcal{E})$ the coefficient $\mu_{e,o}$ may turn out to be negative (McCray, 1966; Zheleznyakov, 1967a,b; Zheleznyakov and Suvorov, 1968; Bratman and Suvorov, 1969; Sazonov, 1970, 1973; Kaplan and Tsytovich, 1973; Tsytovich and Kaplan, 1974). It is immediately clear from (9.45) that for a power-law electron spectrum, $N_e(\mathcal{E}) = K_e \mathcal{E}^{-\gamma}$, negative values of $\mu_{e,o}$ are possible only when $\gamma < -2$, that is, for a function $N_e(\mathcal{E})$ which in some region increases faster than $\mathcal{E}^2$. In the opposite case the integrand in (9.45) is always positive, since the function $p(\nu,\mathcal{E})$ is positive. The region where the function $N_e(\mathcal{E})$ increases with increasing $\mathcal{E}$ can usually not be very large and, at any rate, when $\mathcal{E}$ increases further, one must shift into a region where $N_e(\mathcal{E})$ decreases. A power-law spectrum is therefore in the case under discussion of negative reabsorption of no interest (Zheleznyakov (1967a,b) has considered a spectrum of the form $N_e(\mathcal{E}) = K_e \mathcal{E}^{\gamma'}$, $\gamma' > 2$, $\mathcal{E}_1 < \mathcal{E} < \mathcal{E}_2$; $N_e(\mathcal{E}) = 0$, $\mathcal{E} < \mathcal{E}_1$, $\mathcal{E}_2 < \mathcal{E}$). Of more interest is a spectrum with a rather sharp maximum at an energy $\mathcal{E}_i$, say (the width $\Delta\mathcal{E}$ of the spectrum must satisfy the condition $\Delta\mathcal{E}/mc^2 \ll 3\,eH_\perp \nu^2/4\pi mc\nu_p^3$, which is completely compatible with inequality (9.12)). For such a spectrum we have (Zheleznyakov, 1967a,b)

$$\mu = \mu^{I} = \frac{4\pi}{3\sqrt{3}}\frac{e}{H_\perp}\left(\frac{mc^2}{\mathcal{E}_i}\right)^5 N_{e,i}\,K_{\frac{5}{3}}(Z_i)\ ,\quad Z_i = \frac{4\pi mc\nu}{3eH_\perp}\left(\frac{mc^2}{\mathcal{E}_i}\right)^2 \tag{9.47}$$

when $\mathcal{E}_i^2 \ll \mathcal{E}_*^2 \equiv (mc^2\,\nu/\nu_p)^2$ (see (9.45) and (9.46)). If, however, $\mathcal{E}_i^2 \gg \mathcal{E}_*^2$ (see (9.46)), we have

$$\left.\begin{aligned} \mu = \mu^{II} &= \frac{\sqrt{3}\,e^3 H_\perp}{8\pi m\nu\nu_p}\frac{mc^2}{\mathcal{E}_i^2} N_{e,i}\,\Phi(Z_i)\ , \\ \Phi(Z) &= 2Z\int_Z^\infty K_{\frac{5}{3}}(u)\,du - Z^2 K_{\frac{5}{3}}(Z)\ ,\quad Z_i = \frac{4\pi me\nu_p^3}{3\,eH_\perp \nu^2}\,\frac{\mathcal{E}_i}{mc^2}\ . \end{aligned}\right\} \tag{9.48}$$

In (9.47) and (9.48) $N_{e,i}$ is the electron density with the energy $\mathcal{E}_i$ ($\gg mc^2$) which we are considering.

Expression (9.47) is always positive; when there is no plasma it is valid for all energies, in agreement with what we have said earlier. The function $\Phi(Z_i)$ can be negative and in the corresponding range of Z_i values the coefficient

---

† This remark is valid only if the function $N_e(\mathcal{E})$ is sufficiently smooth so that the expressions used for μ (see (9.17) and (9.45)) are valid. For very 'sharp' functions $N_e(\mathcal{E})$ and anisotropic velocity distributions one may encounter a negative value of μ even in vacuo (see the paper by Zhelesnyakov and Suvorov (1972) and the references quoted there).

$\mu^{II} < 0$. It is negative in a range of the order of 0.7 to $1.3 \times \nu_{max}$, where ν_{max} is the frequency for which $|\mu^{II}|$ is a maximum. At that frequency

$$\left.\begin{aligned} \mu^{II}_{max} &\approx -10^{-2}\,\frac{e^2}{mc}\,\frac{\nu_p^3}{\nu_{max}^4}\,N_{e,i} = -8.5\times 10^{-5}\,\frac{\nu_p^3}{\nu_{max}^4}\,N_{e,i}\ \text{cm}^{-1}\,, \\ \nu_{max} &\approx \left(0.24\,\frac{2\pi mc\nu_p^3}{eH_\perp}\,\frac{\mathcal{E}_i}{mc^2}\right)^{\frac{1}{2}}\,. \end{aligned}\right\} \tag{9.49}$$

At the same time the coefficient μ^I at the maximum of the frequency spectrum, that is at the frequency ν_m (see (5.40)), is equal to

$$\mu^I(\nu_m) \approx 2.4\times 10^{-8}\,\frac{N_{e,i}}{H_\perp}\left(\frac{mc^2}{\mathcal{E}_i}\right)^5 \text{cm}^{-1}\,,\quad \nu_m \approx 0.07\,\frac{eH_\perp}{mc}\left(\frac{\mathcal{E}_i}{mc^2}\right)^2\,. \tag{9.50}$$

Zheleznyakov (1967a,b) and Kaplan and Tsytovich (1973; Tsytovich and Kaplan, 1974) have given some estimates of the negative reabsorption coefficient for a number of cosmic sources.

So far we have considered only the case of quasi-longitudinal propagation where we could neglect the difference $\mu_e - \mu_o$. Zheleznyakov and Suvorov (1968) have considered transverse propagation ($\chi = \frac{1}{2}\pi$) in a plasma and in that case it also turned out that negative reabsorption is possible. Bratman and Suvorov (1969) obtained expressions for μ_e and μ_o for any angle χ between the field and the line of sight. It turns out that the coefficients μ_e can be negative for any χ, but only, of course, for well defined kinds of spectra $N_e(\mathcal{E})$, and not for all frequency ranges. Moreover, they found an expression for the difference $\mu_e - \mu_o$ for quasi-longitudinal wave propagation. This difference is small since

$$|\mu_e - \mu_o| \sim \left\{a\,\frac{\omega_H\omega_p^2}{\omega^3(1-n^2v^2/c^2)} + b\,\frac{\omega_H}{\omega} + d\left[1 - n^2\,\frac{v^2}{c^2}\right]^{\frac{1}{2}}\right\}\mu_{e,o}\,, \tag{9.51a}$$

where a, b, and d are numerical coefficients of order unity. At the maximum of the radiation

$$\omega \sim \omega_H\,\frac{mc^2}{\mathcal{E}}\left(1 - n^2\,\frac{v^2}{c^2}\right)^{-\frac{3}{2}} \approx \omega_H\,\frac{mc^2}{\mathcal{E}}\,\eta^{-3}\,,$$

$$\eta = \left[\left(\frac{mc^2}{\mathcal{E}}\right)^2 + \frac{\omega_p^2}{\omega^2}\right]^{\frac{1}{2}} \approx \left[1 - n^2\,\frac{v^2}{c^2}\right]^{\frac{1}{2}}$$

and, hence, we have in this case

$$|\mu_e - \mu_o| \sim \left\{a\,\frac{\omega_p^2\mathcal{E}}{\omega^2 mc}\,\eta + b\,\frac{\mathcal{E}}{mc^2}\,\eta^3 + d\,\eta\right\}\mu_{e,o}\,. \tag{9.51b}$$

It is clear from conditions (9.46a) and (9.46b) that in the region where the effect of the plasma is important, but not yet too large, $(\omega_p^2/\omega^2)(\mathcal{E}/mc^2)^2 \sim 1$,

$\eta \sim mc^2/\mathcal{E}$ and, hence, $|\mu_e - \mu_o| \sim mc^2/\mathcal{E}$. In the broad, and more important, range of parameter values where $\omega_p^2/\omega^2 \leqslant mc^2/\mathcal{E}$ we have

$$|\mu_e - \mu_o| \sim \eta = \left\{\left(\frac{mc^2}{\mathcal{E}}\right)^2 + \frac{\omega_p^2}{\omega^2}\right\}^{\frac{1}{2}} .$$

In most cases the factor η is small so that even when $|\mu_{e,o}|L \gg 1$, it is difficult to expect that the condition $|\mu_e - \mu_o|L \geqslant 1$ is satisfied. If, nonetheless, this condition will be satisfied for negative $\mu_{e,o}$, in the synchrotron radiation of the source one of the waves will dominate, that is, one should observe in that case complete circular polarization (we refer to Zheleznyakov (1968, 1974) for the general expression for the degree of circular polarization in the case when $|\mu_{e,o}|L \gg 1$).

There can not be any circular polarization in the approximation in which $\mu_e = \mu_o$ and the emissivities $\boldsymbol{\varepsilon}_e = \boldsymbol{\varepsilon}_o$. However, the linear polarization can change also in the case, when the effect of the plasma does not affect the absorption and emission of waves. In fact, if condition (9.36) is not satisfied, one observes not only a rotation of the plane of polarization, but also depolarization of the radiation. The fact is that through the effect of the Faraday rotation by itself the degree of linear polarization decreases by a factor

$$\frac{\sin[\frac{1}{2}(k_e - k_o)L]}{\frac{1}{2}(k_e - k_o)L} ,$$

where $k_{e,o} = (\omega/c)\, n_{e,o}$ and L is the size of the emitting region along the line of sight (see, for instance, Gardner and Whiteoak, 1966; Zheleznyakov, 1968, 1974). The degree of circular polarization from a thick layer with $\mu > 0$ is equal to (Zheleznyakov, 1968, 1974)

$$\Pi_c = \frac{V}{I} = \frac{\mu_e - \mu_o + (\mu_e + \mu_o)\Pi_c^{(o)}}{\mu_e + \mu_o + (\mu_e - \mu_o)\Pi^{(o)}} \approx \frac{\mu_e - \mu_o}{2\mu} + \Pi_c^{(o)} , \tag{9.52}$$

where we used, when going over to the last expression, the relations

$$|\mu_e - \mu_o| \ll \mu_{e,o} \approx \mu \quad \text{and} \quad \Pi_c^{(o)} = \frac{\boldsymbol{\varepsilon}_o - \boldsymbol{\varepsilon}_e}{\boldsymbol{\varepsilon}_o + \boldsymbol{\varepsilon}_e} \ll 1 .$$

The estimate for $|\mu_e - \mu_o|$ had already been given (see (9.51)); it is clear from (9.17), (9.18), and (9.28) that in the region where these formulae are valid $(\mu_e - \mu_o)2\mu \sim \Pi_c^{(o)}$. Moreover, the formula (9.17) for μ_e itself was obtained under the assumption that the radiation had the nature of needle radiation, that is, neglecting terms of order $mc^2/\mathcal{E}$. Furthermore, it is well known that even in vacuo $\Pi_c^{(o)} \sim mc^2/\mathcal{E}$ (see Chapter 5). Combining various estimates we arrive at the conclusion that usually (when $\mu > 0$) the degree of circular polarization, $\Pi_c^{(o)}$ or Π_c, is small and of order $mc^2/\mathcal{E}$ or $\eta = \{(mc^2/\mathcal{E})^2 + \omega_p^2/\omega^2\}^{\frac{1}{2}}$. The occurrence of circular or elliptic polarization of synchrotron radiation is therefore remarkable, as in the simplest cases this radiation is always linearly polarized. Circularly or elliptically polarized synchrotron radiation for a system of radiating electrons with a quasi-isotropic distribution may arise either in the case of not too relativistic energies, or when we take into account the effect of plasma anisotropy. The position changes if we are dealing with synchro-Compton radiation (see the end of Chapter 5 and the references given there).

Under conditions of negative reabsorption there may occur not only changes in the polarization, but also a strong dependence of the reabsorption coefficients $\mu_{\perp,\parallel}$ or $\mu_{e,o}$ on the angle χ between the field and the line of sight. As a result, if the field in the source is inhomogeneous, but not totally random, when $\mu < 0$ radiation will be preferentially amplified in direction with maximal $|\mu|$. Thus, when $|\mu| L > 1$ and, especially when $|\mu| L \gg 1$ separate regions of an inhomogeneous source will look anomalously bright.

In this Chapter we could dwell only upon a relatively small part of the problem of the effect of a cold plasma on synchrotron radiation and its reabsorption. One still needs to analyze in this field a whole range of problems and possibilities; first of all there are the problems of negative reabsorption and the polarization relations under various conditions and their applications to different kinds of sources. Zheleznyakov and Suvorov (1972) have reviewed this field.

Chapter X

ELECTRODYNAMICS OF MEDIA WITH SPATIAL DISPERSION

Spatial dispersion. Normal waves in an anisotropic medium.
Some effects of spatial dispersion in crystal optics.

When earlier we used the electrodynamics of continuous media we assumed that the dielectric permittivity was either constant or dependent on the frequency only (taking temporal dispersion into account). It is, however, well known that there is a rather broad range of phenomena — especially in plasma physics, but also in metal physics and in optics — for the analysis of which one must pay attention also to spatial dispersion, that is, the dependence of the permittivity on the wavevector. In the present chapter we shall consider both general problems of the electrodynamics of media with spatial dispersion and also a few effects of spatial dispersion in optics (for details see Agranovich and Ginzburg, 1966; in what follows we shall to a large extent follow another paper by Agranovich and Ginzburg (1971)). The role played by spatial dispersion in a plasma will be discussed in Chapter 11.

Spatial dispersion occurs as the induction $\mathbf{D}$ in some point $\mathbf{r}$ is determined by the electromagnetic fields $\mathbf{E}$ and $\mathbf{B}$ not only in the same point, but also in the vicinity of that point. We shall make this clear in detail a little later, but before that we shall for the sake of convenience give again the initial field equations and we shall write down the relations connecting $\mathbf{D}$ and $\mathbf{E}$ in linear electrodynamics.

The initial equations have the form

$$\text{curl}\,\mathbf{B} = \frac{1}{c}\frac{\partial \mathbf{D}}{\partial t} + \frac{4\pi}{c}\mathbf{j}_{ext}\ , \quad \text{div}\,\mathbf{D} = 4\pi\rho_{ext}\ ,$$
$$\text{curl}\,\mathbf{E} = -\frac{1}{c}\frac{\partial \mathbf{B}}{\partial t}\ , \qquad \text{div}\,\mathbf{B} = 0\ , \tag{10.1}$$

where $\mathbf{E}$ is the electric field strength, $\mathbf{D}$ and $\mathbf{B}$ are the electric and magnetic inductions (we do not distinguish between the fields $\mathbf{B}$ and $\mathbf{H}$ but, in contrast to other chapters, we use $\mathbf{B}$ as is usual in the electrodynamics of media with spatial dispersion), and $\mathbf{j}_{ext}$ and ρ_{ext} are the current and charge densities of the external sources. The induction $\mathbf{D}$ is given by the relation

$$\frac{\partial \mathbf{D}}{\partial t} = \frac{\partial \mathbf{E}}{\partial t} + 4\pi\mathbf{j}\ ,$$

where $\mathbf{j}$ is the current density produced by the fields $\mathbf{E}$ and $\mathbf{B}$; it is sometimes convenient to introduce also the polarization $\mathbf{P}$ and then $\mathbf{D} = \mathbf{E} + 4\pi\mathbf{P}$.

If there are sufficiently sharp boundaries present it is necessary to use boundary conditions which one can obtain from (10.1) by taking suitable limits. These conditions have the following form

$$E_{1t} = E_{2t}, \quad [\mathbf{n} \wedge (\mathbf{B}_2 - \mathbf{B}_1)] = \frac{4\pi}{c}(\mathbf{i} + \mathbf{i}_{ext}), \qquad (10.2)$$
$$B_{1n} = B_{2n}, \quad D_{2n} - D_{1n} = 4\pi(\sigma + \sigma_{ext}).$$

here $\mathbf{n}$ is the normal to the boundary surface in the direction from medium 1 into medium 2, and indexes n and t correspond to the normal and tangential components, $\mathbf{i}_{ext}$ and σ_{ext} are, respectively, the external current and charge density, while the densities $\mathbf{i}$ and σ can, in particular, be expressed in terms of $\mathbf{D}$ by integrating over the thickness of the surface layer. To be precise

$$\mathbf{i} = \frac{1}{4\pi}\int_1^2 (\partial\mathbf{D}/\partial t)\, d\ell$$

and

$$\sigma = \frac{1}{4\pi}\int_1^2 \left[\mathbf{n} \wedge [\mathbf{D} \wedge \mathbf{n}]\right] d\ell$$

(see Silin and Rukhadze, 1961; Ginzburg and Rukhadze, 1975). The densities $\mathbf{i}$ and σ are, in general, non-vanishing when spatial dispersion is taken into account in which case there are derivatives which occur in the relation between $\mathbf{D}$ and $\mathbf{E}$ (vide infra), In this connection we note that if one usually puts $\mathbf{i} = 0$ and $\sigma = 0$ when neglecting spatial dispersion, then there are no special reasons for detailed assumptions when spatial dispersion is taken into account. We note also that in deriving the boundary conditions (10.2) we assumed that the physical fields $\mathbf{E}$ and $\mathbf{B}$ could not become infinite at the boundary while the induction $\mathbf{D}$ may tend to infinity (this is, for instance, possible when the value of $\mathbf{D}$ is determined by the derivatives of $\mathbf{E}$ along the normal to the boundary).

When one obtains Eqns.(10.1) from the microscopic Maxwell equations one must perform statistical averages in these micro-equations (for details see Robinson, 1973 and the literature quoted there). The vectors $\mathbf{E}$, $\mathbf{D}$ and $\mathbf{B}$ in (10.1) are thus statistical averages. The meaning of these averages is completely clear for media which are in thermodynamic equilibrium. However, the average can also be evaluated in the more general case, such as for metastable states corresponding to overheating or undercooling. Due to the averaging fluctuations are not taken into account in (10.1), but the average fields $\mathbf{E}$, $\mathbf{D}$ and $\mathbf{B}$ can change in any way whatever in space and time — any additional averaging (additional to the statistical averaging) of the fields over $\mathbf{r}$ is not only not necessary but, in general, is impossible in the electrodynamics of media if one takes spatial dispersion consistently into account. Similarly, it, is, in general, impossible to average over t when temporal dispersion is taken into account.

The set of Eqns. (10.1) is not complete and, if at all, without content until we have given a relation which enables us to express $\mathbf{D} = \mathbf{E} + 4\pi\mathbf{P}$ in terms of $\mathbf{E}$. In the framework of linear electrodynamics this relation — which is sometimes called the material equation or the connecting equation — can be written in the general form

$$D_i(\mathbf{r},t) \int_{-\infty}^{t} dt' \int d^3\mathbf{r}' \; \hat{\varepsilon}_{ij}(t,t',\mathbf{r},\mathbf{r}') \, E_j(\mathbf{r}',t') \,. \qquad (10.3)$$

The causality principle has been taken into account in (10.3); as a result the induction at time t is determined only by the fields in the past and the present, that is, at times $t' \leq t$. If the properties of the medium do not change with time, the kernel $\hat{\varepsilon}_{ij}$ can depend only on the difference $\tau = t - t'$. Similarly, if we can take the medium to be spatially uniform, $\hat{\varepsilon}_{ij}$ will depend only on the difference $\mathbf{R} = \mathbf{r} - \mathbf{r}'$. For such media Eqn.(10.3) can be written particularly simply for fields in the form of plane waves, that is, for fields of the form

$$E_i(\mathbf{r},t) = E_i(\omega,\mathbf{k}) \exp\left[i(\mathbf{k}\cdot\mathbf{r}) - i\omega t\right] . \qquad (10.4)$$

Substituting (10.4) into (10.3) we find

$$D_i(\omega,\mathbf{k}) = \varepsilon_{ij}(\omega,\mathbf{k}) \, E_j(\omega,\mathbf{k}) \,, \qquad (10.5)$$

where

$$\varepsilon_{ij}(\omega,\mathbf{k}) = \int_0^\infty d\tau \int d^3\mathbf{R} \exp\left[-i(\mathbf{k}\cdot\mathbf{R}) + i\omega\tau\right] \hat{\varepsilon}_{ij}(\tau,\mathbf{R}) \,. \qquad (10.6)$$

It is clear that for fields which in some way depend on $\mathbf{r}$ and t the quantities $\mathbf{E}(\omega,\mathbf{k})$ and $\mathbf{D}(\omega,\mathbf{k})$ have the meaning of the appropriate Fourier components; for instance,

$$E_i(\omega,\mathbf{k}) = \frac{1}{(2\pi)^4} \int E_i(\mathbf{r},t) \, e^{-i(\mathbf{k}\cdot\mathbf{r}) + i\omega t} \, d^3\mathbf{r} \, dt \,.$$

The tensor $\varepsilon_{ij}(\omega,\mathbf{k})$ describes completely not only the electrical, but also the magnetic properties of the medium, that is, it takes into account the effect of **B** on **D** — or, what amounts to the same, on the induced current $\mathbf{j} = \frac{1}{4\pi}\frac{\partial}{\partial t}(\mathbf{D}-\mathbf{E})$. Indeed, the field equation

$$\text{curl}\,\mathbf{E} = -\frac{1}{c}\frac{\partial \mathbf{B}}{\partial t}$$

in terms of the Fourier components has the form

$$\mathbf{B}(\omega, \mathbf{k}) = \frac{c}{\omega}\left[\mathbf{k} \wedge \mathbf{E}(\omega, \mathbf{k})\right];$$

One can thus assume that in (10.3), and then also in (10.5) and (10.6), the effect of **B** on **D** is taken into account together with the spatial dispersion. If, however, one neglects spatial dispersion, that is, if one puts

$$\varepsilon_{ij}(\omega, \mathbf{k}) = \varepsilon_{ij}(\omega, \mathbf{k} \to 0) = \varepsilon_{ij}(\omega),$$

one must assume for non-ferromagnetic substances in the optical region of the spectrum that $B_i = \mu_{ij} H_j = H_i$, that is, $\mu_{ij} = \delta_{ij}$, where δ_{ij} is the Kronecker delta-function. For ferromagnetics $\mu_{ij}(\omega)$ does no longer always reduce to δ_{ij} even in the optical range, let alone the low-frequency range. As $k \to 0$ it is necessary at low frequencies to introduce the magnetic permeability tensor μ_{ij} also for para- and diamagnetics. This problem — like practically all others touched upon in the present chapter — is either discussed in detail in the book by Agranovich and Ginzburg (1966), or the appropriate literature is cited there (for more recent references we refer to Agranovich and Ginzburg's papers (1971, 1973; Ginzburg, 1973h)). It is sometimes convenient to use instead of the tensor $\varepsilon_{ij}(\omega, \mathbf{k})$ its inverse, that is, the tensor $\varepsilon^{-1}_{ij}(\omega, \mathbf{k})$. In that case we have clearly

$$E_i(\omega, \mathbf{k}) = \varepsilon^{-1}_{ij}(\omega, \mathbf{k})\, D_j(\omega, \mathbf{k}). \tag{10.7}$$

The dielectric permittivity tensor $\varepsilon_{ij}(\omega, \mathbf{k})$ is assumed to be known in the framework of the phenomenological theory. To calculate it for a crystal, or for any other condensed medium, is a problem for the microscopic theory.

One sees easily from Eqns.(10.3), (10.5), and (10.6) that the **k**-dependence of the tensor $\varepsilon_{ij}(\omega, \mathbf{k})$, that is, the spatial dispersion is immediately connected with the fact that the induction at the point **r** is determined by the value of the electric field strength not only at the point **r**, but also in its vicinity. In other words, spatial dispersion is caused by the non-local relation between **D** and **E**. Similarly, the ω-dependence of $\varepsilon_{ij}(\omega, \mathbf{k})$(temporal or frequency dispersion) is caused by the 'non-local' time-dependence of Eqn. (10.3). As the eigenfrequencies ω_i of the medium usually fall into the frequency range considered, the ratio ω_i/ω is of order unity and the temporal dispersion is, in general, large. The position is different in the case of spatial dispersion for dielectrics. This is connected with the fact that the wavelength $\lambda = \lambda_0/n = 2\pi c/n\omega$ in the optical band (let alone for lower frequencies) for dielectrics appreciably exceeds the characteristic size of the neighbourhood of the point $\mathbf{r}$ which makes a significant contribution to the integral (10.3). Indeed, this size is in dielectrics of the order of atomic dimensions or of the order of the lattice constant $a \sim 10^{-8}$ to 10^{-7} cm so that the ratio $a/\lambda \sim 10^{-3}$, that is, very small. At the same time, in an isotropic plasma, for instance, the characteristic size a determining the magnitude of the spatial dispersion is the Debye radius $r_D = (k_B T/8\pi N e^2)^{\frac{1}{2}}$ and in a conducting medium it is the mean free path ℓ, when collisions are taken into account. For those media, therefore, spatial dispersion is, in general, small only when the conditions $r_D \ll \lambda$ and $\ell \ll \lambda$ hold which cannot be satisfied in a number of completely realistic cases (see, in particular, Chapter 11). The smallness of the spatial dispersion considerably simplifies the analysis of the optical effects caused by it, and we shall use this in what follows.

We now make a few remarks about the general properties of the tensor $\varepsilon_{ij}(\omega, \mathbf{k})$. This tensor is, in general, complex, even for real ω and $\mathbf{k}$. At the same time, a real field $\mathbf{E}$ in the medium leads, of course, to a real induction $\mathbf{D}$. From this it follows that the kernel $\hat{\varepsilon}_{ij}(\tau, \mathbf{R})$ in (10.6) is real and, hence, in the general case we have (for complex ω and $\mathbf{k}$)

$$\varepsilon_{ij}(\omega, \mathbf{k}) = \varepsilon_{ij}(-\omega^*, -\mathbf{k}^*), \tag{10.8}$$

or, what amounts to the same,

$$\varepsilon_{ij}(\omega^*, \mathbf{k}^*) = \varepsilon^*_{ij}(-\omega, -\mathbf{k}). \tag{10.9}$$

Using the symmetry properties of kinetic coefficients (see Landau and Lifshitz, 1960, §§ 82, 88) leads to another relation:

$$\varepsilon_{ij}(\omega, \mathbf{k}, \mathbf{B}_{ext}) = \varepsilon_{ji}(\omega, -\mathbf{k}, -\mathbf{B}_{ext}); \tag{10.10}$$

here $\mathbf{B}_{ext}$ is the magnetic induction which is constant in time and which is non-vanishing when there is an external magnetic field or a magnetic structure (ferro- and antiferromagnetics). For the sake of simplicity we shall in both those cases call $\mathbf{B}_{ext}$ the induction of the external magnetic field; hence the index.

The proof of Eqn. (10.10) for the tensors $\varepsilon_{ij}(\omega)$ and $\varepsilon_{ij}(\omega, \mathbf{B}_{ext})$ is given by Landau and Lifshitz (1960). There are no principally new features in the generalization to the case when spatial dispersion is present — as the vectors

$\mathbf{k}$ and $\mathbf{B}_{ext}$ behave in the same way under time reversal one might say that changing from the relation $\varepsilon_{ij}(\omega, \mathbf{B}_{ext}) = \varepsilon_{ji}(\omega, -\mathbf{B}_{ext})$ to (10.10) is almost trivial. We note merely that Eqn. (10.10) is usually proved only for real ω and $\mathbf{k}$. However, in the analyticity domain, which we shall be discussing below, these relations retain their form also for complex ω and $\mathbf{k}$.

A medium is called non-gyrotropic, if for all ω and $\mathbf{k}$

$$\varepsilon_{ij}(\omega, \mathbf{k}) = \varepsilon_{ji}(\omega, \mathbf{k}) . \tag{10.11}$$

We have dropped the argument $\mathbf{B}_{ext}$ in (10.11) as the tensor ε_{ij} is, strictly speaking, in an external magnetic field always asymmetrical (it equals ε_{ji} only when $\mathbf{B}_{ext}$ is replaced by $-\mathbf{B}_{ext}$; we have here in mind just the external field and do not refer to the case of an antiferromagnetic when there is no external field present in which case the tensor ε_{ij} may remain symmetric, even without the change from $\mathbf{B}_{ext}$ to $-\mathbf{B}_{ext}$). We might therefore talk about gyrotropy caused by the magnetic field. In order not to cause confusion we call such a gyrotropy magnetic activity and the corresponding medium a magneto-active medium (see Chapter 11). We shall call in what follows a medium gyrotropic only when the tensor ε_{ij} for it is non-symmetric only when spatial dispersion is taken into account.

It is clear from (10.10) and (10.11) that for a non-gyrotropic medium

$$\varepsilon_{ij}(\omega, \mathbf{k}) = \varepsilon_{ij}(\omega, -\mathbf{k}) . \tag{10.12}$$

In a gyrotropic medium there must thus exist at least one direction which is not equivalent to the directly opposite direction. In other words, only a medium without a centre of symmetry can be gyrotropic. The opposite conclusion is incorrect — a medium can be without a centre of symmetry, but not be gyrotropic, as the satisfying of Eqn.(10.12) can be guaranteed due to the presence of other symmetry elements.

When there is no spatial dispersion, condition (10.12) is satisfied automatically and the medium is always non-gyrotropic. Gyrotropy is thus a spatial dispersion effect and is the most important optical effect of this kind which has been known for a long time. However, if there is no spatial dispersion, but $\mathbf{B}_{ext} \neq 0$, magnetic activity, of course, appears as

$$\varepsilon_{ij}(\omega, \mathbf{B}_{ext}) = \varepsilon_{ji}(\omega, -\mathbf{B}_{ext}) , \tag{10.13}$$

and the tensor ε_{ij} is, in general, for given ω and $\mathbf{B}_{ext}$ non-symmetric (in some cases $\varepsilon_{ij}(\omega, \mathbf{B}_{ext}) = \varepsilon_{ij}(\omega, -\mathbf{B}_{ext})$ which may occur in antiferromagnetics when there is no external magnetic field present; in antiferromagnetics $\mathbf{B}_{ext}$ is the statistically averaged magnetization in a given micro-volume of the crystal, which vanishes only on the average taken over the whole elementary cell of the magnetic structure of the crystal).

The definition of a gyrotropic medium as a medium with a non-symmetric tensor $\varepsilon_{ij}(\omega, \mathbf{k})$ is, of course, somewhat formal. It will, however, become clear from what follows that it is just the non-symmetric feature of the tensor ε_{ij} which leads to those peculiar features (for instance, the rotation of the plane of polarization in normal waves when there is no absorption) which distinguish gyrotropic from non-gyrotropic media.

It is often convenient to split the tensor ε_{ij} into its real and imaginary parts $\mathrm{Re}\,\varepsilon_{ij}$ and $\mathrm{Im}\,\varepsilon_{ij}$ and also into two Hermitean tensors ε'_{ij} and ε''_{ij}

$$\varepsilon_{ij} = \mathrm{Re}\,\varepsilon_{ij} + i\,\mathrm{Im}\,\varepsilon_{ij} , \qquad (10.14)$$

$$\varepsilon_{ij} = \varepsilon'_{ij} + i\,\varepsilon''_{ij} , \quad \varepsilon'_{ij} = (\varepsilon'_{ji})^* , \quad \varepsilon''_{ij} = (\varepsilon''_{ji})^* , \qquad (10.15)$$

where the asterisks, as in other cases, indicate complex conjugates; we note that one sometimes introduces instead of ε''_{ij} the Hermitean conductivity tensor σ_{ij}, defined by

$$\varepsilon''_{ij} = \frac{4\pi\sigma_{ij}}{\omega} . \qquad (10.16)$$

It is true that one also uses the complex conductivity tensor

$$\sigma_{ij} = \sigma'_{ij} + i\,\sigma''_{ij} = -i\,\omega(\varepsilon_{ij} - \delta_{ij})/4\pi ;$$

in that case the quantity σ'_{ij} should occur in (10.16).

When one neglects spatial dispersion and when there is no constant magnetic field present we find clearly from (10.12) and (10.13) that the tensor $\varepsilon_{ij}(\omega)$ is symmetric

$$\varepsilon_{ij}(\omega) = \varepsilon_{ji}(\omega) . \qquad (10.17)$$

Both in this case and in the more general case (10.11) we have clearly

$$\mathrm{Re}\,\varepsilon_{ij} = \varepsilon'_{ij} \quad \text{and} \quad \mathrm{Im}\,\varepsilon_{ij} = \varepsilon''_{ij} .$$

It follows from requirements which are connected with the causality principle that the function $\varepsilon_{ij}(\omega, \mathbf{k})$ in a medium which is in equilibrium (or at least which is stable) does not have any singularities in the upper half-plane and on the real axis of the complex variable ω. Using this fact one can understand a whole number of relations for and properties of the functions

$\varepsilon_{ij}(\omega, \mathbf{k})$. The most important ones of those are the dispersion relations connecting $\mathrm{Re}\,\varepsilon_{ij}(\omega, \mathbf{k})$ with $\mathrm{Im}\,\varepsilon_{ij}(\omega, \mathbf{k})$. Taking spatial dispersion into account introduces here little what is new — it usually reduces to the same relations as for the function $\varepsilon_{ij}(\omega)$, but including the wavevector $\mathbf{k}$ as a parameter. We shall therefore not dwell upon the dispersion relations — this is done in detail by Landau and Lifshitz (1960) for the case without spatial dispersion in an application to an isotropic medium. The generalization to the case of an anisotropic medium and a medium with spatial dispersion can be found in § 1 of the book by Agranovich and Ginzburg (1966).

An important part of the electrodynamics of continuous media is devoted to the study of the propagation of electromagnetic waves produced by sources outside the medium considered. In particular, in crystal optics one studies usually just that kind of problem and most often one is dealing with even a narrower problem — the propagation of plane monochromatic waves in which the electric field is of the form

$$\mathbf{E} = \mathbf{E}_0 e^{i(\mathbf{k}\cdot\mathbf{r}) - i\omega t} ; \tag{10.18}$$

here $\mathbf{E}_0$ is a complex vector which is independent of the space and time coordinates $\mathbf{r}$ and t, while $\mathbf{k}$ and ω are the wavevector and frequency.

One should bear in mind that Eqn.(10.18) with $\mathbf{E}_0$ = constant is not the most general one — sometimes one must consider also a field of the type (10.18), but with $\mathbf{E}_0 = \mathbf{E}_\infty(\mathbf{k}\cdot\mathbf{r})$, $\mathbf{E}_\infty$ = constant. However, this necessity arises only very rarely (case of singular axes in crystals of the lowest crystal systems and a few other cases; see Agranovich and Ginzburg, 1966, § 2). We shall therefore restrict ourselves in what follows to expressions of the form (10.18).

Solutions of the type (10.18) satisfy the homogeneous electromagnetic field equation, that is, Eqns.(10.1) without the external (given) currents and charges $\mathbf{j}_{ext}$ and ρ_{ext}, only if $\mathbf{k}$ and ω are interrelated. This connection is given by the dispersion equation and allows us, for example, to express $\mathbf{k}$ in terms of ω:

$$\mathbf{k} = \frac{\omega}{c}\tilde{n}(\omega, \mathbf{s})\mathbf{s} . \tag{10.19}$$

Here $\tilde{n} = n + i\kappa$ is the complex index of refraction, n is the index of refraction, κ the index of absorption ($\mu = 2\omega\kappa/c$ is the absorption coefficient in terms of intensity), and $\mathbf{s}$ a real unit vector (we consider now only uniform plane waves in which $\mathbf{k} = \mathbf{k}_1 + i\mathbf{k}_2$, $\mathbf{k}_1$ and $\mathbf{k}_2$ being collinear; moreover, in accordance with the statement of the problem which one encounters in optics we have taken the frequency ω to be real). The dispersion equation determines the function $\tilde{n}$ in terms of the coefficients which occur in the field equations, that is, in terms of the permittivity tensor ε_{ij} and to each value of ω and $\mathbf{k}$ there correspond several values $\tilde{n} = \tilde{n}_\ell$ where the index ℓ corresponds to one or other solution — a normal wave. Normal waves (for given ω and $\mathbf{s}$, but different ℓ) differ also in their polarization, that is, in the vector $\mathbf{E}_{0\ell}$ in (10.18) which is determined (apart from a multiplying factor) from the field equations.

The problem of crystal optics† consists formally speaking, thus in the first instance of a study of the functions $\tilde{n}_\ell(\omega, \mathbf{s})$ and $\mathbf{E}_{0\ell}(\omega, \mathbf{s})$. In turn all information about these functions is contained, if we do not mention the field equations, in the complex dielectric permittivity tensor $\boldsymbol{\varepsilon}_{ij}(\omega, \mathbf{k})$. If we in this case neglect spatial dispersion, that is, assume that $\boldsymbol{\varepsilon}_{ij} = \boldsymbol{\varepsilon}_{ij}(\omega)$, the problem reduces to what is usually called 'classical' crystal optics. In this way classical crystal optics is, of course, contained in crystal optics with spatial dispersion as a special (or limiting) case.

In the optics of practically non-absorbing or weakly absorbing crystals the spatial dispersion is weak in that sense that its magnitude is determined, as we already mentioned, by the small parameter

$$ka \sim \frac{a}{\lambda} = \frac{an}{\lambda_0} \ll 1 \,. \tag{10.20}$$

This is just the reason why one can usually neglect spatial dispersion in crystal optics, unless we are dealing with qualitatively new effects (gyrotropy, optical anisotropy of cubic crystals, the appearance of additional normal waves, a non-vanishing group velocity for longitudinal waves, and so on). Moreover, the presence of a small parameter enables us to simplify strongly and to make more precise the study of the effect of spatial dispersion.

Taking spatial dispersion into account in crystal optics and in general in electrodynamics is not something principally new and it can be traced back to the last century. However, it is only in the last 15 to 20 years that crystal optics with spatial dispersion became a more or less independent subject of study. At the same time, and even more strikingly, taking spatial dispersion into account became an organic constituent of part of modern electrodynamics of continuous media, of plasma physics, solid state theory, theory of metals, and so on. The modern structure and position of crystal optics, even if we

---

† We use here this term for the sake of simplicity also when we are dealing with a more general case, namely, the propagation of waves in an arbitrary (in general, anisotropic) medium.

forget about plasma physics, must thereby be based upon electrodynamics taking spatial dispersion into account. In other words, one should start from the relation (10.5) and obtain a whole series of general results, and after that go over as a particular (albeit very important) case to the exposition of classical crystal optics. However, in the optical literature one proceeds, as a rule, in the old-fashioned way — one develops to begin with, and often even exclusively, the classical crystal optics. This is just the reason why it is appropriate to discuss in the present book crystal optics taking spatial dispersion into account.

We now turn to finding all normal electromagnetic waves of the type (10.18) in an infinite uniform medium characterized by the tensor $\boldsymbol{\varepsilon}_{ij}(\omega, \mathbf{k})$. Such waves satisfy Eqns.(10.1) with $\mathbf{j}_{ext} = 0$ and $\rho_{ext} = 0$ from which we get the wave equation

$$\text{curl curl } \mathbf{E} + \frac{1}{c^2}\frac{\partial^2 \mathbf{D}}{\partial t^2} = 0 \,. \tag{10.21}$$

For the plane waves (10.18) Eqn.(10.21) takes the form

$$\mathbf{D} = \frac{c^2}{\omega^2}\left\{k^2\mathbf{E} - \mathbf{k}\,(\mathbf{k}\cdot\mathbf{E})\right\} \,. \tag{10.22}$$

Substituting here the relation (10.5) we find

$$\left\{\frac{\omega^2}{c^2}\,\boldsymbol{\varepsilon}_{ij}(\omega\,,\mathbf{k}) - k^2\,\delta_{ij} + k_i\,k_j\right\} E_j = 0 \,. \tag{10.23}$$

This algebraic set of equations has a non-trivial solution $\mathbf{E} \neq 0$ provided its determinant vanishes, that is, provided

$$\left|\frac{\omega^2}{c^2}\;\boldsymbol{\varepsilon}_{ij}(\omega\,,\mathbf{k}) - k^2\,\delta_{ij} + k_i\,k_j\right| = 0 \,. \tag{10.24}$$

Equation (10.24) is often called the dispersion equation — it establishes a relation between ω and $\mathbf{k}$ for normal waves, and its solution can be written in the form

$$\omega_\ell = \omega_\ell(\mathbf{k}) \;\; ; \;\; \ell = 1, 2, 3, \ldots \tag{10.25}$$

or in the form (10.19). that is, expressing $\mathbf{k}$ in terms of ω,

$$\mathbf{k} = \frac{\omega}{c}\,\tilde{n}_\ell(\omega\,,\mathbf{s})\,\mathbf{s} \;\;,\;\; \tilde{n}_\ell^2 = \frac{c^2 k^2}{\omega^2} \;\;, \tag{10.26}$$

where the index ℓ corresponds to the different normal waves. One can also write Eqn.(10.24) in the form of an equation for $\tilde{n}^2(\omega\,,\mathbf{s})$:

$$(\boldsymbol{\varepsilon}_{ij}\,s_i\,s_j)\,\tilde{n}^4 - \{(\boldsymbol{\varepsilon}_{ij}\,s_i\,s_j)\,\boldsymbol{\varepsilon}_{kk} - (\boldsymbol{\varepsilon}_{ik}\,\boldsymbol{\varepsilon}_{kj}\,s_i\,s_j)\}\,\tilde{n}^2 + |\boldsymbol{\varepsilon}_{ij}| = 0 \,, \tag{10.27}$$

where $\varepsilon_{kk} = \mathrm{Tr}\,\varepsilon_{ij} = \varepsilon_{11} + \varepsilon_{22} + \varepsilon_{33}$ and $|\varepsilon_{ij}|$ is the determinant of the matrix ε_{ij}.

When spatial dispersion is neglected, that is, when $\varepsilon_{ij} = \varepsilon_{ij}(\omega)$, Eqn.(10.27) is often called the Fresnel equation, and it is the basis of classical crystal optics. In that case Eqn.(10.27) always has only two solutions $\tilde{n}_1^2$ and $\tilde{n}_2^2$ whence it is clear that for arbitrary ω and $\mathbf{s}$ in a medium there can propagate only two normal waves for which the vector $\mathbf{E}(\omega, \mathbf{s})$ has the transverse component $\mathbf{E}_\perp = \mathbf{E} - (\mathbf{s} \cdot \mathbf{E})\mathbf{s}$, $\mathbf{s} = \mathbf{k}/k$. Waves with $\mathbf{E}_\perp = 0$, that is, longitudinal waves can in that case exist only for a discrete set of frequencies ω. Indeed, for such waves $\mathbf{D} = 0$ (see (10.22)) and as we have the relation $D_i = \varepsilon_{ij} E_j$, we are led to the conclusion that for $\mathbf{D} = 0$ and $\mathbf{E} \neq 0$ for longitudinal waves ω and $\mathbf{k}$ must in the general case satisfy the relation

$$|\varepsilon_{ij}(\omega, \mathbf{k})| = 0 . \tag{10.28}$$

We must, however, bear in mind that satisfying this equation is necessary, but not sufficient for the occurrence of longitudinal waves. When we neglect spatial dispersion Eqn.(10.28), which can also be obtained from (10.24) takes the form

$$|\varepsilon_{ij}(\omega)| = 0 . \tag{10.29}$$

The frequencies $\omega \equiv \omega_\parallel$ which satisfy Eqn.(10.29) are the frequencies which longitudinal waves can possess. For an isotropic medium we get from (10.29) the condition

$$\varepsilon(\omega) = 0 . \tag{10.30}$$

Of course, one could have arrived directly at this very important condition by considering an isotropic medium without spatial dispersion when $\mathbf{D}(\omega) = \varepsilon(\omega)\,\mathbf{E}(\omega)$. As in a longitudinal wave by definition $\mathbf{E}_\parallel = E\mathbf{k}/k$, in it $\mathbf{D} = 0$ (see (10.22)) and the field $\mathbf{E}$ is non-vanishing only when condition (10.30) is satisfied. When, however, spatial dispersion is taken into account Eqn.(10.28) gives for longitudinal waves a dispersion relation $\omega_\parallel = \omega_\parallel(\mathbf{k})$.

Waves with $\mathbf{E}_\perp \neq 0$ play the main role in crystal optics, as just these waves are most strongly excited by light. When spatial dispersion is taken into account Eqn.(10.27) can for these waves have in some spectral ranges not two, but a larger number of solutions. However, even in the case when new solutions ('additional waves') do not turn up, spatial dispersion leads to a number of new effects amongst which natural optical activity (gyrotropy), and also the optical anisotropy of cubic crystals (it is well known that when spatial dispersion is neglected cubic crystals are optically isotropic) are the most important ones. All these effects were already mentioned and are discussed below, and for their analysis in a phenomenological theory it is necessary to know how the tensor $\boldsymbol{\varepsilon}_{ij}(\omega, \mathbf{k})$ depends on $\mathbf{k}$ for small $\mathbf{k}$.

Before we turn to that problem, we make a few remarks. For instance, we must bear in mind that when we consider the $\mathbf{k}$-dependence of $\boldsymbol{\varepsilon}_{ij}(\omega, \mathbf{k})$ we must remember that Eqn.(10.5), which connects the quantities $\mathbf{D}$ and $\mathbf{E}$ for one and the same value of $\mathbf{k}$, was obtained earlier under the assumption that the medium is spatially uniform. However, crystals are in actual fact not spatially uniform media as, for instance, the lattice sites are not equivalent to other points. Use of the tensor $\boldsymbol{\varepsilon}_{ij}(\omega, \mathbf{k})$, which was introduced under the assumption of a uniform medium (after statistical averaging), for applications to crystals must therefore clearly be limited. An analysis of this problem (see Agranovich and Ginzburg, 1966) enables us to reach the conclusion that the use of Eqn.(10.5) in crystals is justified provided the wavevector $\mathbf{k}$ is small compared to the basis vectors of the elementary cell of the reciprocal lattice, that is, provided $k \ll 1/a$ or $\lambda \gg a$, where a is the lattice constant. These inequalities, the meaning of which is obvious already from purely qualitative considerations, are clearly satisfied in the optical range of wavelengths where $a/\lambda \sim 10^{-3}$. We shall therefore in what follows use the tensor $\boldsymbol{\varepsilon}_{ij}(\omega, \mathbf{k})$ without restrictions in crystal optics.

Another and completely independent problem about the conditions of the applicability of the material Eqn. (10.5) when spatial dispersion is taken into account arises when we change from a crystal of infinite extent which is, of course, an idealization (which, however, was used when we changed from (10.3) to (10.5) and (10.6)) to finite size crystals. As Eqn.(10.3) is an integral relation the presence of boundaries of the crystal have been taken into account in it and the exact boundary conditions for the fields are retained. If the point $\mathbf{r}$ is far from the crystal surface at a distance appreciably longer than the size R of the neighbourhood which makes the main contribution to the value of the induction, the kernel $\hat{\boldsymbol{\varepsilon}}_{ij}$ becomes equal to the kernel $\hat{\boldsymbol{\varepsilon}}_{ij}$ in (10.6) which is used for an infinite crystal. For such points the electric field in the form (10.4) clearly leads to the induction

$$\mathbf{D}(\mathbf{r}, t) = \mathbf{D}(\omega, \mathbf{k})\, e^{i(\mathbf{k}, \mathbf{r}) - i\omega t}$$

which is also in the form of a plane wave and the relation between the amplitudes of $\mathbf{E}$ and $\mathbf{D}$ is just given by Eqn.(10.5). It therefore follows from what we have said that one can use the material equation in the form (10.5) provided the thickness of the crystal is large compared to R. In dielectrics the magnitude of R is usually of the order of a where a is the lattice constant.

One should, moreover, note that when we consider additional waves and at the same time use the tensor $\boldsymbol{\varepsilon}_{ij}(\omega, \mathbf{k})$, and not the general integral relation (10.3), the boundary conditions (10.2) are insufficient. One can in principle obtain additional boundary conditions (10.3), but usually they are introduced less rigorously from different considerations (see Agranovich and Ginzburg, 1966, § 10). One should in connection of the problem of the boundary conditions in the electrodynamics of media with spatial dispersion make yet another remark which one should bear in mind also when there are not additional waves present. The question is the necessity to take into account the possible role of higher derivatives when writing down the actual form of the boundary conditions (10.2). Let us elucidate this using the equation div $\mathbf{D} = 0$ as an example. To obtain the boundary condition at the boundary between the media 1 and 2 we take a limit — we integrate the equation

$$\operatorname{div} \mathbf{D} = \frac{\partial D_x}{\partial x} + \frac{\partial D_y}{\partial y} + \frac{\partial D_z}{\partial z}$$

along a direction normal to the diffuse separating boundary. Taking that direction as the z-axis and taking the limit of a sharp boundary we get the condition

$$D_{2z} - D_{1z} = \boldsymbol{\varepsilon}_2 E_{2z} - \boldsymbol{\varepsilon}_1 E_{1z} = 0 , \tag{10.31}$$

where we have put $\mathbf{D} = \boldsymbol{\varepsilon} \mathbf{E}$ when changing to the second equation, and have taken $\boldsymbol{\varepsilon} = \boldsymbol{\varepsilon}_1$ in the first and $\boldsymbol{\varepsilon} = \boldsymbol{\varepsilon}_2$ in the second medium. Let us now assume that

$$\mathbf{D} = \boldsymbol{\varepsilon} \mathbf{E} + \delta_1 \operatorname{curl} \mathbf{E} + \operatorname{curl} (\delta_{II} \mathbf{E}) , \tag{10.32}$$

where, of course, in a uniform medium $\mathbf{D} = \boldsymbol{\varepsilon} \mathbf{E} + (\delta_I + \delta_{II}) \operatorname{curl} \mathbf{E}$, the material Eqn.(10.32) corresponds to the relation between $\mathbf{D}$ and $\mathbf{E}$ in the case of the simplest (isotropic) gyrotropic medium.

Proceeding now in the usual way we get the boundary condition

$$D_{2n} - D_{1n} = \delta_{II,2} \operatorname{curl}_n \mathbf{E}_2 - \delta_{II,1} \operatorname{curl}_n \mathbf{E}_1 , \tag{10.33}$$

or

$$\boldsymbol{\varepsilon}_2 E_{2n} - \boldsymbol{\varepsilon}_1 E_{1n} + \delta_{I,2} \operatorname{curl}_n \mathbf{E}_2 - \delta_{I,1} \operatorname{curl}_n \mathbf{E}_1 = 0 , \tag{10.34}$$

where we have used here the more general index n rather than z , as it characterizes the direction of the normal. The appearance of additional terms in

(10.33) and (10.34), as compared to (10.31), is clearly explained by the inadmissibility of neglecting the integrals

$$\int_0^{\ell} \left(\frac{\partial D_x}{\partial_x}\right) dz \quad \text{and} \quad \int_0^{\ell} \left(\frac{\partial D_y}{\partial_y}\right) dz ,$$

when using the relation (10.32), even when the thickness of the transition layer between the media $\ell \to 0$.

In the optical band where the spatial dispersion is small (this means in (10.32) that $\delta \sim a \ll \lambda$) the generalization of the boundary conditions such as (10.33) is clearly not of particular interest because of the necessity even when spatial dispersion is neglected to complicate the boundary conditions (10.31) somewhat to take into account that the boundary between two media is not sharp, that there are impurities on such a boundary, and so on (for details see Agranovich and Ginzburg, 1973 and the literature cited there). However, as a matter of principle and in practice for media with a strong spatial dispersion one should take into account possible changes even in the usual boundary conditions such as (10.31). If we write the boundary conditions in electrodynamics in the form (10.2), we can express what we have said here in the form of a statement about the necessity to make the expressions for the surface densities $\mathbf{i}$ and σ more precise, as it is sometimes impossible, when spatial dispersion is taken into account, to assume in general that $\mathbf{i} = 0$ and $\sigma = 0$ (see (10.2), and (10.33) and (10.34)).

We make yet one more remark of a methodological nature, which is connected with the following problem which often occurs. The dispersion equation connects $\mathbf{k}$ and ω and thus $\boldsymbol{\varepsilon}_{ij}(\omega, \mathbf{k})$ in fact depends on ω only. How then does spatial dispersion differ from the temporal one? The answer consists of the fact that the tensor $\boldsymbol{\varepsilon}_{ij}(\omega, \mathbf{k})$ is introduced (say, in (10.5)) not just for normal waves, for which $\mathbf{k}$ and ω are interrelated, but for an arbitrary electromagnetic field with sources (see (10.1)). The wavevector $\mathbf{k}$ and the frequency ω are completely independent in such a field. Let us, for instance, consider a field of the form $\mathbf{E} = \mathbf{E}_0 e^{i(\mathbf{k}\cdot\mathbf{r})}$ with arbitrary $\mathbf{k}$ and frequency $\omega = 0$. Such a field produces an induction $\mathbf{D} = \mathbf{D}_0 e^{i(\mathbf{k}\cdot\mathbf{r})}$ which is connected with the field $\mathbf{E}$ through Eqn. (10.5), where $\boldsymbol{\varepsilon}_{ij} = \boldsymbol{\varepsilon}_{ij}(0, \mathbf{k})$. The situation is similar for any other frequency ω. Hence it is also clear, incidentally, that it is the tensor $\boldsymbol{\varepsilon}_{ij}(\omega, \mathbf{k})$ and not the refractive index $\tilde{n}(\omega, \mathbf{s})$ which is the fundamental quantity determining the electrodynamic properties of the medium.

One does not normally use energy considerations in crystal optics or, at any rate, they are of secondary importance. We shall therefore not go into any detailed discussion of this side of the problem (see Agranovich and Ginzburg, 1966, § 3 and also Chapter 12 of the present book) and restrict ourselves only to the discussion of a single problem.

Poynting's theorem follows in the usual way from the field Eqns.(10.1)

$$\frac{1}{4\pi}\left[\left(\mathbf{E}\cdot\frac{\partial\mathbf{D}}{\partial t}\right)+\left(\mathbf{B}\cdot\frac{\partial\mathbf{B}}{\partial t}\right)\right]=-\frac{c}{4\pi}\,\mathrm{div}\,\left[\mathbf{E}\wedge\mathbf{B}\right]-(\mathbf{j}_{ext}\cdot\mathbf{E})\,. \qquad (10.35)$$

When temporal and especially spatial dispersion is taken into account and also absorption, the use and interpretation of Eqn.(10.35) is in general far from obvious. The same simplicity to which we are accustomed when dispersion and absorption are neglected — when we can put, say, $\mathbf{D}=\boldsymbol{\varepsilon}\mathbf{E}$ and obtain the expression $\boldsymbol{\varepsilon}E^2/8\pi$ for the energy density — is very deceptive since the situation changes when we take dispersion and/or absorption into account. A more detailed examination of this problem can be found in § 3 of the monograph by Agranovich and Ginzburg (1966; see also, for instance, Landau and Lifshitz, 1960; Ginzburg, 1970b; and Barash and Ginzburg, 1976). Now we give the example of a medium without temporal dispersion. We consider thus a gyrotropic medium in which $\mathbf{D}$ and $\mathbf{E}$ are connected through Eqn.(10.32) with $\boldsymbol{\varepsilon}$ and $\delta_{I,II}$ frequency-independent. In that case (10.35) becomes $(\delta=\delta_I+\delta_{II})$

$$\frac{\partial}{\partial t}\left\{\frac{\boldsymbol{\varepsilon}E^2+B^2+\delta(\mathbf{E}\cdot\mathrm{curl}\,\mathbf{E})}{8\pi}\right\}=-\frac{c}{4\pi}\,\mathrm{div}\,\left\{\left[\mathbf{E}\wedge\mathbf{B}\right]-\frac{\delta}{2c}\left[\mathbf{E}\wedge\frac{\partial\mathbf{E}}{\partial t}\right]\right\}$$
$$+\left(\left\{\mathrm{grad}\,(2\delta_{II}-\delta)\right\}\cdot\left[\frac{\mathbf{E}}{8\pi}\wedge\frac{\partial\mathbf{E}}{\partial t}\right]\right)-(\mathbf{j}_{ext}\cdot\mathbf{E})\,. \qquad (10.36)$$

It is clear from this equation, firstly, that taking spatial dispersion into account, when $\delta\neq 0$, leads to the appearance of an additional term $\delta(\mathbf{E}\cdot\mathrm{curl}\,\mathbf{E})$ in the expression for the energy density and a term $-(\delta/2c)[\mathbf{E}\wedge(\partial\mathbf{E}/\partial t)]$ in the expression for the energy flux density. Secondly, there appears in (10.36) a term $A=(1/8\pi)(\{\mathrm{grad}\,(2\delta_{II}-\delta)\}\cdot[\mathbf{E}\wedge(\partial\mathbf{E}/\partial t)])$ which is proportional to $\mathrm{grad}\,(2\delta_{II}-\delta)=\mathrm{grad}\,(\delta_{II}-\delta_I)$ and is thus non-vanishing (localized) only close to the separating boundary of the two media. If $A\neq 0$, which would be true, if $\delta_{II}\neq\delta_I$ there will be a liberation or absorption of energy. The result is, of course, unusual, but in principle well possible, for instance, when some kind of surface waves are excited. Nonetheless, the appearance of the term A is suspicious and the problem arises whether or not it vanishes in the general case by virtue of the following condition being satisfied:

$$\delta_I = \delta_{II} = \tfrac{1}{2}\delta \tag{10.37}$$

Fedorov (1973; see also the literature cited there) assumed that condition (10.37) and the condition which generalizes it to the case of an anisotropic medium follow from the requirement that the Poynting relation (10.36) should have the form of the energy conservation law in the usual form

$$\frac{\partial w}{\partial t} + \operatorname{div} \mathbf{S} = 0 ,$$

where w is the energy density and $\mathbf{S}$ the energy flux density. Such an argument, however, is insufficient when dispersion and absorption are taken into account. It would thus appear (Agranovich and Ginzburg, 1973; Ginzburg, 1973b) that condition (10.37) is not obligatory. It turns out, however, that it is valid in the general form as it follows from the symmetry principle of kinetic coefficients (Agranovich and Yudson, 1973). This example is sufficiently indicative and useful (at any rate the author of the present book thought it to be particularly pertinent to give it, as he himself has here a certain lack of understanding).

The analysis of the boundary problems and of the actual boundary conditions in media with spatial dispersion is still far from completed. This fact is fortunately no obstacle to the possibility of studying a whole range of problems on the basis of using the tensor $\varepsilon_{ij}(\omega, \mathbf{k})$ or related tensors. For instance since for normal waves (and in general when there are no sources) $\operatorname{div} \mathbf{D} = 0$ one sometimes uses in crystal optics instead of the tensor $\varepsilon_{ij}(\omega, \mathbf{k})$ the so-called transverse dielectric permittivity tensor $\varepsilon_{\perp,ij}(\omega , \mathbf{k})$. This tensor connects in normal waves the induction vector with the transverse component $\mathbf{E}_\perp$ of the field $\mathbf{E}$, that is,

$$D_i(\omega , \mathbf{k}) = \varepsilon_{\perp,ij}(\omega , \mathbf{k})\, E_{\perp,j}(\omega , \mathbf{k}). \tag{10.38}$$

We shall not use the tensor $\varepsilon_{\perp,ij}$ in what follows;† its connection with the tensor ε_{ij} has been elucidated elsewhere (Agranovich and Ginzburg, 1966,1971).

Thanks to the fact that in the optical range spatial dispersion is weak for crystals (condition (10.20) is satisfied) in all known cases we can use instead of the relatively complicated function of $\mathbf{k}$ (which the tensor $\varepsilon_{ij}(\omega , \mathbf{k})$ usually is) an expansion of the tensors $\varepsilon_{ij}(\omega , \mathbf{k})$ or $\varepsilon^{-1}_{ij}(\omega , \mathbf{k})$ in series in $\mathbf{k}$, retaining two or three terms. In the theory of optical activity

---

† We indicate transverse and longitudinal quantities by the indexes $\perp$ and $\parallel$. Just as often one uses the indexes ℓ and tr (in particular, we shall do so in Chapter 11; in the present chapter this would be inconvenient as there are so many indexes).

this fact has been known for a long time, and in that case it was in general sufficient to retain in such an expansion only the terms linear if $\mathbf{k}$ so that†

$$\varepsilon_{ij}(\omega, \mathbf{k}) = \varepsilon_{ij}(\omega) + i\gamma_{ij\ell}(\omega)\, k_\ell \tag{10.39}$$

For a more general discussion we can instead of (10.39) use the expansion

$$\varepsilon_{ij}(\omega, \mathbf{k}) = \varepsilon_{ij}(\omega) + i\gamma_{ij\ell}(\omega)\, k_\ell + \alpha_{ij\ell m}(\omega)\, k_\ell k_m \tag{10.40}$$

or the analogue for the inverse tensor:

$$\varepsilon^{-1}_{ij}(\omega, \mathbf{k}) = \varepsilon^{-1}_{ij}(\omega) + i\delta_{ij\ell} k_\ell + \beta_{ij\ell m}(\omega)\, k_\ell k_m, \tag{10.41}$$

where $\mathbf{k} = (\omega/c)\tilde{n}\mathbf{s}$, and $\tilde{n} = n + i\kappa$ is the complex refractive index for waves of frequency ω propagating in the direction $\mathbf{s} = \mathbf{k}/k$.

Using the tensors ε_{ij} and ε^{-1}_{ij} either in the general form or in the form of the expansions (10.40) and (10.41) is equivalent within wide limits and the choice of one or other of them is a question of convenience. Exceptions are the cases where some components of the tensors $\varepsilon_{ij}(\omega)$ or $\varepsilon^{-1}_{ij}(\omega)$ tend to infinity (increases strongly). For instance, if a component of $\varepsilon_{ij}(\omega)$ tend to infinity, the expansion (10.40) for the corresponding components of $\varepsilon_{ij}(\omega, \mathbf{k})$ loses its meaning, if in this case the coefficients $\gamma_{ij\ell}(\omega)$, $\alpha_{ij\ell m}(\omega)$, ... also tend to infinity. Clearly, in such a case one must use the expansion for the tensor $\varepsilon^{-1}_{ij}(\omega, \mathbf{k})$ which becomes especially effective when $\varepsilon^{-1}_{ij}(\omega)$ decreases. Similarly, when in some region the components of $\varepsilon^{-1}_{ij}(\omega)$ increase strongly, one must use (10.40) rather than (10.41).

We illustrate what we have said by an elementary example, choosing the tensor $\varepsilon_{ij}(\omega, \mathbf{k})$ in the form

$$\varepsilon_{ij}(\omega, \mathbf{k}) = \varepsilon(\omega, \mathbf{k})\, \delta_{ij},$$

where

$$\varepsilon(\omega, \mathbf{k}) = \varepsilon_0 - \frac{a}{\omega^2 - \omega_i^2 + \mu k^2}.$$

If we expand this function $\varepsilon(\omega, \mathbf{k})$ in a series in k^2, it will in fact be an expansion in powers of the ratio $\mu k^2/(\omega^2 - \omega_i^2)$. As in this case

---

† In some crystals (the crystal classes C_{3v}, C_{4v}, C_{3h}, and D_{3h}) the tensor $\gamma_{ij} = 0$, notwithstanding the absence of a centre of inversion. In such cases the whole rotation of the plane of polarization is determined by terms in the expansion of the form $i\gamma_{ij\ell mn} k_\ell k_m k_n$, which were dropped in (10.39). Moreover, close to the quadrupole absorption lines the coefficients $\gamma_{ij\ell}(\omega)$ cannot have a resonance character whereas the ω-dependence of the coefficients $\gamma_{ij\ell mn}$ can be of a resonance nature. In that case the role of the terms omitted in (10.39) increases. Molchanov (1966) has found the non-vanishing components of the tensor $\gamma_{ij\ell mn}$.

$$\boldsymbol{\varepsilon}(\omega) = \varepsilon_0 - \frac{a}{\omega^2 - \omega_i^2},$$

we find that as $\omega \to \omega_i$, when $\boldsymbol{\varepsilon}(\omega) \to \infty$, also the other coefficients of the expansion (10.40) will tend to infinity. At the same time the expansion (10.41) for the tensor $\boldsymbol{\varepsilon}^{-1}(\omega, \mathbf{k})$ is, in fact, in powers of the ratio $\mu k^2/\{(\omega^2 - \omega_i^2)\boldsymbol{\varepsilon} - a\}$ so that as $\omega \to \omega_i$ the coefficients in the expansion (10.41) remain finite.

We shall in what follows consider the effects of spatial dispersion in crystal optics using the expansions (10.39) to (10.41). We note therefore, apart from what we have said about the conditions for their use, that as the tensor $\boldsymbol{\varepsilon}_{ij}(\omega, \mathbf{k})$ for all real media depends not only on ω but also on $\mathbf{k}$, the presence of a region of applicability of classical crystal optics is based in fact on the assumption that there exists a limit as $\mathbf{k} \to 0$ which is reached in this case by the tensors $\boldsymbol{\varepsilon}_{ij}(\omega, \mathbf{k})$ and $\boldsymbol{\varepsilon}_{ij}^{-1}(\omega, \mathbf{k})$ while the limits themselves (that is, the tensors $\boldsymbol{\varepsilon}_{ij}(\omega)$ and $\boldsymbol{\varepsilon}_{ij}^{-1}(\omega)$, respectively) are independent of the direction of $\mathbf{k}$ as $\mathbf{k} \to 0$. In other words, the basis of crystal optics is the assumption that the tensors $\boldsymbol{\varepsilon}_{ij}(\omega, \mathbf{k})$ and $\boldsymbol{\varepsilon}_{ij}^{-1}(\omega, \mathbf{k})$ are analytical functions of $\mathbf{k}$ for small $\mathbf{k}$. The validity of this assumption follows to a large extent from the fact that in the framework of classical crystal optics one is able to explain a huge set of experimental data. A rigorous proof of the analyticity of the tensors $\boldsymbol{\varepsilon}_{ij}(\omega, \mathbf{k})$ and $\boldsymbol{\varepsilon}_{ij}^{-1}(\omega, \mathbf{k})$ as functions of $\mathbf{k}$ for small $\mathbf{k}$ can be obtained in the framework of microscopic theories which enable us to find these tensors in explicit form for several crystals. There are several calculations of the tensors $\boldsymbol{\varepsilon}_{ij}(\omega, \mathbf{k})$ and $\boldsymbol{\varepsilon}^{-1}(\omega, \mathbf{k})$ for crystals (see, for instance, Agranovich and Ginzburg, 1966). In all cases known to us, the tensors $\boldsymbol{\varepsilon}_{ij}(\omega, \mathbf{k})$ and $\boldsymbol{\varepsilon}_{ij}^{-1}(\omega, \mathbf{k})$ turned out to be analytical functions of $\mathbf{k}$ as $\mathbf{k} \to 0$, both outside resonance regions and in the neighbourhood of resonance, but taking damping into account. Apparently, there are no grounds to doubt the analyticity of the functions $\boldsymbol{\varepsilon}_{ij}(\omega, \mathbf{k})$ and $\boldsymbol{\varepsilon}_{ij}^{-1}(\omega, \mathbf{k})$ as $\mathbf{k} \to 0$.

Finally, one more remark regarding our initial assumptions. If we neglect damping, expansions such as (10.40) and (10.41) may turn out to be insufficient, when we retain only a few terms such as we have written down, in the following unusual situation. Let, for instance, $\boldsymbol{\varepsilon}_{ij}(\omega, \mathbf{k}) = \boldsymbol{\varepsilon}(\omega, \mathbf{k})\,\delta_{ij}$ with

$$\boldsymbol{\varepsilon}(\omega, \mathbf{k}) = \boldsymbol{\varepsilon}(\omega) + \frac{\rho k^2}{[(\omega - \omega_i)/\omega_i] - \mu k^2} \qquad (10.42)$$

(the index i of ω_i has, of course, nothing whatever to do with the tensor indexes i, j, ...). This kind of expression sometimes approximately describes the behaviour of $\boldsymbol{\varepsilon}(\omega, \mathbf{k})$ in the vicinity of the frequency of a quadrupole absorption line.

As long as the term μk^2 in (10.42) is unimportant we are dealing with an expansion of the form (10.40). However, in the general case

$$[\boldsymbol{\varepsilon}(\omega, \mathbf{k}) - \boldsymbol{\varepsilon}(\omega)]^{-1} = \frac{[(\omega - \omega_i)/\omega_i] - \mu k^2}{\rho k^2},$$

which corresponds to neither (10.40) nor (10.41). One can, however, easily generalize Eqn.(10.42) for any crystal in the spirit of the phenomenological expansion (10.40):

$$\begin{aligned} \boldsymbol{\varepsilon}_{ij}(\omega, \mathbf{k}) &= \boldsymbol{\varepsilon}_{ij}(\omega) + \gamma_{ij\ell}(\omega) k_\ell + \alpha_{ij\ell m}(\omega, \mathbf{k}) k_\ell k_m, \\ \alpha^{-1}_{ij\ell m}(\omega, \mathbf{k}) &= \xi_{ij\ell m}(\omega) + i\eta_{ij\ell mn}(\omega) k_n + \xi_{ij\ell mnp}(\omega) k_n k_p, \end{aligned} \qquad (10.43)$$

Similarly we can replace $\gamma_{ij\ell}(\omega)$ in (10.40) by $\gamma_{ij\ell}(\omega, \mathbf{k})$, and so on. For a non-gyrotropic cubic crystal Eqns.(10.43) and (10.42) are equivalent where

$$\alpha_{ij\ell m} k_\ell k_m = \frac{\rho(\omega) k^2}{[(\omega - \omega_i)/\omega_i] - \mu k^2} \delta_{ij}.$$

If we generalize Eqn.(10.42) somewhat and write it in the form

$$\boldsymbol{\varepsilon}_{ij}(\omega, \mathbf{k}) = \boldsymbol{\varepsilon}_{ij}(\omega) + \frac{\alpha_{ij\ell m} k_\ell k_m}{[(\omega - \omega_i)/\omega\ ] + i\nu + \mu_{\ell m} k_\ell k_m}, \qquad (10.44)$$

at resonance, that is, as $\omega \to \omega_i$ and $\nu = 0$, the tensor $\boldsymbol{\varepsilon}_{ij}(\omega, \mathbf{k})$ becomes as $\mathbf{k} \to 0$ dependent, in general, on $\mathbf{s} = \mathbf{k}/k$, that is, it turns out to be a non-analytical function of $\mathbf{k}$. However, in actual fact we have always $\nu \neq 0$ and the analytical behaviour of $\boldsymbol{\varepsilon}_{ij}(\omega, \mathbf{k})$ as function of $\mathbf{k}$ as $\mathbf{k} \to 0$ is retained, even at resonance. Therefore, the necessity for an expansion such as (10.43) can arise only in the vicinity of a resonance and when we neglect damping. When we take damping into account, however, the use of an expansion such as (10.43) can be justified in general only when we study the effects of spatial dispersion to higher order.

The tensors occurring in Eqns.(10.39) to (10.41) satisfy a number of relations following from the general symmetry properties of the tensor $\boldsymbol{\varepsilon}_{ij}(\omega, \mathbf{k})$ which we discussed earlier. For instance, by virtue of (10.10)

$$\boldsymbol{\varepsilon}_{ij}(\omega) = \boldsymbol{\varepsilon}_{ji}(\omega), \quad \gamma_{ij\ell}(\omega) = -\gamma_{ji\ell}(\omega), \quad \alpha_{ij\ell m}(\omega) = \alpha_{ji\ell m}(\omega), \qquad (10.45)$$

$$\boldsymbol{\varepsilon}^{-1}_{ij}(\omega) = \boldsymbol{\varepsilon}^{-1}_{ji}(\omega) \ , \ \delta_{ij\ell}(\omega) = -\delta_{ji\ell}(\omega) \ , \ \beta_{ij\ell m}(\omega) = \beta_{ji\ell m}(\omega). \quad (10.45)$$

Moreover, one can always choose the tensors $\alpha_{ij\ell m}$ and $\beta_{ij\ell m}$ in such a way that $\alpha_{ij\ell m} = \alpha_{ijm\ell}$ and $\beta_{ij\ell m} = \beta_{ijm\ell}$ (we shall assume in what follows that we have made such a choice). We remind ourselves also that the magnetic induction of the external field $\mathbf{B}_{ext}$ is always, unless a statement to the contrary is made, assumed to be zero.

When there is a centre of symmetry or in general for a non-gyrotropic medium it follows from (10.12) that

$$\gamma_{ij\ell} = 0 \ , \ \delta_{ij\ell} = 0 \ . \quad (10.46)$$

When there is no absorption and when $\mathbf{k}$ is real, the tensor $\text{Im}\,\boldsymbol{\varepsilon}_{ij}(\omega, \mathbf{k}) = 0$ and hence the tensor $\boldsymbol{\varepsilon}_{ij}(\omega, \mathbf{k}) = \text{Re}\,\boldsymbol{\varepsilon}_{ij}(\omega, \mathbf{k})$ is Hermitean. By virtue of (10.45) in this case, all the tensors

$$\boldsymbol{\varepsilon}_{ij}(\omega) \ , \ \boldsymbol{\varepsilon}^{-1}_{ij}(\omega) \ , \ \gamma_{ij\ell}(\omega) \ , \ \delta_{ij\ell}(\omega) \ , \ \alpha_{ij\ell m}(\omega) \text{ , and } \beta_{ij\ell m}(\omega)$$

are real. We mentioned earlier dipole and quadrupole absorption lines. In fact, near dipole lines it is usually sufficient to use the expansions (10.39), (10.40), or (10.41) while Eqns.(10.42) to (10.44) can be met with in first instance in the case of quadrupole lines. We must at the same time emphasize that the above mentioned expansions in series $\mathbf{k}$ are not multipole expansions. Moreover, the use of Eqns.(10.39) to (10.41) is not limited at all to the region of some lines. We must also bear in mind that in an arbitrary optically anisotropic medium and for an arbitrary direction of propagation of light the appearance of resonance lines (or when we neglect absorption — the appearance of a pole in the function $\tilde{n}^2 = n^2$) is not connected with the resonance growth of the components of $\boldsymbol{\varepsilon}_{ij}(\omega, \mathbf{k})$ as is already clear from the formulae of classical crystal optics (see, for instance, (10.27)).

The tensors $\boldsymbol{\varepsilon}_{ij}(\omega, \mathbf{k})$, $\boldsymbol{\varepsilon}^{-1}(\omega, \mathbf{k})$ and, of course, the tensors

$$\boldsymbol{\varepsilon}_{ij}(\omega), \boldsymbol{\varepsilon}^{-1}_{ij}(\omega) \ , \ \gamma_{ij\ell} \ , \ \delta_{ij\ell} \ , \ \alpha_{ij\ell m}, \text{ and } \beta_{ij\ell m}$$

which occur in (10.40) and (10.41) can be significantly simplified when the crystal has symmetry elements. For instance, when there is a centre of symmetry $\gamma_{ij\ell} = \delta_{ij\ell} = 0$ (see (10.46)) and in an isotropic, non-gyrotropic medium the tensor $\alpha_{ij\ell m}$ has only two independent components so that in that case (see also Chapter 11)

$$\boldsymbol{\varepsilon}_{ij}(\omega, \mathbf{k}) = \boldsymbol{\varepsilon}(\omega)\,\delta_{ij} + \alpha_{\perp}(\omega)\,(\delta_{ij} + s_i s_j)\,k^2 + \alpha_{\parallel}(\omega)\,s_i s_j k^2 \ . \quad (10.47)$$

We refer to the books by Agranovich and Ginzburg (1966) and Nye (1957) for a detailed discussion of the well known consequences for the tensors

$$\boldsymbol{\varepsilon}_{ij}(\omega)\ ,\ \gamma_{ij\ell}(\omega)\ ,\ \alpha_{ij\ell m}\ ,\ \ldots$$

which are connected with crystal symmetry. We shall restrict ourselves merely to a few examples.

It follows from Eqn.(10.45) that the tensor $\gamma_{ij\ell}$ (and $\delta_{ij\ell}$) possesses the following properties:

$$\gamma_{xx,\ell} = \gamma_{yy,\ell} = \gamma_{zz,\ell} = 0\ ,\ \gamma_{xy,\ell} = -\gamma_{yx,\ell}\ ,\ \gamma_{yz,\ell} = -\gamma_{zy,\ell}\ ,$$
$$\gamma_{zx,\ell} = -\gamma_{xz,\ell} \qquad (\ell = 1, 2, 3 \equiv x, y, z)\ .$$

In the general case the tensors $\gamma_{ij\ell}$ and $\delta_{ij\ell}$ thus have nine independent components, and we can write them in the form

$$\gamma_{ij\ell} = e_{ijm}\, g'_{m\ell}\ ,\quad \delta_{ij\ell} = e_{ijm}\, f'_{m\ell}\ , \tag{10.48}$$

where e_{ijm} is the completely antisymmetric third rank unit pseudotensor ($e_{123} = 1$, $e_{213} = -1$, $e_{112} = 0$, and so on; $e_{ij\ell}$ does not change under a cyclic permutation) and $g'_{m\ell}$ and $f'_{m\ell}$ are second rank pseudotensors.

Besides the tensors $g'_{m\ell}$ and $f'_{m\ell}$ one sometimes introduces the gyration vectors $\mathbf{g}'$ and $\mathbf{f}'$ defined by the relations

$$g'_m = g'_{m\ell}\, k_\ell\ ,\quad f'_m = f'_{m\ell}\, k_\ell\ ,\quad \mathbf{k} = k\mathbf{s}\ . \tag{10.49}$$

If we neglect in (10.40) and (10.41) terms quadratic in k, we can use the gyration vectors to write those equations in the following form:

$$D_i = \boldsymbol{\varepsilon}_{ij}(\omega, \mathbf{k})\, E_j = \boldsymbol{\varepsilon}_{ij}(\omega)\, E_j - i\,[\mathbf{g}' \wedge \mathbf{E}]_i\ , \tag{10.50a}$$

$$E_i = \boldsymbol{\varepsilon}^{-1}_{ij}(\omega, \mathbf{k})\, D_j = \boldsymbol{\varepsilon}^{-1}(\omega)\, D_j - i\,[\mathbf{f}' \wedge \mathbf{D}]_i\ . \tag{10.50b}$$

Because the spatial dispersion is small it is in most cases fully sufficient to use Eqns.(10.50) for gyrotropic media. We shall therefore in what follows consider not the general expansions (10.40) and (10.41), but Eqn.(10.50) for gyrotropic media and also the following expressions for a non-gyrotropic medium:

$$\boldsymbol{\varepsilon}_{ij}(\omega, \mathbf{k}) = \boldsymbol{\varepsilon}_{ij}(\omega) + \left(\frac{\omega}{c}\right)^2 \alpha_{ij\ell m}(\omega)\, \tilde{n}^2\, s_\ell\, s_m\ , \tag{10.51a}$$

$$\boldsymbol{\varepsilon}^{-1}_{ij}(\omega, \mathbf{k}) = \boldsymbol{\varepsilon}^{-1}_{ij}(\omega) + \left(\frac{\omega}{c}\right)^2 \beta_{ij\ell m}(\omega)\, \tilde{n}^2\, s_\ell\, s_m\ . \tag{10.51b}$$

One can, of course, always bring the tensor $\boldsymbol{\varepsilon}_{ij}(\omega, \mathbf{k})$ to diagonal form by suitably choosing principal axes.† For arbitrary $\mathbf{s}$ the direction of these axes is not the same as that of $\mathbf{s}$ or as that of the axes of the tensor $\varepsilon_{ij}(\omega)$; in those cases where the axes of the tensor $\varepsilon_{ij}(\omega)$ are fixed (that is, when there is no degeneracy, which occurs in cubic and uni-axial crystals) the axes of the tensor $\varepsilon_{ij}(\omega, \mathbf{k})$ are close to the axes of $\varepsilon_{ij}(\omega)$ due to the fact that the $\mathbf{s}$-dependent terms in (10.50) and (10.51) are small.

In crystal optics with spatial dispersion there is, of course, great interest in those principal axes of $\varepsilon_{ij}(\omega, \mathbf{k})$ which have the same direction as $\mathbf{s}$. For rhombic crystals the x-, y-, z-axes are such axes. If, for instance, the vector $\mathbf{s}$ is along the x-axis, the principal values of the tensor $\varepsilon_{ij}(\omega, \mathbf{k})$ are equal to (see here and below Table III in Agranovich and Ginxburg, 1966)

$$\varepsilon_1 \equiv \varepsilon_{xx}(\omega, \mathbf{k}) = \varepsilon_{xx}(\omega) + \left(\frac{\omega}{c}\tilde{n}\right)^2 \alpha_{xxxx},$$

$$\varepsilon_2 \equiv \varepsilon_{yy}(\omega, \mathbf{k}) = \varepsilon_{yy}(\omega) + \left(\frac{\omega}{c}\tilde{n}\right)^2 \alpha_{yyxx},$$

$$\varepsilon_3 \equiv \varepsilon_{zz}(\omega, \mathbf{k}) = \varepsilon_{zz}(\omega) + \left(\frac{\omega}{c}\tilde{n}\right)^2 \alpha_{zzxx}.$$

In tetragonal crystals of the classes D_4, C_{4v}, D_{2d}, and D_{4h} the tensor $\varepsilon_{ij}(\omega, \mathbf{k})$ turns out to be brought onto principal axes if the vector $\mathbf{s}$ is along the x- or y-axis, and the principal values are different. If, however, the vector $\mathbf{s}$ is along the z-axis (along the fourth order axis) we have

$$\varepsilon_1 = \varepsilon_2 = \varepsilon_\perp(\omega) + \left(\frac{\omega}{c}\tilde{n}\right)^2 \alpha_{xxzz}, \quad \varepsilon_3 = \varepsilon_\parallel(\omega) + \left(\frac{\omega}{c}\tilde{n}\right)^2 \alpha_{zzzz}.$$

Forgetting the crystals of other crystal systems we turn to cubic crystals. In that case

$$\left.\begin{aligned}
&\alpha_1 = \alpha_{xxxx} = \alpha_{yyyy} = \alpha_{zzzz}, \quad \alpha_2 = \alpha_{xxzz} = \alpha_{yyxx} = \alpha_{zzyy},\\
&\alpha_3 = \alpha_{xyxy} = \alpha_{yzyz} = \alpha_{zxzx}, \quad \alpha_4 = \alpha_{zzxx} = \alpha_{xxyy} = \alpha_{yyzz},\\
&\varepsilon_{xx} = \varepsilon + \left(\frac{\omega}{c}\tilde{n}\right)^2 (\alpha_1 s_x^2 + \alpha_4 s_y^2 + \alpha_2 s_z^2), \quad \varepsilon_{xy} = 2\left(\frac{\omega}{c}\tilde{n}\right)^2 \alpha_3 s_x s_y,\\
&\varepsilon_{xz} = 2\left(\frac{\omega}{c}\tilde{n}\right)^2 \alpha_3 s_x s_z, \quad \varepsilon_{yy} = \varepsilon + \left(\frac{\omega}{c}\tilde{n}\right)^2 (\alpha_2 s_x^2 + \alpha_1 s_y^2 + \alpha_4 s_z^2),\\
&\varepsilon_{zz} = \varepsilon + \left(\frac{\omega}{c}\tilde{n}\right)^2 (\alpha_4 s_x^2 + \alpha_2 s_y^2 + \alpha_1 s_z^2), \quad \varepsilon_{yz} = 2\left(\frac{\omega}{c}\tilde{n}\right)^2 \alpha_3 s_y s_z.
\end{aligned}\right\} \quad (10.52)$$

† If the tensor $\varepsilon_{ij}(\omega, \mathbf{k})$ is non-Hermitean, we must independently consider the Hermitean tensors ε'_{ij} and ε''_{ij}, $\varepsilon_{ij} = \varepsilon'_{ij} + i\varepsilon''_{ij}$, and the principal axes (or, more precisely, the eigenvectors which in the general case are complex) of these tensors may not coincide. Unless we make a statement to the contrary we are in the text only considering the tensor ε'_{ij} which we assume to be real.

(for the classes O, T_d, and O_h, we have, moreover $\alpha_2 = \alpha_4$; the factor 2 in the expressions for ε_{xy}, ε_{xz}, and ε_{yz} arises because of the summation in (10.51) of the terms proportional to $s_x s_y$ and $s_y s_x$, and so on). From this it is clear that the x-, y-, and z-axis of the cube are the principal axes of the tensor, if the vector $\mathbf{s}$ is directed along any of the three axes (x-, y-, or z-axis). The corresponding second degree surface is then for $\alpha_2 = \alpha_4$ degenerated into a surface (ellipsoid or hyperboloid) of rotation. If the vector $\mathbf{s}$ is directed along one of the space diagonals of the cube

$$(|s_x| = |s_y| = |s_z| = 1/\sqrt{3})$$

we have

$$\varepsilon_{xx} = \varepsilon_{yy} = \varepsilon_{zz} = \varepsilon + \tfrac{1}{3}\left(\frac{\omega}{c}\tilde{n}\right)^2(\alpha_1 + \alpha_2 + \alpha_4)$$

$$|\varepsilon_{xy}| = |\varepsilon_{xz}| = |\varepsilon_{yz}| = 2\left(\frac{\omega}{c}\tilde{n}\right)^2 \frac{\alpha_3}{3} .$$

We now turn finally to some effects of spatial dispersion in crystal optics. We noted already that as spatial dispersion is small, for instance, in optics, in first instance one is interested in such cases only in those problems where spatial dispersion leads to qualitatively new effects or, at any rate, does not just lead only to insignificant corrections to the formulae from classical crystal optics. In agreement with what we have said the following exposition has a fragmentary nature and reduces essentially to a discussion of those effects for the analysis of which is is necessary to take spatial dispersion into account. As we have already emphasized the most important effect of this kind is gyrotropy (natural optical activity). A phenomenological discussion of gyrotropy has, however, usually without using explicitly spatial dispersion,† been given very long ago and has been elucidated in rather great detail in the literature (see, for instance, the books by Landau and Lifshitz, 1960, Agranovich and Ginzburg, 1966, and Nye, 1957). Another important effect of spatial dispersion — the existence of a non-vanishing group velocity of longitudinal waves — is particularly well known in its application to plasmas and will be mentioned in Chapter 11. We shall therefore discuss here only two effects of spatial dispersion: the optical anisotropy of cubic crystals and the appearance of additional (new) normal waves close to resonance (in the anomalous dispersion region).

---

† Formally one can, neglecting the requirements connected with the symmetry of kinetic coefficients, describe a gyrotropic medium by considering a non-symmetric tensor $\varepsilon_{ij}(\omega)$.

Lorentz predicted the optical anisotropy of cubic crystals already in the last century, but it was first observed in 1960 in Cu_2O in the quadrupole transition region at a wavelength $\lambda = 6125\,\text{Å}$ (for references see Agranovich and Ginzburg, 1966, 1971, 1973).

The appearance of optical activity in cubic crystals when spatial dispersion is taken into account follows immediately from Eqn.(10.51). For cubic crystals Eqn.(10.51a), for instance, takes the form (10.52). Outside the resonance region of the function $\boldsymbol{\varepsilon}(\omega)$ we could with the same justification use either of the Eqns.(10.51). If, however, we consider the resonance region of $\boldsymbol{\varepsilon}(\omega)$, spatial dispersion is taken into account correctly only by using the expansion for the tensor $\boldsymbol{\varepsilon}^{-1}(\omega, \mathbf{k})$.

If we do not pay attention to spatial dispersion, the tensor $\boldsymbol{\varepsilon}_{ij}(\omega)$ reduces to a scalar for cubic crystals and as a result we get for the refractive index $\tilde{n}^2 = \boldsymbol{\varepsilon}(\omega)$, which corresponds to a complete independence of the optical properties of the crystal of the direction of propagation of the light and its polarization. However, when spatial dispersion is taken into account the tensor $\boldsymbol{\varepsilon}_{ij}(\omega, \mathbf{k})$ no longer reduce to a scalar (see (10.52)). Substitution of this tensor into Eqn.(10.27) leads to values of $\tilde{n}^2$ which depend both on the direction of propagation of the light and on its polarization and this corresponds to the optical anisotropy of the crystal. As the coefficients $\alpha_{ij\ell m}$ and $\beta_{ij\ell m}$ are of the order of the square of the lattice constant $a \sim 10^{-8}$ to 10^{-7} cm the anisotropy is small (for $\boldsymbol{\varepsilon}_{ij} \sim 1$, $\tilde{n} \sim 1$, this effect is of the order of $(\omega a/c)^2 = (2\pi a/\lambda_0)^2 \sim 10^{-4}$ to 10^{-5}). The dependence of $\tilde{n}^2$ on the direction of $\mathbf{s} = \mathbf{k}/k$ and on the polarization for cubic crystals of different crystal classes is considered in detail by Agranovich and Ginzburg (1966).

We turn now to the problem of new (additional) waves which appear when spatial dispersion is taken into account. The possibility of the appearance of such waves is at once clear from the general dispersion Eqn.(10.27). Indeed, when spatial dispersion is neglected this equation is a quadratic equation in the refractive index $\tilde{n}^2$ (that is, bi-quadratic in the index $\tilde{n}$). The dispersion equation therefore has, as we have already mentioned, only two solutions $\tilde{n}_1^2$ and $\tilde{n}_2^2$ corresponding to two normal waves (we do not now consider longitudinal oscillations; the solutions $\tilde{n}_{1,2}$ and $-\tilde{n}_{1,2}$ correspond to two opposite propagation directions, not to different kinds of wave). However, when spatial dispersion is taken into account the coefficient themselves in the dispersion equation (10.27) depend on $\tilde{n}$ through $\boldsymbol{\varepsilon}_{ij}(\omega, \mathbf{k}(\omega))$, $\mathbf{k}(\omega) = \frac{\omega}{c}\,\tilde{n}(\omega)\,\mathbf{s}$ as for normal waves ω and $\mathbf{k}$ are just connected though the dispersion equation. As a result this equation can, in principle, have any number of roots. In an actual situation, however, the number of roots, at least those which are not too strongly damped, (that is, which correspond to not too strongly damped waves), usually turns out to be relatively small. What we have said is clearly valid in the case of weak spatial dispersion when one encounters only one or two new waves (roots). Moreover, even such new waves are difficult to observe in optics and so far this has been accomplished only indirectly — using Raman scattering of light (see Arganovich and Ginzburg, 1972). Nonetheless it is undoubtedly of interest to consider the new waves which occur when spatial dispersion is taken into account.

If we bear in mind that we shall in what follows study, in particular, the propagation of waves in the vicinity of absorption bands, we shall use an expansion of the inverse dielectric permittivity tensor which in the most general case we can write in the following form:

$$\varepsilon^{-1}_{ij}(\omega,\mathbf{k}) = \tilde{\varepsilon}^{-1}_{ij}(\omega,\mathbf{k}) + i\delta_{ij\ell}(\omega,\mathbf{k})\,\frac{\omega\tilde{n}}{c}\,s_{\ell}\,, \tag{10.53}$$

where $\tilde{\varepsilon}^{-1}_{ij}(\omega,\mathbf{k})$ and $\delta_{ij\ell}(\omega,\mathbf{k})$ are tensors which are even functions of $\mathbf{k}$ ($\tilde{\varepsilon}^{-1}_{ij}(\omega,\mathbf{k}) = \tilde{\varepsilon}^{-1}_{ij}(\omega,-\mathbf{k})$, $\delta_{ij\ell}(\omega,\mathbf{k}) = \delta_{ij\ell}(\omega,-\mathbf{k})$, while the tensor $\delta_{ij\ell}$ is only different from zero in crystals without a centre of inversion). As the spatial dispersion is by assumption small we retain in the expansions of $\tilde{\varepsilon}^{-1}_{ij}(\omega,\mathbf{k})$ and $\delta_{ij\ell}$ only the first terms in the series in $\mathbf{k}$. We thus use Eqn.(10.53) in a form which is equivalent to the one already considered earlier.

For non-longitudinal waves in the medium the induction $\mathbf{D} \neq 0$. It is thus convenient to study the properties of these waves in a system of coordinates for which the z-axis is along $\mathbf{k}$ so that by virtue of the relation $(\mathbf{k}\cdot\mathbf{D}) = 0$ the equations $D_z \equiv D_3 = 0$ holds. In this system of coordinates the equations (which are equivalent to the basic vector Eqn.(10.22)) which determine the non-vanishing components of the induction vector have the following form

$$\begin{aligned} \left(\frac{1}{\tilde{n}^2} - \varepsilon^{-1}_{xx}\right)D_x - \tilde{\varepsilon}^{-1}_{xy}D_y &= i\frac{\omega}{c}\tilde{n}\,\delta_{123}D_y\,, \\ -\tilde{\varepsilon}^{-1}_{yx}D_x + \left(\frac{1}{\tilde{n}^2} - \tilde{\varepsilon}^{-1}_{yy}\right)D_y &= -i\frac{\omega}{c}\tilde{n}\,\delta_{123}D_x\,. \end{aligned} \tag{10.54}$$

We choose the x- and y-axes along the principal axes of the two-dimensional tensor $\varepsilon^{-1}_{ij}(\omega,0) = \varepsilon^{-1}_{ij}(\omega)$ and denote the principal values of that tensor by $1/\tilde{n}^2_{01}$ and $1/\tilde{n}^2_{02}$. The components ε^{-1}_{xy} and ε^{-1}_{yx} turn out to be small quantities of order k^2 for this choice of axes. The condition that the determinant of the set (10.54) must vanish gives an equation which enables us to determine the possible values of $\tilde{n}^2$. If we drop terms of order k^3, k^4, and so on, this equation has the form

$$\left(\frac{1}{\tilde{n}^2}-\tilde{\varepsilon}^{-1}_{xx}\right)\left(\frac{1}{\tilde{n}^2}-\tilde{\varepsilon}^{-1}_{yy}\right) = \delta^2_{123}(\omega, \mathbf{k})\,\frac{\omega^2}{c^2}\,\tilde{n}^2 . \qquad (10.55)$$

If we put in Eqn.(10.55) $\delta_{123} = 0$ and $\mathbf{k} = 0$, its solutions $\tilde{n}^2_{1,2} = \tilde{n}^2_{0,12}$ are, of course, the same as the solutions of the Fresnel equation with $\boldsymbol{\varepsilon}_{ij} = \boldsymbol{\varepsilon}_{ij}(\omega)$. If, however, $\delta_{123} \neq 0$ and $\mathbf{k} \neq 0$, Eqn.(10.55) determines for a given direction of $\mathbf{s}$ even when $\tilde{\varepsilon}^{-1}_{xx} = \tilde{\varepsilon}^{-1}_{xx}(\omega)$ and $\tilde{\varepsilon}^{-1}_{yy} = \tilde{\varepsilon}^{-1}_{yy}(\omega)$ not two but, in general, several values of the index of refraction (Eqn.(10.55) is for non-longitudinal waves in the given approximation the same as (10.27), as should be the case).

The analysis of Eqn.(10.55) is particularly simple for media which are optically isotropic when spatial dispersion is neglected.

For an isotropic medium $\tilde{\varepsilon}^{-1}_{xx}(\omega, 0) = \tilde{\varepsilon}^{-1}_{yy}(\omega, 0) = 1/\boldsymbol{\varepsilon}(\omega)$. One sees easily from (10.55) that in that case for an isotropic medium with $\delta_{123} \neq 0$ one can neglect the $\mathbf{k}$-dependence of the quantities $\tilde{\varepsilon}^{-1}_{xx}$, $\tilde{\varepsilon}^{-1}_{yy}$, and δ_{123} so that we get instead of (10.55) the following equation to determine $\tilde{n}^2$

$$\left(\frac{1}{\tilde{n}^2} - \frac{1}{\boldsymbol{\varepsilon}(\omega)}\right)^2 = \frac{\omega^2}{c^2}\,\delta^2_{123}\,\tilde{n}^2 . \qquad (10.56)$$

This equation has clearly three roots for $\tilde{n}^2$, that is, we find three values $\tilde{n}^2_1$, $\tilde{n}^2_2$, $\tilde{n}^2_3$. It may turn out that all three roots of Eqn.(10.56) correspond to relative long wavelengths and it is then admissible to consider all three solutions in the framework of a macroscopic approach. In particular, in the vicinity of resonance (that is, when $\omega \approx \omega_i$, where $\boldsymbol{\varepsilon}(\omega_i) \to \infty$) one can neglect the ω-dependence of δ_{123} and one may assume the function $\boldsymbol{\varepsilon}(\omega)$ to be known in that frequency range; the solution of Eqn.(10.56) then allows us to find in the resonance region the functions $\tilde{n}^2(\omega)$ for all three solutions and thus to check the validity of the condition $an/\lambda_0 \ll 1$ (see (10.20)). A detailed discussion of the results can be found in § 6.3 of the book by Agranovich and Ginzburg (1966). We shall therefore here only give a few results for the case where we neglect the absorption of the waves. The quantities δ_{123} and $\varepsilon(\omega)$ are then real and in the region $\omega \approx \omega_i$ we can put

$$\boldsymbol{\varepsilon}(\omega) = \boldsymbol{\varepsilon}_0 - \frac{2A\omega_i^2}{\omega^2 - \omega_i^2} \approx \boldsymbol{\varepsilon}_0 - \frac{A}{\xi} , \qquad (10.57)$$

where $\xi = (\omega - \omega_i)/\omega_i$, $A = 2\pi N_{eff}\,e^2/m\omega_i^2$; here e and m are the charge and mass of a free electron, N_{eff}/N is the oscillator strength, where N is the total number of electrons per unit volume and N_{eff} that part of them which 'effectively' determines the optical properties of the medium in the spectral region considered.

The behaviour of the functions $n(\xi)$ obtained by solving Eqn.(10.56) for this case is shown in Fig.10.1. An interesting feature of the dispersion curves shown there is that to the right of the turning point $\xi_m = (\omega_m - \omega_i)/\omega_i$ there exists only a single real solution whereas to the left of it there are three real solutions. We note that the multiple root (that is, the turning point) corresponds to the frequency ω_m which satisfies the equation

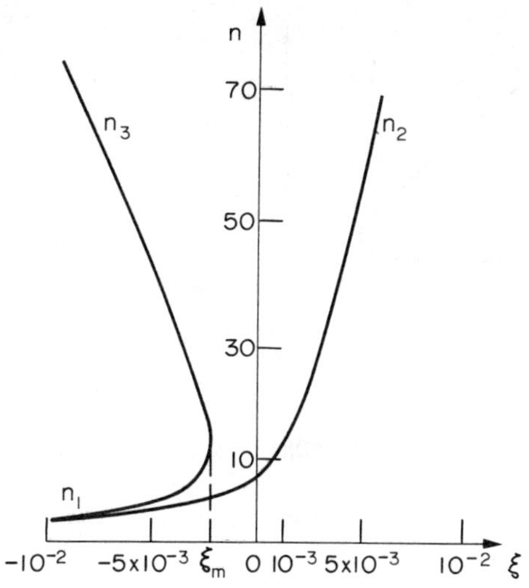

Fig.10.1 The refractive index n_1, n_2, n_3 as function of $\xi = (\omega - \omega_i)/\omega_i$ near the resonance frequency ω_i for the case of a gyrotropic, but isotropic and non-absorbing medium

$$\boldsymbol{\varepsilon}(\omega_m) = \frac{1}{3}2^{\frac{2}{3}}\left(\frac{\omega_m}{c}\delta_{123}\right)^{-\frac{2}{3}}. \tag{10.58}$$

The solution of Eqn.(10.55) is even simpler in the case of an isotropic non-gyrotropic medium. As in that case we have not only the equation $\tilde{\boldsymbol{\varepsilon}}^{-1}_{xx}(\omega, 0) = \tilde{\boldsymbol{\varepsilon}}^{-1}_{yy}(\omega, 0) = 1/\boldsymbol{\varepsilon}(\omega)$, but also $\delta_{123} = 0$, Eqn.(10.55) becomes

$$\frac{1}{\tilde{n}^2} - \frac{1}{\boldsymbol{\varepsilon}(\omega)} = \frac{\omega^2}{c^2}\beta\tilde{n}^2 . \tag{10.59}$$

This equation is obtained from (10.55) by writing

$$\tilde{\boldsymbol{\varepsilon}}^{-1}_{xx}(\omega, \mathbf{k}) = \tilde{\boldsymbol{\varepsilon}}^{-1}_{yy}(\omega, \mathbf{k}) = \frac{1}{\boldsymbol{\varepsilon}(\omega)} + \beta k^2 \equiv \tilde{\boldsymbol{\varepsilon}}^{-1}(\omega) + \frac{\omega^2}{c^2}\beta\tilde{n}^2 .$$

Equation (10.59) determines two values of the refractive index $\tilde{n}^2_{1,2}$ which correspond to the same polarization of light, and

$$\tilde{n}^2_{1,2} = -\frac{1}{2\boldsymbol{\varepsilon}(\omega)\beta'} \pm \left[\left(\frac{1}{2\boldsymbol{\varepsilon}(\omega)\beta'}\right)^2 + \frac{1}{\beta'}\right]^{\frac{1}{2}} , \tag{10.60}$$

where $\beta' = (\omega^2/c^2)\beta$. When $\beta' < 0$ it follows from (10.60) that at a frequency $\omega = \omega_m$, given by the equation

$$\boldsymbol{\varepsilon}(\omega_m) = \frac{1}{2}|\beta'|^{-\frac{1}{2}} , \tag{10.61}$$

the roots $\tilde{n}^2_1$ and $\tilde{n}^2_2$ are equal. Hence, the frequency ω_m corresponds to a turning point. If, however, $\beta' > 0$, no turning point appears. We show in Fig.10.2 the frequency-dependence of the roots $\tilde{n}^2_1$ and $\tilde{n}^2_2$ in the vicinity of the resonance using Eqn.(10.57) and the values $\beta' = \pm 10^{-5}$. Taking damping

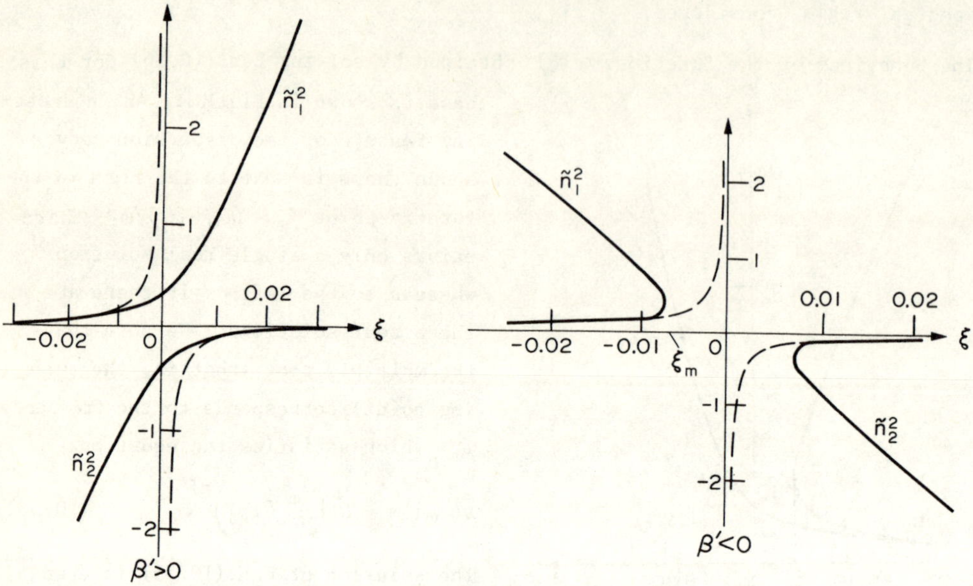

Fig.10.2 $\tilde{n}^2$ as function of $\xi = (\omega - \omega_i)/\omega_i$ close to the frequency ω_i for the case of an isotropic and non-absorbing, but non-gyrotropic medium.
The dotted curves correspond to $\beta = 0$. Along the ordinate axis we have given the values of the quantity $10^{-3}\,\tilde{n}^2$.

into account changes the character of the ω-dependence of $\tilde{n}^2_{1,2}$ considerably. We note also that the situation regarding dispersion for an anisotropic non-gyrotropic medium remains qualitatively the same as for an isotropic medium. The only important difference consists here in that for an anisotropic medium, where, in general, the quantities n^2_{01} and n^2_{02} which satisfy Eqn.(10.55) for $\delta_{123} = 0$ and $\mathbf{k} = 0$, are not the same, the new wave of type 1 appears close to the resonance of n_{01} and the new wave of type 2 close to the resonance of n_{02}. We have elsewhere (Agranovich and Ginzburg, 1971) considered the nature of the dispersion curves and the polarization of the new (additional) waves in an anisotropic (both gyrotropic and non-gyrotropic) medium. There is no reason to discuss here details, as it is rather difficult to observe the new waves although one should nevertheless count on a solution of the corresponding experiments in the future (we forget here the Raman scattering method mentioned earlier which has already enabled one to observe a new wave; see Agranovich and Ginzburg, 1972). Taking spatial dispersion into account is in some cases important in the theory of surface waves in the optical band, or, as one often says, in the theory of surface excitons (see Agranovich and Ginzburg,

1966; Agranovich, 1968; Ginzburg and Kelle, 1973).

At the same time we must once again emphasize that the effects of spatial dispersion in optics are weak — this is caused by the fact that the coefficients $(\omega/c)\delta \sim 2\pi a/\lambda \sim 10^{-2}$ to 10^{-3} and $\beta' = (\omega^2/c^2)\beta \approx 4\pi^2 a^2/\lambda^2 \sim 10^{-4}$ to 10^{-5} are small. It is thus clearly possible to neglect spatial dispersion in most problems. However, it is also very clear why at the present time when measuring techniques have been improved and the field of phenomena and objects studied has been enlarged the value of taking spatial dispersion into account in optics (let alone in plasma physics) is so strongly increasing.

Chapter XI

DIELECTRIC PERMITTIVITY AND WAVE PROPAGATION IN A PLASMA

Dielectric permittivity of a plasma (elementary and kinetic theory). Wave propagation in a uniform plasma and in a magnetized plasma.

A typical plasma is a strongly or fully ionized gas. Often one also calls a weakly ionized gas and the electron gas or electron liquid in semiconductors and metals a plasma (in the latter two cases one speaks in applications to solids about a solid-state plasma). Undoubtedly, there is much in common to all kinds of plasmas. In what follows we shall, however, have in mind only a non-relativistic gaseous plasma for the description of which it is sufficient to restrict ourselves to the classical approximation and, to be precise, we can use the classical non-relativistic kinetic equation. It is just this kind of plasma which one encounters in most cases in astrophysics, the physics of the ionosphere, and when analyzing the problem of thermonuclear fusion using a rarefied plasma (the situation may turn out to be more complex when we consider the heating and compression of pellets of condensed solid hydrogen which are irradiated by lasers or intense electron beams).

The study of plasmas and, in particular, the propagation of waves in a plasma have been the subject of a huge number of papers — tens of thousands of papers. There are so many problems and different special cases in this field that one can only discuss them in a large series of books. This has already been done to a large extent, but we shall here only mention some monographs which contain also extensive bibliographies, namely, the books by Ginzburg (1970b), Zheleznyakov (1970), Kaplan and Tsytovich (1973), Ginzburg and Rukhadze (1975), Tsytovich (1977), Gurevich and Shvartsburg (1973), Akhiezer, Akhiezer, Polovin and Stepanov (1975a,b), and Smirnov (1972)† as well as review articles by Silin and Rukhadze (1961), Silin (1971, 1973), Gould (1974), Erokhin and Moiseev (1973), Gorbunov (1973), and the series edited by Leontovich (1965). We shall for this reason only touch upon a few problems in plasma physics in the present

† (Note by the translator). We may add to this list a few Western monographs, for instance, those by Stix (1962), Spitzer (1962), Thompson (1964), Davidson (1972) and Ichimaru (1973).

book and we do this for two reasons. First of all, we can use plasmas as an example to make more precise and to elucidate a number of the general results about dispersive media (see Chapter 10). Secondly, plasmas are objects of exceptional importance in astrophysics and in the physics of the ionosphere. However, in classical courses of electromagnetic theory plasmas are usually quite insufficiently considered, while special courses on plasma physics are not followed by anything like all students. The exposition of even a few plasma physics problems in the present book may therefore turn out to be useful and perhaps convenient for the readers (some results and remarks referring to plasma physics are already contained in previous chapters, especially in Chapter 7, but, as in other cases, we are not afraid of repeating ourselves). To get acquainted with details about the problems discussed here we refer to the literature cited already (for obvious reasons the discussion in the present chapter is very close both in nature of the exposition and in contents to the monographs by Ginzburg (1970b) and by Ginzburg and Rukhadze (1975)).

We shall consider only two problems in what follows: we evaluate the dielectric permittivity tensor $\boldsymbol{\varepsilon}_{ij}(\omega, \mathbf{k})$ of the plasma and we study the propagation of various normal waves in the plasma. In the general case — when thermal motion is taken into account and when there is an external magnetic field $\mathbf{H}_0 \equiv \mathbf{B}_0$ present — the calculation of the tensor $\boldsymbol{\varepsilon}_{ij}$ and a consistent study of normal waves is rather complicated or at least cumbersome. As the main point of our effort is to elucidate the physical picture and not to give many details we shall start from simple problems, and then move to more complicated ones.

The simplest statement of the problem is the following one: we consider an isotropic plasma (that is, we put $\mathbf{H}_0 = 0$) and neglect thermal motion. In that case (and also when $\mathbf{H}_0 \neq 0$, but still neglecting thermal motion) it is sufficient for the calculation of $\boldsymbol{\varepsilon}_{ij}$ to use the so-called 'elementary theory' which reduces to considering the ordered motion of separate particles (electrons and ions) in an electric field of frequency ω.

For the sake of simplicity we shall assume the ions to have a single charge (in the case of hydrogen this condition is, of course, automatically satisfied), we shall neglect the possibility of negative ions, and, finally, we shall use the condition of quasi-neutrality $N \equiv N_e = N_i$, where N_e and N_i are, respectively, the electron and ion densities; we shall take the electron charge to be $e < 0$. There is, clearly, no volume charge in the plasma under conditions of quasi-neutrality as the charge density is $\rho = e(N - N_i)$. Of course, the condition of quasi-neutrality may be violated, but when we are dealing with large volumes which are typical for astrophysical and ionospheric problems the difference $|N - N_i| \ll N$. We can check this from estimates based on the field equation

$$\operatorname{div} \mathbf{E} = 4\pi\rho = 4\pi e(N - N_i) \tag{11.1}$$

and bearing in mind that when there is a field present in the plasma this produces a current which in most cases will lead very fast to the vanishing of the volume charge (this does not refer to space charges close to external charges introduced into the plasma or to the fields close to the separate particles which constitue the plasma; vide infra).

The current density is in an isotropic plasma equal to

$$\mathbf{j} = e \sum_{n=1} \left(\dot{\mathbf{r}}_n - \dot{\mathbf{r}}_n^{(i)}\right) = -\frac{i\omega(\mathbf{D} - \mathbf{E})}{4\pi} = \mathbf{j}_{cond} - i\omega\mathbf{P}$$

$$= \left(\sigma - i\,\frac{\boldsymbol{\varepsilon} - 1}{4\omega}\,\omega\right)\mathbf{E} == \frac{i\omega}{4\pi}(\boldsymbol{\varepsilon} - 1)\,\mathbf{E}\,. \tag{11.2}$$

Here $\mathbf{j}_{cond} = \boldsymbol{\sigma}\mathbf{E}$ is the conduction current density, $\mathbf{P} = (\varepsilon' - 1)\,\mathbf{E}/4\pi$ is the polarization, $\boldsymbol{\varepsilon} \equiv \boldsymbol{\varepsilon}' + i\varepsilon'' \equiv \boldsymbol{\varepsilon}' + 4\pi\, i\boldsymbol{\sigma}/\omega$ is the complex permittivity ($\varepsilon' = \mathrm{Re}\,\boldsymbol{\varepsilon}$, $\boldsymbol{\sigma}$ is the conductivity) and we have assumed that the electric field $\mathbf{E}$ is monochromatic ($\mathbf{E} = \mathbf{E}_0\, e^{-i\omega t}$). We have written Eqn. (11.2) in such a way that we can illustrate the notations one encounters in the literature.† Quasi-neutrality is taken into account in (11.2) only in the summation over the number of particles (in (11.2) we have, clearly, that $\mathbf{r}_n$ is the radius vector of the nth electron and $\mathbf{r}_n^{(i)}$ the radius vector of the nth ion).

If there is no external magnetic field $\mathbf{H}_0$ and if we neglect collisions and use a local approach (that is, neglect spatial dispersion when we can assume the field $\mathbf{E}$ to be uniform), the equations of motion for the electrons and ions have the form

$$m\ddot{\mathbf{r}}_n = e\,\mathbf{E}_0\, e^{-i\omega t}\;,\quad M\ddot{\mathbf{r}}_n^{(i)} = -\,e\,\mathbf{E}_0\, e^{-i\omega t}\,. \tag{11.3}$$

where m and M are, respectively, the electron and ion masses.

Hence we get, for instance,

$$\mathbf{r}_n = -\frac{e\,\mathbf{E}}{m\omega^2} + \mathbf{r}_n^{(0)}\;,$$

† One should also bear in mind that the field is often put in the form $\mathbf{E} = \mathbf{E}_0\, e^{i\omega t}$ which leads to the complex conjugate quantities. Moreover, the conduction current $\mathbf{j}_{cond}$ is more often denoted by $\mathbf{j}'$ and the total current $-i\omega(\mathbf{D} - \mathbf{E})/4\pi$ is either not introduced or denoted by $\mathbf{j}$.

where $\mathbf{r}_n^{(0)}$ is the electron radius vector when there is no field. From (11.2) and (11.3) one sees easily that

$$\mathbf{P} = \frac{\varepsilon' - 1}{4\pi}\mathbf{E} = -\frac{e^2}{\omega^2}\left(\frac{1}{m}+\frac{1}{M}\right) N\mathbf{E},$$

and the terms with $\mathbf{r}_n^{(0)}$ and $\mathbf{r}_n^{(i,0)}$ drop out, as, by assumption, without a field $\mathbf{P} = 0$. The contribution from the ions is only a correction of order $m/M \leqslant 10^{-3}$ and will be omitted.

We thus have

$$\boldsymbol{\varepsilon} = \boldsymbol{\varepsilon}' = 1 - \frac{4\pi N e^2}{m\omega^2} \equiv 1 - \frac{\omega_{pe}^2}{\omega^2} = 1 - 3.18\times 10^9 \frac{N}{\omega^2}$$
$$= 1 - 8.06\times 10^7 \frac{N}{\nu^2}, \qquad (11.4)$$

where $\omega_{pe} \equiv \omega_p = (4\pi N e^2/m)^{\frac{1}{2}}$ is the electron plasma frequency, $\nu = \omega/2\pi$ and where we have substituted the well known values for the electron parameters ($e = 4.8\times 10^{-10}$ e.s.u., $m = 9.1\times 10^{-28}$ g).

It is completely natural that the conductivity σ is put equal to zero under the assumptions which we have made: As there are no collisions between the electrons and ions they do not transfer their energy to other electrons, ions or molecules and only oscillate under the influence of the field. In the framework of a phenomenological theory one can take the effect of collisions into account by introducing a friction force $m\nu_{eff}\dot{\mathbf{r}}_n$ which is equal to the average change in momentum of the particle per unit time. If we assume that in each collision with an ion or molecule the electron on average loses the whole momentum of its ordered motion, ν_{eff} is the number of collisions per unit time. In actual fact, of course, the change in momentum varies for different collisions and ν_{eff} plays therefore the role of an effective collision number per unit time (frequency). Strictly speaking the quantity ν_{eff} is not yet defined and we have merely made an assumption that the average friction force is proportional to $\dot{\mathbf{r}}_n$. However, in actual fact it is known from the kinetic theory of gases that, say, for collisions with molecules $\nu_{eff} = \pi a^2 N_m \bar{v}$, where a is the effective radius of a molecule, N_m their density, and $\bar{v}$ some average electron thermal velocity. One can obtain better defined expressions for ν_{eff} for collisions of electrons with molecules or ions from kinetic considerations but we can use the quantity ν_{eff} itself also in the elementary theory. Thus, when collisions are taken into account,

$$m\ddot{\mathbf{r}}_n + m\nu_{eff}\dot{\mathbf{r}}_n = e\mathbf{E}_0 e^{-i\omega t} \qquad (11.5)$$

and, proceeding as before, we get

$$\begin{aligned} \varepsilon &= 1 - \frac{4\pi N e^2}{m\omega(\omega + i\nu_{eff})} = \varepsilon' + \varepsilon'' \equiv \varepsilon' + i\frac{4\pi\sigma}{\omega} , \\ \varepsilon' &= 1 - \frac{4\pi N e^2}{m(\omega^2 + \nu^2_{eff})} , \quad \sigma = \frac{1 - \varepsilon'}{4\pi}\nu_{eff} = \frac{N e^2 \nu_{eff}}{m(\omega^2 + \nu^2_{eff})} . \end{aligned} \tag{11.6}$$

In the most important limiting cases we have

$$\varepsilon' = 1 - \frac{4\pi N e^2}{m\omega^2} , \quad \sigma = \frac{N e^2 \nu_{eff}}{m\omega^2} = 2.53 \times 10^8 \frac{N\nu_{eff}}{\omega^2} , \quad \omega^2 \gg \nu^2_{eff} , \tag{11.7}$$

$$\varepsilon' = 1 - \frac{4\pi N e^2}{m\nu^2_{eff}} , \quad \sigma = \frac{N e^2}{m\nu_{eff}} , \quad \omega^2 \ll \nu^2_{eff} . \tag{11.8}$$

In the low frequency limit (11.8) the conductivity is, of course, the same as the one obtained in the elementary theory of the static conductivity ($1/\nu_{eff} = \tau_{eff}$, where τ_{eff} is the effective free flight time which is used together with ν_{eff}).

We have assumed above when deriving the expression for ε that the field $\mathbf{E}_{eff}$ acting upon the electron is equal to the macroscopic (average) field $\mathbf{E}$ as we substituted just that field in the equations of motion (11.3) and (11.5). However, it is well known from the theory of dielectrics that the fields $\mathbf{E}_{eff}$ and $\mathbf{E}$ are, in general, not equal to one another and for the simplest models of dielectrics (see, for instance, Tamm, 1976) $\mathbf{E}_{eff} = \mathbf{E} + \frac{4}{3}\pi\mathbf{P} = \frac{1}{3}(\varepsilon + 2)\mathbf{P}$. In a plasma (and generally in a good conductor) such a formula is inapplicable and at the same time a relatively rigorous discussion of the connection between $\mathbf{E}_{eff}$ and $\mathbf{E}$ is a complicated problem (see Ginzburg, 1970b and the literature cited there). In a gaseous plasma we can assume with good accuracy that

$$\mathbf{E}_{eff} = \mathbf{E} , \tag{11.9}$$

as we have done above and as we shall do in what follows. We note, however, that the problem of $\mathbf{E}_{eff}$ merits, apparently, a more detailed analysis for applications to a dense (metallic) plasma and also when non-linear effects are taken into account. In connection with this remark we emphasize that we are here restricting ourselves always to the linear approximation — linear plasma electrodynamics. In strong fields (the problem when a field can be considered to be strong requires special attention†) the picture is appreciably more

---

† For instance, when there are no collisions the linear approximation is in an isotropic plasma, in general, adequate as long as the velocity of the ordered motion of the electrons is small compared to the thermal velocity, that is, $\dot{r} \sim eE_0/m\omega \ll \bar{v} \sim \sqrt{k_B T/m}$.

complicated and has so far been the subject of many studies (see Silin and Rukhadze, 1961; Leontovich, 1965; Silin, 1971, 1973; Gorbunov, 1973; Gurevich and Shvartsburg, 1973, Gould, 1974; Tsytovich, 1977).

It is perhaps not superfluous to explain why one of the field equations was written above in the form (11.1) and not in the form $\operatorname{div}\mathbf{D} = \operatorname{div}\varepsilon\mathbf{E} = 4\pi\rho_{ext}$ (see, for instance, (10.1)). The fact simply is that we are assuming in (11.1) that there are no external charges and we use the equation $\operatorname{div}\mathbf{D} = \operatorname{div}\mathbf{E} + 4\pi\operatorname{div}\mathbf{P} = 0$. Moreover, we assume that only the electrons and ions which we are considering are the sources of the polarization so that $\operatorname{div}\mathbf{P} = e(N - N_i)$. In other words, the difference between the permittivity ε and unity is caused by the electrons and ions and, if we had written the equation in the form $\operatorname{div}\varepsilon\mathbf{E} = 4\pi e(N - N_i)$, as might have seemed natural at first sight, we would twice have taken the same effect into account.

As the elementary formulae (11.6) to (11.8) have a very wide range of applicability we shall give here a few expressions for the effective collision frequency ν_{eff} (see Ginzburg, 1970b). An exact calculation is, in general, not possible for collisions of electrons with molecules and one widely uses experimental data. If, however, we assume the molecule to be a hard sphere of radius a we have

$$\nu_{eff,m} = \frac{4}{3}\pi a^2 \bar{v} N_m = 8.3\times 10^5 \pi a^2 N_m \sqrt{T}, \quad \omega^2 \gg \nu_{eff}^2, \tag{11.10}$$

$$\nu_{eff,m} = \frac{3\pi}{8}\pi a^2 \bar{v} N_m, \qquad \omega^2 \ll \nu_{eff}^2, \tag{11.11}$$

where $\bar{v} = \sqrt{8k_BT/\pi m}$ is the average absolute magnitude of the electron velocity for an equilibrium velocity distribution of temperature T. The difference between Eqns.(11.10) and (11.11) is minor ($\frac{4}{3} = 1.33$, $\frac{3\pi}{8} = 1.18$), particularly when we bear in mind that we are dealing here with the 'limiting' hard-sphere model. Nonetheless it is clear already from the example that the formulae from the elementary theory are approximate in nature and that the effective collision frequency is a function of the frequency.

For collisions of electrons with ions (for $N = N_i$) we have

$$\nu_{eff,i} = \pi\frac{e^4}{(k_BT)^2}\bar{v}N\,\ell n\left(0.37\frac{k_BT}{e^2N^{\frac{1}{3}}}\right) = \frac{5.5N}{T^{\frac{3}{2}}}\,\ell n\left(220\frac{T}{N^{\frac{1}{3}}}\right), \quad \omega^2 \gg \nu_{eff}^2, \tag{11.12}$$

$$\nu_{eff,i} = \frac{1.6N}{T^{\frac{3}{2}}}\,\ell n\left(324\gamma\frac{T}{N^{\frac{1}{3}}}\right., \quad \gamma \sim 1, \quad \omega^2 \ll \nu_{eff}^2. \tag{11.13}$$

The value (11.13) is approximately three times smaller than the value (11.12) but such a large difference between them is somewhat illusory. The fact is that when $\omega^2 \gg \nu_{eff}^2$ electron-electron collisions do not play a role and Eqn. (11.12) retains its validity. However, when $\omega^2 \ll \nu_{eff}^2$ electron-electron collisions affect $\nu_{eff,i}$ and we must multiply expression (11.13) by 1.73, if they are taken into account. Altogether when electron-electron collisions are taken into account the effective electron-ion collision frequency is in the high-frequency case about twice its value in the low-frequency case.

Even though one can obtain the exact Eqns.(11.12) and (11.13) only as a result of a detailed calculation, one can establish their structure easily from simple considerations. An electron-ion collision leads to an appreciable change in the direction of the electron velocity if it flies past the ion with an impact parameter $p \sim e^2/k_BT$, when its Coulomb energy $e^2/p \sim k_BT$, that is, of the order of the electron kinetic energy. The corresponding cross-section for 'close' collisions $q \sim \pi p^2 \sim \pi e^4/(k_BT)^2$. The collision frequency $\nu_{eff,i}$ is, however, determined not solely by close, but also by 'distant' collisions, and taking these into account leads to the appearance in (11.12) and (11.13) of the logarithmic factor, usually denoted by L. This factor is typical for plasma physics and its appearance is connected with the fact that the Coulomb field decrease slowly (as $1/r$) with distance only up to some Debye screening radius, which equals

$$r_D = \left(\frac{k_BT}{8\pi Ne^2}\right)^{\frac{1}{2}} = 4.9\left(\frac{T(°K)}{N}\right)^{\frac{1}{2}} \text{cm} . \tag{11.14}$$

To be more precise, a given ion of charge e produces in the plasma as the result of the attraction of the electrons and the repulsion of other ions a field with a potential $(e/r)\exp(-r/r_D)$ which is the same as the Coulomb field only when $r \ll r_D$. The field practically vanishes when $r \gg r_D$ and thus, when an electron collides with an ion in a plasma it is just a radius of the order of r_D which plays the role of the maximum impact parameter p_{max}. The Coulomb logarithm is usually written in the form

$$L = \ln\left(\frac{p_{max}}{p_{min}}\right) = \ln\left\{\frac{3k_BT}{2e^2}\left(\frac{k_BT}{8\pi Ne^2}\right)^{\frac{1}{2}}\right\} \approx \frac{3}{2}\ln\left(220\frac{T}{N^{\frac{1}{3}}}\right), \tag{11.15}$$

where we have put $p_{max} = r_D$ and $p_{min} = e^2/\frac{3}{2}k_BT$; we shall consider the screening problem and the derivation of Eqn.(11.14) below (for a different method see, for instance, Ginzburg, 1970b, § 4).

Apart from the provisos already made which determine the region of applicability of the formulae obtained, we must also indicate the limitations connected with the use of a classical theory. If there are no collisions, the quantal limitation when considering the interaction of radiation with free electrons is connected with the condition

$$\hbar\omega \ll mc^2 = 0.51 \times 10^6 \text{ eV} . \qquad (11.16)$$

When this condition is satisfied the scattering of electromagnetic waves by free electrons can for all scattering angles be described classically. Moreover, the refractive index $n = \sqrt{\varepsilon} = \sqrt{\varepsilon'}$ (we assume that $\omega^2 \gg \nu_{eff}^2$) is in fact determined by the scattering of waves by the particles in the medium whence in final reckoning follows the vaildity of Eqn. (11.4). One can verify this also in other ways. When collisions are taken into account and the absorption is calculated (that is, when we take the conductivity $\sigma = \omega\varepsilon''/4\pi$ into account) we must also bear in mind the condition

$$\hbar\omega \ll k_B T = 1.38 \times 10^{-16}\, T\,(^\circ K) \text{ erg} . \qquad (11.17)$$

The meaning of this condition is that the photon energy must be small compared to the kinetic energy of the electrons (see, for instance, Ginzburg, 1970b, § 3 for details about this condition). As we are now dealing with a non-relativistic plasma when

$$k_B T \ll mc^2 , \qquad (11.18)$$

condition (11.17) is considerably more restrictive than condition (11.16), but the requirement (11.17) does not play a role, as we mentioned, when we calculate the quantity ε (for $\omega^2 \gg \nu_{eff}^2$).

For an unlimited applicability of the classical theory it is also necessary that the electron gas stays non-degenerate; this is equivalent to the requirement

$$T \gg T_0 \sim \hbar^2 N^{\frac{2}{3}}/m k_B . \qquad (11.19)$$

The meaning of the degeneracy temperature T_0 is the following: at this temperature the energy $k_B T_0$ is of the order of the zero-point energy

$$\frac{\hbar^2}{m\bar{r}^2} \sim \frac{\hbar^2}{m} N^{\frac{2}{3}} ,$$

connected with the localization of an electron in a volume of order $\bar{r}^3 \sim 1/N$.

In most cases which one encounters (although not by any means always) the conditions (11.16) to (11.19) are satisfied and it turns out that the restrictions connected with the fact that so far we have neglected spatial dispersion are more important. By its very nature spatial dispersion (see Chapter 10) is unimportant only when the field (say, the field of a wave with frequency ω and wavelength $\lambda = 2\pi/k$) changes little over a distance corresponding to the production of a 'response' of the medium, for instance, the production of a polarization **P** as a result of the field **E**. When we take thermal motion into account an electron traverses in a plasma during a period $\tau = 2\pi/\omega$ a distance $\xi \sim \tau\bar{v} \sim \tau\sqrt{k_B T/m}$. It follows from what we have said that we can neglect spatial dispersion, provided $\xi \ll \lambda$, that is, provided

$$\omega \gg k\bar{v} \sim \frac{2\pi}{\lambda}\left(\frac{k_B T}{m}\right)^{\frac{1}{2}} . \tag{11.20}$$

For a wave propagating in the medium

$$E \propto e^{ikz-i\omega t} \propto \exp i\omega\left(\frac{z}{v_{ph}} - t\right) ,$$

and, clearly, the phase velocity of the wave equals

$$v_{ph} = \frac{\omega}{k} . \tag{11.21}$$

From this and (11.20) it is clear that for a wave freely propagating in a plasma we can neglect spatial dispersion, provided

$$v_{ph} \gg \bar{v} \sim \sqrt{k_B T/m} . \tag{11.22}$$

When we use Eqn.(11.3) the neglect of spatial dispersion is reflected in the fact that we write the field in the form $\mathbf{E} = \mathbf{E}_0\, e^{-i\omega t}$, and not in the form

$$\mathbf{E} = \mathbf{E}_0 \exp\left[i(\mathbf{k}\cdot\mathbf{r}_n(t)) - i\omega t\right] .$$

As neglecting spatial dispersion is thus equivalent to neglecting the thermal motion of the particles in the plasma, one speaks of the 'cold' plasma approximation. In other words when one neglects the thermal motion in a plasma, one calls the plasma a cold plasma.

In slight anticipation we note that for waves of a transverse field the condition (11.22) is always well satisfied in a non-relativistic plasma. On the other hand, for a longitudinal field this inequality can easily be violated. The best way to elucidate in detail and accurately under what conditions spatial dispersion can be neglected consists, of course, in analyzing an actual situation using the general expressions in which this dispersion is taken into account. Here we consider only the most widely used and general method which enables us to take temporal and spatial dispersion into account — the kinetic equation method.

We shall describe the state of a plasma by using a distribution function $f(t,\mathbf{r},\mathbf{v})$ which is defined in such a way that the average number dN in a volume $d^3\mathbf{r}\,d^3\mathbf{v} = dx\,dy\,dz\,dv_x\,dv_y\,dv_z$ is equal to $dN = f(t,\mathbf{r},\mathbf{v})d^3\mathbf{r}\,d^3\mathbf{v}$, where $\mathbf{r}$ is the radius vector and $\mathbf{v}$ the velocity of a particle. By definition we have then (this is the normalization condition)

$$\iiint_{-\infty}^{+\infty} f(t,\mathbf{r},\mathbf{v})\,d^3\mathbf{v} = N(\mathbf{r}) , \qquad (11.23)$$

where N is the particle density (for the sake of simplicity we consider here and henceforth electrons, that is, $f \equiv f_e$ and $N \equiv N_e$; in the case of ions or molecules the index e must be replaced by i or m, respectively).

The kinetic equation which determines the function f has the form

$$\frac{\partial f}{\partial t} + (\mathbf{v}\cdot\nabla_{\mathbf{r}} f) + \frac{e}{m}\left(\left\{\mathbf{E} + \frac{1}{c}\left[\mathbf{v}\wedge\mathbf{H}\right]\right\}\cdot\nabla_{\mathbf{v}} f\right) + \mathcal{S} = 0 , \qquad (11.24)$$

where e and m are the charge and mass of the particles considered, $\mathbf{E}$ and $\mathbf{H}$ the electric and magnetic field strengths acting upon the particles (in practice these fields can usually be assumed to be the average macroscopic fields), and

$$\nabla_{\mathbf{r}} f = \frac{\partial f}{\partial x}\mathbf{i} + \frac{\partial f}{\partial y}\mathbf{j} + \frac{\partial f}{\partial z}\mathbf{k} , \quad \nabla_{\mathbf{v}} f = \frac{\partial f}{\partial v_x}\mathbf{i} + \frac{\partial f}{\partial v_y}\mathbf{j} + \frac{\partial f}{\partial v_z}\mathbf{k}$$

(of course, it is not necessary to use only Cartesian coordinates, but we have chosen them here to fix the ideas); the quantity $\mathcal{S}$ in (11.24) is the so-called collision integral which takes into account the change in the function f as a result of collisions of the particles of the kind considered (for instance, electrons) in the volume $d^3\mathbf{r}\,d^3\mathbf{v}$ with particles of all other kinds (that is, with particles of the same kind which are in other volumes of phase space). We can also include in $\mathcal{S}$ terms which determine the change in the function f due to ionization, recombination, and so on.

When there are no fields the distribution functions for electrons and ions are Maxwellian in the equilibrium state (the same is also true for molecules, but we shall not consider them here)

$$f_e \equiv f = f_{00}(v) = N\left(\frac{m}{2\pi k_B T}\right)^{\frac{3}{2}} \exp\left(-\frac{mv^2}{2k_B T}\right) ,$$

$$f_i = f_{i,00}(v) = N_i\left(\frac{M}{2\pi k_B T}\right)^{\frac{3}{2}} \exp\left(-\frac{Mv^2}{2k_B T}\right) . \qquad (11.25)$$

We have assumed here that the electron and ion temperatures are the same. This must, of course, be the case for complete equilibrium.[†] One must, however,

[†] see footnote on next page

bear in mind that the exchange of momentum (relaxation with respect to momentum) in a plasma proceeds appreciably faster than energy exchange (a consequence of the fact that the parameter $m/M \leqslant 10^{-3}$ is small; for details see Ginzburg, 1970b). One sometimes therefore considers a non-isothermal plasma with the Maxwellian distributions (11.25) but with different electron and ion temperatures equal to, respectively, T_e and T_i (in some cases even $T_e \gg T_i$).

If we are interested in problems relating to the realm of the linear theory (in particular, linear electrodynamics) we may assume the electric field $\mathbf{E}$ to be small‡ and, in accordance with that consider the change in the distribution function as a perturbation. In other words we shall look for the distribution function in the form

$$f = f_{00}(\mathbf{v}) + f'(t,\mathbf{r},\mathbf{v}) \quad , \quad |f'| \ll f_{00} \; , \tag{11.26}$$

where we choose the Maxwell distribution (11.25) for the function f_{00} when we try to find the tensor $\varepsilon_{ij}(\omega,\mathbf{k})$ in an equilibrium plasma; we restrict ourselves in what follows to that case.

When there is no external magnetic field $\mathbf{H}_0$ we get then in first approximation

$$\frac{\partial f'}{\partial t} + (\mathbf{v}\cdot\nabla_{\mathbf{r}} f') + \frac{e}{m}(\mathbf{E}\cdot\nabla_{\mathbf{v}} f_{00}) + \mathcal{S} = 0 \; . \tag{11.27}$$

The problem of the form of the collision integral $\mathcal{S}$ in general and in a plasma in particular has been the subject of detailed analysis (see Ginzburg and

---

† Not only must the particles have the same temperature T but for complete equilibrium the electromagnetic radiation must also be thermal (black-body radiation) with the same temperature T. This last requirement is nevertheless usually not satisfied; however, this plays often no role due to the weakness of the interaction between the particles and the radiation. However, one must, of course, check in each actual case whether it is possible to neglect the effect of the radiation on the particle distribution function. As an example of a situation where the particle distribution function will not be in equilibrium when the radiation is not in equilibrium we adduce a gas of relativistic electrons in a strong magnetic field (for instance, near a pulsar). In that case as a result of the large magneto-brems losses the electron distribution function must, in general, be strongly anisotropic (see, for instance, Chapter 4 and Tsytovich and Kaplan, 1974).

‡ The equilibrium distribution function remains Maxwellian in a constant and uniform magnetic field notwithstanding the fact that the trajectories of the individual particles are different in a field and when there is no field (see also below). Moreover, the term with the magnetic field contains in (11.24) the factor v/c. The restriction on the magnetic field is therefore different from that on the electric field.

Rukhadze, 1975, and the literature cited there). Here we restrict ourselves to the simplest case where we can put in the equation for the electron distribution function $S = \{\nu_m(v) + \nu_i(v)\} f'$, where ν_m corresponds to the contribution from the collisions between electrons and molecules, and ν_i from the collisions with ions (see Ginzburg, 1970b, § 4 for a simple interpretation of this expression for S). If we now consider the field $\mathbf{E} = \mathbf{E}_0 e^{i(\mathbf{k}\cdot\mathbf{r})-i\omega t}$ and the function $f' = f'_{\omega,\mathbf{k}}(\mathbf{v})\, e^{i(\mathbf{k}\cdot\mathbf{r})-i\omega t}$, we get from (11.27)

$$-i\left[\omega-(\mathbf{k}\cdot\mathbf{v})\right]f' + \frac{e}{m}(\mathbf{E}\cdot\mathbf{v})\frac{1}{v}\frac{\partial f_{00}(v)}{\partial v} + \nu(v)\, f' = 0\ , \qquad (11.28)$$

where $\nu = \nu_m + \nu_i$ and where we have used the fact that

$$\nabla_{\mathbf{v}} f_{00} = \frac{\mathbf{v}}{v}\ \frac{\partial f_{00}}{\partial v} = -\frac{mv}{k_B T}\ \frac{\mathbf{v}}{v} f_{00}\ .$$

Hence we get

$$f' = -i\,\frac{e}{m}\,\frac{(\mathbf{E}\cdot\mathbf{v})}{\omega-(\mathbf{k}\cdot\mathbf{v})+i\nu}\,\frac{1}{v}\,\frac{\partial f_{00}}{\partial v} = i\,\frac{e}{k_B T}\,\frac{(\mathbf{E}\cdot\mathbf{v})f_{00}}{\omega-(\mathbf{k}\cdot\mathbf{v})+i\nu(v)}\ . \qquad (11.29)$$

Further, we have, by definition, for the current density or, to be more precise, its appropriate Fourier components

$$j_i(\omega,\mathbf{k}) = -\frac{i\omega}{4\pi}\left[\varepsilon_{ij}(\omega,\mathbf{k}) - \delta_{ij}\right] E_j(\omega,\mathbf{k}) = e\int v_i\, f'_{\omega,\mathbf{k}}(\mathbf{v})\, d^3\mathbf{v}\ . \qquad (11.30)$$

Substituting (11.29) into (11.30) we get

$$\varepsilon_{ij}(\omega,\mathbf{k}) = \delta_{ij} + \frac{4\pi e^2}{m\omega}\int \frac{v_i v_j}{\omega-(\mathbf{k}\cdot\mathbf{v})+i\nu(v)}\,\frac{1}{v}\,\frac{\partial f_{00}}{\partial v}\, d^3\mathbf{v}\ . \qquad (11.31)$$

When we take into account both electrons (charge $e = e_1$, mass $m = m_1$) and different kinds of ions (charges e_α, masses m_α, $\alpha = 2, 3, \ldots, \ell$) we find

$$\varepsilon_{ij}(\omega,\mathbf{k}) = \delta_{ij} + 4\pi\sum_{\alpha=1}^{\ell}\frac{e_\alpha^2}{m_\alpha\omega}\int\frac{v_i v_j}{\omega-(\mathbf{k}\cdot\mathbf{v})+i\nu_\alpha(v)}\,\frac{1}{v}\,\frac{\partial f_{00,\alpha}}{\partial v}\, d^3\mathbf{v}\ . \qquad (11.32)$$

It is at once clear from (11.28) to (11.32) that neglecting spatial dispersion, that is, the $\mathbf{k}$-dependence, is equivalent to neglecting the term $(\mathbf{k}\cdot\mathbf{v})$ as compared to $\omega + i\nu(v)$. We thus arrive, when we neglect collisions, at condition (11.20), as should have been expected.

To establish correspondence with the elementary theory we neglect in the denominator in (11.31) the terms $-(\mathbf{k}\cdot\mathbf{v}) + i\nu(v)$ as compared to ω. We then at once obtain the result (11.4)

$$\varepsilon_{ij} = \varepsilon\delta_{ij}\ , \quad \varepsilon = 1 - 4\pi N e^2/m\omega^2,$$

since

$$\int v_i v_j \frac{m}{k_B T} f_{00}\, d^3\mathbf{v} = N\delta_{ij} \quad .$$

On the other hand, if we neglect the term $(\mathbf{k}\cdot\mathbf{v})$, but take the term $i\nu(v)$ into account, we obtain the expression from the elementary theory which we have already given above which includes the appropriate values of the effective collision frequency ν_{eff} (see Ginzburg, 1970, § 6 for detailed calculations). We shall now assume that the collisions are unimportant, but take into account spatial dispersion. It is true that there then arises the problem of how to evaluate the integrals in (11.31) and (11.32) near the pole

$$\omega = (\mathbf{k}\cdot\mathbf{v}) \quad . \tag{11.33}$$

The simplest way to avoid this difficulty is to assume to begin with that the collision frequency ν, though very small, is not completely zero. The pole in the plane of the complex variable $u = (\mathbf{k}\cdot\mathbf{v})/k$ is then shifted to the point $u_0 = (\omega + i\nu)/k$ which lies above the contour of integration (it is essential here that $\nu > 0$), that is, above the real u-axis (the integration over $d^3\mathbf{v}$ in (11.31) and (11.32) reduces to an integration over u and over the velocity components at right angles to the vector $\mathbf{k}$). As a result we can perform the integration, using the relation

$$\lim_{\delta\to 0} \frac{1}{x + i\delta} = \frac{\mathcal{P}}{x} - \pi i\,\delta(x) \quad , \tag{11.34}$$

where $\mathcal{P}$ indicates that the integral over the region near the singularity $x = 0$ is taken in the sense of the principal value integral. Using this method we get from (11.31) as $\nu \to 0$

$$\varepsilon_{ij}(\omega, \mathbf{k}) = \\ = \delta_{ij} + \frac{4\pi e^2}{m\omega}\int v_i v_j \frac{1}{v}\frac{\partial f_{00}}{\partial v}\left\{\frac{\mathcal{P}}{\omega - (\mathbf{k}\cdot\mathbf{v})} - \pi i\,\delta[\omega - (\mathbf{k}\cdot\mathbf{v})]\right\} d^3\mathbf{v} \quad . \tag{11.35}$$

Hence it follows that when there are no collisions there is some absorption (the imaginary part of ε_{ij} is non-vanishing) and responsible for this absorption are only the particles with velocities satisfying condition (11.33). However, condition (11.33) is nothing but the condition for Cherenkov radiation (see (6.55)). We get thus at once a clear physical cause for the absorption even when there are no collisions (at first sight such a result may seem paradoxical, for instance, in view of the elementary theory expounded above). Indeed, under condition (11.33) a particle (say, an electron) emits Cherenkov waves (for those $\omega/k = v_{ph} = v\cos\theta$, where θ is the angle between $\mathbf{k}$ and $\mathbf{v}$).

However, each emission process can be reversed and, hence, for the same condition on ω and $\mathbf{k}$ the wave must cause the inverse effect — Cherenkov absorption connected with the transfer of the appropriate energy and momentum from the wave to the particle. What we have said is, of course, only a repeat of the interpretation of collisionless damping in an isotropic plasma which we discussed in Chapter 7. Collisionless damping of plasma waves is often called Landau damping as he was the first to elucidate it (Landau, 1946) when solving the kinetic equation with initial conditions. One can treat the collisionless absorption not only through the 'Cherenkov interpretation' but also differently, and especially simply in the case when $\cos\theta = 1$, when $\omega/k = v_{ph} = v$; in that case the wave and the particle move in the same direction and the phase of the wave 'at the particle' does not change so that the particle is accelerated all the time and thus obtains energy from the wave.

In an isotropic (and non-gyrotropic) medium we can write the expression for $\boldsymbol{\varepsilon}_{ij}(\omega, \mathbf{k})$ in the form

$$\boldsymbol{\varepsilon}_{ij}(\omega, \mathbf{k}) = \left(\delta_{ij} - \frac{k_i k_j}{k^2}\right) \boldsymbol{\varepsilon}_{tr}(\omega, k) + \frac{k_i k_j}{k^2} \boldsymbol{\varepsilon}_{\ell}(\omega, k) \ . \tag{11.36}$$

Indeed, we have in this case only two tensors δ_{ij} and $k_i k_j$ at our disposal and the arrangement of the terms is determined by the requirement that only the tensor $\boldsymbol{\varepsilon}_{\ell}(\omega, k)$ must 'operate' for a longitudinal field. By definition, the vector $\mathbf{E}$ is for a longitudinal field directed along the wavevector $\mathbf{k}$, that is, $\mathbf{E}_{\ell} \equiv \mathbf{E}_{\|} = E\mathbf{k}/k$. For such a field we have in the case (11.36)

$$\mathbf{D}(\omega, \mathbf{k}) = \boldsymbol{\varepsilon}_{\ell}(\omega, \mathbf{k})\, \mathbf{E}_{\ell} \ . \tag{11.37}$$

On the other hand, for a transverse field $\mathbf{E}_{tr} \equiv \mathbf{E}_{\perp}$, which satisfies the condition $(\mathbf{k} \cdot \mathbf{E}_{tr}) = 0$, we have

$$\mathbf{D}(\omega, \mathbf{k}) = \boldsymbol{\varepsilon}_{tr}(\omega, \mathbf{k})\, \mathbf{E}_{tr} \ . \tag{11.38}$$

Both from general considerations (see Chapter 10) and from the actual expressions for $\boldsymbol{\varepsilon}_{ij}$ (vide infra) it is clear that

$$\boldsymbol{\varepsilon}_{tr}(\omega, 0) = \boldsymbol{\varepsilon}_{\ell}(\omega, 0) = \boldsymbol{\varepsilon}(\omega) \ , \tag{11.39}$$

where $\boldsymbol{\varepsilon}(\omega)$ is the permittivity of an isotropic medium in which the spatial dispersion is neglected (see, for instance, (11.4) or (11.6)). The general expressions for $\boldsymbol{\varepsilon}_{tr}$ and $\boldsymbol{\varepsilon}_{\ell}$ in a plasma, obtained by using (11.31) and (11.32) are, for instance, given by Ginzburg and Rukhadze (1975). We just indicate here a few formulae which correspond to important limiting cases (the plasma is assumed to be strongly ionized).

At 'high' frequencies when

$$\omega \gg k v_{T,e,i} \ , \quad \omega \gg \nu_{eff} \ , \tag{11.40}$$

we have[†]

$$\varepsilon_{tr} = 1 - \frac{\omega_{pe}^2}{\omega^2}\left[1 - i\,\frac{\nu_{eff}}{\omega} - i\sqrt{\tfrac{1}{2}\pi}\,\frac{\omega}{k v_{Te}}\exp\left(-\frac{\omega^2}{2k^2 v_{Te}^2}\right)\right] , \tag{11.41}$$

$$\varepsilon_{\ell} = 1 - \frac{\omega_{pe}^2}{\omega^2}\left[1 - i\,\frac{\nu_{eff}}{\omega} + 3\,\frac{k^2 v_{Te}^2}{\omega^2} - i\sqrt{\tfrac{1}{2}\pi}\,\frac{\omega^3}{k^3 v_{Te}^3}\exp\left(-\frac{\omega^2}{2k^2 v_{Te}^2}\right)\right] , \tag{11.42}$$

The appearance here of exponential factors is caused by our use of the Maxwellian distribution function (11.25); from this it is already clear that for a non-equilibrium plasma the corresponding parts of ε_{tr} and ε_{ℓ} can differ radically from expressions (11.41) and (11.42).

As $k \to 0$ these expressions (provided $\omega \gg \nu_{eff}$) are the same as (11.6), as should be the case. Let now

$$k v_{Te} \gg (\omega \, , \nu_e) \ , \quad k v_{Ti} \gg (\omega \, , \nu_i) \ , \tag{11.43}$$

where ν_e is the electron collision frequency and ν_i the ion collision frequency (one must take into account all significant collisions of a given particle with all others; writing $k v_{Ti} \gg (\omega \, , \nu_i)$ means that $k v_{Ti} \gg \omega$ and $k v_{Ti} \gg \nu_i$, and so on). In that case

$$\varepsilon_{tr} = 1 + i\sqrt{\tfrac{1}{2}\pi}\,\frac{\omega_p^2}{\omega k v_{Te}} \ ,$$

$$\varepsilon_{\ell} = 1 + \sum_{\alpha} \frac{\omega_{p\alpha}^2}{k^2 v_{T\alpha}^2}\left(1 + i\sqrt{\tfrac{1}{2}\pi}\,\frac{\omega}{k v_{T\alpha}}\right) \approx 1 + \frac{1}{k^2 r_D^2} \ . \tag{11.44}$$

Here

$$\omega_{p\alpha}^2 = \frac{4\pi N_\alpha e^2}{m_\alpha} \ , \quad \alpha = e \, , i \tag{11.45}$$

and, when $T_e = T_i = T$, r_D is the Debye radius (11.14); in the more general case

$$r_{D\alpha} = \left(\frac{k_B T_\alpha}{4\pi N_\alpha e^2}\right)^{\frac{1}{2}} = \frac{v_{T\alpha}}{\omega_{p\alpha}} \ . \tag{11.46}$$

When $T_e = T_i$ (isothermal plasma) and $N_i = N_e \equiv N$

---

† We have used here the notation $v_T = v_{Te} = \sqrt{k_B T/m}$ and $v_{Ti} = \sqrt{k_B T/M}$, where M is the ion mass (we assume that there is only one kind of ions). If $T_e \neq T_i$, we must write $v_{T\alpha} = \sqrt{k_B T_\alpha / m_\alpha}$, $\alpha = e \, , i$, and so on. We neglect terms of order m/M and therefore the contributions from the ions do clearly not appear in (11.41) and (11.42) (we must, of course, take the contribution from the ions into consideration in the experession for ν_{eff}).

$$\frac{1}{r_D^2} = \frac{1}{r_{De}^2} + \frac{1}{r_{Di}^2} , \quad r_D = \left(\frac{k_B T}{8\pi N e^2}\right)^{\frac{1}{2}} . \tag{11.47}$$

We consider an 'external' charge at rest with charge density $\rho_{ext} = e\delta(\mathbf{r})$ which is introduced into the plasma. The potential ϕ of the field is determined by the equations

$$\operatorname{div} \mathbf{D} = 4\pi e\delta(\mathbf{r}) , \ \mathbf{E} = \mathbf{E}_\ell = -\nabla\phi , \ \mathbf{D}(\omega, \mathbf{k}) = \boldsymbol{\varepsilon}_\ell(\omega, \mathbf{k}) \ \mathbf{E}(\omega, \mathbf{k}) , \tag{11.48}$$

whence we get in the case (11.44)

$$\phi(0,\mathbf{k}) = \frac{4\pi e}{k^2 \boldsymbol{\varepsilon}_\ell(0,\mathbf{k})} , \quad \phi(\mathbf{r}) = \frac{1}{(2\pi)^3} \int \phi(0,\mathbf{k}) \, e^{i(\mathbf{k}\cdot\mathbf{r})} d^3\mathbf{k} =$$

$$= \frac{e \times \exp(-r/r_D)}{r} , \quad \boldsymbol{\varepsilon}_\ell(0,\mathbf{k}) = 1 + \frac{1}{k^2 r_D^2} . \tag{11.49}$$

Within well defined limits one can consider each ion and electron in the plasma to be external in relation to all other particles and Eqn.(11.49) reflects the fact that the Coulomb field of each particle in the plasma is screened by the other particles. We have already mentioned that this screening is important when one considers collisions. The problem of taking the screening of plasma particles by one another into account reduces thus to finding an expression for the collision integral $\mathfrak{S}$ and the subsequent solution of the kinetic equation (see Leontovich, 1965; Ginzburg and Rukhadze, 1975).

From the example of the limiting case (11.43) or, roughly speaking, in the static limit as $\omega \to 0$ the role of spatial dispersion in a plasma becomes especially clear. If we had used in that case the equation $\varepsilon = 1 - \omega_p^2/\omega^2$ from the elementary theory we would simply have arrived at the result $\varepsilon \to \infty$, $E_\ell \to 0$, while in reality the field penetrates into the plasma to a distance r_D.

Another not less important peculiarity of a plasma as a medium with spatial dispersion is the occurrence of longitudinal waves. We now turn to the problem of these and other normal waves which can propagate in an isotropic plasma.

The general dispersion equation which determines the connection between ω and $\mathbf{k}$ for waves propagating in the medium (when there are no external charges or currents) has the form

$$\left| \frac{\omega^2}{c^2} \boldsymbol{\varepsilon}_{ij}(\omega,\mathbf{k}) - k^2\delta_{ij} + k_i k_j \right| = 0 . \tag{10.24}$$

In an isotropic medium when $\boldsymbol{\varepsilon}_{ij}$ is given by Eqn.(11.36) the dispersion equation splits up into an equation for longitudinal waves (when $\omega \neq 0$)

$$\boldsymbol{\varepsilon}_\ell(\omega, k) = 0 \tag{11.50}$$

and an equation for transverse waves

$$k^2 \equiv \frac{\omega^2}{c^2}\tilde{n}^2(\omega) = \frac{\omega^2}{c^2}\boldsymbol{\varepsilon}_{tr}(\omega, k) . \tag{11.51}$$

Here $\tilde{n}(\omega) \equiv \tilde{n}_{tr}(\omega)$ is the complex refractive index for transverse waves. In an isotropic medium due to degeneracy there corresponds to the single value of $\tilde{n}_{tr}(\omega)$ two possible independent polarization states; moreover Eqn. (11.51) can, in principle, have several roots $\tilde{n}_{tr,j}(\omega)$. One must remember, however, that the connection between k and $\tilde{n}$ is simply the definition of the quantity $\tilde{n}$

$$k(\omega) = \frac{\omega}{c}\tilde{n}(\omega) = \frac{\omega}{c}(n + i\kappa) ,$$

the meaning of which is clear from the expression for the field of a plane wave

$$\mathbf{E} = \mathbf{E}_0 e^{ikz-i\omega t} = \mathbf{E}_0 \exp\left\{i\omega\left(\frac{\tilde{n}}{c}z - t\right)\right\} = \mathbf{E}_0 \exp\left\{-\frac{\omega}{c}\kappa z + i\omega\left(\frac{n}{c}z - t\right)\right\} ,$$

where the z-axis is chosen along the direction of propagation of the wave.

Solving Eqn. (11.50) for k we can write that relation also in the form

$$k(\omega) = \frac{\omega}{c}\tilde{n}_{\ell}(\omega),$$

where $\tilde{n}_{\ell}(\omega)$ is the refractive index for a longitudinal wave or longitudinal waves, as Eqn. (11.50) can in principle have several roots for a given frequency ω.

The simplest way to obtain the results (11.50) and (11.51) is not directly from Eqns. (10.24) or (10.27), but starting from the original wave Eqn. (10.22), (10.23). For the longitudinal field $\mathbf{E}_{\ell} = E\mathbf{k}/k$ we then get at once (for $\omega \neq 0$) the equation

$$\mathbf{D} = \boldsymbol{\varepsilon}_{\ell}(\omega, k)\,\mathbf{E}_{\ell} = 0 , \tag{11.52}$$

from which condition (11.50) follows immediately — otherwise there is no non-trivial solution for $\mathbf{E}_{\ell}$. Similarly for the transverse field

$$\frac{\omega^2}{c^2}\mathbf{D} = \frac{\omega^2}{c^2}\boldsymbol{\varepsilon}_{tr}(\omega, k)\,\mathbf{E}_{tr} = k^2\mathbf{E}_{tr} . \tag{11.53}$$

which leads to condition (11.51).

When we neglect spatial dispersion Eqn. (11.51) takes on the familiar form (see (11.39))

$$k^2 = \frac{\omega^2}{c^2}\tilde{n}^2(\omega) = \frac{\omega^2}{c^2}(n^2 - \kappa^2 + 2\,in\,\kappa) = \frac{\omega^2}{c^2}\boldsymbol{\varepsilon}(\omega) , \tag{11.54}$$

and there is only one, multiple root $\tilde{n}(\omega) \equiv \tilde{n}_{tr}(\omega) \equiv \tilde{n}_{1,2}(\omega)$; we have already mentioned that the multiplicity of the root is connected with polarization degeneracy in an isotropic medium (we do not distinguish here the roots $\pm\tilde{n}_{1,2}$ — they correspond to different directions of wave propagation).

By virtue of (11.54) and (11.2)

$$\varepsilon' = n^2 - \kappa^2 , \quad \frac{4\pi\sigma}{\omega} = 2n\kappa , \quad n = \left[\tfrac{1}{2}\varepsilon' + \left\{\tfrac{1}{4}\varepsilon'^2 + \left(\frac{2\pi\sigma}{\omega}\right)^2\right\}^{\frac{1}{2}}\right]^{\frac{1}{2}} ,$$

$$\kappa = \left[-\tfrac{1}{2}\varepsilon' + \left\{\tfrac{1}{4}\varepsilon'^2 + \left(\frac{2\pi\sigma}{\omega}\right)^2\right\}^{\frac{1}{2}}\right]^{\frac{1}{2}} , \tag{11.55}$$

where the square root inside the braces is always assumed to be positive (for instance, for $\sigma = 0$ and $\varepsilon' < 0$ this root equals $\frac{1}{2}|\varepsilon'| = -\frac{1}{2}\varepsilon'$).

We must for $\varepsilon(\omega)$ in (11.54) take for transverse waves in the plasma Eqn. (11.6) whence we get simple formulae in the limiting cases.

For instance, when

$$|\varepsilon'| \gg \frac{4\pi\sigma}{\omega} , \tag{11.56}$$

we have

$$\left.\begin{aligned} &\text{if } \varepsilon' > 0 , \quad n \approx \sqrt{\varepsilon'} = \left[1 - \frac{\omega_p^2}{\omega^2 + \nu_{eff}^2}\right]^{\frac{1}{2}} , \\ &\kappa \approx \frac{2\pi\sigma}{\omega\sqrt{\varepsilon'}} = \frac{\omega_p^2 \nu_{eff}}{2\omega(\omega^2 + \nu_{eff}^2)\sqrt{\varepsilon'}} , \quad \omega_p^2 = \frac{4\pi N e^2}{m} ; \end{aligned}\right\} \tag{11.57}$$

$$\left.\begin{aligned} &\text{if } \varepsilon' < 0 , \quad n \approx \frac{2\pi\sigma}{\omega\sqrt{(-\varepsilon')}} = \frac{\omega_p^2 \nu_{eff}}{2\omega(\omega^2 + \nu_{eff}^2)\sqrt{(-\varepsilon')}} , \\ &\kappa \approx \sqrt{-\varepsilon'} = \left[\frac{\omega_p^2}{\omega^2 + \nu_{eff}^2} - 1\right]^{\frac{1}{2}} . \end{aligned}\right\} \tag{11.58}$$

If, however,

$$|\varepsilon'| \ll \frac{4\pi\sigma}{\omega} , \tag{11.59}$$

we have

$$n \approx \kappa \approx \left(\frac{2\pi\sigma}{\omega}\right)^{\frac{1}{2}} \approx \left[\frac{\omega_p^2 \nu_{eff}}{2\omega(\omega^2 + \nu_{eff}^2)}\right]^{\frac{1}{2}} . \tag{11.60}$$

According to (11.57) the refractive index $n < 1$; when the number of collisions is small, so that $\omega^2 \gg \nu_{eff}^2$, this case is in practice realized when

$$\omega > \omega_p \quad \left(\frac{4\pi N e^2}{m}\right)^{\frac{1}{2}} = 5.64 \times 10^4 \sqrt{N} . \tag{11.61}$$

If $n < 1$, $v_{ph} = c/n > c$ and Cherenkov radiation and thus absorption by separate particles is impossible. Hence it is clear that the whole of the damping is connected solely with collisions (we have in mind waves which propagate and are not damped in the medium as $\nu_{eff} \to 0$; see condition (11.61) and the remark which follows). The result (11.41) does not contradict what we have just said, as it is obtained in the non-relativistic approximation and using a Maxwellian velocity distribution. In this distribution there are formally also particles with velocities $v > c$ and this leads to the appearance of an exponentially weak damping of transverse waves in (11.41). A relativistic calculation leads, of course, to a complete absence of collisionless damping of transverse waves in a plasma. Nonetheless, Eqn.(11.41) was written out in full, as the expression for ε_{tr} is suitable not only for considering normal waves (we emphasize this important fact yet again; see Chapter 10). Moreover, for transverse waves the factor

$$\exp(-\omega^2/2k^2 v_T^2) = \exp(-c^2/2n^2(\omega)v_T^2)$$

is so negligibly small that the corresponding damping is all the same practically equal to zero.

For transverse waves in an isotropic plasma taking spatial dispersion into account therefore does not play a role and Eqns.(11.57) are adequate and they are widely used in radio-astronomy and the theory of propagation of radio-waves in the ionosphere.

Even in the case of weak true absorption, that is, under condition (11.56), the field in the plasma may be strongly damped — this occurs when $\varepsilon' < 0$. In that case, say, when $\nu_{eff} \to 0$ (see (11.58))

$$n = 0\ ,\ \kappa = \sqrt{-\varepsilon'} = \left[\frac{\omega_p^2}{\omega^2} - 1\right]^{\frac{1}{2}}\ ,\ E = E_0 \exp\left[-\frac{\omega}{c}\sqrt{|\varepsilon'|}\, z\right]\ , \qquad (11.62)$$

where E_0 is the field at $z = 0$ (at the inner boundary of the plasma) and the region close to the boundary is here not considered in detail. Clearly, the case (11.62) can easily be realized and it corresponds to a frequency $\omega < \omega_p$. In this case we are physically dealing with a complete internal reflection of a wave from a plasma layer (for details see Ginzburg, 1970b). In general, both for a plasma and for a more general kind of medium we must bear in mind that absorption of waves (the transfer of wave energy into heat or ordered motion of particles) and the damping are different things. In other words one can express the same by noting that for given waves non-absorbing media may be either transparent (a collisional plasma for transverse waves with $\omega > \omega_p$) or non-transparent (the same plasma for $\omega < \omega_p$).

The dispersion law for normal waves, that is, the relation between k and ω in these waves is often expressed not in terms of the refractive index $\tilde{n}$, but directly. For transverse waves in a collisionless plasma

$$k^2 = \frac{\omega^2}{c^2} n^2(\omega) = \frac{\omega^2 - \omega_p^2}{c^2}$$

or

$$\omega^2 = \omega_p^2 + k^2 c^2\ . \qquad (11.63)$$

It is convenient to show the function $\omega(k)$ in a figure; such a function is shown for the case (11.63) in Fig. 11.1. Of course, for $\omega^2 \gg \omega_p^2$ the effect of the medium (plasma) is unimportant and $\omega = ck$, as in vacuo.

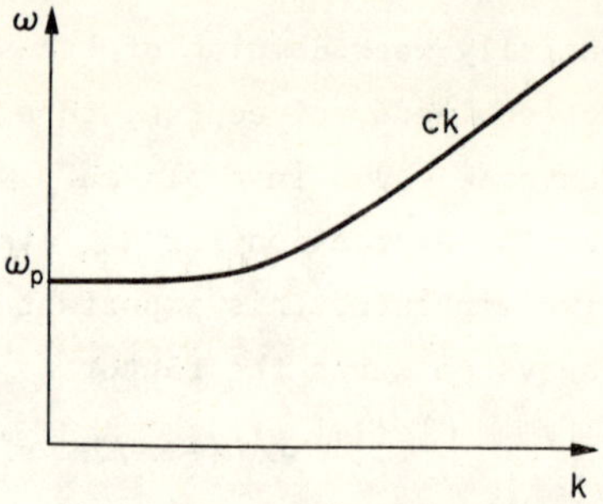

Fig. 11.1 Connection between ω and k (spectrum, dispersion relation) for transverse waves in an isotropic collisionless plasma

We now turn to a consideration of longitudinal waves. Until relatively recently the possibility of the propagation of longitudinal electromagnetic or, more precisely, electrostatic waves in a medium was neglected. There were apparently two reasons for this. Firstly, when one, as was usual, neglects spatial dispersion, the condition for the existence of longintudinal waves in an isotropic media has the form (see (11.5) and (11.39))

$$\varepsilon(\omega) = 0 \; . \tag{11.64}$$

This equation determines discrete frequencies ω_ℓ which are independent of $\mathbf{k}$ and thus longitudinal waves somehow do not manifest themselves explicitly. Secondly, in practically all media, bar plasmas, the roots ω_ℓ of Eqn.(11.64) are complex with a rather large imaginary part, that is, the corresponding longitudinal oscillations are strongly damped. The study of longitudinal waves thus turned out to be connected with the development of the physics of a gaseous plasma although now such waves are considered also in condensed media.

As taking spatial dispersion into account is necessary in the case of longitudinal waves to establish a connection between ω and k even in first approximation, it is expedient to start at once from the dispersion Eqn.(11.50). For the case (11.40), that is, for high-frequency longitudinal waves† we then have, using (11.42) and putting $\omega = \omega' + i\gamma$,

$$\omega'^2 = \omega_p^2 + 3\,v_T^2 k^2 = \omega_p^2\left(1 + 3k^2 r_{D,e}^2\right) ,$$
$$\gamma = -\sqrt{\tfrac{1}{8}\pi}\,\frac{\omega_p}{k^3 r_{D,e}^3}\exp\left(-\frac{1}{2k^2 r_{D,e}^2} - \tfrac{3}{2}\right) - \tfrac{1}{2}\nu_{\rm eff} \; , \tag{11.65}$$

where we have assumed the damping to be weak, that is, we assumed that

---

† Such waves are also called high-frequency Langmuir waves.

$$\mathrm{Im}\,\omega \equiv \gamma \ll \omega' = \mathrm{Re}\,\omega \ . \tag{11.66}$$

For the sake of simplicity we shall drop the prime on ω' and there is thus practically no difference made between ω and ω'.

We note that the condition $\omega^2 \gg (kv_T)^2$ (see (11.40)) means that

$$k^2 r_{D,e}^2 = \left(\frac{2\pi r_{D,e}}{\lambda}\right)^2 \ll 1 \ , \quad r_{D,e} = \left(\frac{k_B T}{4\pi N e^2}\right)^{\frac{1}{2}} \ . \tag{11.67}$$

Hence it is clear from (11.65) that, if (11.67) holds, in zeroth approximation $\omega^2 = \omega_p^2$ and the collisionless damping is exponentially small. This is connected with the fact that in this case the condition for Cherenkov absorption (11.33) is satisfied only in the 'tail' of the Maxwell velocity distribution, where the number of particles is exponentially small. Under the influence of the wave the particle distribution changes and this proceeds relatively easily in the 'tail' region of the distribution. As a result the collisionless absorption changes and can altogether vanish, if the wave appropriately changes the distribution function (the absorption vanishes if the distribution function f_0 ceases to depend on u, which is the component of the particle velocity **v** along the wavevector **k**). The effect of the wave on the distribution function and the related change in the absorption of the wave are, clearly, non-linear processes. We shall not consider here the non-linear theory of the propagation of longitudinal waves (see, for instance, Ginzburg, 1970b; Tsytovich, 1977 and the literature cited there).

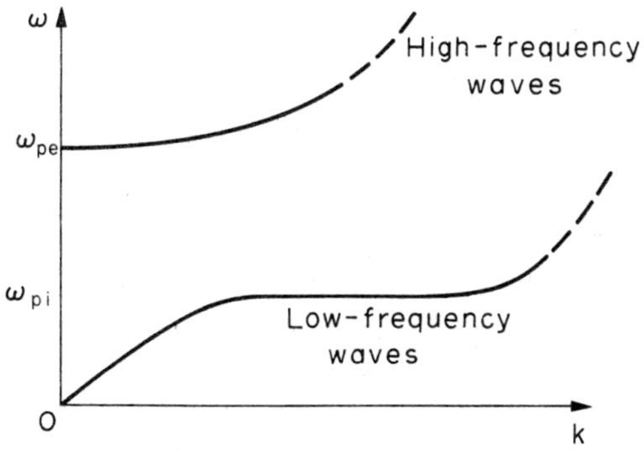

Fig. 11.2 Relation between ω and k for longitudinal waves in an isotropic collisionless plasma. The curves are drawn schematically. The dashed curves correspond to the region of strong damping of the waves and for the low-frequency waves the weakly damped region exists only for a non-isothermal plasma with $T_e \gg T_i$

In terms of the refractive index we can write Eqn.(11.65) in the form (we put $\nu_{eff} = 0$)

$$\left.\begin{aligned} k &= \frac{\omega^2}{c^2}\tilde{n}_3 = \frac{\omega}{c}(n_3 + i\kappa_3) \ , \quad n_3^2 = \frac{1-\omega_p^2/\omega^2}{3v_T^2/c^2} \ , \\ \kappa &= \frac{1}{6}\sqrt{\tfrac{1}{2}\pi}\ \frac{c^5}{n_3^4 v_T^5}\exp\left(-\frac{c^2}{2n_3^2 v_T^2} - \frac{3}{2}\right) \ . \end{aligned}\right\} \tag{11.68}$$

When k increases, that is, when the collisionless absorption increases and when condition (11.67), or even (11.66) is violated, Eqns.(11.65) and (11.68) become already unsuitable. One can emphasize, however, that there exists in the whole of the high-frequency region (11.40) a single branch of longitudinal plasma waves, the behaviour of which is given by Eqns.(11.65) for weak damping. This is the upper branch in Fig.11.2 and the dashed line there indicates the strong damping region. The lower branch in Fig.11.2 corresponds to the low-frequency longitudinal waves which are weakly damped when there are no collisions (condition (11.66)) only in a non-isothermal plasma with $T_e \gg T_i$ and in the wavelength range $k^2 r_{D,i}^2 \ll 1$ (see Ginzburg and Rukhadze, 1975). In that case we have in the region $1/r_{D,e} \leqslant k \ll 1/r_{D,i}$ in zeroth approximation $\omega = \omega_{pi}$, where $\omega_{pi} = \sqrt{(4\pi N_i e^2/M)}$ is the ion plasma (Langmuir) frequency. In the region of even longer wavelength, when $k r_{D,e} \ll 1$ we have for the low-frequency waves $\omega \approx \sqrt{k_B T_e/M}\,k$ which corresponds to isothermal sound in a gas with temperature T_e, but with particles of mass M. When collisions are neglected longitudinal waves are always strongly damped in an isothermal plasma. The presence of collisions and, especially, a large number of neutral particles (that is, when we change to a weakly ionized plasma) changes the picture in the region of very long wavelengths (such that the mean free path $\bar{v}/\nu_{eff}$ is small compared to the wavelength, that is, $k\bar{v} \ll \nu_{eff}$). In that case we are dealing already with waves related to ordinary sound which may be weakly damped.

We now turn to a consideration of a magnetized plasma and, to be precise, a uniform plasma in a uniform and constant magnetic field $\mathbf{H}_0$.

In the framework of the elementary theory taking the magnetic field $\mathbf{H}_0$ into account reduces to adding the Lorentz force to Eqn.(11.5), as a result of which we have

$$m\ddot{\mathbf{r}}_n + m\nu_{eff}\dot{\mathbf{r}}_n = e\mathbf{E}_0 e^{-i\omega t} + \frac{e}{c}[\dot{\mathbf{r}}_n \wedge \mathbf{H}_0]\,. \tag{11.69}$$

Solving this equation and the analogous equation for the ions we find $\dot{\mathbf{r}}_n$ and $\dot{\mathbf{r}}_n^{(i)}$, and after that the current

$$\mathbf{j} = e\sum_{n=1}(\dot{\mathbf{r}}_n - \dot{\mathbf{r}}_n^{(i)})\,.$$

At the same time, by definition, $j_i = -(i\omega/4\pi)(\varepsilon_{ij} - \delta_{ij})E_j$ and by comparing the two expressions we find ε_{ij}. We have elsewhere (Ginzburg, 1970b, § 10) given the corresponding calculations in detail and here we shall merely quote the results:

$$\left.\begin{aligned}
\varepsilon_{xx} = \varepsilon_{yy} &= 1 - \tfrac{1}{2}\omega^2\left(\frac{1}{\omega^2 + \omega\omega_H + i\omega\nu_{eff}} + \frac{1}{\omega^2 - \omega\omega_H + i\omega\nu_{eff}}\right) = \\
&= 1 - \frac{\omega_p^2(\omega + i\nu_{eff})}{\omega\{(\omega + i\nu_{eff})^2 - \omega_H^2\}}, \\
\varepsilon_{zz} &= 1 - \frac{\omega_p^2}{\omega(\omega + i\nu_{eff})}, \\
\varepsilon_{xy} = -\varepsilon_{yx} &= i\frac{\omega_p^2\omega_H}{\omega(\omega + \omega_H + i\nu_{eff})(\omega - \omega_H + i\nu_{eff})}, \\
\varepsilon_{xz} = \varepsilon_{zx} &= \varepsilon_{yz} = \varepsilon_{zy} = 0 .
\end{aligned}\right\} \quad (11.70)$$

It is important that we have chosen here a right-handed Cartesian system of coordinates with the z-axis along the field $\mathbf{H}_0$. Moreover, the gyro-frequency ω_H for electrons (charge $e < 0$) is taken to be positive:

$$\omega_H = \frac{|e|H_0}{mc} = -\frac{eH_0}{mc} = 1.76 \times 10^7 H_0 \text{ s}^{-1}, \quad \lambda_H = \frac{2\pi c}{\omega_H} = \frac{1.07 \times 10^4}{H_0} \text{ cm}. \quad (11.71)$$

Of course, the role of the ions has been neglected in (11.70). While this can always be done when there is no magnetic field by neglecting terms of order N_i/m as compared to N_e/m, in a magnetized plasma the role of ions is usually unimportant only provided

$$\omega \gg \Omega_H = \frac{|e|H_0}{Mc} = 1.76 \times 10^7 \frac{m}{M} H_0 \text{ s}^{-1}, \quad (11.72)$$

where Ω_H is the ion gyrofrequency.

Inequality (11.72) turns sometimes out to be insufficient and the role of the ions may be important even at higher frequencies. For instance, when waves propagate across the field — when they propagate at an angle α between $\mathbf{k}$ and $\mathbf{H}_0$, which is equal to $\frac{1}{2}\pi$, the effect of ions can be neglected only when $\omega \gg \sqrt{\omega_H \Omega_H} = \sqrt{M/m}\,\Omega_H \gg \Omega_H$.

Waves which can be considered neglecting the effect of the ions are called high-frequency waves. Waves and, in general, fields with a frequency

$$\omega \ll \Omega_H \quad (11.73)$$

we shall call low-frequency waves and fields.

The reason why the magnetic field $\mathbf{H}_0$ at low frequencies (satisfying condition (11.73)) radically changes the 'response' of the plasma to an external field $\mathbf{E} = \mathbf{E}_0 e^{-i\omega t}$ is clear already from the equations of motion. These reduce in the elementary theory to (11.69) and the equation

$$M\ddot{\mathbf{r}}_n^{(i)} + M\nu_{eff}^{(i)}\dot{\mathbf{r}}_n^{(i)} = -e\mathbf{E}_0 e^{-i\omega t} - \frac{e}{c}[\dot{\mathbf{r}}_n^{(i)} \wedge \mathbf{H}_0] , \tag{11.74}$$

where the term proportional to $\nu_{eff}^{(i)}$ takes the collisions of a given ion with all other particles into account (see also below). If, for the sake of simplicity, we forget about collisions, it is clear from (11.69) and (11.74) that when (11.73) holds the Lorentz force provides the main terms. This force does, however, not contain the particle mass and in the appropriate approximation the induced electron and ion velocities, $\dot{\mathbf{r}}_n$ and $\dot{\mathbf{r}}_n^{(i)}$, are the same so that the current $\mathbf{j} = e\sum_n (\dot{\mathbf{r}}_n - \dot{\mathbf{r}}_n^{(i)}) \to 0$. At the same time, when we neglect the contribution from the ions the current does not tend to zero as $\omega \to 0$ (see (11.70)). For a more detailed analysis of the role of the ions it is especially convenient to write down the expression for ε_{ij}. For instance, for a mixture of electrons and ions of mass M and if we neglect collisions we can write

$$\varepsilon_{xx} = \varepsilon_{yy} = 1 - \frac{\omega_{pe}^2}{\omega^2 - \omega_H^2} - \frac{\omega_{pi}^2}{\omega^2 - \Omega_H^2} . \tag{11.75}$$

In the high- and low-frequency limits (11.72) and (11.73) we get from this approximately:

$$\text{when} \quad \omega \gg \Omega_H , \qquad \varepsilon_{xx} = \varepsilon_{yy} \approx 1 - \frac{\omega_{pe}^2}{\omega^2 - \omega_H^2} \tag{11.76}$$

$$\text{when} \quad \omega \ll \Omega_H , \quad \varepsilon_{xx} = \varepsilon_{yy} \approx 1 + \frac{\omega_{pe}^2}{\omega_H^2} + \frac{\omega_{pi}^2}{\Omega_H^2} =$$

$$= 1 + \frac{4\pi N mc^2}{H_0^2} + \frac{4\pi N M c^2}{H_0^2} = 1 + \frac{4\pi\rho_M c^2}{H_0^2} \tag{11.77}$$

where $\rho_M = mN + MN \approx NM$ is the mass density of the plasma considered; of course, the value of ρ_M is determined by the ions and, clearly, the contribution from the ions is the determining one.

Very often

$$\frac{4\pi\rho_M c^2}{H_0^2} \gg 1 . \tag{11.78}$$

Under such conditions one can write

$$\varepsilon_{xx} = \varepsilon_{yy} = \frac{4\pi\rho_M c^2}{H_0^2} = \frac{c^2}{v_A^2} , \quad v_A = \frac{H_0}{\sqrt{4\pi\rho_M}} , \tag{11.79}$$

where v_A is the magnetohydrodynamic (or Alfvén) velocity — the velocity of waves in a medium with permittivity (11.79) for the case where these waves are polarized in the xy-plane, that is, the electrical field $\mathbf{E}$ in the wave is at right angles to the field $\mathbf{H}_0$ (one can easily check this, for instance, from the general expressions given above).

The low-frequency region borders in the case of a magnetized plasma upon the magnetohydrodynamics region and under well-defined conditions coincides with it. One can arrive at the magnetohydrodynamic equations, although not rigorously, but in fact rather convincingly, by simply adding the equations of motion (11.69) and (11.74) for electrons and ions. One must solely bear in mind that, for instance, in a pure electron-ion plasma the average 'friction' force on an electron due to ions equals, because of the action = reaction law the average 'friction' force on an ion due to the electrons (we are, in fact, dealing here with the change in the average momenta of the electrons and ions; collisions between electrons and between ions do not play a role in the approximation considered). For this reason the friction force drops out when we add the equations of motion and after multiplying by N we find

$$\rho_M \dot{\mathbf{v}} = \frac{1}{c}[\mathbf{j} \wedge \mathbf{H}_0] , \quad \mathbf{j} = e \sum_{n=1}^{N} (\dot{\mathbf{r}}_n - \dot{\mathbf{r}}_n^{(i)}) , \tag{11.80}$$

where we must understand by $\mathbf{v}$ the velocity of the plasma 'as a whole' or, practically, the ion velocity (strictly speaking, we consider a regime where $m\ddot{\mathbf{r}}_n \ll M\ddot{\mathbf{r}}_n^{(i)}$); making some rather obvious generalizations we get from (11.80) a more general magnetohydrodynamic equation which contains the gradient of the pressure p,

$$\rho_M \frac{d\mathbf{v}}{dt} = \frac{1}{c}[\mathbf{j} \wedge \mathbf{H}] - \nabla p . \tag{11.81}$$

It is inappropriate to dwell here in detail upon magnetohydrodynamics and we wished only to note its connection with the problem of the evaluation of the permittivity tensor in a magnetized plasma (for details see Landau and Lifshitz, 1960; Leontovich, 1965; Ginzburg and Rukhadze, 1975; Tsytovich, 1977, and the literature cited there).

The effect of a magnetic field on the properties of a plasma is, in general, small, if

$$\omega \gg \omega_H = 1.76 \times 10^7 H_0 \, s^{-1} . \tag{11.82}$$

Of course, in that case inequality (11.72) is automatically satisfied and this is true also for the even stricter condition $\omega \gg \sqrt{\omega_H \Omega_H}$ which determines that the role of the ions is small. In the case (11.82) one can to first approximation assume the plasma to be isotropic and the tensor (11.70) reduces to the tensor $\varepsilon_{ij} = \varepsilon(\omega)\,\delta_{ij}$ where $\varepsilon(\omega)$ is given by Eqn. (11.6).

One should, however, remember that even when condition (11.82) is satisfied the role of the magnetic field may turn out to be important, especially when one is dealing with effects which are absent when $H_0 = 0$. As an example we mention the rotation of the polarization plane when transverse electromagnetic waves propagate in a plasma along the magnetic field (Faraday effect; see Chapter 9 and below). As the angle of rotation of the polarization vector (that is, the electric vector of the field in the wave) in a magnetic field with a given direction increases with increasing distance (it is an integral effect), the rotation may turn out to be appreciable even in a field which is completely negligible from other points of view.

The elementary theory is in applications to a magnetized plasma valid only for a cold plasma, that is, when we neglect the thermal motion of the particles, in first instance, of the electrons. This means in a collisionless plasma that the following inequalities must hold:

$$\frac{k_z v_T}{\omega} \ll 1 \ , \quad \frac{k_\perp v_T}{\omega} \ll 1 \ , \tag{11.83}$$

where k_z is the component of the wavevector $\mathbf{k}$ along the field (along the z-axis) and $k_\perp$ the component of $\mathbf{k}$ at right angles to the field $\mathbf{H}_0$ (we bear in mind that $v_T = \sqrt{k_B T/m}$). We have already in essence discussed the first condition (see (11.20) and (11.40)) as the field $\mathbf{H}_0$ does not change the motion of the electrons along it. However, the projection of the electron trajectory on the plane perpendicular to $\mathbf{H}_0$ is a circle of radius $r_H = v_\perp/\omega_H$. As a result of thermal motion $v_\perp \sim v_T$ and $\bar{r}_H \sim v_T/\omega_H$, whence it is clear that the second of conditions (11.83) has the form $2\pi v_T/\lambda_\perp \omega_H \ll 1$, where $\lambda_\perp = 2\pi/k_\perp$. The application of the elementary theory which is equivalent to neglecting spatial dispersion is thus possible when two requirements are satisfied: the phase velocity of the waves $v_{ph} = \omega/k_z$ must be much larger than the thermal velocity v_T and the wavelength $\lambda_\perp$ must be large compared to the Larmor radius $\bar{r}_H \sim v_T/\omega_H$. In the Earth's ionosphere $T \sim 300$ to $1000\ ^\circ$K, $v_T \sim 10^7$ cm/s, $\omega_H \sim 10^7\ s^{-1}$, $\bar{r}_H \sim 1$ cm and in the short wavelength band ($\omega \sim 10^8\ s^{-1}$, $\lambda = 2\pi c/\omega \sim 20$ m) the ratio $v_T/\omega \sim 0.1$ cm. From this it is rather obvious that in the radio-wave band the role of spatial dispersion is normally small in the ionosphere. One can easily make similar estimates for the solar corona, the interstellar medium, and so on and so forth. However, it seems to us not to be appropriate to fix the ideas in this way as it is impossible to legislate for all possibilities while the propagation of waves in a magnetized plasma in the general case is characterized by rather a high degree of complexity. It is useful to bear in mind that in a study of each actual problem in plasma physics one must carefully mention what approximations have been used and estimate their accuracy, in particular, when one takes into account thermal motion (and thereby also spatial dispersion).

One can use the kinetic equation (11.24) to obtain an expression for the tensor $\varepsilon_{ij}(\omega, \mathbf{k})$ in a magnetized plasma taking thermal motion into account or, as one says, for a 'hot' plasma. In the linear approximation when one can use perturbation theory (see (11.26) and (11.27)) we are led at once to an equation which differs from (11.27) only by the addition of the term $(e/m)[\mathbf{v} \wedge \mathbf{H}_0]\nabla_{\mathbf{v}} f'$. We have here taken into account that for an isotropic unperturbed distribution function $f_{00}(v)$ the gradient $\nabla_{\mathbf{v}} f_{00} = (\partial f_{00}/\partial v)(\mathbf{v}/v)$ and $[\mathbf{v} \wedge \mathbf{H}_0]\nabla_{\mathbf{v}} f_{00} = 0$. From this it is clear that the Maxwellian velocity distribution remains the equilibrium distribution also when there is a magnetic field present. At first sight such a conclusion may look suspicious as in a magnetic field the electron moves along a helix and when there is no field it moves along a straight line. There is here nevertheless no contradiction whatever as the probability to find given values of the velocity $\mathbf{v}$ are the same in both cases for a system of electrons (or, in other words, the number of electrons with velocities in the range $\mathbf{v}$, $\mathbf{v} + \mathbf{dv}$ is the same in both cases).

The expression for the tensor $\varepsilon_{ij}(\omega, \mathbf{k})$ in a Maxwellian (equilibrium) plasma can, for instance, be found in the monograph by Ginzburg and Rukhadze (1975). If one can neglect the thermal motion (in the simplest case inequalities (11.83) should be satisfied for this) the tensor $\varepsilon_{ij}(\omega, \mathbf{k})$ reduces to the tensor (11.70), as one should have expected. Perhaps the most specific feature of the kinetic theory is the existence of collisionless absorption. It occurs when the following conditions are satisfied (the z-axis is taken along $\mathbf{H}_0$):

$$\omega = |s\omega_H + k_z v_z|, \quad \omega = |s'\Omega_H + k_z v_z|, \quad s, s' = 0, \pm 1, \pm 2, \pm 3, \ldots \qquad (11.84)$$

When $s, s' = 0$ we are dealing with Cherenkov absorption and when $s \neq 0$ and $s' \neq 0$ with magneto-brems absorption for electrons and ions, respectively. We have already discussed this problem in Chapter 7 and we shall not discuss it now in detail. Apart from collisionless absorption it is particularly important to take thermal motion into account in a magnetized plasma as in an isotropic plasma, in those cases where we are dealing with waves which propagate with a velocity comparable to the thermal velocity v_{Te} or v_{Ti}.

The analysis of the propagation of different waves in a magnetized plasma is distinguished by a certain cumbersomeness even for a cold plasma (that is, when we neglect thermal motion and thus also spatial dispersion). We shall only consider the propagation of waves in a cold plasma and even then very briefly (the books by Ginzburg (1970b), Ginzburg and Rukhadze (1975), and the review series edited by Leontovich (1965) consider in detail wave propagation both in a cold and in a hot plasma; see also the large number of other books and papers quoted there). However, expressions which do not need the actual form of the tensor $\varepsilon_{ij}(\omega, \mathbf{k})$ are equally valid, of course, for cold and for hot plasmas. For instance, the dispersion Eqn.(10.27) which determines the refractive index $\tilde{n}$ for normal waves in an anisotropic medium remains, of course, valid in the general case (it is true, assuming that we can introduce the tensor $\varepsilon_{ij}(\omega, \mathbf{k})$). However, for a magnetized plasma we can go further by using the symmetry properties of the tensor $\varepsilon_{ij}(\omega, \mathbf{k})$. For example, in the case of a uniform field $\mathbf{H}_0$ directed along the z-axis there occur simplifications which are obvious from (11.70) and which remain true also for a hot plasma (we are thinking, for instance of the relation $\varepsilon_{xx} = \varepsilon_{yy}$).

As for a cold plasma $\varepsilon_{ij}(\omega, \mathbf{k}) = \varepsilon_{ij}(\omega)$ the dispersion Eqn.(10.27) is a quadratic equation in $\tilde{n}^2$. The corresponding roots $\tilde{n}^2_{1,2}$ correspond to the ordinary (index 2) and the extra-ordinary (index 1) waves which can propagate in a cold plasma. Moreover, in a cold plasma there can occur a longitudinal wave or, more exactly, longitudinal oscillations (when spatial dispersion is neglected the group velocity $v_{gr} = d\omega/dk$ vanishes for the longitudinal wave). However, while in an isotropic plasma the wavevector $\mathbf{k}$ in the longitudinal wave can be directed arbitrarily, in a magnetized plasma this wavevector can be only in the direction of the field $\mathbf{H}_0$. This all reduces therefore simply to the fact that when there is a magnetic field present the well known expression

$$\varepsilon_{zz} = 1 - \frac{\omega_p^2}{\omega(\omega + i\nu_{eff})}$$

remains unchanged and the field does not affect the longitudinal wave (in this wave $\mathbf{E} = \mathbf{E}_\ell = \mathrm{E}\mathbf{k}/k$) propagating along the field. The problem of the limiting transition from a magnetized plasma to an isotropic one is quite instructive, in particular, in connection with the possibility of the existence of a plasma wave which can propagate only in one direction (in a magnetized plasma) or in any direction (isotropic plasma). A consistent study of this limiting transition is possible only when we take spatial dispersion into account, even though the crux of the matter can be explained without it (see Ginzburg, 1970b, § 12).

We restrict ourselves here to only one remark concerning longitudinal waves. Normal waves in an anisotropic medium, in particular in a magnetized plasma, are, in general, neither longitudinal nor transverse. It should therefore not cause special surprise that one of the normal waves (1 or 2) can under special circumstances turn out to be almost longitudinal. Moreover, in the case of a cold plasma the wave 1 or 2 may be even strictly longitudinal and is formally so when $\tilde{n}^2_{1,2} \to \infty$. One might call (see Agranovich and Ginzburg, 1966) such longitudinal waves fictitious longitudinal waves as these waves are, when spatial dispersion is taken into account, in general, not strictly longitudinal. What we have said is, of course, insufficient to understand the essence of the matter but here we want only to explain that when in the literature longitudinal waves in a magnetized plasma are mentioned, one usually has in mind normal waves in the region of the poles $\tilde{n}^2_{1,2} \to \infty$ (see also below). There is thus no contradiction whatever with the above statement that 'true' longitudinal waves in a magnetized plasma propagate only along the field $\mathbf{H}_0$.

We shall give an expression for $\tilde{n}^2_{1,2}(\omega)$ for high-frequency waves (condition (11.72)) in a cold plasma, that is, when we use the tensor (11.70). Substituting (11.70) into the dispersion Eqn.(10.27) or, more simply, writing out the dispersion equation anew, but now for the case considered (see Ginzburg, 1970b, § 11) we get

$$\tilde{n}^2_{1,2} = (n + i\kappa)_{1,2} = $$
$$= 1 - \frac{2v(1+is-v)}{2(1+is)(1+is-v) - u\sin^2\alpha \pm \{u^2\sin^4\alpha + 4u(1+is-v)^2\cos^2\alpha\}^{\frac{1}{2}}}, \tag{11.85}$$

where

$$v = \frac{\omega_p^2}{\omega^2} = \frac{4\pi N e^2}{m\omega^2}, \quad u = \frac{\omega_H^2}{\omega^2} = \frac{e^2 H_0^2}{m^2 c^2 \omega^2}, \quad s = \frac{\nu_{eff}}{\omega}, \tag{11.86}$$

and α is the angle between $\mathbf{k}$ and $\mathbf{H}_0$.

When there are no collisions

$$\tilde{n}^2_{1,2} = 1 - \frac{2v(1-v)}{2(1-v) - u\sin^2\alpha \pm \{u^2\sin^4\alpha + 4u(1-v)^2\cos^2\alpha\}^{\frac{1}{2}}}$$
$$= 1 - \frac{2\omega_p^2(\omega_p^2 - \omega_p^2)}{2(\omega^2-\omega_p^2)\omega^2 - \omega_H^2\omega^2\sin^2\alpha \pm \{\omega^4\omega_H^4\sin^4\alpha + 4\omega_H^2\omega^2(\omega^2-\omega_p^2)^2\cos^2\alpha\}^{\frac{1}{2}}}. \tag{11.87}$$

Even Eqn.(11.87), let alone (11.85), is so cumbersome that one studies it either in particular cases or graphically. In the case of longitudinal propagation, $\alpha = 0$, we have

$$\left.\begin{aligned} \tilde{n}_1^2 \equiv \tilde{n}_+^2 &= 1 - \frac{v}{1-\sqrt{u}} = 1 - \frac{\omega_p^2}{\omega(\omega-\omega_H)}\,, \\ \tilde{n}_2^2 \equiv \tilde{n}_-^2 &= 1 - \frac{v}{1+\sqrt{u}} = 1 - \frac{\omega_p^2}{\omega(\omega+\omega_H)}\,. \end{aligned}\right\} \qquad (11.88)$$

The normal waves 1, 2 (or ±) are in this case transverse, and their polarization is circular. The direction of rotation of the electric vector **E** in the + wave is the same as the direction of rotation of an electron in the magnetic field $\mathbf{H}_0$ (and is independent of the direction of propagation, that is, the same, if the wavevector **k** in the wave is directed along the field $\mathbf{H}_0$ or in the direction $-\mathbf{H}_0$). It is thus natural that when the frequency ω in the + wave approaches the electron gyro-frequency ω_H, there occurs resonance (see (11.88), where we have put $\nu_{eff} = 0$; the + or 1 wave is called the extra-ordinary wave). If a wave of arbitrary polarization is incident upon an anisotropic medium, in a uniform medium without spatial dispersion it splits into two normal waves which propagate independently of one another. In particular, for longitudinal (along the field $\mathbf{H}_0$) propagation of the waves in a cold plasma an incident (transverse) linearly polarized wave splits in two waves, ±, with circular polarization. As $n_+ \neq n_-$ the phases of the ± waves along the ray change differently and we get a rotation of the plane of polarization (Faraday effect). It is clear that the difference of the phases of the − (or 2) and + (or 1) waves which occurs after passing along a path L is equal to

$$\Delta\phi = \frac{\omega}{c}(n_2 - n_1)L\,.$$

One easily verifies that such a phase advance corresponds to the rotation of the polarization plane over an angle

$$\Psi = \tfrac{1}{2}\Delta\phi = \frac{\omega}{2c}(n_2 - n_1)L \approx \frac{\omega_p^2\omega_H L\cos\alpha}{2c\omega^2} = 0.93\times 10^6\,\frac{NH_0 L\cos\alpha}{\omega^2}\,; \qquad (11.89)$$

here we have as illustration substituted the values of $n_{1,2}$ from (11.88) for the limiting case $|n_{1,2} - 1| \ll 1$, $\omega \gg \omega_H$ which is often met with under cosmic conditions.

For transverse propagation ($\alpha = \frac{1}{2}\pi$) we get from (11.87)

$$\tilde{n}_1^2 = 1 - \frac{\omega_p^2(1-\omega_p^2/\omega^2)}{\omega^2 - \omega_H^2 - \omega_p^2}\,, \quad \tilde{n}_2^2 = 1 - \frac{\omega_p^2}{\omega^2}\,. \qquad (11.90)$$

In the ordinary wave (2) the polarization vector (the vector **E**) is directed along $\mathbf{H}_0$ and it is from this clear that $\tilde{n}_2^2$ for this wave has the same value as $\tilde{n}^2$ in an isotropic plasma. In the extra-ordinary wave (1) the vector **E** describes an ellipse in a plane at right angles to $\mathbf{H}_0$, that is, with respect to $\mathbf{k}$ it has both longitudinal and transverse components (we remind ourselves once again that in the case (11.90) the vector $\mathbf{k}$ itself is perpendicular to the field $\mathbf{H}_0$).

If we speak of the diagrams of the functions $\tilde{n}_{1,2}(\omega, \omega_p, \omega_H, \alpha; \nu_{eff} = 0)$ in the general case, they are to a considerable extent characterized by the zeroes and poles of these functions

$$\left.\begin{aligned} \tilde{n}^2_{1,2} &= 0; \; v_{20} \equiv \frac{\omega_p^2}{\omega_{20}^2} = 1 \, , \; v^{\pm}_{10} \equiv \frac{\omega_p^2}{(\omega^{\pm}_{10})^2} = 1 \pm \sqrt{u} = 1 \pm \frac{\omega_H}{\omega^{\pm}_{10}} \, , \\ \tilde{n}^2_{1,2} &= \infty; \; v_{1\infty} \equiv \frac{\omega_p^2}{\omega_{1\infty}^2} = \frac{1-u}{1-u\cos^2\alpha} - \frac{\omega_{1\infty}^2 - \omega_H^2}{\omega_{1\infty}^2 - \omega_H^2\cos^2\alpha} \, , \quad \text{for} \quad u < 1 \, , \end{aligned}\right\} \tag{11.91}$$

$$v_{2\infty} \equiv \frac{\omega_p^2}{\omega_{2\infty}^2} = \frac{u-1}{u\cos^2\alpha - 1} = \frac{\omega_H^2 - \omega_{2\infty}^2}{\omega_H^2\cos^2\alpha - \omega_{2\infty}^2} \, , \quad \text{for} \quad u > 1 \, . \tag{11.92}$$

As, by definition $v = \omega_H^2/\omega^2 > 0$, the root v_{10} is, of course, fictitious when $u = \omega_H^2/\omega^2 > 1$. For the same reason when $u > 1$, but $u\cos^2\alpha < 1$, the function $\tilde{n}_{1,2}$ has no poles,

As an illustration we have given in Figs.11.3 and 11.4 a few graphs of the functions $\tilde{n}^2_{1,2}(v)$ for different values of u and α. We have used as variable the quantity $v = \omega_p^2/\omega^2$; this may turn out to be artificial. In fact, such graphs are very convenient when we are dealing with explanations for the way $\tilde{n}^2_{1,2}$ depends on the electron density N (indeed, $v = 4\pi N e^2/m\omega^2$). Some other graphs, for instance, of $\tilde{n}^2_{1,2}(\omega/\omega_p)$, turn out to be also convenient; an example of these is given in Fig.11.5. As there is a region of small ω-values in this graph it is necessary to remind ourselves once again that Eqn. (11.85) and the subsequent ones refer to the high-frequency case $\omega \gg \Omega_H$. When the frequency decreases the effect of the ions starts to become discernible and yet another branch of oscillations appears. We restrict ourselves here by giving Fig.11.6 in which we show the functions $\tilde{n}^2_{1,2}(\omega)$, taking ions into account.

We discuss yet one more important limiting case

$$\omega_p^2 \gg \omega^2 \, , \quad \omega_p^2 \gg \omega_H^2 \, , \quad \omega^2 \ll \omega_H^2\cos^2\alpha \, , \quad \omega \gg \Omega_H \, , \tag{11.93}$$

or

$$v \gg 1 \, , \quad v \gg u \, , \quad u\cos^1\alpha \gg 1 \, .$$

Under the conditions (11.93) we have according to (11.87)

$$\tilde{n}_1^2 = -\frac{v}{\sqrt{u}\cos\alpha} \, , \quad \tilde{n}_2^2 = \frac{c^2k^2}{\omega^2} \approx \frac{v}{\sqrt{u}\cos\alpha} = \frac{\omega_p^2}{\omega\,\omega_H\cos\alpha} = \frac{4\pi c N|e|}{\omega H_0\cos\alpha} \, , \tag{11.94}$$

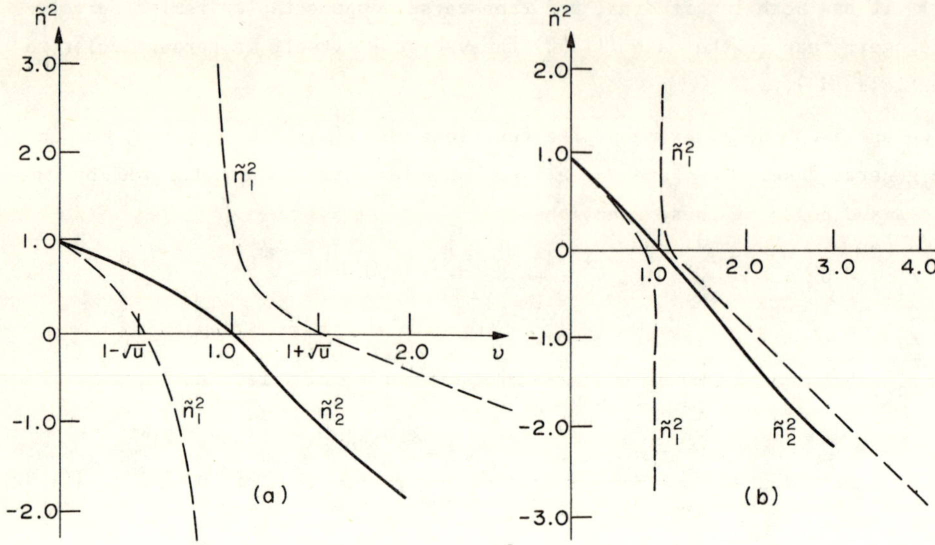

Fig. 11.3 The functions $\tilde{n}^2_{1,2}(v)$ for $\nu_{eff} = 0$ and $\alpha = 45°$. (a) $u = \omega_H^2/\omega^2 = \frac{1}{4}$; (b) $u = 0.01$

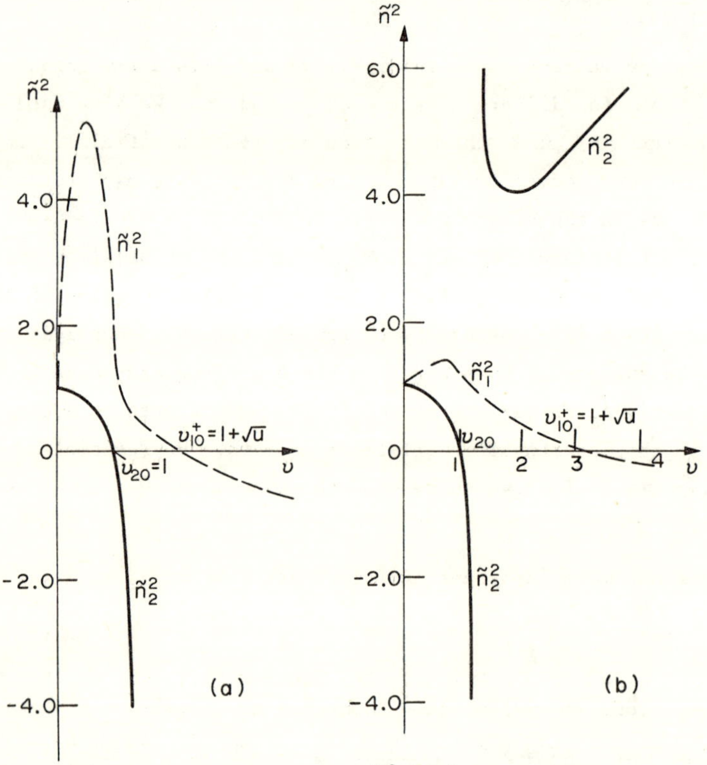

Fig. 11.4 The functions $\tilde{n}^2_{1,2}(v)$ for $\nu_{eff} = 0$ and $\alpha = 20°$. (a) $u = \omega_H^2/\omega^2 = 1.08$; (b) $u = 4$

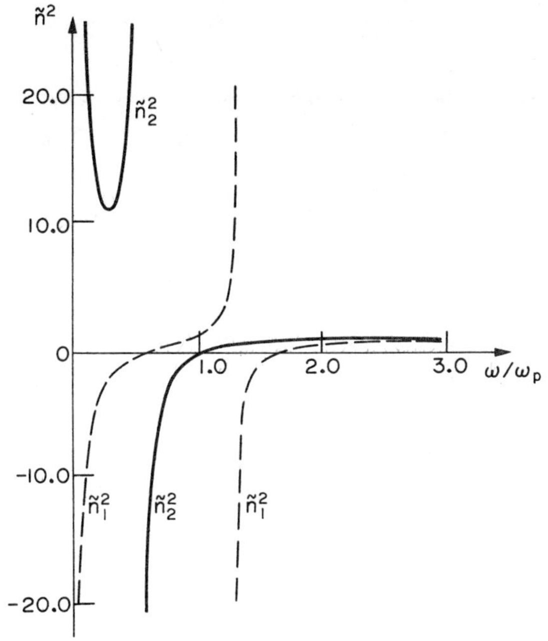

Fig. 11.5 The functions $\tilde{n}^2_{1,2}(\omega/\omega_p)$ for $\omega_H^2/\omega_p^2 = u/v = 1$ and $\alpha = 45°$

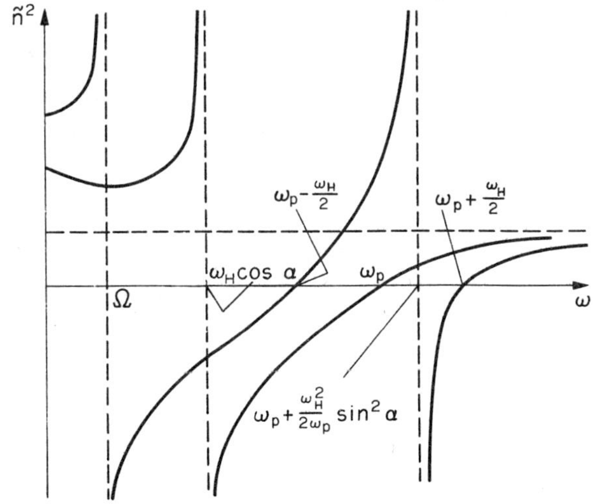

Fig. 11.6 The functions $\tilde{n}^2_{1,2}(\omega)$ for a cold, collisionless plasma taking ions into account. We assume that $\omega_p^2 \gg \omega_H^2$ and the values of the roots and the poles of the functions $\tilde{n}^2_{1,2}(\omega)$ are indicated approximately.

or

$$\omega = \frac{c^2 k^2 \omega_H \cos\alpha}{\omega_p^2} = \frac{c H_0 k^2 \cos\alpha}{4\pi N|e|} . \qquad (11.95)$$

Under the conditions considered here the wave 1 does not propagate at all (this wave is damped, as $\tilde{n}_{1,2}^2 < 0$) and for the wave 2 the value $\tilde{n}_2^2 \gg 1$. One encounters such waves, for instance, in the Earth's magnetosphere ('atmospheric whistlers') and in solid state plasmas ('helical waves' in metals which are in a magnetic field).

Apart from everything else, the neglecting of the spatial dispersion is the less legitimate the smaller the phase velocity of the waves considered $v_{ph} = \omega/k$. If $v_{ph} \leqslant v_T$, the effects of spatial dispersion are large. It is clear in this connection that spatial dispersion in the first instance must be taken into account in the region of the poles of the functions $\tilde{n}_{1,2}^2(\omega)$ which are obtained for a cold plasma. For the 'atmospheric whistlers' and the 'helical waves' the role of the spatial dispersion is also larger than for waves with a large phase velocity because of the condition $\tilde{n}_{1,2}^2 \gg 1$ which means that $v_{ph} = c/n_{1,2}$ is much smaller.

Finally, one more remark: a reminder that in the foregoing we have considered only a uniform plasma. Meanwhile under actual conditions a plasma is always non-uniform — either there are boundaries or the plasma properties (such as the electron density) change from point to point. A lot of attention is therefore paid to the propagation of waves in a non-uniform plasma. Of large value for plasma physics is also the discussion of non-equilibrium plasmas (for instance, a cold plasma penetrated by electron or ion beams), the study of various non-linear phenomena and effects, relativistic considerations, and so on. We have not mentioned plasma physics in a wider context and only considered calculations of the dielectric permittivity of a plasma and wave propagation in it; we have thus in the present chapter only touched upon a very small part of that field of problems.

Chapter XII

THE ENERGY-MOMENTUM TENSOR IN MACROSCOPIC ELECTRODYNAMICS

The energy-momentum tensor in macroscopic electrodynamics. Applications of the energy and momentum conservation laws to the radiation of electromagnetic wave (photons) in a medium. Forces acting on the medium.

We have already discussed in Chapter 10 the Poynting theorem and the energy conservation law in macroscopic electrodynamics (as applied to a simple model medium). Agranovich and Ginzburg (1966) considered the energy conservation law for the case where an arbitrary temporal and spatial dispersion was taken into account within wide limits and this problem has also often been discussed in general in the literature. It is impossible to say the same about the momentum conservation law in macroscopic electrodynamics. There are apparently two reasons for this. Firstly, one encounters the momentum of the electromagnetic field appreciably more seldom than the energy. Secondly, the problem of the momentum of the field in a medium has to a certain extent turned out to be confused — it is connected with the choice for the expression for the energy-momentum tensor of the electromagnetic field in a medium. This problem has been discussed for more than 60 years up to the most recent times (see Brevik, 1970a,b; Møller, 1972; Skobel'tsyn, 1973; Robinson, 1973, 1975; Ginzburg, 1973c; Walker and Lahoz, 1975; Ginzburg and Ugarov, 1976; and the literature cited there; we use the last named paper in what follows). It is therefore opportune to give a discussion of the momentum conservation law in macroscopic electrodynamics.

In order to avoid complications which are not directly related to the problems which we wish to explain we shall consider a non-magnetic, dispersionless medium at rest. The field equations then have the form, well known to the reader, but it is convenient to write them down once more ($\mathbf{B} = \mathbf{H}$, and ε is independent of ω and $\mathbf{k}$):

$$\mathrm{curl}\,\mathbf{H} = \frac{4\pi}{c}\mathbf{j} + \frac{\varepsilon}{c}\frac{\partial \mathbf{E}}{\partial t}, \tag{12.1}$$

$$\mathrm{curl}\,\mathbf{E} = -\frac{1}{c}\frac{\partial \mathbf{H}}{\partial t}, \tag{12.2}$$

$$\mathrm{div}\,\varepsilon\mathbf{E} = 4\pi\rho, \tag{12.3a}$$

$$\mathrm{div}\,\mathbf{H} = 0. \tag{12.3b}$$

We take the scalar product of Eqn.(12.1) with $\mathbf{E}$ and that of Eqn.(12.2) with $\mathbf{H}$. Subtracting the equations obtained from one another and using the identity $(\mathbf{E}\cdot\mathrm{curl}\,\mathbf{H}) - (\mathbf{H}\cdot\mathrm{curl}\,\mathbf{E}) = -\mathrm{div}\,[\mathbf{E}\wedge\mathbf{H}]$, we get

$$\frac{1}{8\pi}\frac{\partial}{\partial t}(\varepsilon E^2 + H^2) = -(\mathbf{j}\cdot\mathbf{E}) - \mathrm{div}\,\mathbf{S}, \tag{12.4}$$

that is, we obtain Poynting's theorem which in the present case we can without particular complications interpret as the energy conservation law (the energy density is $w = (\varepsilon E^2 + H^2)/8\pi$ and the energy flux density is $\mathbf{S} = (c/4\pi)[\mathbf{E}\wedge\mathbf{H}]$).†

We now take the vector product of Eqn.(12.1) and $\mathbf{H}$ and of Eqn.(12.2) and $\varepsilon\mathbf{E}$ and adding the expressions obtained we find

$$\frac{1}{4\pi}\left\{[\mathbf{H}\wedge\mathrm{curl}\,\mathbf{H}] + \varepsilon[\mathbf{E}\wedge\mathrm{curl}\,\mathbf{E}]\right\} = -\frac{1}{c}[\mathbf{j}\wedge\mathbf{H}] - \frac{\varepsilon}{4\pi c}\frac{\partial}{\partial t}[\mathbf{E}\wedge\mathbf{H}].$$

We add now the expression $-\rho\mathbf{E}$ to both sides of this equation and on the left hand side use Eqn.(12.3a) to change it to $-\mathbf{E}(\mathrm{div}\,\varepsilon\mathbf{E})/4\pi$. As a result we find

$$\frac{1}{4\pi}\left\{[\mathbf{H}\wedge\mathrm{curl}\,\mathbf{H}] + \varepsilon[\mathbf{E}\wedge\mathrm{curl}\,\mathbf{E}] - \mathbf{E}\,\mathrm{div}\,\varepsilon\mathbf{E}\right\} + \frac{1}{4\pi c}\frac{\partial}{\partial t}[\mathbf{E}\wedge\mathbf{H}] =$$

$$= -\left\{\rho\mathbf{E} + \frac{1}{c}[\mathbf{j}\wedge\mathbf{H}] + \frac{\varepsilon-1}{4\pi c}\frac{\partial}{\partial t}[\mathbf{E}\wedge\mathbf{H}]\right\}. \tag{12.5}$$

On the right hand side we have here the Lorentz force density

$$\mathbf{f}^L = \rho\mathbf{E} + \frac{1}{c}[\mathbf{j}\wedge\mathbf{H}]$$

and the volume force density

$$\mathbf{f}^A = \frac{\varepsilon-1}{4\pi c}\frac{\partial}{\partial t}[\mathbf{E}\wedge\mathbf{H}], \tag{12.6}$$

which is sometimes called the Abraham force. The minus sign on the right hand side of (12.5) is connected with the fact that the sum $\mathbf{f}^L + \mathbf{f}^A$ is the force on the medium, while Eqn.(12.5) determines the balance of the forces and momentum as referring to the field, where

$$\mathbf{g}^A = \frac{1}{4\pi c}[\mathbf{E}\wedge\mathbf{H}] = \frac{\mathbf{S}}{c^2} \tag{12.7}$$

is the momentum density of the field (it is just this expression which is the same for the vacuum and for a medium at rest which corresponds to the energy-momentum tensor in Abraham's form; vide infra).

For the sake of simplicity we shall to begin with assume the medium to be uniform (in that case $\varepsilon = \mathrm{constant}$; the case of a non-uniform medium and the

---

† We note that when there is dispersion and absorption the interpretation of the separate terms in the Poynting relation is more complicated (see Agranovich and Ginzburg, 1966, § 3; Ginzburg, 1970b, § 22; and, for details, Barash and Ginzburg, 1976).

possible dependence of ε on the density of the medium are considered below). One can then very easily transform Eqn.(12.5) into the standard form (it is convenient for the transition from (12.5) to (12.8) to use the identity $[\mathbf{a} \wedge \mathrm{curl}\,\mathbf{a}] = \frac{1}{2}\nabla a^2 - (\mathbf{a}\cdot\nabla)\mathbf{a}$)

$$\frac{\partial \sigma_{\alpha\beta}}{\partial x_\beta} - \frac{\partial g^A_\alpha}{\partial t} = f_\alpha \;, \quad f_\alpha = f^L_\alpha + f^A_\alpha \;, \quad \alpha, \beta = 1, 2, 3 \;, \tag{12.8}$$

where $\sigma_{\alpha\beta}$ is the Maxwell stress tensor;

$$\sigma_{\alpha\beta} = \frac{1}{4\pi}\left\{\varepsilon E_\alpha E_\beta + H_\alpha H_\beta - \tfrac{1}{2}(\varepsilon E^2 + H^2)\,\delta_{\alpha\beta}\right\} . \tag{12.9}$$

The momentum conservation law (12.8) thus follows from the field equations without further assumptions. If we combine this law and the energy conservation law (12.4) in a single four-dimensional relation — the energy-momentum conservation law — we are at the same time led to an expression for the energy-momentum tensor T_{ik}:

$$T^{(A)}_{ik} = \begin{pmatrix} \sigma_{\alpha\beta} & -ic\,\mathbf{g}^A \\ -\frac{i}{c}\mathbf{S} & w \end{pmatrix}, \quad w = \frac{\varepsilon E^2 + H^2}{8\pi} \;, \quad \mathbf{S} = \frac{c}{4\pi}[\mathbf{E}\wedge\mathbf{H}] = c^2\mathbf{g}^A \;, \tag{12.10}$$

$$\frac{\partial T^{(A)}_{ik}}{\partial x_k} = f_i \;, \quad f_\alpha = f^L_\alpha + f^A_\alpha \;, \quad f_4 = \frac{i}{c}(\mathbf{j}\cdot\mathbf{E}) \;, \tag{12.11}$$

$$(i, k = 1, 2, 3, 4;\quad \alpha, \beta = 1, 2, 3;\quad x_4 = ict) .$$

The tensor (12.10) is the energy-momentum tensor suggested by Abraham for a uniform medium at rest; for a moving medium this tensor looks somewhat more complicated (vide infra).

The energy-momentum tensor used by Minkovski under the same assumptions as in the case (12.10) and (12.11) has the form

$$T^{(M)}_{ik} = \begin{pmatrix} \sigma_{\alpha\beta} & -ic\,\mathbf{g}^M \\ -\frac{i}{c}\mathbf{S} & w \end{pmatrix}, \quad \mathbf{g}^M = \frac{\varepsilon}{4\pi c}[\mathbf{E}\wedge\mathbf{H}] = \varepsilon\,\mathbf{g}^A \;, \tag{12.12}$$

$$\frac{\partial T^{(M)}_{ik}}{\partial x_k} = f^L_i \;, \quad f^L_\alpha = \rho E_\alpha + \frac{1}{c}[\mathbf{j}\wedge\mathbf{H}]_\alpha \;, \quad f^L_4 = f_4 = \frac{i}{c}(\mathbf{j}\cdot\mathbf{E}) . \tag{12.13}$$

It is completely obvious that at least from a formal point of view the conservation laws (12.11) and (12.13) are identical — they differ merely in the different splitting up into terms of the same sum. To be precise, if we transfer the Abraham force (12.6) from the right-hand to the left-hand side of Eqn.(12.11) and combine it with $\partial T_{ik}^{(A)}/\partial x_k$ we get at once the expression $\partial T_{ik}^{(M)}/\partial x_k$ and we can take for the energy-momentum tensor the Minkovski tensor. Such ambiguity in the choice for an expression for the energy-momentum tensor is nonetheless surprising in that it is very general in nature and occurs even in field theory in vacuo (see, for instance, Landau and Lifshitz, 1975, § 32). Moreover, the field in a medium is not a closed system — only the system consisting of the field and the medium is 'closed' and the medium is characterised by its own energy-momentum tensor $T_{ik}^{(med)}$. There is a conservation law $\partial T_{ik}/\partial x_k = 0$ valid for the total tensor $T_{ik} = T_{ik}^{(med)} + T_{ik}^{(e.m.)}$, where $T_{ik}^{(e.m.)}$ is the energy-momentum tensor of the field (for instance, the tensor (12.10)); however the tensor T_{ik} is not uniquely defined and even less so are its parts $T_{ik}^{(med)}$ and $T_{ik}^{(e.m.)}$. The force density is altogether a different thing; it is, at least in principle, a unique and measurable quantity. In this connection the fate of the 'controversy' about the Abraham and Minkovski tensors will ultimately be solved as a result of the choice of an expression for the force. The Abraham force (12.6) is genetically connected with the force due to the magnetic field (Lorentz force) acting upon the displacement current. It is in reality impossible to question this force notwithstanding that it has as yet not been reliably measured directly.† In that way the problem would be solved 'in favour' of the Abraham tensor. Skobel'tsyn (1973) has also shown in detail that the various expressions one meets with in the literature which contradict the choice of the Abraham tensor are without a foundation, and this was done also by Brevik (1970a,b). We restrict ourselves here to one of the arguments in favour of choosing Minkovski's tensor rather than Abraham's. This is that when one chooses the Minkovski tensor one finds for the case of a quasi-monochromatic wave train in any system of reference for the flux of the field energy in a transparent medium $\mathbf{S} = w\mathbf{v}_{gr}$, where w is the energy density and $\mathbf{v}_{gr}$ the group velocity. This just analogous to the relation $\sigma_{\alpha\beta} = -g_\alpha^M v_{gr,\beta}$ for the Minkovski tensor (see Agranovich and Ginzburg, 1966, § 3.2, and the literature cited there). When, on the other hand, one chooses the Abraham tensor such relations do not hold and this is taken somehow to be a disadvantage or difficulty. Indeed, as Skobel'tsyn (1973) proved especially and in

---

† The appropriate possibilities have been discussed by Brevik (1970a,b) and Skobel'tsym (1973). Undoubtedly, for one's 'peace of mind' it would be justified to measure the Abraham force. This has been done relatively recently (Walker and Lahoz, 1975a,b) and the validity of Eqn.(12.6) was confirmed, albeit only with a relatively low accuracy.

detail, this is all again connected with the presence of the volume force $\mathbf{f}^A$ when we use the Abraham tensor. In a moving medium this force does work on the medium and the relation $\mathbf{S} = w\mathbf{v}_{gr}$ therefore can and should not be satisfied. The situation is here completely analogous to the one for a medium at rest† when the relation $\mathbf{S} = w\mathbf{v}_{gr}$ is violated when there is absorption and, in general, when there are some energy sources or sinks in the medium. When applying this to the momentum density flux $g_\alpha v_{gr,\beta}$ what we have said refers already to the case of a transparent medium at rest, as the relation $\sigma_{\alpha\beta} = -g_\alpha v_{gr,\beta}$ can hold only when there are not volume forces. This last requirement is at once satisfied by the Minkovski tensor (we assume that there are no charges or currents) for which $\partial T^{(M)}_{ik} / \partial x_k = 0$.

All we have said allows us to take the Abraham tensor to be the 'correct' one, but it seems to us that one can label the Minkovski tensor as being 'incorrect' only in a somewhat formal approach. In reality in most situations the results obtained using the Abraham and Minkovski tensors are completely identical. This makes it possible not only to use in appropriate cases the Minkovski tensor but even to consider its application to be fully suitable, if in that way one reaches a certain simplification. One should therefore hardly label the Minkovski tensor $T^{(M)}_{ik}$ to be 'erroneous'; rather it is an auxiliary concept which can be used fully. This does not imply any loss of 'prestige' for the more fundamental and, when it is convenient, 'truer' energy-momentum tensor of an electromagnetic field in a medium, $T^{(A)}_{ik}$.

An analysis of the problem of the energy and momentum conservation laws when electromagnetic waves (photons) are emitted in a medium confirms and illustrates just what we have said. Indeed, let us consider what the momentum of a wave train is equal to in a medium, using the Abraham and Minkovski tensors and after that turn in both cases to the conservation laws.

We consider the propagation in the medium of a plane wave of the form

$$\left.\begin{aligned} \mathbf{E} &= \tfrac{1}{2}\left(\mathbf{E}_0\, e^{i(\mathbf{k}\cdot\mathbf{r})-i\omega t} + \mathbf{E}_0^*\, e^{-i(\mathbf{k}\cdot\mathbf{r})+i\omega t}\right), \\ \mathbf{H} &= \tfrac{1}{2}\left(\mathbf{H}_0\, e^{i(\mathbf{k}\cdot\mathbf{r})-i\omega t} + \mathbf{H}_0^*\, e^{-i(\mathbf{k}\cdot\mathbf{r})+i\omega t}\right). \end{aligned}\right\} \tag{12.14}$$

If the wave is quasi-monochromatic, $\mathbf{E}_0$ and $\mathbf{H}_0$ are 'slowly' varying (as compared to the period $2\pi/\omega$) functions of the time t. However, for the sake of simplicity we neglect dispersion and we shall thus assume $\mathbf{E}_0$ and $\mathbf{H}_0$ to be

† In such a case the volume force acting upon the medium does no work, as this work equals the product of the force and the velocity of the medium.

constants, but the wave train to have a cross-section of unit area and length L (for an account of dispersion see, for instance, Agranovich and Ginzburg, 1966, § 3). Substituting (12.14) into the field Eqns.(12.1) and (12.2) with real ε = constant† and $\mathbf{j} = 0$, we find

$$\mathbf{E}_0 = -\frac{c}{\varepsilon\omega}[\mathbf{k}\wedge\mathbf{H}_0] \quad , \quad \mathbf{H}_0 = \frac{c}{\omega}[\mathbf{k}\wedge\mathbf{E}_0] \; . \tag{12.15}$$

Hence we get, from the condition for the existence of a non-trivial solution, the dispersion equation

$$k \equiv \frac{\omega}{c} n = \frac{\omega}{c}\sqrt{\varepsilon} \; . \tag{12.16}$$

Moreover we get for the time-averaged (that is, averaged over the high frequency) quantities (see (12.10) and (12.13))

$$\left.\begin{aligned} \bar{w} &= \frac{\overline{\varepsilon E^2 + H^2}}{8\pi} = \frac{1}{16\pi}\left\{\varepsilon(\mathbf{E}_0\cdot\mathbf{E}_0^*) + (\mathbf{H}_0\cdot\mathbf{H}_0^*)\right\} = \frac{n^2}{8\pi}(\mathbf{E}_0\cdot\mathbf{E}_0^*) \, , \\ \bar{\mathbf{S}} &= \frac{c}{4\pi}\,[\overline{\mathbf{E}\wedge\mathbf{H}}] = \frac{1}{16\pi}\left\{[\mathbf{E}_0^*\wedge\mathbf{H}_0] + [\mathbf{E}_0\wedge\mathbf{H}_0^*]\right\} = \frac{cn}{8\pi}(\mathbf{E}_0\cdot\mathbf{E}_0^*)\,\frac{\mathbf{k}}{k} \, , \\ \overline{\mathbf{g}^A} &= \frac{\bar{\mathbf{S}}}{c^2} \quad , \quad \overline{\mathbf{g}^M} = n^2\,\overline{\mathbf{g}^A} \, , \end{aligned}\right\} \tag{12.17}$$

$$\mathbf{G}^{(A)} = \overline{\mathbf{g}^A} L = \frac{\bar{w}L}{cn}\frac{\mathbf{k}}{k} = \frac{\mathcal{H}}{cn}\frac{\mathbf{k}}{k} \quad , \tag{12.18}$$

$$\mathbf{G}^{(M)} = \overline{\mathbf{g}^M} L = \frac{n\bar{w}L}{c}\frac{\mathbf{k}}{k} = \frac{\mathcal{H}n}{c}\frac{\mathbf{k}}{k} \quad , \tag{12.19}$$

where $\mathbf{G}^{(A,M)}$ and $\mathcal{H} = \mathcal{H}^{(A)} = \mathcal{H}^{(M)} = \bar{w}L$ are, respectively, the momentum and energy of the wave train. The relation (12.19) is exactly the same as the relation between the energy and momentum of a 'photon in a medium' obtained through quantization of the field in a medium (see Chapters 6 and 7). Indeed, the energy of a photon in a medium is equal to $\mathcal{H} = \hbar\omega$ and its momentum equals $\mathbf{G} = (\hbar\omega n/c)(\mathbf{k}/k)$, that is, $\mathbf{G} = (\mathcal{H} n/c)(\mathbf{k}/k)$. We showed in Chapter 7 that the use of (12.19) enables us to use the energy and momentum conservation laws to find the condition for Cherenkov radiation; the same refers to the Doppler formula. In both cases we get, of course, in a classical calculations, when we have not made a definite choice for $\mathcal{H}$ (which therefore does not contain the quantum constant $\hbar$) only classical formulae which do not take into account recoil — the change in the motion of the emitting particle. To find more general formulae taking recoil into account we must use the conservation laws

---

† If we wanted to be precise we should have added that we assume the medium not only to be non-absorbing (ε real), but also transparent (condition $\varepsilon > 0$).

already when applying them to a separate photon in a medium. To be precise, when a single photon is emitted we must use the relations (see (7.1) and (7.2))

$$E_0 - E_1 = \hbar\omega \quad , \quad \mathbf{p}_0 - \mathbf{p}_1 = \frac{\hbar\omega n}{c}\frac{\mathbf{k}}{k} , \tag{12.20}$$

where $E_{0,1}$ and $\mathbf{p}_{0,1}$ are the energies and momenta of the particle (radiator), respectively in the initial state 0 and the final state 1. Because of what we have said it is clear (and is, of course, confirmed by calculations) that when one 'obtains' quanta (photons in a medium) with energy $\hbar\omega$ and momentum $(\hbar\omega n/c)(\mathbf{k}/k)$ through standard quantization one must have recourse to the Minkovski energy-momentum tensor. If, however, we use the Abraham tensor, we find both classically (see (12.18)) and also quantally and taking only the momentum $\mathbf{G}^{(A)}$ into account a completely wrong result for the momentum of the 'photon'. In reality, however, the use of the Abraham tensor leads, as we should expect, to a correct result, but one must calculate also the action of the Abraham force on the medium when the radiation is emitted (the same applies to the absorption process). It is, indeed, necessary to do this as the force $\mathbf{f}^{(A)}$ (see (12.6)) does not vanish when a wave train is emitted (or, for instance, when it enters the medium). We are here interested not in the force itself, but in the momentum related to the force which in the emission of a wave train equals

$$\mathbf{F}^{(A)} = \frac{n^2-1}{4\pi c}\int \frac{\partial}{\partial t}[\mathbf{E}\wedge\mathbf{H}]\, d^3\mathbf{r}\, dt = \frac{n^2-1}{16\pi c}\left\{[\mathbf{E}_0\wedge\mathbf{H}_0^*] + [\mathbf{E}_0^*\wedge\mathbf{H}_0]\right\}L =$$

$$= \frac{(n^2-1)n}{8\pi c}(\mathbf{E}_0\cdot\mathbf{E}_0^*)\, L\frac{\mathbf{k}}{k} = \frac{(n-1)\,\mathcal{H}\,\mathbf{k}}{cnk} , \tag{12.21}$$

where we have dropped the oscillating terms and hence we are dealing with a time-averaged quantity.†

We note that we can right from the start consider a more or less arbitrary wave train and calculate and afterwards compare the integral quantities

$$\mathcal{H} = \int w\, d^3\mathbf{r}\, dt \quad , \quad \mathbf{G}^{(A,M)} = \int \mathbf{g}^{A,M} d^3\mathbf{r}\, dt \quad \text{and} \quad \mathbf{F}^{(A)} .$$

The relations between these quantities remain the same as in (12.17) to (12.19), (12.21) for a train with sharp boundaries.

It is clear that by virtue of (12.17) to (12.19)

$$\mathbf{G}^{(A)} + \mathbf{F}^{(A)} = \mathbf{G}^{(M)} = \frac{\mathcal{H}n}{c}\frac{\mathbf{k}}{k} . \tag{12.22}$$

---

† The force acts only as long as the wave (train) enters the medium or is emitted by the source. During the propagation, however, of a train of a given length in a uniform medium the force momentum $\mathbf{F}^{(A)}$ vanishes.

In the field of applications of the energy and momentum conservation law usually only two points are important: firstly, what is the energy and momentum which is lost (or gained in absorption) the emitting particle or 'system'; secondly, what is the field energy emitted in a given direction. On the other hand, the problem of the distribution or redistribution of the momentum of the radiation is unimportant from this point of view. In the case considered the particle loses a momentum $-\mathbf{G}^{(M)}$, the field in the medium gains a momentum $\mathbf{G}^{(A)}$ and the medium obtains the force momentum $\mathbf{F}^{(A)} = \mathbf{G}^{(M)} - \mathbf{G}^{(A)}$. For a medium — 'grainy matter' — under the influence of the force $\mathbf{f}^A$ the particles in the medium are accelerated and $\mathbf{F}^{(A)} = \mathbf{G}^{(med)}$ is the momentum of the medium (the sum of the momenta of the 'grains', the density of which we assume to be constant). In the general case, however, the state of the medium is determined by the appropriate equations of motion, for instance, the equations of elasticity theory or the equations of hydrodynamics in which the volume force density equals $\mathbf{f}^A$ and can contain in principle also other terms. In that case it is, of course, impossible to assume that $\mathbf{F}^{(A)} = \mathbf{G}^{(med)} = \mathbf{g}^{(med)}L$, where $\mathbf{g}^{(med)}$ is the density of the momentum of the medium (it is just this fact which has been correctly emphasized by Skobel'tsyn (1973)).† It is thus, in general, also impossible to state that the Minkovski momentum density $\mathbf{g}^M$ equals $\mathbf{g}^A + \mathbf{g}^{med}$. However, we saw in the relation between the integral quantities — the momenta and the force momentum $\mathbf{F}^{(A)}$ — that the result (12.21) is completely independent of the properties of the medium and remains true also if we assume (in general, incorrectly) that $\mathbf{g}^M = \mathbf{g}^A + \mathbf{g}^{med}$. The use of the

---

† A similar situation also occurs in the case of sound quanta — phonons. The propagation of sound is not accompanied by a transport of mass and in that connection the momentum of sound waves vanishes (we neglect here a relativistic effect — the fact that a train of source waves with energy $\mathcal{H}$ has a mass $\mathcal{H}/c^2$ and thus possesses a momentum $(\mathcal{H}/c^2)v_s$, where v_s is the sound velocity). After quantization we find thus that the sound quanta (phonons) have an energy $\hbar\omega$ and a zero momentum (we neglect again the momentum $(\hbar\omega/c^2)v_s$). The statement, however, that the momentum of a phonon (say, when it is emitted by an electron) equals $\hbar\mathbf{k} = (\hbar\omega/v_s)(\mathbf{k}/k)$ means, indeed, that when a phonon is emitted the lattice as a whole obtains a momentum $\hbar\mathbf{k}$ (Umklapp processes are neglected here). When we apply the conservation laws in the case of emission, absorption, or scattering of sound, nothing is changed, however, if we assume as is usually done, that the phonons not only have an energy $\hbar\omega$, but also a momentum $\hbar\mathbf{k} = (\hbar\omega/v_s)(\mathbf{k}/k)$. Incidentally, the field momentum according to Abraham $\mathbf{G}^{(A)} = \mathcal{H}/cn = (\mathcal{H}/c^2)(c/n)$ (see (12.18)), that is, it has the same meaning as the 'true' phonon momentum $(\mathcal{H}/c^2)v_s$, as the velocity of the electromagnetic pulse is equal to c/n (we neglect dispersion).

Minkovski tensor in this case is thereby in reality justified as it not only leads to the correct result, but also leads directly to the goal without considering the action of the volume force. It is true that taking the action of this force into account in the framework of the classical approach is always simple (vide infra) but it turned out that apparently it is rather cumbersome quantum-mechanically. As far as we know such a quantum-mechanical consideration has not yet been given. For those non-stationary problems for the solution of which there are clear advantages or even the necessity to apply the Abraham tensor, the corresponding quantal analysis would be justified (although, of course, not necessary as long as the problem is classical which is probably the case for any realistic situation when one considers the problem of measuring the Abraham force). As to the use of the energy and momentum conservation laws dicussed above (and particularly in Chapter 7) in the emission of 'photons in a medium' it seems to us that the problem of the nature and meaning of such a discussion may be assumed to be completely clear already in the light of the remarks made here (it is also perhaps more convenient to turn to an earlier paper (Ginzburg, 1973c) where the quantization of the field in a medium and the problem of the energy-momentum tensor are discussed in one spot rather than to Chapters 6 and 7).

We discussed above the problems of the energy-momentum tensor and the forces in a medium for the simplest case. In particular, we assumed the medium to be at rest, uniform, and unmagnetized; we also neglected the possibility that the dielectric permittivity ε might depend on the density of the medium ρ.

Let us now drop all these assumptions and write the field equations in the form

$$\operatorname{curl}\mathbf{H} = \frac{4\pi}{c}\mathbf{j} + \frac{1}{c}\frac{\partial\mathbf{D}}{\partial t}\,, \tag{12.23}$$

$$\operatorname{curl}\mathbf{E} = -\frac{1}{c}\frac{\partial\mathbf{B}}{\partial t}\,, \tag{12.24}$$

$$\operatorname{div}\mathbf{D} = 4\pi\rho_e\,, \tag{12.25}$$

$$\operatorname{div}\mathbf{B} = 0\,. \tag{12.26}$$

We have here denoted the charge density by ρ_e, to save the notation ρ for the density of the medium. If in the frame of reference in which the medium is at rest $\mathbf{D} = \varepsilon\mathbf{E}$, $\mathbf{B} = \mu\mathbf{H}$ (as before we neglect dispersion), we get for a slowly moving medium (see Tamm, 1976, § 111)

$$\mathbf{D} = \varepsilon\mathbf{E} + \left(\varepsilon - \frac{1}{\mu}\right)\left[\frac{\mathbf{u}}{c}\wedge\mathbf{B}\right]\,, \quad \mathbf{E} = \frac{\mathbf{D}}{\varepsilon} - \left(1 - \frac{1}{\varepsilon\mu}\right)\left[\frac{\mathbf{u}}{c}\wedge\mathbf{B}\right]\,, \tag{12.27}$$

$$\mathbf{H} = \frac{1}{\mu}\mathbf{B} + \left(\varepsilon - \frac{1}{\mu}\right)\left[\frac{\mathbf{u}}{c} \wedge \mathbf{E}\right] = \frac{1}{\mu}\mathbf{B} + \left(1 - \frac{1}{\varepsilon\mu}\right)\left[\frac{\mathbf{u}}{c} \wedge \mathbf{D}\right], \tag{12.27}$$

where we have assumed that the velocity of the medium $\mathbf{u}$ with respect to the 'laboratory' frame of reference is small, that is, we have neglected terms of order u^2/c^2. However, even when we use more general relations between $\mathbf{D}, \mathbf{B}$, $\mathbf{H}$ and $\mathbf{E}$ the initial Eqns.(12.23) to (12.26) also remain valid.

Taking the scalar product of (12.23) and (12.24) with $\mathbf{E}$ and $\mathbf{H}$, respectively, and then proceeding as usual (see (12.4)) we get the relation

$$\frac{1}{4\pi}\left\{\left(\frac{\partial \mathbf{D}}{\partial t} \cdot \mathbf{E}\right) + \left(\frac{\partial \mathbf{B}}{\partial t} \cdot \mathbf{H}\right)\right\} = -(\mathbf{j} \cdot \mathbf{E}) - \operatorname{div} \mathbf{S}, \quad \mathbf{S} = \frac{c}{4\pi}[\mathbf{E} \wedge \mathbf{H}]. \tag{12.28}$$

Taking the vector product of (12.23) and (12.24) with $\mathbf{B}$ and $\mathbf{D}$, respectively, and proceeding then as we did to get Eqn.(12.5) we find

$$\frac{1}{4\pi}\left\{[\mathbf{D} \wedge \operatorname{curl} \mathbf{E}] + [\mathbf{B} \wedge \operatorname{curl} \mathbf{H}] - \mathbf{E} \operatorname{div} \mathbf{D}\right\} = -\mathbf{f}^L - \frac{1}{4\pi c}\frac{\partial}{\partial t}[\mathbf{D} \wedge \mathbf{B}], \tag{12.29}$$
$$\mathbf{f}^L = \rho_e \mathbf{E} + \frac{1}{c}[\mathbf{j} \wedge \mathbf{B}].$$

Equations (12.28) and (12.29) are generalizations of (12.4) and (12.5) and represent the energy and momentum conservation laws following from the field Eqns.(12.23) to (12.26). To be more precise, we are dealing with conservation laws which are related to the energy and momentum conservation laws. However, the clarification of this relation requires an additional analysis and additional assumptions. Indeed, this was already made clear above — when there are a number of terms present in a conservation law it is impossible without further assumptions to interpret this one or that one from them. At the same time when we make the problem more precise the use of the conservation laws enables us, of course, to obtain useful results. As an example we shall use Eqn.(12.28) to find an expression for the force density $\mathbf{f}_m$ which acts upon the medium considered. To do this we evaluate the derivative

$$\frac{\partial}{\partial t}\frac{(\mathbf{D} \cdot \mathbf{E}) + (\mathbf{B} \cdot \mathbf{H})}{8\pi} \equiv \frac{\partial w^M}{\partial t}, \tag{12.30}$$

where $w^M = [(\mathbf{D} \cdot \mathbf{E}) + (\mathbf{B} \cdot \mathbf{H})]/8\pi$ can so far only be considered to be a particular notation.

We use Eqns.(12.27) to evaluate the derivatives $\partial \mathbf{E}/\partial t$ and $\partial \mathbf{H}/\partial t$. In that case we must also somehow make precise the values $\partial \varepsilon/\partial t$ and $\partial \mu/\partial t$. We shall assume that in each element of the medium ε can change only because the density ρ changes. In that case

$$\frac{d\boldsymbol{\varepsilon}}{dt} \equiv \frac{\partial\boldsymbol{\varepsilon}}{\partial t} + (\mathbf{u}\cdot\nabla\boldsymbol{\varepsilon}) = \frac{\partial\boldsymbol{\varepsilon}}{\partial\rho}\frac{d\rho}{dt} = -\frac{\partial\boldsymbol{\varepsilon}}{\partial\rho}\,\rho\,\mathrm{div}\,\mathbf{u}\ , \tag{12.31}$$

where we have used also the continuity equation

$$\frac{\partial\rho}{\partial t} + \mathrm{div}\,\rho\,\mathbf{u} = 0.$$

By virtue of (12.27) and (12.31) and the analogous expression for $d\mu/dt$ we find easily

$$\begin{aligned}\frac{\partial\mathbf{E}}{\partial t} &= \frac{1}{\boldsymbol{\varepsilon}}\frac{\partial\mathbf{D}}{\partial t} - \left(1-\frac{1}{\boldsymbol{\varepsilon}\mu}\right)\left[\frac{\mathrm{u}}{\mathrm{c}}\wedge\frac{\partial\mathbf{B}}{\partial t}\right] - \mathbf{D}\left\{\left(\mathbf{u}\cdot\nabla\frac{1}{\boldsymbol{\varepsilon}}\right) - \frac{1}{\boldsymbol{\varepsilon}^2}\left(\frac{\partial\boldsymbol{\varepsilon}}{\partial\rho}\rho\right)\mathrm{div}\,\mathbf{u}\right\},\\ \frac{\partial\mathbf{H}}{\partial t} &= \frac{1}{\mu}\frac{\partial\mathbf{B}}{\partial t} + \left(1-\frac{1}{\boldsymbol{\varepsilon}\mu}\right)\left[\frac{\mathbf{u}}{\mathrm{c}}\wedge\frac{\partial\mathbf{D}}{\partial t}\right] - \mathbf{B}\left\{\left(\mathbf{u}\cdot\nabla\frac{1}{\mu}\right) - \frac{1}{\mu^2}\left(\frac{\partial\mu}{\partial\rho}\rho\right)\mathrm{div}\,\mathbf{u}\right\}.\end{aligned} \tag{12.32}$$

Here, as everywhere else, we assumed the velocity $\mathbf{u}$ to be constant or, to be more precise, we neglected all derivatives of $\mathbf{u}$ with respect to the time or coordinates, but took into account the divergence $\mathrm{div}\,\mathbf{u}$ which arises when we use the continuity equation. By virtue of (12.32) and again using the relation (12.27) and neglecting terms of order u^2/c^2 we find

$$\begin{aligned}\frac{\partial w^M}{\partial t} &= \frac{1}{4\pi}\left\{\left(\frac{\partial\mathbf{D}}{\partial t}\cdot\mathbf{E}\right)+\left(\frac{\partial\mathbf{B}}{\partial t}\cdot\mathbf{H}\right)\right\} + \frac{1}{8\pi}\,(\mathbf{u}\cdot\nabla\boldsymbol{\varepsilon})\,E^2 + \\ &+ \frac{1}{8\pi}\left(\frac{\partial\boldsymbol{\varepsilon}}{\partial\rho}\rho\right)E^2\,\mathrm{div}\,\mathbf{u} + \frac{1}{8\pi}(\mathbf{u}\cdot\nabla\mu)\,H^2 + \frac{1}{8\pi}\left(\frac{\partial\mu}{\partial\rho}\rho\right)H^2\,\mathrm{div}\,\mathbf{u}\ .\end{aligned} \tag{12.33}$$

Combining (12.33) and (12.28) we finally find

$$-\frac{\partial w^M}{\partial t} = (\mathbf{j}\cdot\mathbf{E}) + (\mathbf{f}_m\cdot\mathbf{u}) + \mathrm{div}\left\{\mathbf{S} - \frac{\mathbf{u}}{8\pi}\left[\left(\frac{\partial\varepsilon}{\partial\rho}\rho\right)E^2 + \left(\frac{\partial\mu}{\partial\rho}\rho\right)H^2\right]\right\}\ , \tag{12.34}$$

$$\mathbf{f}_m = -\frac{E^2}{8\pi}\nabla\varepsilon - \frac{H^2}{8\pi}\nabla\mu + \frac{1}{8\pi}\nabla\left\{\left(\frac{\partial\varepsilon}{\partial\rho}\rho\right)E^2\right\} + \frac{1}{8\pi}\nabla\left\{\left(\frac{\partial\mu}{\partial\rho}\rho\right)H^2\right\}\ . \tag{12.35}$$

It is natural to interpret this relation as the energy conservation law where w^M is the energy density of the field and $\mathbf{f}_m$ the force acting upon the medium (the work done by it is $(\mathbf{f}_m\cdot\mathbf{u})$); the addition to the energy flux of a term proportional to $\mathbf{u}$ cannot cause surprise, but the problem of the accuracy of the corresponding expression needs further analysis. We shall not be interested in this problem as the aim of the derivation was to obtain an expression for the force $\mathbf{f}_m$ which acts upon a medium at rest when $\mathbf{u} = 0$. However, it is not possible to put $\mathbf{u} = 0$ at once, as in that case the work done by the force $(\mathbf{f}_m\cdot\mathbf{u})$ also vanishes. Equation (12.35) is, of course, the same as the one normally obtained (see Landau and Lifshitz, 1960; Tamm, 1976), but as the result of considering displacements of elements of the medium in the field. Such a derivation is in general equivalent to the one given but it refers immediately only to the static case.

One should not think, although this might look so at first sight, that the given derivation determines the density of the force acting upon the medium unqiuely. Indeed, let us take for a moving medium as energy density and momentum density the following expressions:

$$w^A = \frac{1}{8\pi}\left\{(\mathbf{D}\cdot\mathbf{E}) + (\mathbf{B}\cdot\mathbf{H})\right\} - \left(\frac{\mathbf{u}/c}{4\pi(1-u^2/c^2)}\cdot\left\{[\mathbf{D}\wedge\mathbf{B}] - [\mathbf{E}\wedge\mathbf{H}]\right\}\right), \qquad (12.36)$$

$$\mathbf{g}^A = \frac{\mathbf{S}^A}{c^2} = \frac{1}{4\pi c}\left\{[\mathbf{E}\wedge\mathbf{H}] - \frac{\mathbf{u}/c^2}{1-(u^2/c^2)}\left\{(\mathbf{u}\cdot[\mathbf{D}\wedge\mathbf{B}]) - (\mathbf{u}\cdot[\mathbf{E}\wedge\mathbf{H}])\right\}\right\}. \qquad (12.37)$$

It is just this kind of expression which is obtained (see Pauli, 1958; Skobel'tsyn, 1973) for a moving medium through relativistic transformations if we take for the energy-momentum tensor in the medium at rest the Abraham expression (12.10).

It is clear that

$$\frac{\partial w^M}{\partial t} + (\mathbf{f}_m\cdot\mathbf{u}) = \frac{\partial w^A}{\partial t} + (\mathbf{f}_m\cdot\mathbf{u}) + (\mathbf{f}^A\cdot\mathbf{u}), \qquad (12.38)$$

$$\mathbf{f}^A = \frac{\partial}{\partial t}(\mathbf{g}^M - \mathbf{g}^A) = \frac{1}{4\pi c}\frac{\partial}{\partial t}\left\{[\mathbf{D}\wedge\mathbf{B}] - [\mathbf{E}\wedge\mathbf{H}] + \frac{\mathbf{u}\{(\mathbf{u}\cdot[\mathbf{D}\wedge\mathbf{B}]) - (\mathbf{u}\cdot[\mathbf{E}\wedge\mathbf{H}])\}}{c^2\{1-(u^2/c^2)\}}\right\} \qquad (12.39)$$

If we are, therefore, speaking only about the satisfying of the conservation law (12.28), we can equally well take the volume force density to be equal to $\mathbf{f}_m$ as to $\mathbf{f}_m + \mathbf{f}^A$. The same holds also in respect of the choice of an expression for the momentum density on the basis of the conservation law (12.29); both the Minkovski expression $\mathbf{g}^M = [\mathbf{D}\wedge\mathbf{B}]/4\pi c$ and the Abraham expression (12.37) are compatible with (12.29). The only difference consists in that in the Abraham version apart from other forces there acts on the medium also a force with density $\mathbf{f}^A$, which is not present in the Minkovski version, so to say, due to a corresponding change in the field momentum density in the medium The problem about whether the force $\mathbf{f}^A$, which for an isotropic and unmagnetized medium and when $u^2/c^2 \ll 1$ has the form (12.6), is real must be answered experimentally or on the basis of a microscopic theory. We have already mentioned that there is no reason to doubt the presence of the force $\mathbf{f}^A$ from either point of view.

In conclusion we must note that also in the general case the relation $\mathbf{G}^M \equiv \int \mathbf{g}^M d^3\mathbf{r} = \mathbf{G}^A + \mathbf{F}^A$ (see (12.22) is retained, that is, the total momentum transferred to the medium by a radiator in it can be evaluated using the Minkovski expression for the momentum density $\mathbf{g}^M$ (for details see Ginzburg and Ugarov, 1976, 1977). As in the earlier considered particular case of an isotropic and unmagnetized medium at rest this conclusion does, clearly, not contradict at all the acknowledgement of the reality of the Abraham force $\mathbf{f}^A$.

Chapter XIII

FLUCTUATIONS AND VAN DER WAALS FORCES

Fluctuations in an electric circuit. Thermal radiation in a medium. Molecular (van der Waals) forces between macroscopic bodies. Interaction between electrons and the field in an empty resonator.

When we expounded the electrodynamics of continuous media in Chapter 10 we emphasized that the fields considered are statistically averaged fields. Automatically we excluded thereby fluctuation effects. At the same time it is well known that various fluctuations and, in particular, electromagnetic fluctuations and effects connected with it play a very large role in physics and astrophysics. It is sufficient to remind ourselves of fluctuations in electric circuits, of fluctuations of the electromagnetic field in resonators (both empty ones, and ones filled with a medium), and of molecular (van der Waals) forces between condensed bodies (the calculation of these forces is closely connected with the problem of electromagnetic fluctuations). The scattering of electromagnetic waves (radio-waves, light, X-rays) in a medium is also a fluctuation effect — one may say that one is dealing with the scattering by the fluctuations of the dielectric permittivity tensor $\varepsilon_{ij}(\omega, \mathbf{k})$. It is thus perfectly natural in the concept of this book to elucidate in the present and the following chapter a few problems connected with fluctuations and with the scattering of waves. It is, however, necessary to emphasize that we shall not concentrate on the general problems from the theory of electromagnetic fluctuations (see Landau and Lifshitz, 1960, 1969; Abrikosov, Gor'kov and Dzyaloshinskii, 1965; Sitenko, 1967, 1977; Rytov, 1966; Fain, 1972) but only on a few special cases, which are very interesting from a physical point of view.

The first of these problems is that of fluctuations in a linear electric circuit with lumped capacitance C, self-induction L, and resistance R (Fig. 13.1). We assume the size of the circuit ℓ to be very small as compared to the wavelength $\lambda = 2\pi c/\omega$

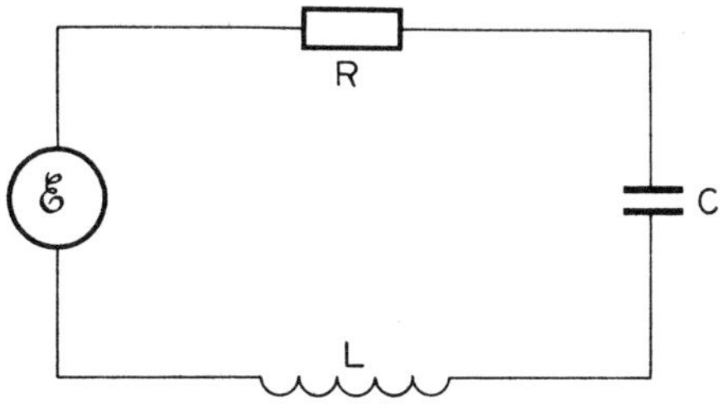

Fig. 13.1 Electric LCR-circuit

corresponding to the frequency range ω considered. Under such conditions the quasi-stationary approximation (see, for instance, Landau and Lifshitz, 1960; Tamm, 1976) is valid for the fields and currents in the circuit; in the framework of this approximation the current is the same in all parts of the circuit and equal to

$$J(t) = \int_{-\infty}^{+\infty} J_\omega e^{-i\omega t} d\omega .$$

If the quantities C, L, and R are independent of the frequency, the current will be given by the equation

$$L\ddot{q} + R\dot{q} + \frac{q}{C} = L\dot{J} + RJ + \frac{1}{C}\int J\,dt = \mathcal{E}(t) , \qquad (13.1)$$

where $q = \int J\,dt$ is the charge on the capacitor and $\mathcal{E}(t)$ the electromotive force (e.m.f.) imposed on (or, rather, included in) the circuit.† If at least one of the quantities L, C, or R is frequency-dependent (a direct analogue to temporal dispersion in a medium) Eqn. (13.1) holds only for the Fourier components:

$$\left. \begin{aligned} &\left\{- i\omega L(\omega) + R(\omega) - \frac{1}{i\omega C(\omega)}\right\} J_\omega = \mathcal{E}_\omega , \\ &J_\omega = \frac{1}{2\pi}\int_{-\infty}^{+\infty} J(t)\, e^{i\omega t} dt , \\ &\mathcal{E}_\omega = \frac{1}{2\pi}\int_{-\infty}^{+\infty} \mathcal{E}(t)\, e^{i\omega t} dt. \end{aligned} \right\} \qquad (13.2)$$

If we introduce the impedance (complex resistivity) of the circuit

$$Z(\omega) = R - i\left(\omega L - \frac{1}{\omega C}\right) , \qquad (13.3)$$

we have from (13.2)

$$\mathcal{E}_\omega = Z(\omega)\, J_\omega . \qquad (13.4)$$

If there is no external e.m.f. the statistical average of the current J is, of course, zero. At the same time it is clear that in the circuit as an effect or, rather, as the result of thermal motions there appear and disappear all the time fluctuating currents. As usual in such cases such currents can be characterized by a correlation function

$$\phi(t' - t) = \phi(\tau) = \overline{J(t)\, J(t + \tau)} \qquad (13.5)$$

where we have assumed uniformity in time (hence the dependence on $\tau = t' = t$ only; the circuit would be non-uniform in time, if its parameters were time-dependent) and the bar indicates a statistical and, if needs be, a quantum-mechanical average (for details see Landau and Lifshitz, 1960, 1969; for the sake of simplicity we write equations such as (13.5) only down in their classical case). For random (fluctuating) currents J which do not tend to zero as $|t| \to \infty$, one must exercize care in using Fourier components, but for the square integrable quantities used in what follows it is unnecessary to take such precautions (Landau and Lifshitz, 1969; Rytov, 1966). Therefore, substituting the expression

$$J(t) = \int J_\omega e^{-i\omega t} \, d\omega$$

into (13.5) we get

$$\phi(\tau) = \iint_{-\infty}^{+\infty} \overline{J_\omega J_{\omega'}} e^{-i\omega t - i\omega' t'} \, d\omega \, d\omega' .$$

However, the right-hand side of this formula depends on τ only if there is an appropriate δ-function present and, hence, we can write

$$\phi(\tau) = \int_{-\infty}^{+\infty} (J^2)_\omega \, e^{-i\omega\tau} \, d\omega \ , \quad (J^2)_\omega = \frac{1}{2\pi} \int_{-\infty}^{+\infty} \phi(\tau) \, e^{i\omega\tau} \, d\tau \ , \tag{13.6}$$

where the quantity $(J^2)_\omega$ is defined as follows:

$$\overline{J_\omega J_{\omega'}} = (J^2)_\omega \, \delta(\omega + \omega') \ .$$

For the mean square of the current we have

$$\overline{J^2} = \phi(0) = \int_{-\infty}^{+\infty} (J^2)_\omega \, d\omega = 2 \int_0^{\infty} (J^2)_\omega \, d\omega \ . \tag{13.7}$$

By measuring the strength of the fluctuating current $J(t)$ in the circuit we can find $\phi(\tau)$ and, hence, the spectral density of the mean square fluctuations $(J^2)_\omega$. For a circuit in thermodynamic equilibrium (at temperature T) the quantity $(J^2)_\omega$ follows from theory — obtaining the appropriate expression is now our aim. Instead of $(J^2)_\omega$ we can with equal success look for the spectral density of the mean square 'random' e.m.f.

$$(\mathcal{E}^2)_\omega = |Z(\omega)|^2 \, (J^2)_\omega \ . \tag{13.8}$$

Indeed, in a thermodynamic context fluctuating currents in a circuit do not differ at all from currents flowing under the influence of an 'external' e.m.f.. Therefore, if we use the relation (13.4), where according to (13.3) $Z(-\omega) = Z^*(\omega)$, we are led to (13.8).

The value of $(\mathcal{E}^2)_\omega$ in thermodynamic equilibrium can at once be obtained by using the general so-called fluctuation-dissipation theorem which connects the fluctuations (for instance, the quantities $\overline{J^2}$ and $(J^2)_\omega$) with the dissipative properties of the system (in the case of a circuit — with its resistance R). We shall not add here to the derivations given in the textbooks (Landau and Lifshitz, 1960, 1969; Rytov, 1966). However, an elementary discussion is sufficient; this is given below and it leads not only at once to its goal but also elucidates the essential features of the whole physical contents of the fluctuation-dissipation theorem. Furthermore, departing from the circuit we obtain in fact appreciably more general results.

Let us therefore turn to obtaining an expression for $(\mathcal{E}^2)_\omega$ in an equilibrium circuit. We start from the statement that for any circuit

$$(\mathcal{E}^2)_\omega = R(\omega)\, f(\omega, T) \ , \tag{13.9}$$

where $f(\omega, t)$ is a universal function of the frequency ω and the temperature T, that is, a function which is independent of the circuit parameters L, C, and R.

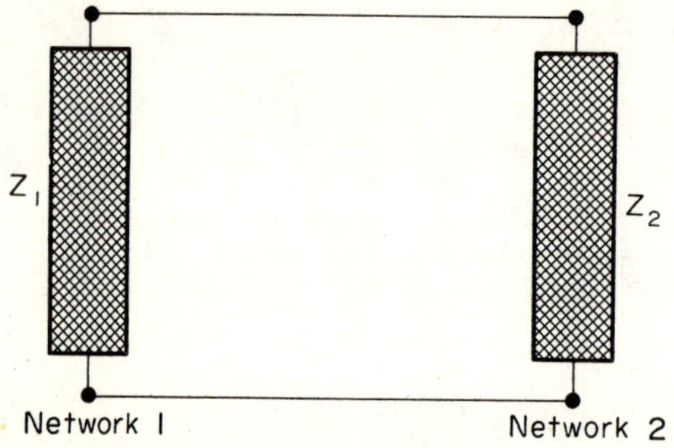

Fig. 13.2 Circuit consisting of two networks in series with impedances Z_1 and Z_2

To prove relation (13.9) we consider two circuits in series — a network forming a closed circuit which is schematically shown in Fig. 13.2 (see, for instance, Gorelik, 1951; Ginzburg, 1952a); the idea of the proof goes in fact back to Nyquist (1928) who considered a two-wire line). As a result of fluctuations in the whole of the circuit there flows a current $J(t)$ and the fluctuating e.m.f. in each of the networks is equal to, respectively, $\mathcal{E}_1 = Z_1 J$ and $\mathcal{E}_2 = Z_2 J$, where $Z_{1,2}$ are the impedances of the networks. In thermodynamic equilibrium the average power P_{12} transferred from network 1 to network 2 must equal the power P_{21} transferred from network 2 to network 1, that is,

$$P_{12} = \int \frac{R_2(\mathcal{E}_1^2)_\omega \, d\omega}{|Z|^2} = P_{21} = \int \frac{R_1(\mathcal{E}_2^2)_\omega \, d\omega}{|Z|^2} \ , \tag{13.10}$$

Here $Z(\omega) = Z_1 + Z_2$ is the impedance of the circuit considered which consists of two networks in series; one easily finds the expressions used for the power using the well known relation $P = \overline{RJ^2}$. Equation (13.10) must hold for any networks, in particular, when for fixed R_1 and R_2, but varying self-inductions $L_{1,2}$ and capacitances $C_{1,2}$ the impedance Z of the whole circuit is changed. This is, however, possible only when the integrands are equal, that is, when $R_2(\mathcal{E}_1^2)_\omega = R_1(\mathcal{E}_2^2)_\omega$. Thus

$$\frac{(\mathcal{E}_1^2)_\omega}{R_1} = \frac{(\mathcal{E}_2^2)_\omega}{R_2} = \frac{(\mathcal{E}^2)_\omega}{R} = f(\omega\,,\,T)$$

is a quantity which may depend on ω and T, but not on the circuit parameters (see Gorelik, 1951 for more details). To find the function $f(\omega,T)$ we now choose any circuit and, of course, the simplest one is a weakly damped LC-circuit with $R\to 0$.

Beforehand we write down an expression for the average electric and magnetic energies $\bar{U}$ and $\bar{K}$ in a circuit with arbitrary L, C, and R:

$$\bar{U} = \frac{\overline{q^2}}{2C} = \frac{1}{2C}\int_{-\infty}^{+\infty} \frac{1}{\omega^2}(J^2)_\omega\, d\omega =$$

$$= \frac{1}{C}\int_0^\infty \frac{(\mathcal{E}^2)_\omega\, d\omega}{\omega^2|Z(\omega)|^2} = \int_0^\infty \frac{CRf(\omega\,,\,T)\,d\omega}{R^2C^2\omega^2 + (LC\omega^2-1)^2}\,, \tag{13.11}$$

$$\bar{K} = \frac{\overline{LJ^2}}{2} = \tfrac{1}{2}L\int_{-\infty}^{+\infty}(J^2)_\omega\, d\omega = \int_0^\infty \frac{\omega^2C^2LRf(\omega\,,\,T)\,d\omega}{R^2C^2\omega^2+(LC\omega^2-1)^2}\,. \tag{13.12}$$

If $R\to 0$, the resulting circuit is completely analogous to an undamped harmonic oscillator described by the equation $m\ddot{x}+kx=0$ with eigenfrequency $\omega_i = \sqrt{k/m} = 1/\sqrt{LC}$ (see (13.1)). However, it is well known that for an oscillator its average energy at a temperature T is given by (it is unnecessary to elucidate this)

$$\bar{W} = \bar{U}+\bar{K} = 2\bar{U} = 2\bar{K} = \overline{\tfrac{1}{2}m\dot{x}^2} + \overline{\tfrac{1}{2}kx^2} = \frac{\overline{q^2}}{2C} + \overline{\tfrac{1}{2}LJ^2} =$$

$$= \tfrac{1}{2}\hbar\omega_i + \frac{\hbar\omega_i}{\exp[\hbar\omega/k_BT]-1} = \tfrac{1}{2}\hbar\omega_i \coth\frac{\hbar\omega_i}{2k_BT}\,. \tag{13.13}$$

Of course, the internal energy of the resistance R itself has here not been taken into consideration.

We note further that for small R the integrals (13.11) and (13.12) have a steep maximum at $\omega=\omega_i=1/\sqrt{LC}$ and we can therefore put (strictly speaking this is valid as $R\to 0$; $\alpha = CR/\sqrt{LC}$)

$$\bar{U} = f(\omega,T)\int_0^\infty \frac{\alpha\, d\eta}{\alpha^2\eta^2+(\eta^2-1)^2} = \bar{K} = f(\omega,T)\int_0^\infty \frac{\alpha\eta^2\,d\eta}{\alpha^2\eta^2+(\eta^2-1)^2} = \tfrac{1}{2}\pi f(\omega,T)\,. \tag{13.14}$$

The integrals occurring here are evaluated exactly (most simply using the residue theorem), but to obtain the result it is sufficient to take into account that as $\alpha\to 0$, both integrals reduce to $\int_0^\infty \alpha\, d\eta/[\alpha^2+4(\eta-1)^2] \approx \tfrac{1}{2}\pi$. Comparing

(13.13) and (13.14) and bearing in mind that the frequency ω_i is arbitrary, we find for $f(\omega, T)$

$$f(\omega,T) = \frac{\hbar\omega}{2\pi} \coth \frac{\hbar\omega}{2k_BT} . \tag{13.15}$$

Finally, we get from (13.9) Nyquist's formula

$$(\mathcal{E}^2)_\omega = \frac{\hbar\omega}{2\pi} R(\omega) \coth \frac{\hbar\omega}{2k_BT} = \frac{R}{\pi}\left(\frac{1}{2}\hbar\omega + \frac{\hbar\omega}{\exp(\hbar\omega/k_BT) - 1}\right),$$

$$\mathcal{E}^2 = \frac{\hbar}{\pi}\int_0^\infty R(\omega)\,\omega \coth \frac{\hbar\omega}{2k_BT}\, d\omega . \tag{13.16}$$

In the classical case when $\hbar\omega \ll k_BT$ we have

$$(\mathcal{E}^2)_\omega = \frac{R}{\pi} k_BT , \quad \hbar\omega \ll k_BT . \tag{13.17}$$

The derivation given here is completely rigorous or, at any rate, not less conclusive that the derivation based upon the general fluctuation-dissipation theorem. Moreover, one can consider it as a derivation of the theorem itself which for the case of a random variable x has the form (see, for instance, Landau and Lifshitz, 1960; Abrikosov, Gor'kov, and Dzyaloshinskii, 1965)

$$(x^2)_\omega = \frac{\hbar\alpha''}{2\pi} \coth \frac{\hbar\omega}{2k_BT} , \tag{13.18a}$$

where α'' is the imaginary part of the quantity α which determines the response of the system to an external perturbation and to the same extent to a random (fluctuating) perturbation Q. To be precise

$$x_\omega = \alpha(\omega)\, Q_\omega$$

and

$$\overline{x_\omega x_{\omega'}} = \alpha(\omega)\,\alpha(\omega')\,\overline{Q_\omega Q_{\omega'}} = (x^2)_\omega\, \delta(\omega+\omega') = |\alpha|^2 (Q^2)_\omega\, \delta(\omega+\omega') .$$

Comparing this expression with (13.18a) we find

$$(Q^2)_\omega = \frac{\hbar\alpha''}{2\pi|\alpha|^2} \coth \frac{\hbar\omega}{2k_BT} . \tag{13.18b}$$

However, clearly Eqn.(13.18b) is equivalent to (13.16), if $\alpha = i\omega/Z(\omega)$ and, hence, $\alpha'' = \mathrm{Im}\,\alpha = \omega R/|Z|^2$. One can see easily and directly (see, for instance, Landau and Lifshitz, 1960, § 89) that for an electric circuit α has just this meaning. It is clear from what has been said that, the other way round, we can start from Nyquist's Eqn.(13.16) and at the same time making the meaning of the parameter α and the quantities x and Q more precise obtain

Eqn.(13.18) for a rather wide class of random physical quantities.

We now make a few remarks referring to the properties of an LCR circuit. In the classical limit $\hbar\omega \ll k_B T$ the function $f(\omega, T) = k_B T/\pi$, and it is clear from (13.11), (13.12), and (13.14) that we can for any values of L, C, and R write

$$\bar{U} = \bar{K} = \tfrac{1}{2} k_B T \, . \tag{13.19}$$

Of course, we should have expected such a result, as in the present case it is equivalent to the statistical theorem about the equipartition of energy over degrees of freedom. It is true that we have assumed above that L, C, and R are independent of ω but in the opposite case it is, in general. difficult to speak of a circuit with a single degree of freedom and, in any case, the classical equipartition law does not have to be satisfied.

It is important that in the quantum-mechanical case equipartition does, in general, not hold, not even when L, C, and R are constant. Of course, as $R \to 0$, Eqns.(13.13) hold and, in particular, $\bar{U} = \bar{K}$. In general, as a first approximation Eqns.(13.13) are valid provided

$$\frac{R}{L} \ll \frac{1}{\sqrt{LC}} \, , \tag{13.20}$$

which guarantees that the damping is weak.† If, however, inequality (13.20) does not hold, we have $\bar{U} \neq \bar{K}$, and we can use (13.11), (13.12), and (13.15) to obtain general expressions for $\bar{U}$ and $\bar{K}$. We see that $\bar{U}$ and $\bar{K}$ depend on two parameters, namely $\alpha = CR/\sqrt{LC}$ and $\beta = \hbar/k_B T \sqrt{LC}$.

Condition (13.20) is equivalent to $\alpha \ll 1$ and then $\beta = \hbar\omega_i/k_B T$, where $\omega_i = 1/\sqrt{LC}$ is the frequency of the circuit. Let, for example, the circuit be strongly damped (for details see Ginzburg, 1952a), that is,

$$\frac{R}{L} \gg \frac{1}{\sqrt{LC}} \, , \tag{13.21}$$

or, what comes to the same, $\alpha \gg 1$. In that case

$$\left.\begin{aligned} \bar{U} &= \tfrac{1}{2} k_B T \, , \quad \text{when} \quad \frac{\hbar}{RC} \ll k_B T \, , \text{ that is, when } \beta \ll \alpha \, ; \\ \bar{U} &= \frac{\pi}{6} \frac{k_B T}{\hbar/RC} k_B T \, , \quad \text{when} \quad \frac{\hbar}{RC} \gg k_B T \end{aligned}\right\} \tag{13.22}$$

---

† The eigenfrequency of the LCR-circuit follows from Eqn.(13.1) with $\mathcal{E}(t) = 0$ and equals

$$\omega_i = -i \frac{R}{2L} \pm \left[\frac{1}{LC} - \left(\frac{R}{2L}\right)^2\right]^{\frac{1}{2}} \, . \tag{13.21}$$

$$\bar{K} = \tfrac{1}{2}k_BT\,, \quad \text{when} \quad \frac{\hbar R}{L} \ll k_BT\,, \quad \text{that is, when} \quad \beta \ll \frac{1}{\alpha}\,;$$

$$\bar{K} = \frac{\pi}{6}\,\frac{k_BT}{\hbar R/L}\,k_BT\,, \quad \text{when} \quad \frac{\hbar R}{L} \gg k_BT\,. \qquad (13.23)$$

The condition for 'classicism' is thus completely different for the electric energy $\bar{U}$ from the corresponding condition for the magnetic energy $\bar{K}$, and so on. It is also interesting that a circuit far from classical behaviour for small R (the condition $\hbar\omega_i = \hbar/\sqrt{LC} \gtrsim k_BT$ being satisfied) for sufficiently increasing R and unvarying L and C becomes classical as far as $\bar{U}$ is concerned (that is, $\bar{U} \to \frac{1}{2}k_BT$) whereas $\bar{K} \to 0$. We only need remind ourselves that the requirement of quasi-stationarity of the circuit, which we started from, imposes well known limitations on the quantities L, C, and R (in particular, it is impossible simply to let $R \to \infty$ as in that case the circuit turns out to be open and formally $(\mathcal{E}^2)_\omega \to \infty$, although $(J^2)_\omega \to 0$). We do not know, however, whether such limitations affect the analysis of the corresponding actual problems at all or seriously interfere with the transition to the limiting cases $\hbar/RC \ll k_BT$ and $\hbar R/L \gg k_BT$ (see (13.23)).

The cause of the above-noted situation — the different behaviour of $\bar{U}$ and $\bar{K}$ — is completely obvious. A classical system may retain its characteristic features even for strong damping — it 'remains itself'. For instance, the oscillations of a pendulum (oscillator) are very strongly damped when the viscosity of the medium surrounding it increases, but it remains all the same a pendulum. If, on the other hand, we have a quantum harmonic oscillator with frequency ω_i, this system has, when there is no damping, energy levels which lie at a distance $\hbar\omega_i$ from one another. When the damping increases (as the result, say, of collisions or of interaction with radiation) the levels spread out and finally overlap. It is also perfectly obvious that when the overlap of the levels is large the system possesses a clearly pronounced continuous spectrum and has little in common with the quantum harmonic oscillator. Also, in various quantum systems, depending on the nature of their energy spectrum, the average total energy and the average kinetic or potential energy are, in general, completely different.

The harmonic oscillator plays in physics an exceptionally large role by no means solely because such a system is often met with (pendulum, molecular oscillations, and so on). It is even more important that an extra-ordinarily wide class of problems, connected with the discussion of small (linear) perturbations and waves in media with 'distributed constants', that is, in tne electrodynamics of continuous media, in acoustics, and so on, reduces to some extent to the oscillator problem. This is also true of the electromagnetic field in vacuo — the Hamiltonian method expounded in Chapters 1 and 6 for electrodynamics in vacuo or in a medium is a clear illustration of this. Besides, the expansion into waves, by no means only plane waves, goes beyond the limits of the Hamiltonian method and has, as we have said, an extremely wide range of applications. It is in this connection at once clear that the study of electrical fluctuations in an electric circuit which was made above can be generalized not only to discrete systems (mechanical oscillator, discrete chains, and so on) but also to continuous media. While referring to Landau and Lifshitz (1960) or Rytov (1966) for details we shall give here merely a few expressions which relate to the fluctuations of an electromagnetic field in a medium.

We can indicate the presence of fluctuations by writing the relation between $\mathbf{D}$ and $\mathbf{E}$ in the framework of linear electrodynamics in the form

$$D_i(\omega, \mathbf{r}) = \int \hat{\varepsilon}_{ij}(\omega, \mathbf{r}, \mathbf{r}')\, E_j(\omega, \mathbf{r}')\, d^3\mathbf{r}' + K_i(\omega, \mathbf{r}) \,. \tag{13.24}$$

We have here already made the transition to the Fourier components with respect to ω, and otherwise this relation differs from (10.3) only through the addition of a fluctuating electric induction $\mathbf{K}(\omega, \mathbf{r})$ which takes into account the appearance of fluctuations in $\mathbf{D}$ even when there is no average field $\mathbf{E}$. When there are no external sources the basic field equations for $\mathbf{E}$ and $\mathbf{B}$ take the form

$$\begin{aligned} \mathrm{curl}_i\, \mathbf{B}(\omega, \mathbf{r}) &= -\frac{i\omega}{c} \int \hat{\varepsilon}_{ij}(\omega, \mathbf{r}, \mathbf{r}')\, E_j(\omega, \mathbf{r}')\, d^3\mathbf{r}' - \frac{i\omega}{c} K_i(\omega, \mathbf{r}') \,, \\ \mathrm{curl}\, \mathbf{E}(\omega, \mathbf{r}) &= \frac{i\omega}{c} \mathbf{B}(\omega, \mathbf{r}) \,. \end{aligned} \tag{13.25}$$

The relation (13.24) clearly takes into account the possibility of spatial dispersion and we can therefore without losing generality assume that $\mathbf{B} = \mathbf{H}$ (see Chapter 10). When we neglect spatial dispersion we can (and sometimes must) introduce the magnetic permeability and in that case we must when considering fluctuations introduce also a fluctuating magnetic induction $L(\omega)$ (see Landau and Lifshitz, 1960, § 90).

According to the fluctuation-dissipation theorem we have in equilibrium

$$\begin{aligned} \overline{K_i(\omega, \mathbf{r})\, K_j(\omega, \mathbf{r}')} &\equiv (K_i(\mathbf{r})\, K_j(\mathbf{r}'))_\omega = \\ &= i\hbar \coth \frac{\hbar\omega}{2k_B T} \left\{ \hat{\varepsilon}^*_{ji}(\omega, \mathbf{r}', \mathbf{r}) - \hat{\varepsilon}_{ij}(\omega, \mathbf{r}, \mathbf{r}') \right\} , \end{aligned} \tag{13.26}$$

where the bar indicates statistical averaging. When we neglect spatial dispersion and put $\mathbf{B}=\mathbf{H}$ we have $\hat{\varepsilon}_{ij}(\omega,\mathbf{r},\mathbf{r}')=\varepsilon_{ij}(\omega)\,\delta(\mathbf{r}-\mathbf{r}')$ and

$$(K_i(\mathbf{r})\,K_j(\mathbf{r}'))_\omega = i\hbar\left\{\varepsilon^*_{ji}(\omega)-\varepsilon_{ij}(\omega)\right\}\delta(\mathbf{r}-\mathbf{r}')\coth\frac{\hbar\omega}{2k_BT} = 2\hbar\varepsilon''_{ij}(\omega)\,\delta(\mathbf{r}-\mathbf{r}')\coth\frac{\hbar\omega}{2k_BT}\,. \tag{13.27}$$

Solving Eqn.(13.25) in order to find the fields $\mathbf{B}$ and $\mathbf{E}$ produced by the fluctuating term $\mathbf{K}$, and then obtaining the quadratic expressions, we can use the fluctuation-dissipation relations (13.26) and (13.27) to express the result in terms of ε''_{ij} and after that in terms of other suitable quantities. For instance, in a transparent medium $\varepsilon''\to 0$, but the presence of a δ-function in (13.27) guarantees the correct limiting transition, say, to the equation (see Landau and Lifshitz, 1960, § 91)

$$(\mathbf{E}(\mathbf{r})\cdot\mathbf{E}(\mathbf{r}'))_\omega = \frac{\hbar\omega^2}{\pi c^2}\,\frac{\sin(n\tilde{r}\omega/c)}{\tilde{r}}\coth\frac{\hbar\omega}{2k_BT}\,,$$

$$(\mathbf{E}^2)_\omega = \frac{\hbar\omega^3}{\pi c^3}\,n\coth\frac{\hbar\omega}{2k_BT}\,, \tag{13.28}$$

where $\tilde{r} = |\mathbf{r}-\mathbf{r}'|$ and the refractive index $n=\sqrt{\varepsilon}=\sqrt{\varepsilon'}$, as $\varepsilon''=0$.

Hence it is also straightforward to obtain equations for the equilibrium density of electromagnetic energy in a transparent, dispersive medium (in such cases one usually speaks of thermal radiation). Indeed, this density equals †

$$w_\omega = \frac{1}{8\pi}2\,(E^2)_\omega\frac{d(\omega n^2)}{d\omega}+\frac{1}{8\pi}2\,(H^2)_\omega = \frac{\hbar\omega^3}{4\pi^2c^3}\left(n\frac{d(\omega n^2)}{d\omega}+n^3\right)\coth\frac{\hbar\omega}{2k_BT} = \left(\frac{1}{2}\hbar\omega+\frac{\hbar\omega}{\exp(\hbar\omega/k_BT)-1}\right)\frac{\omega^2n^2}{\pi^2c^3}\,\frac{d(\omega n)}{d\omega}\,; \tag{13.29}$$

we have used here Eqn.(13.28) and the fact that $(H^2)_\omega=\varepsilon(E^2)_\omega$ for transverse normal waves in an isotropic medium (and these are the only ones considered now).

---

† We refer to the literature (for instance, Vainshtein, 1957; Landau and Lifshitz, 1960; Agranovich and Ginzburg, 1966; Ginzburg, 1970b) for a discussion of general expressions of the kind

$$\bar{w} = \left\{\frac{d(\omega\varepsilon)}{d\omega}\,E^2+H^2\right\}/\,8\pi\,.$$

In the text we have, however, used quantities such that

$$\overline{E^2} = \int_{-\infty}^{+\infty}(E^2)_\omega\,d\omega = 2\int_0^\infty(E^2)_\omega\,d\omega\,.$$

On the other hand, we can derive Eqn.(13.29) much more simply and directly by assuming that each 'field oscillator' (of index α) has an average energy

$$w_\alpha = \left(\frac{1}{2}\hbar\omega_\alpha + \frac{\hbar\omega_\alpha}{\exp(\hbar\omega_\alpha/k_BT) - 1}\right) = \frac{1}{2}\hbar\omega_\alpha \coth\frac{\hbar\omega_\alpha}{2k_BT}, \tag{13.30}$$

while the number of such oscillators in an interval $d\omega$ equals (a factor 2 derives from the two polarization directions)

$$\frac{2\,dk_x\,dk_y\,dk_z}{(2\pi)^3} = \frac{8\pi k^2\,dk}{(2\pi)^3} = \frac{\omega^2 n^2}{\pi^2 c^3}\frac{d(\omega n)}{d\omega}\,d\omega, \tag{13.31}$$

as

$$\frac{dk}{d\omega} = \frac{d(\omega n/c)}{d\omega}, \quad k = \frac{\omega}{c}\,n(\omega).$$

The expressions obtained for the transverse field in a transparent uniform medium (in general, without spatial dispersion as we consider only two waves) can easily be generalized to the case of an arbitrary transparent medium in which normal waves with refractive index $n_\ell(\omega,\mathbf{s})$, $\ell = 1, 2, 3, \ldots$ can propagate. To be precise, in that case

$$w_\ell(\omega,\mathbf{s})\,d\omega\,d^2\Omega = \frac{\omega^2 n_\ell^2\,w(\omega)}{(2\pi c)^3}\left|\frac{\partial(\omega n_\ell)}{d\omega}\right| d\omega\,d^2\Omega, \tag{13.32}$$

where $w(\omega)$ is the function (13.20) with $\omega_\alpha = \omega$ which in the classical limit equals k_BT and $\mathbf{s} = \mathbf{k}/k$ is a unit vector corresponding to an element of solid angle $d^2\Omega$. We have already met an application of Eqn.(13.32) in Chapter 6 (see also the literature cited there). There is no doubt whatever that we can also obtain this formula by starting from the fluctuation-dissipation theorem (13.26) by consistently applying it to an arbitrary medium and taking the limit as $\varepsilon''_{ij} \to 0$. For a transparent medium this method is relatively cumbersome and clearly not suitable for the problem. Undoubtedly the usefulness of Eqn.(13.26) and the whole fluctuation approach with random inductions and so on (see, for instance, (13.25)) is connected with the possibility to study in that way absorbing media and the transition to the transparent case serves usually only as a control.

The energy of the electromagnetic field in an absorbing medium† is, in general,

† The concept of the energy of the electromagnetic field in an absorbing medium needs, in general, be made more precise as it, at the least, is ambiguous (see Agranovich and Ginzburg, 1966; Ginzburg 1970b; Barash and Ginzburg, 1976). However, the field energy in an absorbing medium can, first of all, be introduced by giving a precise model of the medium and, secondly, and this is here particularly important, in thermodynamic equilibrium there is no dissipation in the medium and the internal energy of the electromagnetic field has a completely well defined meaning (see also below).

only a small part of the total (or of the free) energy of the thermal motion in the medium. We have here, of course, in view the electromagnetic field in the long wavelength region or in the region of typical distances which are much larger than the atomic size a. If we consider the whole field, in final reckoning a normal substance has only electromagnetic energy (apart from the nuclear energy which remains unchanged unless the isotopic constitution of the substance changes). This energy is, however, first of all mainly electrostatic (Coulomb) energy and, secondly, so to speak 'concentrated' at sizes of the order of atomic sizes ($a \sim 10^{-8}$ to 10^{-7} cm) and, thirdly, should be calculated quantum-mechanically. The relatively long-wavelength field with $\lambda \gg a$ carries, as indicated, in general a small amount of energy. The problem of the role played by this energy is perhaps not altogether clear and needs a further analysis at least for 'non-standard' media (for instance, chemical compounds with layer or filamentary structures). There are now, however, two statements of the problem known when one must take the electromagnetic fluctuations in a continuous absorbing medium into account. This is, firstly, the problem of the forces between macroscopic bodies acting at distances $\ell \gg a$ (such forces are usually called molecular or van der Waals forces). The second problem is related to the previous one — the question of thermal radiation by macroscopic bodies. It is true that if the wavelength of the emitted waves $\lambda \ll \ell$, where ℓ is a characteristic dimension of the body such as the radius of the heated sphere or cylinder, one can usually apply the geometric optics approximation as well as the classical theory of temperature radiation (Kirchhoff's law and so on; see, for instance, Landau and Lifshitz, 1969, § 60), but, if $\lambda \geqslant \ell$ (which may happen for antennae, for heated bodies in waveguides or resonators, and so on) we need more complete electrodynamic calculations. We shall not dwell here upon the corresponding class of problems (see Rytov, 1966), but briefly touch upon the problem of the molecular forces.

If we state the problem relatively simply, the question is how to find the force between two half-spaces 1 and 2 which are filled with media with permittivities $\varepsilon_1(\omega)$ and $\varepsilon_2(\omega)$. The distance (gap) between the media is ℓ and the gap itself may be filled with a medium (a gas or liquid, say) of permittivity $\varepsilon_3(\omega)$. We called this statement of the problem a relatively simple one bearing in mind the natural possibility of a number of generalizations — changing to an anisotropic medium, media with spatial dispersion, a set of plates (layers), surfaces which are not planes, and so on. As far as the possibility of a complete quantitative analysis is concerned, even the problem posed

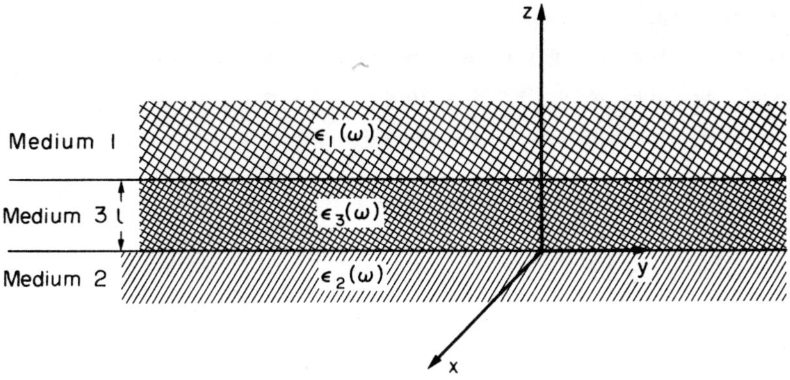

Fig. 13.3 Two half-spaces 1 and 2 (with permittivities $\varepsilon_1(\omega)$ and $\varepsilon_2(\omega)$), separated by a gap 3 filled with a medium of permittivity $\varepsilon_3(\omega)$

here is already complicated or, more precisely, cumbersome. To make this clear we give an expression for the force F acting on unit area of each of the plates (and formally the half-spaces) 1 and 2 which are separated by the gap 3 (Fig. 13.3)

$$\left.\begin{aligned}
F(\ell,T) &= \frac{k_B T}{\pi c^3} \sum_{m=0}^{\infty}{}' \varepsilon_3^{\frac{3}{2}} \omega^3 \int_1^{\infty} p^2 \{\ \} \, dp \\
\{\ \} &= \left\{\left[\frac{(s_1+p)(s_2+p)}{(s_1-p)(s_2-p)} \exp\left(\frac{2p\omega_m}{c} \ell \sqrt{\varepsilon_3}\right) - 1\right]^{-1}\right. \\
&\quad + \left.\left[\frac{(s_1+p\varepsilon_1/\varepsilon_3)(s_2+p\varepsilon_2/\varepsilon_3)}{(s_1-p\varepsilon_1/\varepsilon_3)(s_2-p\varepsilon_2/\varepsilon_3)} \exp\left(\frac{2p\omega_m}{c} \ell \sqrt{\varepsilon_3}\right) - 1\right]^{-1}\right\} \\
s_1 &= \sqrt{\left\{\frac{\varepsilon_1}{\varepsilon_3} - 1 + p^2\right\}}, \quad s_2 = \sqrt{\left\{\frac{\varepsilon_2}{\varepsilon_3} - 1 + p^2\right\}}, \quad \omega_m = \frac{2\pi m k_B T}{\hbar},
\end{aligned}\right\} \tag{13.33}$$

where the prime on the summation sign indicates that the term with $m = 0$ must be multiplied by $\frac{1}{2}$; moreover, all complex permittivities ε_1, ε_2, and ε_3 are taken for the imaginary frequency $i\omega_m$ (a positive value of F corresponds to attraction and a negative value to repulsion between the bodies). In the limiting case of a small distance between the bodies 1 and 2 (this means that for the wavelengths which are important in the problem $\lambda_c \sim 2\pi c/\omega\sqrt{\varepsilon_3} \gg \ell$; moreover, we have assumed that $k_B T \ell/\hbar c \ll 1$ by virtue of which we have put $T = 0$) Eqn. (13.33) reduces to

$$F = \frac{\hbar}{16\pi^2 \ell^3} \int_0^{\infty}\int_0^{\infty} x^2 \left[\frac{(\varepsilon_1+\varepsilon_3)(\varepsilon_2+\varepsilon_3)}{(\varepsilon_1-\varepsilon_3)(\varepsilon_2-\varepsilon_3)} e^x - 1\right]^{-1} dx\, d\xi, \qquad \varepsilon_{1,2,3} = \varepsilon_{1,2,3}(i\xi). \tag{13.34}$$

The general expression (13.33) can also be somewhat simplified in another limiting case — for wide gaps where $\ell \gg \lambda_c$. We restrict ourselves here to an even more special case of two well conducting media (in the limit — perfect conductors) separated by an empty wide gap. In that case

$$F = \frac{\pi^2}{240} \frac{\hbar c}{\ell^4} . \qquad (13.35)$$

In the most important case of a vacuum gap (that is, when $\varepsilon_3 = 1$) Eqn. (13.33), although apparently not simplified, was obtained by Lifshitz (1956), one could say straightforwardly by calculating the fluctuating fields $\mathbf{E}_\omega$ and $\mathbf{H}_\omega$ in the gap and subsequently calculating the stress tensor,† using the theorem (13.27). The calculations were then so cumbersome they were not even reproduced in the relevant Landau and Lifshitz volume (1960; see § 92) where, as a rule, all important calculations are given. Dzyaloshinskii, Lifshitz, and Pitaevskii (1961; see also Abrikosov, Gorkov, and Dzyaloshinskii, 1965, Chapter 6) obtained a generalization of the result to the case of a gap filled with a medium (that is Eqn. (13.33)) using quantum field theoretical methods (or, one sometimes calls them when they are applied to statistical physics, the methods of the quantum many-body theory). The efficiency and fruitfulness of these methods have been proved. However, they do not at all prevent a tendency to obtain this or that result by simpler means. Leaving alone the methodological side of the problem, one can in general say that more transparent and simpler, and therefore less cumbersome methods turn out to be preferable when one must go over to more complex problems; they give a means of checking, and so on. In our opinion this is just the situation in the case of calculating the forces between macroscopic bodies.

Indeed, as was noted by van Kampen, Nijboer, and Schram (1968) the result (13.34) (Ninham, Parsegian, and Weiss (1970) and Gerlach (1971) showed later that the same is true also for Eqn. (13.33)) can be obtained relatively more simply than was done by Dzyaloshinskii, Lifshitz, and Pitaevskii (1961) as follows.‡ We assume that all media 1, 2, 3 are transparent. The internal energy W and the free energy $\mathfrak{F}$ of the system can then be written in the form

---

† The force F is equal to the σ_{zz} component of the stress tensor, if we take the z-axis at right angles to the gap.

‡ In fact van Kampen, Nijboer and Schram (1968), Ninham, Parsegian, and Weiss (1970), and Gerlach (1971) proceed just as stated in the text, but oddly enough there is no mention that they consider directly only transparent media.

$$\left.\begin{aligned} \bar{W} \equiv W &= \mathfrak{F} - T\frac{\partial \mathfrak{F}}{\partial T} = \sum_{\alpha} w_{\alpha}(\omega_{\alpha}, T) \ , \ w_{\alpha} = \tfrac{1}{2}\hbar\omega_{\alpha} \coth \frac{\hbar\omega_{\alpha}}{2k_B T} \ , \\ \mathfrak{F} &= \sum_{\alpha} \left\{ k_B T \, \ell n \left(1 - \exp\left(-\frac{\hbar\omega_{\alpha}}{k_B T} \right) \right) + \tfrac{1}{2}\hbar\omega_{\alpha} \right\} , \end{aligned}\right\} \qquad (13.36)$$

where ω_{α} are the eigenfrequencies; when evaluating the force $F = -\partial\mathfrak{F}/\partial\ell$ only the frequencies ω_{α} which depend on ℓ are important. Such frequencies which correspond to 'surface' oscillations in the gap can be found easily. Substituting them into (13.36) leads to (13.33) or, in particular, to (13.34). To illustrate this we derive the last formula.†

The wave equation in all three uniform regions 1, 2, and 3 (see Fig.13.3) has the form

$$\nabla^2 \mathbf{E} + \frac{\varepsilon\omega^2}{\omega^2} \mathbf{E} = 0 \ .$$

We look for its solution in the form

$$\mathbf{E} = \mathbf{E}_0(z) \exp\left[i(k_x x + k_y y)\right] \ .$$

We then get for $\mathbf{E}_0(z)$ the equation

$$\frac{d^2\mathbf{E}_0(z)}{dz^2} - K^2\mathbf{E}_0(z) = 0 \quad , \quad K^2 = k^2 - \frac{\varepsilon\omega^2}{c^2} \quad , \quad k^2 = k_x^2 + k_y^2 \ .$$

The solution $\mathbf{E}_0(z)$ which is of interest to us is a 'surface' type of solution, that is, one which is localized near the gap; it clearly has the form $A\exp(-K_1 z)$ in region 1, $B\exp(-K_3 z) + C\exp(K_3 z)$ in region 3, and $D\exp(K_2 z)$ in region 2. At the boundaries (for $z = 0$ and $z = \ell$) these solutions must be joined using the electromagnetic boundary conditions that is, the requirement that the quantities $\varepsilon E_{0z}(z)$, $E_{0x}(z)$, and $E_{0y}(z)$ be continuous.

Since we wish, as we have said, to limit ourselves to the static case we must put $K^2 = k^2$ (formally this is obtained by letting $c \to \infty$) and we must not consider the magnetic field (that is the reason why we have not written down the

---

† We note that this method, when applied to two perfect conductors, separated by an empty gap, had been used already rather a long time ago (Casimir, 1948; Boyer, 1970) and led at once to Eqn.(13.35). In this case the role of the zero-point oscillations of the field was particular translucent — the force is

$$F = -\frac{\partial W}{\partial \ell} = -\frac{\partial}{\partial \ell} \sum_{\alpha} \tfrac{1}{2}\hbar\omega_{\alpha}(\ell)$$

where $\omega_{\alpha}(\ell)$ are the frequencies of the eigenoscillations of the electromagnetic field in the gap (one assumes that $T = 0$)

corresponding equation). Finally, from the condition div $\mathbf{E} = 0$ which is valid in each of the three regions it follows that the component $E_{oy}(z)$ is proportional to $dE_{oz}(z)/dz$ (we have for the sake of convenience chosen the x-axis along $\mathbf{k}$ which does not limit the generality). The quantities εE_{oz} and dE_{oz}/dz must thus be continuous at the 1-3 and 3-2 boundaries. As a result we obtain four homogeneous equations for the amplitudes A, B, C, and D. The condition for the existence of a non-trivial solution of this system of equations has the form

$$\mathcal{D}(\omega_\alpha) = \frac{(\varepsilon_1 + \varepsilon_3)(\varepsilon_2 + \varepsilon_3)}{(\varepsilon_1 - \varepsilon_3)(\varepsilon_2 - \varepsilon_3)} e^{2k\ell} - 1 = 0 , \tag{13.37}$$

where ε_1, ε_2, and ε_3 are functions of ω.

The dispersion Eqn. (13.37) connects $k = (k_x^2 + k_y^2)^{\frac{1}{2}}$ with $\omega = \omega_\alpha$, that is, determines the eigenfrequencies of the waves in the gap.

When $T = 0$ we have according to (13.36)

$$W = \sum_\alpha \tfrac{1}{2}\hbar\omega_\alpha = \frac{1}{(2\pi)^2}\int_0^\infty 2\pi k \, dk \, \{\ \} ,$$
$$\{\ \} = \frac{1}{2\pi i}\oint \tfrac{1}{2}\hbar\omega \frac{\partial}{\partial\omega}[\ln \mathcal{D}(\omega)]\, d\omega , \tag{13.38}$$

where we have used a well known theorem from the theory of analytical functions — where we express as a contour integral the sume of values of some function (in this case $\frac{1}{2}\hbar\omega$) at all zeroes $\omega = \omega_\alpha$ of another function (in this case $\mathcal{D}(\omega)$). Moreover, it is important that we must not take into account the poles of the function $\mathcal{D}(\omega)$, as the corresponding values ω_∞ are independent of ℓ (see Barash and Ginzburg, 1975 for details about the applicability of the calculations). We have also used the fact that the roots ω_α of Eqn. (13.37) depend on k as a parameter and that the number of such roots in an interval dk is equal to $2\pi k\, dk/(2\pi)^2$; as in (13.36) there occurs the sum over all roots we must integrate over k. The force $F = -\partial W/\partial\ell$ is determined by the value of the derivative

$$\frac{\partial}{\partial\ell}\frac{\partial\mathcal{D}/\partial\omega}{\mathcal{D}} = -\frac{2k\,\partial\mathcal{D}/\partial\omega}{\mathcal{D}^2} ,$$

as by virtue of (13.37) $\partial\mathcal{D}/\partial\ell = 2k(\mathcal{D} - 1)$. Finally we can write

$$-\frac{\omega\,\partial\mathcal{D}/\partial\omega}{\mathcal{D}^2} = \frac{\partial}{\partial\omega}\left(\frac{\omega}{\mathcal{D}}\right) - \frac{1}{\mathcal{D}}$$

and use that relation when integrating over ω in (13.38). Introducing the variables $x = 2k\ell$ and $\xi = -i\omega$ and taking as integration contour the imaginary ω-axis we get at once Eqn. (13.34). We note that we can find the values of the functions $\varepsilon(i\xi)$ for the imaginary frequency $\omega = i\xi$ as we know the value of $\varepsilon'' = \mathrm{Im}\,\varepsilon$ for the real frequency ω; we are thinking here of the formula (see Landau and Lifshitz, 1960, § 62)

$$\varepsilon(i\xi) = 1 + \frac{2}{\pi}\int_0^\infty \frac{x\varepsilon''(x)}{x^2 + \xi^2}\,dx \,.$$

We have mentioned that Ninham, Parsegian, and Weiss (1970) and Gerlach (1971) obtained also the general result (13.33) in a similar way, but taking retardation into account (that is, by finding the frequencies $\omega_\alpha(k)$ in the gap exactly — for $K^2 = k^2 - \varepsilon\omega^2/c^2$ rather than for $K^2 = k^2$; vide supra). Of course, at the end of the calculation the permittivities were assumed to be arbitrary, corresponding to real media. Apart from a general maxim that 'victors should not be judged' one can justify the derivation of Eqns.(13.33) and (13.34) for absorbing media on the basis of the general Eqn.(13.36) for transparent media by the following arguments. Firstly, the permittivities ε_1, ε_2, and ε_3 occur in (13.33) as functions. Secondly, the function $\varepsilon(\omega)$ is always real on the imaginary axis (see Landau and Lifshitz, 1960, § 62). The result obtained for transparent media (ε_1, ε_2, and ε_3 are real positive quantities for the real frequency ω), must therefore clearly be the same as the appreciably more general one which is applicable to absorbing media. However, this would hardly confirm such a conclusion unless it had been obtained earlier without additional assumptions. Both for this reason and also bearing in mind other related or similar problems one must somehow consistently generalize the expansion in eigenoscillations with frequencies ω_α which we used here to absorbing media. This can, indeed, be done (Barash and Ginzburg, 1972, 1975; Barash, 1975a,b).

The possibility of the above-mentioned generalization is by itself already obvious from the example of the electric circuit considered above — its internal energy was found also for $R \neq 0$; it is equal to (see (13.11), (13.12), (13.15), and (13.30))

$$\bar{W} = \bar{U} + \bar{K} = \frac{1}{\pi}\int^\infty \frac{CRw(\omega, T)\,d\omega}{R^2C^2\omega^2 + (LC\omega^2 - 1)^2} + \frac{1}{\pi}\int^\infty \frac{C^2LRw(\omega, T)\,\omega^2\,d\omega}{R^2C^2\omega^2 + (LC\omega^2 - 1)^2}\,,$$

$$w(\omega, T) = \tfrac{1}{2}\hbar\omega\coth\frac{\hbar\omega}{2k_BT}\,. \tag{13.39}$$

To generalize this result we must transform it to the general form of an expansion in eigenfrequencies, though not those of the circuit considered but those of an auxiliary circuit. The fact is that we are interested in induced oscillations in the circuit under the influence of some e.m.f. $\mathcal{E}_\omega e^{-i\omega t}$ (in the present case we are dealing with the fluctuating e.m.f.). It is clear from (13.1) that we then have

$$q_\omega = \frac{\mathcal{E}_\omega}{-L\omega^2 - i\omega R + 1/C} = \frac{L^{-1}\mathcal{E}_\omega}{\omega_1^2 - \omega^2}, \quad \omega_1^2(\omega) = \frac{1}{LC} - i\frac{R}{L}\omega . \qquad (13.40)$$

However, the frequencies $\omega_1(\omega)$ are the eigenfrequencies for the circuit (and this is the auxiliary circuit)

$$L\ddot{q} + \frac{\omega}{\omega_1} R\dot{q} + \frac{q}{C} = 0, \qquad (13.41)$$

which differs from (13.1) by the replacement of R by $(\omega/\omega_1)R$ (to avoid confusion we explain once more that the solutions $\exp[-i\omega_1(\omega)t]$ in which the frequency ω, as in (13.40), is considered to be a parameter, satisfy Eqn. (13.41)). In terms of the frequencies $\omega_1(\omega)$ we can write Eqn.(13.39) in the form

$$\bar{W} = -\frac{i}{\pi}\int_{-\infty}^{+\infty} \frac{w(\omega,T)}{\omega_1^2(\omega) - \omega^2}\,\omega\, d\omega + \frac{i}{2\pi}\int_{-\infty}^{+\infty} \frac{w(\omega,T)}{\omega_1^2(\omega) - \omega^2}\frac{d\omega_1^2(\omega)}{d\omega}\, d\omega . \qquad (13.42)$$

One can easily check directly that Eqns.(13.39) and (13.42) are identically the same.

Of course, in the case of a circuit nothing whatever is changed, but in the case of generalizations (for instance, oscillations in a gap) it is necessary to perform some calculations starting from Eqn.(13.42) and using, mainly, the frequencies $\omega_\alpha(\omega)$ which are the analogues of the frequency $\omega_1(\omega)$. The last method is well known (see Vainshtein, 1957, §§ 100 to 102 and Barash and Ginzburg, 1975), Let the whole system be immersed in some auxiliary resonator with perfectly conducting walls and consider the frequency ω to be a parameter, while the eigenfrequencies of the resonator $\omega_\alpha(\omega)$ are determined from the homogeneous field equations

$$\left.\begin{aligned} \mathrm{curl}\,\mathbf{H}_{\omega_\alpha(\omega)}(\omega,\mathbf{r}) &= -\frac{i\omega_\alpha(\omega)}{c}\int \hat{\varepsilon}(\omega,\mathbf{r},\mathbf{r}')\,\mathbf{E}_{\omega_\alpha(\omega)}(\omega,\mathbf{r}')\,d^3\mathbf{r}', \\ \mathrm{curl}\,\mathbf{E}_{\omega_\alpha(\omega)}(\omega,\mathbf{r}) &= \frac{i\omega_\alpha(\omega)}{c}\,\mathbf{H}_{\omega_\alpha(\omega)}(\omega,\mathbf{r}), \end{aligned}\right\} \qquad (13.43)$$

where $\hat{\varepsilon}$ is the linear operator occurring in (13.24) and where we have omitted the tensor indexes i, j to simplify the notation. It is clear from (13.43) that the auxiliary resonator is the same as the actual one (that is, the system considered) when $\omega_\alpha = \omega$ and when there are no external and fluctuating sources. The eigenfunctions $\mathbf{E}_{\omega_\alpha(\omega)}(\omega, \mathbf{r})$ and $\mathbf{H}_{\omega_\alpha(\omega)}(\omega, \mathbf{r})$ of the auxiliary resonator have a number of properties (such as orthogonality) which enable us to find particularly simply the induced solutions which depend on $\mathbf{K}(\omega, \mathbf{r})$ for the actual problem which is described by Eqns.(13.25). The average internal energy of the system is in this case simply of the form

$$\bar{W} = -\frac{i}{\pi}\sum_\alpha \int_{-\infty}^{+\infty} \frac{w(\omega, T)}{\omega_\alpha^2(\omega) - \omega^2}\,\omega\, d\omega + \frac{i}{2\pi}\sum_\alpha \int_{-\infty}^{+\infty} \frac{w(\omega, T)}{\omega_\alpha^2(\omega) - \omega^2}\,\frac{d\omega_\alpha^2(\omega)}{d\omega}\, d\omega, \quad (13.44)$$

which immediately generalizes Eqn.(13.42). This last fact makes it already clear that as $\varepsilon'' \to 0$ (or $R \to 0$) Eqn.(13.44) goes over into (13.36) for $\bar{W}$ (the same is, of course, also true for $\mathfrak{F}$). Moreover, Eqn.(13.44) is valid also for absorbing media and it is clear from (13.43) that it can be used also when we take into account anisotropy and spatial dispersion. When applying it to the gap problem (see Fig.13.3) for an isotropic medium without spatial dispersion one can check (Barash and Ginzburg, 1972, 1975; Barash, 1975a,b; Barash, 1978) that from Eqn.(13.44) or, to be more precise, from the analogous expression for the free energy $\mathfrak{F}$ Eqn.(13.33) follows.† When solving yet more complicated problems, for instance, when the media 1 and 2 are anisotropic (Barash, 1978) the advantage of the method described here of an expansion in eigenfrequencies becomes even more striking. One may think that just such an approach would dominate when one solves a whole number of related problems about molecular forces for various media under different geometrical conditions or when calculating the free energy in an absorbing medium, and so on.

In conclusion we shall discuss yet another electrodynamic fluctuation problem — that of the effect of fluctuating voltages in a resonator on electrons passing through it (Ginzburg and Fain, 1957a,b).

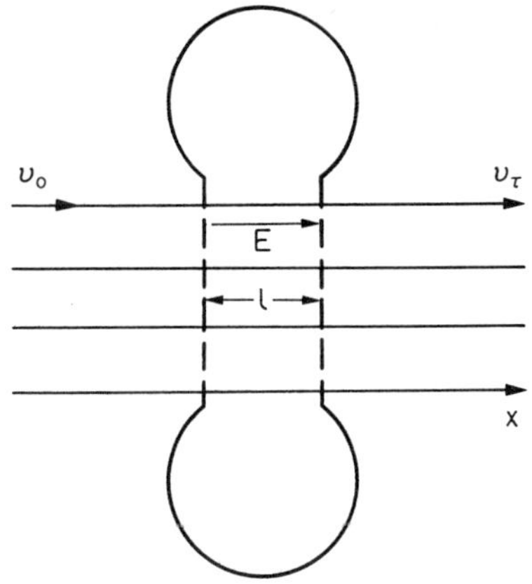

Fig. 13.4 The problem of the flight of electrons through an empty resonator

---

† As $T \to 0$ the free energy becomes the same as the internal energy. In the problem of the van der Waals forces, however, one is usually interested only in the case as $T \to 0$ (see Lifshitz, 1956; Landau and Lifshitz, 1960; Abrikosov, Gor'kov, and Dzyaloshinskii, 1965). This is the reason why we did not write down the expression for $\mathfrak{F}$.

Let a non-relativistic electron with initial energy $K_0 = \frac{1}{2}mv_0^2$ enter the resonator at time $t = 0$ and leave it at time τ with energy $K_\tau = \frac{1}{2}mv_\tau^2$. The field in the resonator is assumed to be uniform and in the direction of the electron velocity — along the x-axis (such a case corresponds to a resonator of a well defined shape, for instance, such as shown schematically in Fig. 13.4). If

$$E = E_1 \cos\omega t + (E_2 + E_0)\sin\omega t ,$$

we have

$$m\ddot{x} = eE \ , \ v_\tau \equiv \dot{x}(\tau) = v_0 + \frac{e}{m}\left\{E_1 \sin\omega\tau + (E_2 + E_0)(1 - \cos\omega\tau)\right\} . \tag{13.45}$$

We shall further assume that E_1 and E_2 are random variables so that $\overline{E}_1 = \overline{E}_2 = 0$ and $\overline{E_1^2} = \overline{E_2^2} = \overline{V^2}/\ell$, where ℓ is the thickness of the resonator (the path traversed by the electrons in the resonator) and $\overline{V^2}$ is the mean square of the fluctuating voltage at the 'plates' of the resonator. Up to terms of order e^2 we then have $(\overline{K}_0 = \frac{1}{2}mv_0^2)$

$$\begin{aligned} \overline{K}_\tau - \overline{K}_0 &= \frac{e^2}{2m\omega^2}\left\{\frac{4\overline{V^2}}{\ell^2}\sin^2\tfrac{1}{2}\omega\tau + E_0^2(1-\cos\omega\tau)^2\right\} + \frac{2ev_0}{\omega}E_0\sin^2\tfrac{1}{2}\omega\tau , \\ \overline{(\Delta K_\tau)^2} &\equiv \overline{K_\tau^2} - (\overline{K}_\tau)^2 = \overline{(K_\tau - K_0)^2} - (\overline{K_\tau - K_0})^2 = \frac{4e^2v_0^2}{\omega^2}\,\frac{\overline{V^2}}{\ell^2}\sin^2\tfrac{1}{2}\omega\tau . \end{aligned} \tag{13.46}$$

The term of order e^4 which we have dropped here is unimportant, if the electron velocity changes little, that is, if $v_0 - v_\tau \ll v_0$; in the same approximation the time taking to fly through the resonator is $\tau = \ell/v_0$ and

$$\overline{(\Delta K_\tau)^2} = e^2\,\overline{V^2}\left(\frac{\sin\frac{1}{2}\omega\tau}{\frac{1}{2}\omega\tau}\right)^2 . \tag{13.47}$$

We have here assumed that the time dependence of V is unimportant. If that would not be the case we have instead of (13.47)

$$\overline{(\Delta K_\tau)^2} = 2e^2\int_0^\infty (V^2)_\omega\left(\frac{\sin\frac{1}{2}\omega\tau}{\frac{1}{2}\omega\tau}\right)^2 d\omega \ , \quad \overline{V^2} = 2\int_0^\infty (V^2)_\omega\, d\omega . \tag{13.48}$$

Of course, in the classical case it is possible that $(V^2)_\omega = 0$, but in the thermodynamic equilibrium state there is always electromagnetic radiation in the resonator and

$$(V^2)_\omega = \frac{R(\omega)\ k_BT/\pi}{R^2C^2\omega^2 + (LC\omega^2 - 1)^2} , \tag{13.49}$$

where we have used the classical Nyquist Eqn.(13.17) for $(\mathcal{E}^2)_\omega$ while $L(\omega)$, $C(\omega)$, and $R(\omega)$ are the self-induction, capacitance, and resistance of the circuit which is equivalent to the given resonator when one calculates the voltage V. We here assume the circuit to be of the kind shown in Fig. 13.1 (connected in series) so that $Z = R - i(\omega L - 1/\omega C)$, $J_\omega = \mathcal{E}_\omega / Z(\omega)$, and $V_\omega = J_\omega / i\omega C = \mathcal{E}_\omega / i\omega C Z$. When we use the quantal Nyquist Eqn.(13.16) we get

$$\overline{(\Delta K_\tau)^2} = e^2 \int_0^\infty \frac{(2/\pi)\{\frac{1}{2}\hbar\omega + \hbar\omega/[\exp(\hbar\omega/k_B T) - 1]\}}{R^2 C^2 \omega^2 + (LC\omega^2 - 1)^2} \left(\frac{\sin\frac{1}{2}\omega\tau}{\frac{1}{2}\omega\tau}\right)^2 d\omega\,, \qquad (13.50)$$

or, for a weakly damped (high-Q) resonator with eigenfrequency $\omega_i = 1/\sqrt{LC}$

$$\overline{(\Delta K_\tau)^2} = \frac{e^2}{C(\omega_i)} \left[\frac{1}{2}\hbar\omega_i + \frac{\hbar\omega_i}{\exp(\hbar\omega_i/k_B T) - 1}\right] \left(\frac{\sin\frac{1}{2}\omega_i\tau}{\frac{1}{2}\omega_i\tau}\right)^2 . \qquad (13.51)$$

As the motion of the electrons was assumed earlier to be classical, the result given here is valid only within certain limitations. If we recall, however, that the electron wavelength $\lambda = 2\pi\hbar/mv \leqslant 10^{-8}$ cm for $K = \frac{1}{2}mv^2 \geqslant 10$ eV, it is clear that the classical approach is practically always adequate for resonators through which electron beams pass. At the same time in a whole number of papers (see Ginzburg and Fain, 1957a,b for references) the whole problem considered above was solved, and with the same result, using quantum mechanics both for the radiation and for the motion of the electrons. It is clear from what we have said that this is, in general, superfluous (see, Ginzburg and Fain, 1957a,b, for some additional provisos).

Chapter XIV

SCATTERING OF WAVES IN A MEDIUM

Scattering of electromagnetic waves (light) in a medium. Linewidth in the emission spectrum and in the spectrum of scattered light. Combinational (Raman) scattering of light involving formation of polaritons (real excitons). Scattering by free electrons and in a plasma. Transition scattering.

A medium is called homogeneous if its statistically averaged properties are independent of the coordinates (this definition has everywhere been tacitly assumed in the foregoing). These were the conditions under which we introduced, for instance, the tensor $\varepsilon_{ij}(\omega,\mathbf{k})$ which connects the average induction with the average field (by averages we understand here statistically averaged values). In a homogeneous medium the average fields (waves) propagate without any scattering; this is clear from the solutions of the field equations — it also followed directly, in particular, in Chapters 10 and 11 when we found the normal waves. However, when we take fluctuations into account the medium is, of course, no longer homogeneous and the waves are scattered. In principle the scattering by the thermal fluctuations of the permittivity does not differ from the scattering by inhomogeneities in ε_{ij} produced by external sources. As in the presence of spatial dispersion the introduction itself of the tensor $\varepsilon_{ij}(\omega,\mathbf{k})$ assumes the homogeneity of the medium, one needs a special analysis for the discussion of scattering when spatial dispersion is taken into account. When we discuss in what follows scattering with the formation of polaritons we shall consider that problem, but otherwise we shall assume that in the medium under consideration there is no spatial dispersion. The general theory of scattering of electromagnetic waves in an isotropic medium, neglecting spatial dispersion and mainly for the case when the frequency of the scattered radiation is changed little compared with the frequency of the incident radiation has been given by Landau and Lifshitz (1960, Chapter 14; see also Fabelinskii, 1968). As usual, we shall therefore restrict ourselves to only briefly remind the reader of some general results and after that turn to some additional remarks.

In most cases the scattered field $\mathbf{E}', \mathbf{H}'$ is weak compared with the field $\mathbf{E}_0, \mathbf{H}_0$ in the incident (to be scattered) wave (an exception is, for instance, the region of critical opalescence, apart from scattering by various turbid media such as emulsions). Under such conditions — and we shall restrict ourselves to those — we can apply perturbation theory. In fact, we can write the relation between the total induction $\mathbf{D}(\omega, \mathbf{r}) = \mathbf{D}_0 + \mathbf{D}'$ and the field $\mathbf{E}(\omega, \mathbf{r}) = \mathbf{E}_0 + \mathbf{E}'$ in the form

$$D_i = \varepsilon_{ij} E_j \approx \varepsilon_{ij}^{(0)} E_j + \delta\varepsilon_{ij} E_{0,j} = \varepsilon_{ij}^{(0)} E_{0,j} + \varepsilon_{ij}^{(0)} E'_j + \delta\varepsilon_{ij} E_{0,j} =$$
$$= D_{0,i} + \varepsilon_{ij}^{(0)} E'_j + \delta\varepsilon_{ij} E_{0,j}, \tag{14.1}$$

where $\varepsilon_{ij}(\omega, \mathbf{r}) = \varepsilon_{ij}^{(0)}(\omega) + \delta\varepsilon_{ij}(\omega, \mathbf{r})$, and the fluctuating part of the permittivity $\delta\varepsilon_{ij}$ is assumed to be small of the same order as the field E'_j; in an isotropic medium $\varepsilon_{ij}^{(0)} = \varepsilon\delta_{ij}$ and assuming the medium to be transparent the tensors $\varepsilon_{ij}^{(0)}$ and $\delta\varepsilon_{ij} = \delta\varepsilon_{ji}$ can also be assumed to be real (of course, even in an isotropic medium one can not, in general, reduce the tensor $\delta\varepsilon_{ij}$ to a scalar). Under all those simplifying assumptions it is convenient to express $\mathbf{E}'$ in terms of $\mathbf{D}'$ and to write

$$\mathbf{E}' = \frac{1}{\varepsilon}\mathbf{D}' - \frac{\mathbf{C}}{\varepsilon}, \quad C_i = \delta\varepsilon_{ij} E_{0,j}. \tag{14.2}$$

We now turn to the wave equation for the field $\mathbf{E}$, that is, to the equation (see (10.21))

$$\text{curl curl}\,\mathbf{E} - \frac{\omega^2}{c^2}\mathbf{D} = 0.$$

Substituting the relation (14.2) into this equation and using the fact that by assumption

$$\text{curl curl}\,\mathbf{E}_0 - \frac{\omega^2}{c^2}\mathbf{D}_0 = 0,$$

we get

$$\nabla^2\mathbf{D}' + \frac{\omega^2}{c^2}\varepsilon(\omega)\mathbf{D}' = -\,\text{curl curl}\,\mathbf{C}, \tag{14.3}$$

as $\text{div}\,\mathbf{D} = \text{div}\,\mathbf{D}_0 = \text{div}\,\mathbf{D}' = 0$.

The solution of Eqn. (14.3) for an incident wave $\mathbf{E}_0 = \mathbf{E}_{00}\exp[i(\mathbf{k}_e \cdot \mathbf{r})]$ and observation in the direction $\mathbf{k}_s$ at large distance R_0 from the scattering volume (but in a region with the same value of $\varepsilon(\omega)$) leads to the result

$$\mathbf{E}' = -\frac{\exp(ik_s R_0)}{4\pi\varepsilon R_0}\left[\mathbf{k}_s \wedge [\mathbf{k}_s \wedge \mathbf{G}]\right], \tag{14.4}$$

$$G_i = \int (\delta \boldsymbol{\varepsilon}_{ij} E_{00j})\, e^{i(\mathbf{q}\cdot\mathbf{r})}\, d^3\mathbf{r} \ ,$$
$$\mathbf{q} = \mathbf{k}_e - \mathbf{k}_s \ , \quad k_e = k_s = \frac{\omega}{c}\sqrt{\boldsymbol{\varepsilon}(\omega)} \ , \tag{14.4}$$

where $\mathbf{k}_e$ and $\mathbf{k}_s$ are the wavevectors of the incident and scattered waves, respectively.

Here we have, strictly speaking, assumed the scattering to proceed without change in frequency and we therefore did not distinguish the frequency of the incident light† ω_e from the frequency ω_s of the scattered light; for the same reason $k_e = k_s$. In fact, however, nothing is changed if $\omega_e \neq \omega_s$ as long as we can neglect the frequency dependence of $\boldsymbol{\varepsilon}$ and $\delta\boldsymbol{\varepsilon}_{ij}$ in the frequency range $\Delta\omega \sim (\omega_e - \omega_s)$. In other words, we can assume that the field $\mathbf{E}'$ is the total field of the scattered light provided $\Delta\omega \ll \omega_e$. The evaluation of $\overline{|\mathbf{E}'|^2}$, where the bar indicates time-averaging, allows us therefore to find the total intensity (for all frequencies) of the scattered radiation. We have according to (14.4) for the intensity of the scattered light $I = \bar{S} R_0^2 = \frac{c\sqrt{\boldsymbol{\varepsilon}}}{8\pi} \overline{|\mathbf{E}'|^2} R_0^2$, relative to the intensity of the incident light $I_0 = \frac{c\sqrt{\boldsymbol{\varepsilon}}}{8\pi} |\mathbf{E}_0|^2$ and per unit

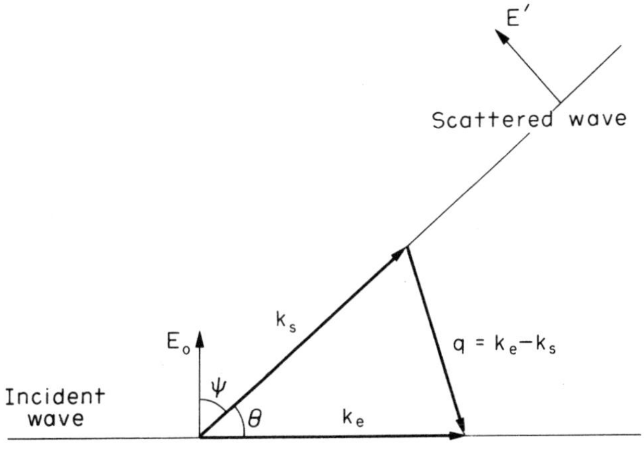

Fig. 14.1 Scattering in a medium. The field in the scattered wave is linearly polarized as shown in the figure only when the fluctuations $\delta\boldsymbol{\varepsilon}_{ij} = \delta\boldsymbol{\varepsilon}\,\delta_{ij}$

† We speak here of light only for the sake of simplicity; essentially we are clearly considering the scattering of electromagnetic waves in all wavebands, provided only that $\lambda \gg a$, where $a \sim 10^{-7}$ to 10^{-8} cm is the atomic size.

solid angle,

$$\frac{I}{I_0} = \frac{1}{16\pi^2}\left(\frac{\omega}{c}\right)^4 \frac{\overline{|G|^2}}{|\mathbf{E}_0|^2} \sin^2\psi , \tag{14.5a}$$

where ψ is the angle between $\mathbf{G}$ and $\mathbf{k}_s$ (the wavevector of the scattered light) and where we have used the fact that $k_e^2 = k_s^2 = (\omega^2/c^2)\varepsilon$.

In a number of cases, for instance, for a gas of spherically symmetric molecules and also for a number of liquids we can assume that the fluctuations $\delta\varepsilon_{ij}$ are scalar, that is, put $\delta\varepsilon_{ij} = \delta\varepsilon\,\delta_{ij}$. Under such conditions (see (14.4) and (14.5a) and also Fig.14.1)

$$\frac{I}{I_0} = \frac{1}{16\pi^2}\left(\frac{\omega}{c}\right)^4 \overline{|\delta\varepsilon_{\mathbf{q}}|^2} \sin^2\psi , \quad \mathbf{G} = \delta\varepsilon\,\mathbf{E}_{00}, \tag{14.5b}$$
$$\delta\varepsilon_{\mathbf{q}} = \int \delta\varepsilon(\mathbf{r})\, e^{i(\mathbf{q}\cdot\mathbf{r})}\, d^3\mathbf{r} ,$$

where $\overline{|\delta\varepsilon_{\mathbf{q}}|^2} \equiv ((\delta\varepsilon)^2)_{\mathbf{q}}$, but the latter notation which is analogous to the one used in Chapter 13 is less convenient.

We remind ourselves that we have assumed the incident field to be of the form $\mathbf{E}_0 = \mathbf{E}_{00}\exp[i(\mathbf{k}_e\cdot\mathbf{r})]$; when $\mathbf{E}_{00} = \text{const}$, this corresponds to a linearly polarized wave. One sees easily that for natural (unpolarized) light of intensity I_0

$$\frac{I}{I_0} = \frac{1}{32\pi^2}\left(\frac{2\pi}{\lambda_0}\right)^4 \overline{|\delta\varepsilon_{\mathbf{q}}|^2}\,(1+\cos^2\theta) , \quad \frac{\omega}{c} = \frac{2\pi}{\lambda_0} , \tag{14.6}$$

where θ is the scattering angle, that is, the angle between $\mathbf{k}_e$ and $\mathbf{k}_s$. Under conditions when $\overline{|\delta\varepsilon_{\mathbf{q}}|^2}$ is independent of $\mathbf{q}$ (vide infra) the intensity I can immediately be integrated over the angles (the element of solid angle $d^2\Omega = \sin\theta\, d\theta\, d\phi$) and as a result we get for the so-called extinction coefficient

$$h = \frac{\int I\, d^2\Omega}{I_0} = \frac{\omega^4}{6\pi c^4}\,\frac{\overline{|\delta\varepsilon_{\mathbf{q}}|^2}}{V} . \tag{14.7}$$

The scattering volume of the medium V occurs here because, by definition h is the ratio of the intensity of the light scattered in all directions to the intensity of the incident light where the scattering volume is taken to be unity; in other words, as a result of the scattering the intensity of the incident light changes as $dI = -hI\,dz$. Of course, the extinction coefficient h is for polarized light equal to the same expression (14.7) — this result, besides the general considerations, is obtained from (14.5b) by integrating over the angles.

By definition

$$\overline{|\delta\varepsilon_{\mathbf{q}}|^2} = \overline{\iint \delta\varepsilon(\mathbf{r}_1)\,\delta\varepsilon^*(\mathbf{r}_2)\exp[i(\mathbf{q}\cdot\{\mathbf{r}_1-\mathbf{r}_2\})]\,d^3\mathbf{r}_1 d^3\mathbf{r}_2} .$$

When averaging we must take into account that when we consider all frequencies of the scattered light at once the fluctuations $\delta\boldsymbol{\varepsilon}(\mathbf{r}_1)$ and $\delta\boldsymbol{\varepsilon}(\mathbf{r}_2)$ are usually correlated only over distances $|\mathbf{r}_1 - \mathbf{r}_2| \sim a \sim 10^{-7}$ to 10^{-8} cm. For light with $\lambda \gg a$ we can thus in the expression for $\overline{|\delta\boldsymbol{\varepsilon}_{\mathbf{q}}|^2}$ put $\exp[i(\mathbf{q}\{\mathbf{r}_1 - \mathbf{r}_2\})] = 1$. In a uniform medium the average $\overline{\delta\boldsymbol{\varepsilon}(\mathbf{r}_1)\,\delta\boldsymbol{\varepsilon}(\mathbf{r}_2)}$ can depend only on the difference $\mathbf{r} = \mathbf{r}_2 - \mathbf{r}_1$. Therefore, changing to the variables $\frac{1}{2}(\mathbf{r}_1 + \mathbf{r}_2)$ and $\mathbf{r} = \mathbf{r}_2 - \mathbf{r}_1$ we can easily show that under the conditions discussed

$$\overline{|\delta\boldsymbol{\varepsilon}_{\mathbf{q}}|^2} = V\int\overline{(\delta\boldsymbol{\varepsilon})^2}\,d^3\mathbf{r} = V^2\,\frac{\overline{(\int\delta\varepsilon\,d^3\mathbf{r})^2}}{V^2} \equiv V^2\,\overline{(\delta\varepsilon)_V^2}\,, \tag{14.8}$$

where V is the volume of the scattering region of the medium; of course, one should have expected that $\overline{|\delta\boldsymbol{\varepsilon}_{\mathbf{q}}|^2}$, and thus also the intensity I (see (14.5b)) is proportional to the volume V (it is convenient to introduce the quantity $\overline{(\delta\boldsymbol{\varepsilon})_V^2}$, as it is proportional to $1/V$).

The role played by the temperature fluctuations is small to first approximation (for details see Fabelinskii, 1968) and therefore

$$\overline{(\delta\boldsymbol{\varepsilon})^2} \approx \left(\frac{\partial\boldsymbol{\varepsilon}}{\partial\rho}\right)_T^2 \overline{(\delta\rho)_V^2}\,, \quad \overline{(\delta\rho)^2} = \frac{k_B T\rho}{V}\left(\frac{\partial\rho}{\partial p}\right)_T; \tag{14.9}$$

whence we get from (14.5), (14.7), and (14.8) the formula due to Einstein (1910):

$$\frac{I}{I_0} = \frac{V}{16\pi^2}\left(\frac{\omega}{c}\right)^4\left(\rho\,\frac{\partial\boldsymbol{\varepsilon}}{\partial\rho}\right)_T^2 \beta_T k_B T\sin^2\psi\,, \quad h = \frac{\omega^4}{6\pi c^4}\left(\rho\,\frac{\partial\boldsymbol{\varepsilon}}{\partial\rho}\right)_T^2 \beta_T k_B T\,, \tag{14.10}$$

where $\beta_T = \frac{1}{\rho}\left(\frac{\partial\rho}{\partial p}\right)_T$ is the isothermal compressibility.

For uncondensed gases $\boldsymbol{\varepsilon} - 1 = \text{const}\times\rho$, $\rho\left(\frac{\partial\boldsymbol{\varepsilon}}{\partial\rho}\right)_T = \boldsymbol{\varepsilon} - 1 \approx 2(n-1) \approx 4\pi\alpha N$ and $\beta_T \approx 1/k_B T N$ (N is the particle density in the gas and α the molecular polarizability), and we thus get from (14.10) the Rayleigh formula

$$h = \frac{2\omega^4(n-1)^2}{3\pi c^4 N} = \frac{8\pi\omega^4\alpha^2 N}{3c^4}\,. \tag{14.11}$$

The result (14.11) is equivalent to the one obtained by assuming that each molecule scatters completely independently of the others. Indeed, in the field of the wave an isotropic molecule acquires a dipole moment

$$\mathbf{p} = \alpha\mathbf{E}_0 = \alpha\mathbf{E}_{00}\,e^{-i\omega t}$$

and, hence, per unit time it scatters an energy (see, for instance, (1.85) and (3.1))

$$\int I\,d^2\Omega = \frac{\alpha^2\omega^4}{3c^3}\,|\mathbf{E}_{00}|^2\,;$$

As in this case $I_0 = (c/8\pi)|\mathbf{E}_{00}|^2$, we get at once (14.11) for $h = \int I\, d^2\Omega/I_0$. In gases or, strictly speaking, in sufficiently rarefied (ideal) gases the scattering by density fluctuations turns out to be equivalent to the independent (incoherent) scattering by separate particles (molecules) by virtue of the fact that under such conditions the position of the particles is uncorreated and $\overline{|\delta N|^2} = N$.

We have here for the sake of convenience only briefly repeated (and, as stated, following mainly Landau and Lifshitz, 1960) the method for calculating the intensity of the scattered light. To elucidate the spectral composition or, as one often says, the line width in the spectrum of the scattered light we must clarify the time-dependence of the fluctuations $\delta\boldsymbol{\varepsilon}_{ij}(\mathbf{r}, t)$. We prefer here, however, to start with the problem of the line width both for the emission and for the scattering of light in rarefied gases, that is, assuming the particles to emit and scatter independently of one another. In this way, going later over to scattering in a condensed medium, we illuminate a number of points which are usually insufficiently emphasized in the literature (we use for what follows an earlier paper (Ginzburg, 1972d)).

We start by defining the line width for the emission of light using the widely employed example of a classical damped oscillator. The appropirate equation of motion has the form

$$\ddot{x} + \gamma\dot{x} + \omega_0^2 x = 0 . \tag{14.12}$$

We shall take x_0 to be the initial displacement of the oscillator at time $t = 0$, that is, we use the solution

$$\begin{aligned} x(t) &= x_0\, e^{-\frac{1}{2}\gamma t} \cos(\omega_k t + \phi) , \quad \omega^2 = \omega_0^2 - \tfrac{1}{4}\gamma^2 , \quad t \geq 0 ; \\ x(t) &= 0 , \quad t < 0 , \end{aligned} \tag{14.13}$$

where ϕ is an arbitrary phase. Expanding the oscillation (14.13) in a Fourier integral

$$x(t) = \int_{-\infty}^{+\infty} x_\omega\, e^{-i\omega t}\, d\omega , \quad x_\omega = \frac{1}{2\pi} \int_{-\infty}^{+\infty} x(t)\, e^{i\omega t}\, dt ,$$

we get

$$x_\omega = -\frac{x_0}{4\pi} \left\{ \frac{e^{-i\phi}}{-\frac{1}{2}\gamma - i(\omega_k - \omega)} + \frac{e^{i\phi}}{\frac{1}{2}\gamma + i(\omega_k + \omega)} \right\} . \tag{14.14}$$

The intensity (power) of the dipole radiation is well known to be proportional to $(e\ddot{x})^2$, where e is the particle charge. Clearly, therefore, the spectral density of the intensity $I(\omega)$ is proportional to $\omega^4 |x_\omega|^2$. We shall also assume the phase to be arbitrary and we average over the phase, bearing in mind that one observes the radiation from a collection of oscillators with arbitrary phases. We have then

$$I(\omega) = A\omega^4 \overline{|x_\omega|^2} \equiv A\omega^4 (x^2)_\omega$$
$$= \frac{A x_0^2 \omega^4}{16\pi^2} \left\{ \frac{1}{(\omega_k - \omega)^2 + \frac{1}{4}\gamma^2} + \frac{1}{(\omega_k + \omega)^2 + \frac{1}{4}\gamma^2} \right\} =$$
$$= \frac{A x_0^2 \omega^4 (\omega_0^2 + \omega^2)}{8\pi^2 \left[(\omega^2 - \omega_0^2)^2 + \gamma^2 \omega^2 \right]} , \qquad (14.15)$$

where A is a proportionality coefficient and the averaging is indicated by a bar on top. If, as is usually the case, except in the radio-band,

$$\gamma \ll \omega_0 , \qquad (14.16)$$

we can write to a good approximation

$$I(\omega) = \frac{A x_0^2 \omega_0^4}{16\pi^2 \left[(\omega \quad \omega_0)^2 + \frac{1}{4}\gamma^2 \right]} = \frac{\gamma I_0 / 2\pi}{(\omega - \omega_0)^2 + \frac{1}{4}\gamma^2} , \quad I_0 = \int_0^\infty I(\omega)\, d\omega . \qquad (14.17)$$

One normally uses Eqn.(14.17) for $I(\omega)$; its meaning is completely clear. In the case of the more general Eqn.(14.15) it is clear (vide infra) that x_0^2 cannot in a simple way be expressed in terms of $\overline{x^2} = \int (x^2)_\omega \, d\omega$ and the assumption that x_0 is constant for all oscillators is thus totally arbitrary and unreasonable (vide infra). Both for this reason and also with the aim of a further exposition we consider now a more realistic problem, namely the same oscillator affected by a random force $f(t)$:

$$\ddot{x} + \gamma \dot{x} + \omega_0^2 x = f(t) = \int_{-\infty}^{+\infty} f_\omega e^{-i\omega t} \, d\omega . \qquad (14.18)$$

Hence

$$x_\omega = \frac{f_\omega}{-\omega^2 + \omega_0^2 - i\gamma\omega} . \qquad (14.19)$$

The role of the 'force' f can, for instance, be played by collisions which maintain the mean square amplitude of the oscillations of the oscillator at some unchanging average level. If we assume that $f(t) = \sum_m a_m \delta(t - t_m)$, we have

$$f_\omega = \frac{1}{2\pi} \sum_m a_m \exp(i\omega t_m)$$

and in the case of random (uncorrelated) collisions we get the average value

$$(f^2)_\omega \equiv \frac{1}{4\pi^2} \sum_m a_m^2 .$$

Further, in that case

$$\overline{x^2} = \int_{-\infty}^{+\infty} (x^2)_\omega \, d\omega = (f^2)_\omega \int_{-\infty}^{+\infty} \frac{d\omega}{(\omega^2 - \omega_0^2)^2 + \gamma^2\omega^2} = \frac{\pi(f^2)_\omega}{\gamma\omega_0^2} . \qquad (14.20)$$

The average $\overline{x^2}$ and, thus, the average values of the potential energy $\frac{1}{2}m\omega_0^2\overline{x^2}$ and of the kinetic energy $\frac{1}{2}m\overline{\dot{x}^2}$ turn thus out to be constants for a given $(f^2)_\omega$ = const (in thermal equilibrium these average values equal $\frac{1}{2}k_BT$). The use of Eqn.(14.19) rather than (14.14) is thus not only simpler and more convenient, but also more meaningful. Taking what we have said into account we get at once from (14.19)

$$I(\omega) = A\omega^4(x^2)_\omega = \frac{A\omega^4(f^2)_\omega}{(\omega^2 - \omega_0^2)^2 + \gamma^2\omega^2} . \qquad (14.21)$$

Of course, when condition (14.16) holds, Eqn.(14.21) changes into (14.17). In the general case, however, the spectral density (14.21) of the intensity is obtained, as we have already mentioned, under more meaningful and natural assumptions that Eqn.(14.15). As to the broadening of the emission and absorption lines under real conditions, rather than the simplest model discussed here, there are many possibilities and variants in that respect (see Chen and Takeo, 1957; Sobel'man, 1972).

We shall now consider the same harmonic oscillator but as a scatterer rather than as a spontaneous light emitter. We assume that the incident light is monochromatic, that is, that the field of the incident wave has the form

$$E(t) = \int_{-\infty}^{+\infty} E_\omega e^{-i\omega t} \, d\omega = E_0 e^{-i\omega_e t} , \quad E_\omega = E_0 \, \delta(\omega - \omega_e) , \qquad (14.22)$$

where the frequency ω_e lies far from resonance.

What will be the spectral composition of the scattered light when the damping of the scattering oscillator is taken into account or under conditions when the spontaneous emission of the oscillator undergoes collision broadening (in the latter case one obtains under the simplest assumptions Eqn.(14.17) with $\gamma = 2/\tau$, where τ is the mean time between collisions; see Born, 1933; Chen and Takeo, 1957; Sobel'man, 1972)?

A number of authors gave in their time the following answer to this problem: the width of the scattering line will be the same as in the case of an emission line; the author got often a similar answer, for instance, in examinations. All the same, one sees easily that under the assumptions made the scattered light will be monochromatic, that is, there will practically be no width at all. Indeed, the equation of motion of the oscillator in the field (14.22) which is assumed to be along the x-axis has the form

$$\ddot{x} + \gamma\dot{x} + \omega_0^2 x = f(t) + \frac{e}{m} E_0 e^{-i\omega_e t} . \tag{14.23}$$

Hence

$$x_\omega = \frac{(e/m) E_0 \delta(\omega - \omega_e) + f_\omega}{-\omega^2 + \omega_0^2 + i\gamma\omega} , \tag{14.24}$$

and far from resonance, that is, when $|\omega_e - \omega| \gg \gamma$, and also under the assumption that collisions occur not too often, the term proportional to $\delta(\omega - \omega_e)$ will dominate absolutely, as the random force f has a wide spectrum.

What we have said is, of course, clear also without any spectral expansion: when scattering the light the oscillator performs forced oscillations with the frequency equal to the frequency of the inducing force (scattered wave). Collisions, however, lead, as long as we can neglect their duration $\Delta\tau$, to a change in the amplitude and the phase of the eigenoscillations of the oscillator which have a frequency $\omega_k = (\omega_0^2 - \frac{1}{4}\gamma^2)^{\frac{1}{2}}$, which is assumed to be appreciably different from the frequency ω_e of the incident wave. During a time $\Delta\tau$, when it is a different system which is scattering, the scattering is changed and this may lead, in particular, to a depolarization of the scattered light (McTague and Birnbaum, 1971). Broadening appears, of course, also when one approaches the resonance (although this is in principle completely obvious, one needs separately consider also scattering not of a monochromatic wave, but of alternating pulses).

Far from resonance and when we neglect the duration of the collisions as compared to the mean free flight time the broadening of the scattered line is connected only with the motion of the scatterer. First of all, the normal Doppler broadening occurs (Born, 1933; Chen and Takeo, 1957; Sobel'man, 1972):

$$I(\Omega) = \text{const} \times e^{-\Omega^2/b^2} , \quad \Omega = \omega_e - \omega , \quad b^2 = \frac{8 k_B T \omega_e^2 \sin^2 \frac{1}{2}\theta}{Mc^2} , \tag{14.25}$$

where θ is the scattering angle and T the temperature of the scattering gas, consisting of particles (a system of oscillators) with mass M; clearly Ω is the difference of the frequencies of the incident and the scattered waves.

Moreover, there exists a broadening which is generically connected with the Doppler broadening, while the corresponding intensity is proportional to the square of the pressure. We have earlier (Ginzburg, 1941) considered this effect for a rarefied gas when the mean free path $\ell \gg \lambda_0/2\sin\frac{1}{2}\theta$, $\lambda_0 = 2\pi c/\omega_e$. In the wings of the line (where $\Omega \gg b$) the intensity $I(\Omega) = \mathrm{const} \times p^2/\Omega^6$, where p is the pressure. This broadening is caused by the fact that in collisions the velocity component of the atom (oscillator) in the direction of observation is changed. The Doppler shift of the frequency is thus also changed, that is, the derivative of the phase suffers a discontinuity; in other words, the scattered wave consists of sections with different frequencies, although with a continuous phase. Of course, the Fourier expansion of such a wave has an additional wing, the intensity of which increases with increasing pressure.†

The pressure range for which $\ell \sim \lambda_0/2\sin\frac{1}{2}\theta$ is an intermediate one and is difficult to analyze. If, however, $\ell \ll \lambda_0/2\sin\frac{1}{2}\theta$ (compressed gas) one can use, as in the discussion of scattering in condensed media, a phenomenological approach; to be precise, the Rayleigh scattering is described as scattering by sound and entropy waves (vide infra). In that case the problem of the broadening of the scattering line has been considered already a long time ago (Leontovich, 1931; Ginzburg, 1944b, 1945; Landau and Lifshitz, 1960; Fabelinskii, 1968). We shall return a little later to the scattering in a condensed medium but now we turn to a consideration of the broadening of combinational (Raman) scattering lines in gases.

The normally used classical model which serves for the description of combinational scattering of light by a molecule is an oscillator (with generalized coordinate x, say, which is proportional to the distance between two nuclei in a diatomic molecule) modulating the electron polarizability $\alpha(x)$ of the molecule; the dipole moment of the molecule induced by the incident field is then equal to (for details see Born, 1933; Sushchinskii, 1969; Gorelik and Sushchinskii, 1969)

$$p(t) = \alpha(x)\,E = \alpha(x)\,E_0\,e^{-i\omega_e t}\,, \quad \alpha(x) = \alpha(0) + \left(\frac{d\alpha}{dx}\right)_0 x\,. \qquad (14.26)$$

In some approximation we can describe the change in the coordinate x by Eqn. (14.18). From (14.19) and (14.26) we have then

---

† If the gas consists of different kinds of atoms (with different masses) we must include in the expression for the intensity also a term proportional to Ω^{-4}. Moreover, an additional broadening arises when there are collisions which transfer atoms (molecules) to states with different polarizabilities. We note, finally, that in the case of degenerate levels of the scattering molecule one should add to the Rayleigh (coherent) scattering, the scattering connected with transitions of the molecule to other sublevels of the level considered. In fact, we are dealing here with combinational (Raman) scattering which is accompanied by line broadening (vide infra).

$$p_\omega = \alpha(0)\,E_0\,\delta(\omega-\omega_e) + \left(\frac{d\alpha}{dx}\right)_0 \frac{f_\Omega}{-\Omega^2+\Omega_0^2-i\gamma\Omega} = p_{\omega_e} + p_\Omega \ , \quad \Omega = \omega_e - \omega \ , \qquad (14.27)$$

where we have now denoted for the sake of uniformity the oscillator frequency ω_0 in (14.18) by Ω_0.

The first term in the expression for p_ω corresponding to Rayleigh scattering is now not of interest to us. We can thus write the spectral density of the combinational scattering in the form

$$I(\Omega) = A\omega_e^4\,\overline{|p_\Omega|^2} = \frac{(\gamma\Omega_0^2/\pi)I_0}{(\Omega^2-\Omega_0^2)^2+\gamma^2\Omega^2} \ , \quad I_0 = \int_{-\infty}^{+\infty} I(\Omega)\,d\Omega \ , \qquad (14.28)$$

where we have put $\omega = \omega_e$ which is valid provided $\Omega \ll \omega_e$; we have also assumed that $\overline{|f_\Omega|^2} \equiv (f^2)_\Omega$ = constant.

The frequency region $\Omega < 0$ corresponds to the red satellite and the region $\Omega > 0$ to the violet satellite. If $\Omega_0 \gg \gamma$ we have for each of the satellites

$$I(\Omega) = \frac{\gamma I_0/4\pi}{(\Omega-\Omega_0)^2+\frac{1}{4}\gamma^2} \ , \quad \Omega_0 \gg \gamma \ , \quad \Omega = \omega_e - \omega \ , \qquad (14.29)$$

where I_0 is the total intensity of both satellites.

In the case of emission or absorption lines the condition $\omega_0 \gg \gamma$ is always satisfied in optics (see (14.16)) and the general Eqn.(14.21) is therefore in the optical band not really useful and Eqn.(14.17) is always adequate for this kind of broadening. In the case of scattering, however, the range of applicability of Eqn.(14.28) is appreciably wider at the frequency Ω_0 may be small, as occurs for some oscillations (modes), for instance, when one approaches a second-order phase transition point.†

We obtained above for the width of the emisssion line not only Eqn.(14.21), but also Eqn.(14.15). If, as is sometimes done, we proceed in similar fashion in the application to combinational scattering lines, that is, if we do not introduce the random force $f(t)$, but write in (14.26)

$$x = x_0\,e^{-\frac{1}{2}\gamma t}\cos(\Omega_k t + \phi) \ , \quad \Omega_k^2 = \Omega_0^2 - \tfrac{1}{4}\gamma^2 \ , \qquad (14.30)$$

---

† See Ginzburg, 1963. It is necessary to note that when in that paper we consider scattering near second-order phase transition points or their analogues (see also some of the references cited there) an appreciable latitude was allowed in applications to solids because we neglected the role of shear deformations (Levanyuk, 1974; Ginzburg and Levanyuk, 1974).

the Fourier expansion for

$$p(t) = \left(\frac{d\alpha}{dx}\right)_0 x E_0 e^{-i\omega_e t}$$

leads to a formula like (14.15):

$$I(\Omega) = \frac{A' \omega_e^4 (\Omega_0^2 - \Omega^2)}{(\Omega^2 - \Omega_0^2)^2 + \gamma^2 \Omega^2} . \qquad (14.31)$$

It is just this kind of expression which is sometimes indicated in the literature, and it is then assumed to be more exact than Eqn.(14.28). However, we have seen above that the situation is in fact the opposite and one should use Eqn.(14.28) rather than (14.31) in the framework of the model chosen here.

As the result (14.28) obtained for the width of the combinational scattering lines is, as we said, analogous to Eqn.(14.21) for the emission line width, one might get the impression that essential differences between them are restricted to the case of Rayleigh scattering. However, we have seen that such a conclusion would be too hasty and in fact refers only to the simplest cases, in particular to the oscillator model discussed here which to some extent describes the scattering in gases. When we turn, however, to any kind of scattering in a condensed medium there are, in general, considerable differences in the expressions for the absorption (emission) and scattering line widths.

When light is scattered in sufficiently rarefied gases there is a characteristic independence (incoherence) of the scattering by different volumes or, one may assume, by different molecules (scattering oscillators). In dense gases and in condensed media one cannot take the scattering in different points to be independent, particularly not when analyzing the spectral composition of the scattered light. In those cases an adequate approach, the use of which goes back to the earlier mentioned paper by Einstein (1910), is the idea of the scattering by the spatial Fourier components of the fluctuations of the dielectric permittivity,† or, in fact, the related consideration of different plane waves which propagate in the medium. It was just in that way that above we performed right from the start the calculation of the intensity of the scattered light (see (14.4) to (14.7)). However, now we shall no longer assume the frequency of the scattered light to be the same as that of the incident light.

† One can especially distinguish the interesting problem of the narrowing of the lines of combinational scattering of light in gases when one goes over to large pressures when one can still consider scattering by separate molecules (Alekseev and Sobel'man, 1969).

For the sake of convenience we repeat the notation we use. The wavevectors of the incident and the scattered light are denoted by $\mathbf{k}_e$ and $\mathbf{k}_s$ and the corresponding frequencies by ω_e and ω_s (before we often omitted the index s). Assuming that the medium is transparent for the frequencies ω_e and ω_s we can take all quantities $\mathbf{k}_e$, $\mathbf{k}_s$, ω_e, and ω_s to be real. The scattering wave, say, of the Fourier component of the fluctuating change in the dielectric permittivity $\delta\boldsymbol{\varepsilon}$ (we consider Rayleigh scattering, neglecting anisotropy; for details see Ginzburg, 1944b, 1945; Landau and Lifshitz, 1960; Fabelinskii, 1968) is then characterized by a frequency Ω and a wavevector $\mathbf{q}$ which are equal to

$$\Omega = \omega_e - \omega_s \,, \quad \mathbf{q} = \mathbf{k}_e - \mathbf{k}_s \,. \tag{14.32}$$

If the change in frequency Ω is small, we have $k_s \approx k_e = 2\pi n/\lambda_0 = \omega_e n(\omega_e)/c$ and

$$q \equiv \frac{2\pi}{\Lambda} = \frac{4\pi n}{\lambda_0} \sin\tfrac{1}{2}\theta = \frac{2\omega_e n(\omega_e)}{c} \sin\tfrac{1}{2}\theta \,, \tag{14.33}$$

where $n(\omega_e)$ is the refractive index at frequency $\omega_e \approx \omega_s$ and θ is the scattering angle.

Under the conditions discussed here the intensity I of the scattered light in the volume V per unit angle is given by Eqn. (14.5b).

The spectral composition of the scattered light is determined by the kinetics of the fluctuations $\delta\boldsymbol{\varepsilon}_{\mathbf{q}}$ and, to be precise,

$$I(\Omega) = A\,\overline{|\delta\boldsymbol{\varepsilon}_{\mathbf{q},\Omega}|^2} \equiv A\left((\delta\boldsymbol{\varepsilon})^2_{\mathbf{q}}\right)_\Omega \,,$$

$$\delta\boldsymbol{\varepsilon}_{\mathbf{q},\Omega} = \frac{1}{2\pi}\int_{-\infty}^{+\infty} \delta\boldsymbol{\varepsilon}_{\mathbf{q}}(t)\, e^{-i\Omega t}\, dt \,. \tag{14.34}$$

The meaning of the quantity $((\delta\boldsymbol{\varepsilon})^2_{\mathbf{q}})_\Omega$ is clear from what was said in Chapter 13, that is,

$$\overline{|\delta\boldsymbol{\varepsilon}_{\mathbf{q}}|^2} = \int_{-\infty}^{+\infty} \left((\delta\boldsymbol{\varepsilon})^2_{\mathbf{q}}\right)_\Omega d\Omega \,,$$

and the bar indicates a statistical average.

As we mentioned, to a rather good approximation

$$\delta\boldsymbol{\varepsilon}_{\mathbf{q}} = \left(\frac{\partial\boldsymbol{\varepsilon}}{\partial\rho}\right)_T \delta\rho_{\mathbf{q}} \,,$$

where ρ is the density; the density fluctuations $\delta\rho$ in turn can be expanded in terms of the pressure fluctuations δp and the entropy fluctuations δS

$$\overline{(\delta\rho)^2} = \left(\frac{\partial\rho}{\partial p}\right)_S^2 \overline{(\delta p)^2} + \left(\frac{\partial\rho}{\partial S}\right)_p^2 \overline{(\delta S)^2} .$$

Adiabatic (isentropic) density fluctuations, proportional to δp, change with time according to the laws of hydrodynamics, while the kinetics of isobaric fluctuations, proportional to δS, are determined by the equations of heat conduction. We shall not discuss in detail how to get all the appropriate formulae (see Fabelinskii, 1968, and the literature cited there) but we shall all the same make a few remarks in this connection.

If we assume both viscosity coefficients η and ζ as well as the heat conduction coefficient κ to be equal to zero, sound propagates in a liquid without absorption and the entropy fluctuations do not dissipate. Under such conditions one observes in the spectrum of the scattered light a triplet of unbroadened lines — in the centre a line with an unshifted frequency $\omega_s = \omega_e$ (here $\Omega = \omega_e - \omega_s = 0$) and the Mandel'shtam-Brillouin doublet $\Omega = \pm\Omega_0$ with $\Omega_0 = uq = (2un\omega_e/c)\sin\frac{1}{2}\theta$, where u is the sound velocity for the frequency Ω_0. In quantum language the appearance of the satellites $\Omega = \pm\Omega_0$ is described as light scattering accompanied by the emission of a phonon of energy $\hbar\Omega_0$ and momentum $\hbar\mathbf{q} = (\hbar\Omega_0/u)\,\mathbf{q}/q$ (red satellite) or the absorption of such a phonon (violet satellite). Of course, when applying classical theory we always assume that $\hbar\Omega_0 \ll k_B T$.

If we do not neglect viscosity and heat conductivity, the sound is damped and the entropy fluctuations are dissipated, as a result of which all lines of the triplet are broadened. The kinetics of the isobaric fluctuations is then determined by the heat conduction equation

$$\frac{\partial T}{\partial t} - \chi\nabla^2 T = f_T(t,\mathbf{r}) \quad , \quad \chi = \frac{\kappa}{\rho c_p} \quad , \tag{14.35}$$

where f_T are the random 'forces' caused by the thermal motions in the liquid; the fluctuations of the temperature T are for given pressure proportional to the fluctuations in the entropy S and in final reckoning lead to fluctuations in the density ρ and the permittivity $\boldsymbol{\varepsilon}$ (vide supra).

We thus get from (14.34) and (14.25)

$$I_{isob}(\Omega) = A'(\Omega)\,\overline{(f^2_{T,\mathbf{q}})_\Omega} = \frac{(\gamma/2\pi)\,I_{0,isob}}{\Omega^2 + \frac{1}{4}\gamma^2} ,$$
$$\gamma = 2\chi q^2 = 4\left(\frac{\omega_e n}{c}\right)^2 \chi(1-\cos\theta) \quad , \quad I_{0,isob} = \int_{-\infty}^{+\infty} I_{isob}(\Omega)\,d\Omega , \tag{14.36}$$

where, as before, we have assumed that the frequency-dependence of the quantity

$\overline{(f^2_{T,\mathbf{q}})_\Omega}$ is unimportant.

In the case of the Mandel'shtam-Brillouin components, which correspond to scattering by adiabatic fluctuations we shall not take into account any subtleties connected with the dispersion of the sound (see Landau and Lifshitz, 1959, Chapter 8) and we shall therefore use the following equation for the pressure:

$$\frac{\partial^2 p}{\partial t^2} - u^2 \nabla^2 p - \Gamma \nabla^2 \frac{\partial p}{\partial t} = f_p(t, \mathbf{r}) , \qquad \Gamma = \frac{1}{\rho}\left\{\frac{4}{3}\eta + \zeta + \frac{\kappa}{c_p}\left(\frac{c_p}{c_V} - 1\right)\right\} . \tag{14.37}$$

Hence

$$\left.\begin{aligned} I_{ad} &= \frac{(\gamma/\pi)\,\Omega_0^2\, I_{0,ad}}{(\Omega^2 - \Omega_0^2)^2 + \gamma^2\Omega^2} , \\ \Omega_0 &= uq = \frac{2u\omega_e n}{c}\sin\tfrac{1}{2}\theta , \quad \gamma = \Gamma q^2 , \\ I_{0,ad} &= 2I_0^{MB} = \int_{-\infty}^{+\infty} I_{ad}(\Omega)\, d\Omega , \end{aligned}\right\} \tag{14.38}$$

where I_0^{MB} is the total intensity of a single satellite; for narrow lines (when $\gamma \ll \Omega_0$) we have for each of the satellites

$$\left.\begin{aligned} I^{MB}(\Omega) &= \frac{(\gamma/2\pi)\, I_0^{MB}}{(\Omega - \Omega_0)^2 + \frac{1}{4}\gamma^2} , \quad I_0^{MB} = \int_{-\infty}^{+\infty} I^{MB}(\Omega)\, d\Omega , \\ \tfrac{1}{2}\gamma &= \frac{q^2}{2\rho}\left\{\frac{4}{3}\eta + \zeta + \frac{\kappa}{c_p}\left(\frac{c_p}{c_V} - 1\right)\right\} , \quad q^2 = 2\left(\frac{n\omega_e}{c}\right)^2 (1 - \cos\theta) . \end{aligned}\right\} \tag{14.39}$$

As to the intensities $I_{0,isob}$ and $I_{0,ad}$, their sum is in the simplest case given by Eqn.(14.10) and their ratio is given by $I_{0,ad}/(I_{0,ad} + I_{0,isob}) = c_V/c_p$ that is, $I_{0,isob}/I_{0,ad} = c_p/c_V - 1$ (see Landau and Lifshitz, 1960; Fabelinskii, 1968).

The formulae obtained above are, apart from the notation, the same as well known expressions (see Landau and Lifshitz, 1960, for instance, where the quantity $\frac{1}{2}\gamma$ in (15.36) and (14.39) is denoted by γ; Leontovich, 1931; Ginzburg, 1941; Fabelinskii, 1968). We gave the derivation of the equations nonetheless in order to emphasize the fact which normally remains hidden that one uses forced solutions and not the solutions of the homogeneous equations of motion (we are here dealing with Eqns.(14.35) and (14.37)). Moreover, if we were interested in the propagation of sound in the liquid, we would have used in the approximation considered the equation

$$\frac{\partial^2 p}{\partial t^2} - u^2 \nabla^2 p - \Gamma \nabla^2 \frac{\partial p}{\partial t} = 0 , \qquad (14.40)$$

the solution of which takes for a monochromatic plane wave with real **q** the form

$$p = p_0 e^{i(\mathbf{q}\cdot\mathbf{r}) - i\Omega_q t} = p_0 e^{-\frac{1}{2}\gamma t} e^{i(\mathbf{q}\cdot\mathbf{r}) - i\Omega'_q t}$$
$$\Omega_q = \Omega'_q - \tfrac{1}{2} i\gamma , \ \gamma = \Gamma q^2 , \ \Omega'_q = (\Omega_0^2 - \tfrac{1}{4}\gamma^2)^{\frac{1}{2}} , \ \Omega_0^2 = u^2 q^2 . \qquad (14.41)$$

If, however, we assume that the frequency Ω_q is real, what corresponds to another possible statement of the problem, the wavevector **q** will be complex as there follows from Eqn.(14.40) only a general relation (dispersion relation)

$$\Omega^2 - u^2 q^2 - i\,\Gamma \Omega_q q^2 = 0 . \qquad (14.42)$$

In the case of light scattering, however, both quantities Ω and **q** in (14.32) are real because $\mathbf{k}_e$, $\mathbf{k}_s$, ω_e , and ω_s are real. Such 'sound' waves can propagate in a medium only because they are forced solutions of Eqn.(14.37) and forced solutions do, of course, not have to obey the dispersion equation. When absorption of sound is taken into account it is, strictly speaking, incorrect to speak of light scattering involving the absorption or emission of a phonon — it is not a sound wave which can freely propagate in the given medium which is absorbed or emitted, but some forced sound perturbation with frequency Ω and wavevector **q** given by (14.32). What we have said does, in general, not impede the use of scattering line widths measurements to determine the supersonic absorption coefficient. Indeed, determining the quantity γ from (14.36) or (14.37) we thereby find the coefficients Γ or γ also for sound propagation (see (14.41)). The situation is so simple, however, only because we are dealing here with the neglect of the sound dispersion, that is, the dependence of the frequency-dependence of the viscosity and heat conduction coefficients. When the absorption is strong and, in general, in most cases it is impossible to proceed in that way and the determination of the supersonic velocity and damping (that is, a study of the dispersion equation $F(\Omega_q, q) = 0$ for sound propagation) through the light scattering method may turn out to be difficult. A similar situation arises also in other cases, for instance, for combinational scattering of light in crystals (and, in general, in a condensed medium) when various excitations, such as excitons, polaritons, or magnons, are formed. This class of problems has latterly been the topic of a large number of studies. Here we dwell only on the combinational scattering line width involving the formation of polaritons (real excitons) as the corresponding discussion

(Agranovich and Ginzburg, 1972) is rather closely connected with the preceding part of the chapter.

It is usual to call excitons which propagate in crystals and which are considered taking retardation into account polaritons or, more rarely, real excitons;† in fact this means that we are dealing with 'normal' electromagnetic waves or photons in a medium (for details see Agranovich and Ginzburg, 1966, 1971; Agranovich, 1968). The scattering of light involving the formation of polaritons (and, to be precise, one polariton) while neglecting the damping of the polaritons is combinational (Raman) scattering in which a 'normal' electromagnetic wave — a polariton of frequency Ω and wavevector $\mathbf{q}$, satisfying condition (14.32) — is emitted (or absorbed). In other words, the process discussed is completely analogous to the scattering involving the formation of the Mandel'shtam-Brillouin satellites in liquids (or solids), but with the polaritons (real excitons) replacing the phonons.

For the sake of simplicity we restrict ourselves to an optically isotropic medium‡ and we shall neglect spatial dispersion. The optical properties of the medium are then characterized by the dielectric permittivity $\boldsymbol{\varepsilon}(\omega) = \boldsymbol{\varepsilon}'(\omega) + i\boldsymbol{\varepsilon}''(\omega)$. As above, in the case of Rayleigh scattering, we shall assume the medium to be transparent for incident and scattered waves with frequencies ω_e and ω_s. This means that $\boldsymbol{\varepsilon}(\omega_e)$ and $\boldsymbol{\varepsilon}(\omega_s)$ are real quantities, that is, that we may put $\boldsymbol{\varepsilon}''(\omega_e) = \boldsymbol{\varepsilon}''(\omega_s) = 0$. As far as the scattering wave of frequency $\Omega = \omega_e - \omega_s$ is concerned, it is, in general, not possible to neglect its absorption.

If the wave with frequency Ω propagates freely in the given medium its dispersion relation has the form

$$\frac{c^2q^2}{\Omega^2} \equiv (n + i\kappa)^2 = \boldsymbol{\varepsilon}(\Omega) = \boldsymbol{\varepsilon}'(\Omega) + i\boldsymbol{\varepsilon}''(\Omega) \ . \tag{14.43}$$

---

† The introduction of the term 'real exciton' is connected with the fact that one can also consider other excitons, such as Coulomb and mechanical excitons (see Agranovich and Ginzburg, 1966, 1971; Agronovich, 1968). We emphasize also that the terminology in this field is not established and one should bear this in mind when getting acquainted with the literature.

‡ The 'three-photon' process discussed here (we are dealing with the 'interaction' of three waves or three photons in a medium with frequencies ω_e, ω_s, and Ω) is possible only in a medium without a symmetry centre, but amongst such media there are also non-gyrotropic crystals of the class $T_d \equiv \bar{4}3m$ (such as ZnS and ZnSe) which are optically isotropic when one neglects higher-order spatial dispersion effects (this means that $\boldsymbol{\varepsilon}_{ij}(\omega, \mathbf{k}) = \boldsymbol{\varepsilon}(\omega)\delta_{ij}$).

This equation is, of course, the usual expression connecting Ω and $\mathbf{q}$ for the propagation of transverse electromagnetic waves in an isotropic medium. By virtue of (14.43) the normal (free) waves propagating in the medium in an arbitrary direction (z-direction) are the following

$$\left.\begin{aligned} E &= E_0 \exp\left\{-\frac{\Omega}{c}\kappa z - i\left(\Omega t - \frac{\Omega}{c} n z\right)\right\}, \\ n &= \left\{\tfrac{1}{2}\varepsilon' + \left[\left(\tfrac{1}{2}\varepsilon'\right)^2 + \left(\tfrac{1}{2}\varepsilon''\right)^2\right]^{\frac{1}{2}}\right\}^{\frac{1}{2}}, \\ \kappa &= \left\{-\tfrac{1}{2}\varepsilon' + \left[\left(\tfrac{1}{2}\varepsilon'\right)^2 + \left(\tfrac{1}{2}\varepsilon''\right)^2\right]^{\frac{1}{2}}\right\}^{\frac{1}{2}}. \end{aligned}\right\} \quad (14.44)$$

As a result of the presence of absorption (that is, when $\varepsilon''(\Omega) \neq 0$) the normal waves (polaritons) are absorbed and, for instance, if the frequency Ω is real, the wavevector $\mathbf{q}$ in the normal waves is complex. However, when light is scattered with the formation of polaritons, those have real Ω and $\mathbf{q}$ by virtue of (14.32). This apparent contradiction† is removed when we remember that scattering is a forced process‡ and that the dispersion Eqn. (14.43) does not apply to the polaritons formed in the scattering involving the formation of polaritons when we neglect absorption. When absorption is taken into account, however, it is not a free polariton, but some polariton-like wave which is formed. This does not hinder, of course, the use of combinational scattering of light to study polaritons. The situation is in this respect analogous to that discussed earlier for the case of Rayleigh scattering in liquids. To be precise, we get for the scattering involving the formation of polaritons a formula for the line width $I(\Omega, q)$ in which occur the same parameters which determine also the propagation of the normal electromagnetic waves — the polaritons. For further details and the formula itself for the scattering

---

† The fact that one found here some difficulties is clear from a number of papers cited by Agranovich and Ginzburg (1972). For instance, in one of them an attempt was made to connect Ω and q in the maximum of the combinational scattering line through the relation $c^2q^2/\Omega^2 = n^2$; in another paper the relation $c^2q^2/\Omega^2 = \varepsilon'(\Omega)$ is discussed. In both cases this was done in order to have a real quantity on the right-hand side of the dispersion relation. Such an approach does not lead to agreement with observations and is mainly incorrect in essence as Ω and $\mathbf{q}$ referring to a polariton which is formed in a scattering process are not at all related through a dispersion relation.

‡ We have here in mind any scattering process, including spontaneous ones and not only the so-called induced scattering arising when waves of a large intensity are scattered (Bloembergen, 1967; Fabelinskii, 1968; Sushchinskii, 1969; Gorelik and Sushchinskii, 1969; Starunov and Fabelinskii, 1970).

line width we refer to the paper by Agranovich and Ginzburg (1972). We note here merely the fact that in the paper referred to no random 'forces' $f(\mathbf{r}, t)$ are introduced, the consideration of which is particularly convenient for the classical approach to the scattering problem. Moreover, in the equation for the polariton field in that paper there occurs explicitly a 'force' which takes into account the effect on the medium of the electric fields of the incident and the scattered waves. Such an approach which is equivalent to considering the energy of the interaction of the incident and the scattered waves with the sound or exciton wave which is formed (absorbed) as a result of the scattering is natural in those cases when it is necessary or expedient to evaluate the intensity in the framework of quantum theory.

The examples given above may be thought to demonstrate the specific points of the scattering line width problem as compared to the discussion of the width of an absorption line of light or sound which is determined by the homogeneous equations for the propagation of the appropriate waves. For example, the polariton absorption line is formed when in the crystal a free wave of frequency Ω is absorbed (of course, one must vary the frequency Ω for the occurrence of a line). This all reduces therefore to determining the absorption index $\kappa(\Omega)$ which occurs if the dispersion relation (14.43).

The problem of the line width of scattered light (particularly, when we have in mind also induced Rayleigh and combinational scattering (Bloembergen, 1967; Starunov and Fabelinskii, 1970), let alone the scattering of electromagnetic waves in a plasma and by relativistic particles; vide infra) has many facets and is an important one. In the past it has remained somewhat in the shadow because of the purely experimental difficulties — the absence of suitable sources of monochromatic light which in particular prevented a wide range of studies of scattering line widths. Now, however, with the use of lasers such obstacles have disappeared and this has already led to an impressive range of various studies of light scattering in all possible media. In particular, the spectral composition (width) of scattering lines has been studied most often and, probably, such a tendency will be maintained and strengthened. It is therefore useful to have the above remarks in mind.

We now discuss the case which is particularly important from the point of view of astrophysical and ionospheric applications — the scattering of electromagnetic waves in a plasma.

We start by reminding ourselves how a separate free electron scatters (see, for instance, Heitler, 1947, § 5; Landau and Lifshitz, 1975, § 78). We shall assume the electron to be non-relativistic ($K = \frac{1}{2}mv^2 \ll mc^2 = 5.1 \times 10^5$ eV) and the scattering to be classical ($\hbar\omega \ll mc^2$). The case of relativistic electrons and arbitrary frequencies of the radiation is discussed in Chapter 16. In the electric field of the wave

$$\mathbf{E} = \mathbf{E} \cos\left[(\mathbf{k}\cdot\mathbf{r}) - \omega t)\right] , \tag{14.45}$$

and we find, when we use the equation of motion

$$m\ddot{\mathbf{r}} = e\mathbf{E} \tag{14.46}$$

and neglect in (14.45) the phase $(\mathbf{k}\cdot\mathbf{r})$ as compared to ωt, that the electron acquires a velocity

$$\mathbf{v} \equiv \dot{\mathbf{r}} = -\frac{c\mathbf{E}_0}{m\omega} \sin\omega t + \mathbf{v}_0 . \tag{14.47}$$

In order that not only the velocity v_0 without a field, but also the induced velocity be non-relativistic, the condition

$$\frac{eE_0}{mc\omega} \ll 1 \tag{14.48}$$

must be satisfied, and we shall here assume that to be the case.

Just because we have assumed that $v \sim \sqrt{[v_0^2 + (eE_0/m\omega)^2]} \ll c$ we can use Eqn. (14.46) where we have neglected the Lorentz force $\frac{e}{c}[\mathbf{v} \wedge \mathbf{H}]$. To be precise, this is so, if also $H_0 \sim E_0$. For a plane wave in vacuo, of course, $H_0 = E_0$, but in a medium with refractive index n we have $H_0 = nE_0$ and when $n \gg 1$ the relative role of the magnetic field increases. The same may happen in waveguides where for well-defined oscillations (modes), or in some points, one can also possibly observe the inequality $H_0 \gg E_0$. At any rate, we neglect here the effect of the Lorentz force. We note that the condition for classical behaviour $\hbar\omega \ll mc^2$ automatically guarantees also the possibility to neglect the radiative friction force† as we did in (14.46). Finally, when $v \ll c$ (to be more precise, when $v \ll c/n$) the phase $kr \sim \omega nvt/c \ll \omega t$ which justifies replacing the field (14.45) by $\mathbf{E} = \mathbf{E}_0 \cos\omega t$.

Taking what we have said into account we start from (14.46) and can restrict ourselves to the dipole approximation, where

---

† The condition $\hbar\omega \ll mc^2$ is equivalent to the inequality $\hbar\omega \ll c/(\hbar/mc) \ll c/r_e$, where $r_e = e^2/mc^2 = 2.82 \times 10^{-13}$ cm is the classical electron radius. Under these conditions the radiative friction force is very weak (see Chapter 2).

$$\mathbf{r} = -\frac{e\mathbf{E}_0}{m\omega^2}\cos\omega t \quad , \quad \ddot{\mathbf{p}} = e\ddot{\mathbf{r}} = \frac{e^2}{m}\mathbf{E} = \frac{e^2}{m}\mathbf{E}_0 \cos\omega t \ . \tag{14.49}$$

Hence, using Eqn.(6.28) we get the time-average of the intensity of the radiation scattered into the solid angle $d^2\Omega$:

$$I = \left(\frac{e^2}{mc^2}\right) E_0^2 \frac{cn}{8\pi} \sin^2\psi = I_0 r_e^2 \sin^2\psi = I_0 \, d\sigma . \tag{14.50}$$

as the intensity of the incident radiation is

$$I_0 = \frac{cn}{4\pi}\overline{E^2} = \frac{cn}{8\pi}E_0^2$$

(as in (14.5) the angle between $\mathbf{E}_0$ and the wavevector $\mathbf{k}$ of the scattered wave is denoted by ψ in contrast to Eqn.(6.28) where we used the notation θ). Clearly, the total cross-section for scattering is

$$\sigma = \int d\sigma = \frac{\int I \, d^2\Omega}{I_0} = \tfrac{8}{3}\pi r_e^2 \tag{14.51}$$

(the cross-section $\sigma_T = \frac{8}{3}\pi r_e^2 = \frac{8\pi}{3}\left(\frac{e^2}{mc^2}\right)^2 = 6.65\times 10^{-25}\ \text{cm}^2$ is called the Thomson cross-section). For unpolarized light $d\sigma = \frac{1}{2} r_e^2 (1 + \cos^2\theta)$ but, of course, as before $\sigma = \sigma_T$ (here θ is the scattering angle).

It is clear from the calculation given here that the refractive index of the medium n in which the scattering by a free electron takes place drops out of the expressions for $d\sigma$ (we note that it is more convenient in vacuo to start at once from the expression for $\ddot{\mathbf{p}}$ in (14.49) and to use the well known formulae for the instantaneous intensity $I = (\ddot{\mathbf{p}}^2/4\pi c^3)\sin^2\psi$ and $\int I \, d^2\Omega = (2/3c^3)(\ddot{\mathbf{p}})^2$).

It is clear from (14.7) and (14.51) that the extinction coefficient for a gas of independently scattering electrons of density N equals

$$h = \sigma_T N . \tag{14.52}$$

Assuming the gas to be ideal and evaluating the fluctuations $\delta\boldsymbol{\varepsilon}$ for a gas of free electrons we get the same result from Eqn.(14.10) if we bear in mind that in that case

$$\varepsilon = 1 - \frac{4\pi N e^2}{m\omega^2} \quad , \quad \rho\frac{\partial\boldsymbol{\varepsilon}}{\partial\rho} = N\frac{\partial\boldsymbol{\varepsilon}}{\partial N} = -\frac{4\pi e^2}{m\omega^2} \quad \text{and} \quad \beta_T = \frac{1}{k_B T N} \ .$$

The effect of the ions is here, clearly, completely negligible, not only in the expression for $\boldsymbol{\varepsilon}$ but also in that for the compressibility β_T. In other words we assume that the ions do not contribute to the scattering either directly, or indirectly.

When can one proceed in this way? Of course, first of all the contribution from the ions to ε_{ij} must be small. In an isotropic plasma it is small when $\omega \gg kv_{T,e,i}$ and $\omega \gg \nu_{eff}$ (see (11.40)); in a rarefied, isotropic plasma only the inequality $\omega \gg kv_T$, $v_T = \sqrt{k_B T/m}$ is important. For transverse waves this condition is always satisfied, but it is not at all sufficient for taking the fluctuations that are important in the problem to be independent. Indeed, fluctuations are only independent in volumes at distances from one another $r \gg \ell$, where ℓ is the correlation radius. In an uncharged (neutral) gas the mean free path plays the role of ℓ and in a rarefied plasma it is the Debye radius r_D (see (11.14)).

On the other hand, the phases of the scattered waves differ less that π only for scattering volumes which are smaller than the wavelength $\lambda = \lambda_0/n$. For larger volumes or larger distances between the scattering volumes the phase of the scattered waves is the whole time 'disturbed' due to the thermal motion, that is, the time-dependence of the fluctuations. For the scattered light which is not resolved into a spectrum (or even for a sufficiently wide spectral band) scattering by volumes at distances $r \gg \ell$ is incoherent. Hence it is clear that for $\lambda \gg \ell$ or, to be more precise, when (see (14.33); θ is the scattering angle)

$$\frac{2\pi}{q} = \frac{\lambda}{2\sin\frac{1}{2}\theta} \ll \ell\,, \tag{14.53}$$

the correlation of the fluctuations at distances of the order of ℓ does not play a role in the sense that the phase of the scattered waves is thus 'disturbed' at appreciably smaller distances, at which the gas behaves as being nearly ideal.†

We have noted and used this fact already earlier in applications to a gas of neutral particles. For a plasma, however, the scattering by single electrons only, neglecting the role of the ions completely, is therefore allowed only, provided (see (11.14))

$$\frac{\lambda}{2\sin\frac{1}{2}\theta} = \frac{\pi c}{n\omega\,\sin\frac{1}{2}\theta} \ll r_D = \left(\frac{k_B T}{8\pi N e^2}\right)^{\frac{1}{2}} = 4.9\left(\frac{T(^\circ K)}{N(cm^{-3})}\right)^{\frac{1}{2}} \text{ cm}\,. \tag{14.54}$$

---

† This statement is obvious in the case of a neutral gas — at distances less than the mean free path ℓ the particles do not interact at all. In a gaseous plasma which we are considering the interaction between particles is also weak by virtue of the condition $e^2 N^{\frac{1}{3}} \ll k_B T$ which also guarantees that the inequality $r_D \gg N^{-\frac{1}{3}}$ is satisfied, where $N^{-\frac{1}{3}}$ is the average distance between the particles.

If we are not concerned with scattering over vanishingly small angles, inequality (14.45) is in optics ($\lambda \sim 10^{-3}$ cm = 10 μm) normally always satisfied (we have in mind densities N in regions whence the radiation can in general escape under astronomical conditions). For these or other reasons one can use Eqn.(14.52) when inequality (14.54) is satisfied. In the other limiting case†

$$\frac{\lambda}{2\sin\frac{1}{2}\theta} \gg r_D \tag{14.55}$$

volumes with dimensions much larger than the correlation radius scatter incoherently and it is no longer possible to consider the plasma to be an ideal gas of electrons. In general, one must proceed here in accordance with the general theory expounded above — that is, evaluate $\delta\boldsymbol{\varepsilon}_{\mathbf{q}}$, or, to find the spectrum, determine the quantity $\delta\boldsymbol{\varepsilon}_{\mathbf{q},\Omega}$ (see (14.5b) and (14.34)). One can find such calculations in the books by Sitenko (1967, 1977; see also Zheleznyakov, 1970; Kaplan and Tsytovich, 1973; Tsytovich and Kaplan, 1974; Tsytovich, 1977). Something can, however, be said also on a more elementary level.

Just as the density fluctuations in a liquid can be split into adiabatic and isobaric ones, the fluctuations $\delta\boldsymbol{\varepsilon}$ in an isotropic plasma can be split into $\delta\boldsymbol{\varepsilon}_n$ in which only the density of the plasma is changed, but not its charge, and $\delta\boldsymbol{\varepsilon}_e$ connected with changes in the charge. In other words, the $\delta\boldsymbol{\varepsilon}_n$ fluctuations are similar to the fluctuations caused by sound waves and in them the electrons and ions are 'attached' to one another and no charge arises. On the other hand, one can expand the $\delta\boldsymbol{\varepsilon}_e$ fluctuations in terms of the high-frequency plasma waves — in them the ions are fixed and the electrons oscillate with frequencies close to $\omega_{pe} \equiv \omega_p = (4\pi Ne^2/m)$. Up to terms of order m/M the two kinds of fluctuations are statistically independent and thus

$$\overline{(\delta\boldsymbol{\varepsilon})^2} = \overline{(\delta\boldsymbol{\varepsilon}_n)^2} + \overline{(\delta\boldsymbol{\varepsilon}_e)^2} \ . \tag{14.56}$$

In the range (14.55) under consideration, but provided the mean free path is large compared to λ one may think that the neutral $\delta\boldsymbol{\varepsilon}_n$ fluctuations will be close to the fluctuations in the corresponding neutral gas with a density of all particles equal to 2N. It is just the total density which is here important as it determines the pressure which is independent of the particle mass

---

† In the case of a gas of neutral particles we must replace here r_D by the mean free path $\ell \sim v_T/\nu_{eff} \sim 1/\pi a^2 N_m$, where $\pi a^2 \sim 10^{-15}$ cm^2 is the cross-section of the molecules (atoms) and N_m their density.

in thermal equilibrium ($p = 2Nk_BT$). In that case the compressibility is thus $\beta_T = \frac{1}{2}Nk_BT$, but as before $\rho\frac{\partial\varepsilon}{\partial\rho} = -4\pi e^2/m\omega^2$. From this and from what we have said earlier one sees easily that the intensity of the waves scattered by the $\delta\varepsilon_n$ fluctuations differs from the intensity of the waves scattered by free electrons merely by a factor $\frac{1}{2}$, that is

$$h_n = \tfrac{1}{2}\sigma_T N . \qquad (14.57)$$

A more rigorous theory (Sitenko, 1967, 1977) confirms this conclusion (Ginzburg and Zheleznyakov, 1959a, 1965; Ginzburg, Zheleznyakov, and Eidman, 1962). It is rather interesting and also important that longitudinal waves with wavelengths $\lambda \gg r_D$ are scattered by the $\delta\varepsilon_n$ fluctuations exacly like transverse waves. Indeed, when waves are scattered by a volume small compared to the wavelength λ, the orientation of the vector $\mathbf{E}_0$ relative to $\mathbf{k}$ is unimportant. In the longitudinal wave with field $\mathbf{E} = \mathbf{E}_0 \sin\omega t$ the intensity of the scattered transverse waves (we are thus dealing with the transformation of longitudinal waves into transverse ones as a result of scattering (Ginzburg and Zhelexnyakov, 1959a, 1965; Ginzburg, Zheleznyakov, and Eidman, 1962)) is thus given by Eqn. (14.50) with an extra factor $\frac{1}{2}N$ (we consider scattering per unit volume)

$$I_\perp = \left(\frac{e^2}{mc^2}\right)^2 \frac{cn(\omega)}{8\pi} N E_0^2 \sin^2\psi ,$$

$$\int I_\perp d^2\Omega = \frac{1}{6}\left(\frac{e^2}{mc^2}\right)^2 cn(\omega)\, E_0^2 . \qquad (14.58)$$

We did not introduce here the extinction coefficient as we were not concerned about the energy flux in the longitudinal wave (this point is unimportant in this case as one is usually interested just in the way the intensity $I_\perp$ depends on E_0^2).

In the short-wavelength (high-frequency) case when condition (14.54) is satisfied the shape of the spectrum of the scattered light is the Doppler shape (see (14.25), but with M replaced by m), In that case the characteristic width of the spectrum is $\Delta\omega \sim \omega_e\sqrt{(k_BT/mc^2)}$. When long-wavelength waves are scattered (condition (14.55)) we have for the scattering by the $\delta\varepsilon_n$ fluctuations which is the analogue of the Rayleigh scattering in liquids or gases $\Delta\omega \sim \omega_e\sqrt{(k_BT/Mc^2)}$. Indeed, the $\delta\varepsilon_n$ fluctuations dissipate with speeds of the order of the ion velocity $v_{Ti} \sim \sqrt{k_BT/M}$ and, as always, $\Delta\omega \sim (v/c)\omega_e$, where v is the characteristic velocity of the motion.

As we pointed out, the $\delta\varepsilon_e$ fluctuations are connected with plasma waves and have a characteristic frequency ω_p. The scattering by these fluctuations is similar to combinational scattering and leads to the appearance of satellites with frequencies differing from the frequency of the scattered light by $\Omega \sim \omega_p$. In quantum language we are dealing here with the creation and absorption by the incident wave of a single plasmon of energy $\hbar\omega \sim \hbar\omega_p$. The width of the satellites is determined by the damping of the corresponding plasma waves — waves with frequency $\Omega = \omega_e - \omega_s$ (in (14.25) we wrote $\Omega = \omega_e - \omega$, that is, we dropped the index s). The intensity of the satellite depends on the magnitude of the charge fluctuations. When $\lambda \gg r_D$ the plasma waves are weakly damped (and the satellites are thus narrow), but the mean square of the charge fluctuations which is proportional to the quantity $(\delta\varepsilon_e)^2$ decreases steeply — it contains an additional factor $(r_D/\lambda)^2$ as compared to the corresponding expression for $\lambda \ll r_D$. Such a result is physically rather clear — at distances larger than r_D the electrons are still 'attached' to the ions and with increasing λ the thermal motion find it 'more difficult' to separate the electrons from the ions and to produce charge fluctuations or, put differently, to produce plasma waves with an appreciable amplitude.† For long wavelengths $\lambda \gg r_D$ the intensity of combinational scattering involving the creation or absorption of plasma waves (plasmons) turns thus out to be small while the extinction coefficient for transverse waves is given by Eqn.(14.57). Such a situation is in reality realized for the scattering of radio-waves in the ionosphere (we are speaking here about the incoherent scattering method for radio-location in the ionosphere).

On the whole the problem of scattering of electromagnetic waves in a plasma is, of course, far from exhausted by the remarks we have made here. Particular attention also merits taking into account the effect of a magnetic field, the case of a non-isothermal plasma, the transformation of transverse waves into longitudinal ones, of longitudinal waves into transverse ones, and so on. However, we shall not enlarge here on this theme (see Sitenko, 1967, 1977; Zheleznyakov, 1970; Kaplan and Tsytovich, 1973; Tsytovich. 1977; and the literature cited in those books.

---

† We indicate here the result; the detailed derivation of it can be found in Sitenko's book (1967, §§ 4 and 5): for an isothermal electron-ion plasma the mean square charge fluctuations are equal to

$$\overline{|\rho_{\mathbf{k}}|^2} = \frac{2Ne^2k^2}{k^2 + 8\pi Ne^2/k_BT} = \frac{2Ne^2k^2r_D^2}{k^2r_D^2 + 1}, \quad k = \frac{2\pi}{\lambda}$$

Clearly, $\overline{|\rho_{\mathbf{k}}|^2} = 2Ne^2$ when $\lambda \ll r_D$ and $\overline{|\rho_{\mathbf{k}}|^2} = 8\pi^2Ne^2r_D^2/\lambda^2$ when $\lambda \gg r_D$.

In concluding this chapter we shall consider yet one more unusual kind of scattering which can be called transition scattering (Ginzburg and Tsytovich, 1974b, 1978). When various perturbations and waves propagate in a medium there occur in a number of cases also what one may call dielectric permittivity waves. In the simplest case this means that the permittivity changes as

$$\boldsymbol{\varepsilon} = \boldsymbol{\varepsilon}_0 + \boldsymbol{\varepsilon}_1 \cos\left[(\mathbf{k}_0 \cdot \mathbf{r}) - \omega_0 t - \phi_0\right] . \tag{14.59}$$

In an isotropic plasma, when $\boldsymbol{\varepsilon} = 1 - 4\pi N e^2/m\omega^2$ and when a longitudinal (plasma) wave propagates in which the field $\mathbf{E} = \mathbf{E}_0 \cos\left[(\mathbf{k}_0 \cdot \mathbf{r}) - \omega_0 t\right]$ changes with frequency $\omega_0 = \omega_{pe} = \sqrt{(4\pi N_0 e^2/m)}$, we have thus to a first approximation

$$\boldsymbol{\varepsilon}(\omega) = 1 - \frac{\omega_{pe}^2}{\omega^2} + \frac{e k_0 E_0}{m\omega^2} \sin\left[(\mathbf{k}_0 \cdot \mathbf{r}) - \omega_0 t\right] . \tag{14.60}$$

In fact, by virtue of the equation $\mathrm{div}\,\mathbf{E} = 4\pi e(N - N_0)$ and the condition for the wave being longitudinal, $(\mathbf{k}_0 \cdot \mathbf{E}_0) = k_0 E_0$, we get at once (14.60). The plasma wave clearly changes here the original electron density N_0 and this leads to a change in the density

$$\delta N = N - N_0 = - \frac{k_0 E_0}{4\pi e} \sin\left[(\mathbf{k}_0 \cdot \mathbf{r}) - \omega_0 t\right]$$

and thereby to $\delta\boldsymbol{\varepsilon} = - 4\pi e^2 \delta N/m\omega^2$.

In the example given here of (14.60) we are dealing with one of the well known non-linear effects in a plasma (Ginsburg, 1970b; Tsytovich, 1977) and the permittivity wave is also connected with an electric field wave. This is, however, not at all necessary. For instance, in a longitudinal acoustic (sound) wave in a liquid or a neutral gas the electric field is practically and even strictly equal to zero, but by virtue of the change in the density the permittivity $\boldsymbol{\varepsilon}$ changes and may have the form (14.59).

What happens if permittivity waves are incident on a charge e? If that charge moves with a constant velocity $\mathbf{v}$ we are dealing with one form of transition radiation (see Chapter 7). Indeed, the charge moves in this case in a non-uniform medium with permittivity (14.59) and, hence, it must radiate. One may thus say that there occurs a transformation of the permittivity wave into electromagnetic waves and this transformation is realized through transition radiation. Nonetheless we prefer the term 'transition scattering' as it arises also in the case of a fixed charge. To be precise, the situation is the following. For a moving charge (when $v \neq 0$) radiation occurs both when $\omega_0 = 0$, $\mathbf{k}_0 \neq 0$ and when $\omega_0 \neq 0$, $\mathbf{k}_0 = 0$ (see Chapter 7). However, for a fixed charge ($v = 0$) the emission of transverse waves or, in other words, the transformation (scattering) of permittivity waves into transverse ones occurs only when $\omega_0 \neq 0$, $\mathbf{k}_0 \neq 0$. The mechanism of this transition scattering (the term 'transition' is caused by its relation, already noted, with the transition radiation) consists in the fact that the permittivity wave modulates the permittivity ε of the medium near the charge causing thereby the occurrence of a variable polarization and, hence, the emission of electromagnetic waves. The scattering charge can in that case remain completely fixed because in the present case it is surrounded by a field — clearly the longitudinal (Coulomb) field $\mathbf{E}$ which causes the additional polarization $\delta\mathbf{P} = \frac{\delta\varepsilon}{4\pi}\mathbf{E}$.

Without discussing the calculations (see Ginzburg and Tsytovich, 1974b, 1978) we give the result for the time-averaged integral intensity (power) of the transition scattering in one particular case. Let the permittivity wave (14.59) be incident upon a fixed (pinned) charge e and let the emitted (scattered) transverse waves have an appreciable shorter wavelength than the permittivity wave. This means that the magnitude of the wavevector in the transverse waves $k = 2\pi/\lambda = (\omega_0/c)\sqrt{\varepsilon(\omega_0)}$ is small compared to k_0; this should be the case if the velocity of the permittivity waves

$$v_{\varepsilon} = \omega_0/k_0 \ll v_{ph} = \omega_0/k = c/\sqrt{\varepsilon_0}\,,$$

the phase velocity of the transverse electromagnetic waves (the frequency of these waves is, of course, equal to the frequency ω_0 of the permittivity wave).

Under such conditions

$$\frac{dW}{dt} = \int I\, d^2\Omega = \frac{e^2\,\omega_0^4\,\varepsilon_1^2\sqrt{\varepsilon_0(\omega_0)}}{3\,c^2\,k_0^2\,\varepsilon_0^2(0)}\,. \tag{14.61}$$

The appearance here of the quantity $\varepsilon_0(0) \equiv \varepsilon_0(\omega = 0)$ is connected with the fact that the scattering charge is fixed and that the static polarization of the medium produced by it depends just on the value $\varepsilon_0(0)$. For comparison we remind ourselves that a free charge e of mass M in a medium with refractive index $n = \sqrt{\varepsilon_0(\omega_0)}$ scatters an electromagnetic wave of frequency ω_0 and amplitude E_0 in such a way that the corresponding power equals (see, for instance, (14.58))

$$\frac{dW}{dt} = \int I\, d^2\Omega = \frac{e^4\sqrt{\varepsilon_0(\omega_0)}}{3M^2\,c^3}\,E_0^2\,. \tag{14.62}$$

Of course, as $M\to\infty$, the scattering (14.62) vanishes completely, while the transition scattering is independent of the mass of the particle.

Transition scattering is very important for the scattering of longitudinal (plasma) waves by electrons and ions in a plasma (Tsytovich, 1977; Ginzburg and Tsytovich, 1978). However, in that case we must take simultaneously into account also the normal scattering as the mass of the electrons and ions is finite, while there is an electric field in a longitudinal wave. Moreover, one must take spatial dispersion and the motion of the charges into consideration. As a result it turns out, for instance, that the total power when a plasma wave is scattered by an electron which is non-moving (when there are no waves) in the plasma is $\frac{1}{4}$ the power for Thomson scattering in vacuo, that is $\sigma = \frac{1}{4}\sigma_T$. On the other hand, for the scattering of plasma waves by ions the transition scattering is the main scattering mechanism.

Chapter XV

COSMIC RAY ASTROPHYSICS

Introductory remarks. Model for the propagation of cosmic rays.
General characteristics of the problems involved.
Ionization energy losses.
Beam instability and plasma effects in cosmic rays.
Transfer equation in the diffusion approximation.
Simplifications of the transfer equations in the case of the proton-nuclear and electron components. Some estimates.

Over the last 30 to 35 years the appearance of astronomy (and in particular of astrophysics) has changed appreciably. Of course, this period is, generally speaking, not a short one compared to present 'norms' characterizing the development of science. The history of astronomy stretches, however, over thousands of years and the profound change during the last three decades or so is particularly striking. The important fact is that only in this last period astronomy became or, one should rather say, gradually turned into all-wavelength astronomy whereas earlier (practically up to 1945) almost all astronomical information reached us in the optical, and most often even in a yet narrower, the visible, band of electromagnetic waves. However, at present radio-astronomy (mainly in the centimetre, decimetre, and meter wavelength bands) in its general level has overtaken and to some extent outdistanced optical astronomy. X-ray astronomy, born only in 1962 (we have in mind extra-solar X-ray astronomy; X-ray emission from the Sun was observed already in 1948) is now in the stage of rapid flowering and growth. Far-infra-red and gamma-astronomy has also started to develop. Indeed, in this way the whole spectrum of electromagnetic waves either turned out to be mastered, or started to be studied. In the plane characteristic for the present stage in the development of astronomy we must add to this cosmic ray astrophysics (in this case high-energy charged particles — the cosmic rays — carry the information) and also the idea of neutrino astronomy and the astronomy of gravitational waves.

This is not the place to characterize in detail the processes, originating in astronomy or to dwell upon the many achievements (see, for instance, Kraus, 1966; Hayakawa, 1969; Pacholczyk, 1970; Ginzburg, 1970a; Wolfendale, 1970;

and also Ginzburg and Syrovatskii, 1964, 1969; Ginzburg, Sazonov, and Syrovatskii, 1968; Weekes, 1969; Zheleznyakov, 1970; Kaplan and Tsytovich, 1973; Ozernoi, Prolutskii, and Rozental', 1978). We wish only to elucidate the reason why the mechanisms of X-ray and gamma-emission, the analysis of the propagation and of the transformation of the chemical composition of the cosmic rays when they wander through interstellar space, neutrino physics, and many other problems came into astronomy's field of view. Of course, in the teaching of theoretical physics (and not only to future astronomers, but also to an appreciably wider circle of specialists) one should find some reflection of the whole gamut of topics which are important for astronomy and cosmic studies. In the present book Chapters 5 and 9 had an 'astronomical bias' apart from separate remarks in other chapters. The present and the next two chapters are devoted to a number of processes which occur in the Universe (but, of course, not only in the Universe) and we shall restrict ourselves only to that part of the problems which refers to so-called high-energy astrophysics. This term is itself not yet well established but one meets it rather often and it refers here to cosmic ray astrophysics†, X-ray astronomy, gamma-astronomy, and neutrino astronomy (in that case we have in mind processes involving cosmic neutrinos with relatively high energies — larger than, say, 0.1 to 1 MeV). However, we shall not consider neutrino astronomy (see Weekes, 1969; Wolfendale, 1970; Ozernoi, Prilutskii, and Rozental', 1978) and we restrict ourselves to a few problems connected with cosmic ray astrophysics (present chapter), with X-ray astronomy (Chapter 16), and with gamma astronomy (Chapter 17). We must, however, bear in mind that these problems are interrelated and in the majority of cases it is here not easy to, and one should not, draw any sharp boundaries.

Only charged particles (protons, nuclei, electrons, and positrons) of cosmic origins and only those with sufficiently high energies, are called cosmic rays. A clear agreement about the terminology does not exist, but usually only particles with kinetic energies $E_k > 100\,\text{MeV}$ are called cosmic rays and softer, but still fast particles are called sub-cosmic rays.

The main quantity characterizing the cosmic rays is their intensity J (sometimes this quantity is also called the flux in a given direction)‡. By

---

† According to a well-established tradition cosmic ray astrophysics is called also the problem of the origin of cosmic rays.

‡ See footnote on next page

definition J is the number of particles per unit solid angle passing per unit time through unit area perpendicular to the line of sight. The unit of J is the quantity.

$$\frac{\text{number of particles}}{\text{cm}^2.\text{s. sterad}} = 10^4 \frac{\text{number of particles}}{\text{m}^2.\text{s.sterad}} .$$

The flux of particles of the ith kind with intensity J_i equals

$$F_{i\Omega} = \int J_i \cos\theta \, d^2\Omega ,$$

where θ is the angle between the normal to the area and the direction of the particle velocity and $d^2\Omega$ is an element of solid angle. For isotropic radiation the particle flux F_i from a hemi-sphere of directions equals

$$F_i = 2\pi \int_0^{\frac{1}{2}\pi} J_i \cos\theta \sin\theta \, d\theta = \pi J_i . \tag{15.1}$$

The density N_i of particles with velocities v_i is in the case of isotropic radiation equal to

$$N_i = \frac{4\pi}{v_i} J_i . \tag{15.2}$$

One is usually dealing not with mono-energetic particles but with a distribution of particles over energies (that is, as one says, with an energy spectrum). The main quantity is then the spectral (differential) intensity $J_i(E)$ such that $J_i(E)\,dE$ is the intensity of particles with their total energy E lying in the interval $E, E+dE$. The intensity of particles with energies larger than E (the integral intensity) equals

$$J_i(>E) = \int_E^{\infty} J_i(E) \, dE . \tag{15.3}$$

In the case of an isotropic distribution of particles with mass M_i we have

$$N_i(>E) = 4\pi \int \frac{J_i(E)}{v} dE , \quad E = \frac{M_i c^2}{\sqrt{(1 - v^2/c^2)}} . \tag{15.4}$$

The kinetic energy density of isotropic cosmic rays is equal to

$$w_i = \int E_k N_i(E) \, dE = \int \frac{4\pi}{v} E_k J_i(E) \, dE . \tag{15.5}$$

---

† Intensity (especially in the case of electromagnetic radiation) is also the term for the energy flux per unit solid angle. Such an 'energy' intensity is equal to $I = E_k J$, where E_k is the kinetic energy of the particles or the photon energy $h\nu = \hbar\omega$ (we have here in mind mono-energetic particles or photons). We note that following a widely spread notation, we denote here and henceforth the particle energy by E and not by $\mathcal{E}$ as we did in earlier chapters.

One can also introduce the energy intensity

$$I_i = \int E_k J_i(E)\, dE \,, \tag{15.6}$$

but it is rarely used.

For ultra-relativistic particles we have

$$I_i = \frac{c}{4\pi} \int E_k N_i(E)\, dE = \frac{cw_i}{4\pi} \,. \tag{15.7}$$

Of course, the use of the total energy $E = E_k + Mc^2$ is convenient only in the relativistic region, but it is just that one which is studied for the cosmic rays at the Earth. For soft cosmic and sub-cosmic rays one used more often the kinetic energy E_k. Moreover, it is convenient to use for nuclei not only the total energy E or the kinetic energy E_k, but also the total energy per nucleon $\epsilon = E/A$ or the kinetic energy per nucleon $\epsilon_k = E_k/A$, where A is the atomic weight or, to be precise, the mass number of the nucleus. Finally, Eqns.(15.2) and (15.4) have been written down assuming the particle distribution to be isotropic because the cosmic rays at the earth, if we eliminate the effect of the Earth's magnetic field, are to a high degree isotropic.

The degree of anisotropy of the cosmic rays is defined as follows:

$$\delta = \frac{J_{max} - J_{min}}{J_{max} - J_{min}}, \tag{15.8}$$

where J_{max} and J_{min} are, respectively, the maximum and minimum cosmic ray intensities as far as direction is concerned (we have assumed here that $J(\theta)$ has only one maximum, say, in the direction $\theta = 0$; in other words, we assume a dependence such as $J(\theta) = J_0 + J_1 \cos\theta$, so that $\delta = J_1/J_0$). The degree of anisotropy of the cosmic rays has not yet been established reliably; for all cosmic rays with energies larger than 100 GeV[†] the value $\delta \leq 10^{-4}$. The anisotropy can thus be shown up only as the result of special studies, while in all other cases we can with justice assume the cosmic rays to be completely isotropic (we remind ourselves once more that we have assumed that the effect of the Earth's magnetic field has been eliminated). The problem of the study of the primary cosmic rays, that is, cosmic rays outside the atmosphere or taking its effect into account, is thus in fact the determination of the functions $J_i(E)$ for all components of the cosmic rays — for the protons and nuclei (the

---

[†] This means that $J_0 = J_0(E > 100\,\text{GeV})$ and $J_1 = J_1(E > 100\,\text{GeV})$; for larger E and especially for $E > 10^{16}$ to 10^{17} eV the degree of anisotropy may be larger than indicated.

proton-nuclear component) and for the electron-positron component. However, the fraction of positrons in the electron-positron component is very small for $E > 1$ GeV (a few per cent) and they have as yet not been reliably distinguished from the electrons. This is also not done in the overwhelming majority of cases and one measures the intensity $J_e(E)$ of the whole electron-positron component which is simply called the electron component. In the case of the proton-nuclear component the separation of the nuclei into charges, let alone separation of isotopes, is far from always done; one often, therefore, considers the total cosmic ray intensity $J_{c.r.}(E)$ or in practice their total proton-nuclear component as the fraction of electrons (that is, the ratio $J_e(E)/J_{c.r.}(E)$) is of the order of one per cent, and, moreover, electrons are relatively easily separated.

We do not plan to give here a detailed introduction to cosmic rays (see Ginzburg and Syrovatskii, 1964a, 1966b, 1973; Hayakawa, 1969; Weekes, 1969; Ginzburg, 1970a; Hobart, 1973; Ginzburg and Ptuskin, 1976a,b; Ozernoi, Prilutskii, and Rozental', 1978), and we restrict ourselves to only a few remarks including an indication of some characteristic values of such quantities as J and w. At the Earth (outside the influence of the Earth's magnetic field) we can as a guideline take for all cosmic rays the following values:†

$$\left.\begin{aligned} J_{c.r.} &\equiv J \sim 0.2 \text{ to } 0.3 \,\frac{\text{particles}}{\text{cm}^2.\text{s.sterad}}, \\ N_{c.r.} &\sim \frac{4\pi J}{c} \sim 10^{-10}\,\frac{\text{particles}}{\text{cm}^3}, \\ w_{c.r.} &\sim 10^{-12}\,\frac{\text{erg}}{\text{cm}^3} \sim 1\,\frac{\text{eV}}{\text{cm}^3}, \\ I &\sim \frac{cw_{c.r.}}{4\pi} \sim 10^{-3}\,\frac{\text{erg}}{\text{cm}^2.\text{s.sterad}}. \end{aligned}\right\} \qquad (15.9)$$

From Table 15.1 one can reach some idea about the chemical composition of the relativistic cosmic rays ($\epsilon = E/A > 2.5$ GeV/nucleon), where we have split the nuclei into the traditionally used groups (for instance, the L-group contains

---

† We must bear in mind that the energy spectrum of the cosmic rays at the Earth has a maximum corresponding to an energy $E_k \sim 250$ MeV for protons. The cited values $\int_{E_k = 100\,\text{MeV}}^{\infty} J(E)\,dE$ and similar integrals therefore converge. However, their magnitude changes with the solar activity cycle because the contribution from slower particles changes and it is thus to some extent sensitive to the lower integration limit.

the Li, Be, and B nuclei). One should bear in mind that the errors in the values in this table are not less than tens of per cents — the data change when new material is accumulated and when more refined experimental methods are used. The last allow us in a number of cases to obtain information not only about groups of nuclei, but also about separate nuclei. Moreover, recently relatively very rare nuclei with $Z \geq 83$ (that is, heavier than lead) have been observed, the quantity of which in cosmic rays is 8 orders of magnitude less than of all nuclei of group H.

Group of nuclei	Atomic Number	Intensity $J(>\epsilon = 2.5$ GeV/Nucleon) particles/m^2.s.sterad.	Abundance relative to H group nuclei: in cosmic rays	Abundance relative to H group nuclei: average in the Universe
p	1	1300	650	3000 to 7000
α	2	94	47	250 to 1000
L	3 to 5	2.0	1	10^{-5}
M	6 to 9	6.7	3.3	2.5 to 10
H	≥ 10	2.0	1	1
VH	≥ 20	0.5	0.26	0.05
VVH	≥ 30	$\sim 10^{-4}$	$\sim 10^{-4}$	$\sim 10^{-4}$

The most important feature characterizing the composition of the cosmic rays which is clear from Table 15.1 consists in the presence of a rather considerable flux of L nuclei, notwithstanding their negligible amount on average in nature. This peculiarity confirmed also separately for Li, Be, and B nuclei, and also for other rare nuclei (for instance the ^{3}He nucleus), indicates the appreciable role played by the transformation of the chemical composition of the cosmic rays when they propagate in interstellar space and, possibly, also in the sources (that is, in the regions where the cosmic rays originate or, in other words, are accelerated).

The energy dependence (energy spectrum) of the cosmic rays is commonly written in the form

$$J_A(>\epsilon) = \int_\epsilon J_A(\epsilon)\, d\epsilon = K_A \epsilon^{-(\gamma-1)}, \quad J_A(\epsilon) = (\gamma-1) K_A \epsilon^{-\gamma}, \qquad (15.10)$$

where, as already stated, $\epsilon = E/A$ is the energy per nucleon and the index A indicates that we are dealing with nuclei (or group of nuclei) of average atomic number A; moreover, we have introduced similar quantities $J(>E)$ and $J(E)$ for all cosmic rays.

In practice the spectrum is not a power-law, that is, the index γ is energy dependent. However, and this is rather important, in a very wide range of energies the approximation of the spectrum in the form (15.10) turns out to be a good one. For instance, in the energy range 2×10^9 eV $< E < 3 \times 10^{15}$ eV according to a number of data $\gamma = 2.7 \pm 0.2$. Apparently we can now assume the value $\gamma = 2.7$ or $\gamma = 2.6$ to be the best. To give some ideas we give, as an example, the following cosmic ray spectrum in the range 10^{10} eV $< E < 10^{15}$ eV :

$$J(>E) = (5.3 \pm 1.1)\, 10^{-10} \left(\frac{E(\text{eV})}{6 \times 10^{14}}\right)^{-(\gamma-1)} \approx$$

$$\approx \left(\frac{E(\text{eV})}{10^9}\right)^{-1.6} \frac{\text{particles}}{\text{cm}^2.\text{s.sterad}}, \quad \gamma = 2.62 \pm 0.05 . \qquad (15.11)$$

In the low-energy region $E_k < 10^9$ to 10^{10} eV the index γ changes and the spectrum depends strongly on the level of solar activity. We shall not discuss that region. We note only that the very important problem (in the framework of theories about the origin of the cosmic rays) of the shape of the spectrum in the low-energy region and, in particular, of the presence of a maximum in the energy spectrum far from the Sun (beyond the limits of the solar system) has not yet been elucidated. Apparently, down to energies $\epsilon_k \sim 100$ MeV/nucleon there is not yet a maximum in the spectrum of the galactic cosmic rays. At an energy $E \sim 10^{15}$ eV there is a more or less steep bend or, at any rate, a change in the spectrum and for $E > 10^{15}$ eV Eqn.(15.11) does not hold, and the following spectrum is closer to reality

$$J(>E) = (2.0 \pm 0.8)\, 10^{-10} \left(\frac{E(\text{eV})}{10^{15}}\right)^{-(\gamma-1)} \frac{\text{particles}}{\text{cm}^2.\text{s.sterad}}, \qquad (15.12)$$

$$\gamma = 3.2 \pm 0.2 ;$$

according to other data the factor (2.0 ± 0.8) in (15.12) must be replaced by (3.74 ± 0.20) and $\gamma = 3.16 \pm 0.1$. When $E \sim 10^{15}$ eV the spectra (15.11) and (15.12) agree ('join') within the limits of the attainable accuracy, as should be the case. Possibly, the spectrum for $E \geqslant 10^{18}$ eV flattens again, but this assumption has not yet been proved. We show in Fig.15.1 the integral cosmic ray spectrum (for $E > 10^{10}$ eV; we remind ourselves that γ is the index of the differential spectrum; see (15.10)).

The maximum energy of observed cosmic rays is of the order of 10^{20} to 10^{21} eV. In that region the cosmic ray spectrum must, in general, 'steepen' suddenly as a result of the considerable losses which cosmic rays undergo at such large energies when they interact with the radiation which is present in the interstellar and intergalactic space. However, so far such a 'steepening' has not

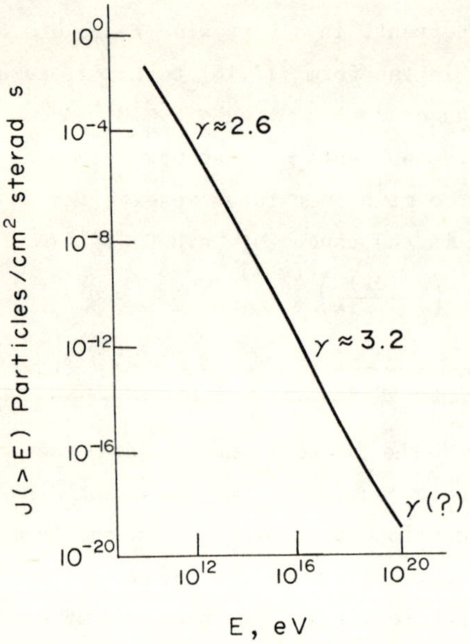

Fig. 15.1 The integral cosmic ray spectrum at the Earth. γ is the power index for the differential spectrum; for a power-law integral spectrum $J(>E) = \text{const.}\ E^{-(\gamma-1)}$

been observed and the problem of the spectrum and thus of the origin of the cosmic rays with very high energies is still open.

The values of γ given here refer to all cosmic rays, but until recently it was assumed that the chemical composition of the cosmic rays, at least up to energies of 100 to 1000 GeV was energy-independent. And it was also assumed in this way that in that region the index $\gamma \approx 2.7$ referred also to all groups of nuclei. In 1972 there appeared indications that the chemical composition already in the energy range up to 100 GeV/nucleon depended, albeit weakly, on the energy — we are referring here to the decrease in the fraction of secondary nuclei (such as Li, Be, and B) which are formed as the result of fragmentation of heavier nuclei when the energy increases. It is also possible that for nuclei of the H group (mainly iron nuclei) the index γ is somewhat smaller than for those of the M group (C, N, O, and F nuclei). The corresponding data are discussed by Ptuskin (1974) and Júliusson (1974). In the region of still higher energies $E > 10^{12}$ eV (or, to be more precise, $\epsilon > 10^{11}$ eV/nucleon) the changes in the chemical composition may turn out to be more substantial, but there are as yet no reliable data about this.

The electron component of the cosmic rays has been studied in less detail than the proton-nuclear component. The spectrum in the energy range up to 1 GeV is particularly sensitive to processes on the Sun and in the solar system and it is itself rather complex here. For $E \equiv E_e > 1$ GeV the power-law approximation fits already better and, for instance, in the range $5 < E < 50$ GeV the spectrum is a power-law spectrum with index $\gamma = 2.7 \pm 0.1$ (to be true, according to other data $\gamma = 3.0 \pm 0.2$). At lower energies the index γ decreases and in the energy region $E_e > 50$ to 100 GeV the data so far contradict each other, although it is possible that $\gamma \approx 3$ up to energies $E_e \sim 500$ to 1000 GeV (there is practically no information about the electrons at even higher energies).

To give some ideas we give the following differential electron specturm:

$$J_e(E) = 1.27\times10^{-2}\,E^{-(2.7\pm0.1)}\,\frac{\text{electrons}}{\text{cm}^2.\text{s.sterad.GeV}}\,, \qquad (15.13)$$

$$5 < E < 300\ \text{GeV}\,,$$

where the electron energy is measured in GeV; hence

$$J_e(>E) \approx 10^{-2}\left(\frac{E(Ev)}{10^9}\right)^{-1.7}, \quad 5\times10^9 < E < 2\times10^{11}\ \text{eV} \qquad (15.14)$$

(we have written Eqn.(15.14) in a form which facilitates its comparison with the spectrum (15.11); it is clear from what we have said that the difference in the indexes γ in (15.11) and (15.14) cannot yet be taken as being significant).

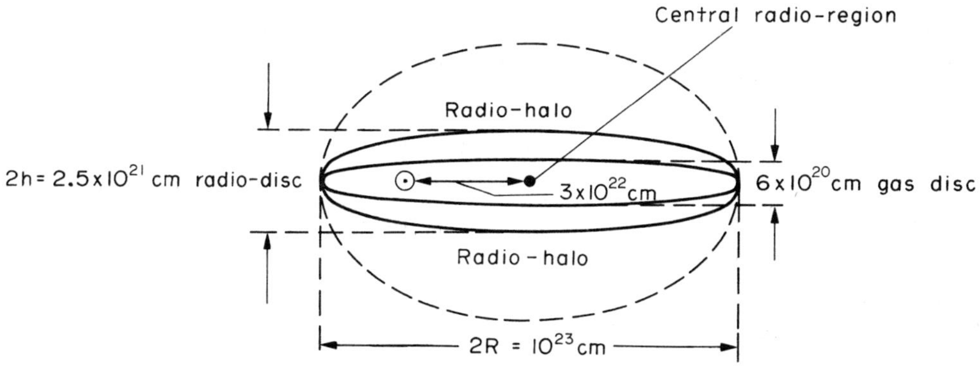

Fig. 15.2 Sketch of the radio-structure of the Galaxy

We emphasize once again that all magnitudes given here are only approximate in character and that one can find in the literature values of the intensity which differ by factors 2 to 3 but this is totally unimportant for our discussion. Another point is the problem how, in principle, to extrapolate all data to the limits of the solar system and into the important regions of the Galaxy.

The radius of curvature of particles with energies $E \gg Mc^2$ which move in a uniform magnetic field H equals (see (4.24); we assume that the particle moves at right angles to the field)

$$r = \frac{E(\text{eV})}{300\,H\,(\text{Oe})}\ \text{cm}\,. \qquad (15.15)$$

In the solar system, in general, $H \leqslant 3\times10^{-5}$ Oe and, moreover, the field is non-uniform. Cosmic rays with energies $E > 10^{12}$ eV and, in practice also for $E > 10^{10}$ eV can clearly not be contained in the solar system. Bearing in mind, however, that the shape of the cosmic ray spectrum at the Earth changes little up to energies $E \sim 10^{15}$ eV, we have all reasons to assume that this spectrum (for $E > 10^{10}$ to 10^{11} eV) is characteristic at least for that region of the Galaxy which adjoins the solar system. However, the conditions in the neighbourhood of the Sun is, apparently, very typical for huge parts of the Galaxy. In this connection radio-astronomy data about the non-thermal radio-emission with a continuous spectrum which comes undoubtedly from synchrotron radiation are particularly important for us. In that way we obtain direct information about the electron component of the cosmic rays in the Galaxy and also far beyond its limits (normal and radio-galaxies, quasars). They are just the radio-data which created, one might say, cosmic ray astrophysics as there is no doubt now about the fact that cosmic rays occur practically everywhere in the Universe and sometimes play a large energetic and dynamic role — this refers primarily to rarefied regions of the galaxies (the halo in normal galaxies, radio-emitting regions, clouds, and outbursts in radio-galaxies). In respect of our own Galaxy, radio-astronomical observations indicate the existence of a radio-disc, a central radio-region and, possibly, a radio-halo (Fig.15.2). Moreover, on the 'radio-map' of the Galaxy one can see the shells of supernovae and various 'inhomogeneities' — some changes or other in intensity (in particular, connected with the spiral structure). It is in this way completely clear that at least the electron component of the cosmic rays fills huge regions of the Galaxy and that, indeed, the circumsolar region is not distinguished at all.

The 'Archilles heel' of cosmic ray astrophysics right from its birth has been and to some extent remains so far a certain lack of definiteness (or, if one wants, lack of uniqueness) connected with using information about the electron components of the cosmic rays. As we saw in Chapter 5 the corresponding transition is realized through making some assumptions about the coefficients κ_H and κ_e:

$$\kappa_H = \frac{w_H}{w_{c.r.}} \quad , \quad \kappa_e = \frac{w_{c.r.}}{w_e} \quad , \tag{15.16}$$

where $w_H = H^2/8\pi$ is the magnetic field energy density, $w_{c.r.}$ the energy density of all cosmic rays, and w_e the energy density of the electron component (in Chapter 5 we considered instead of the energy densities the total energy, but this is in general equivalent, if we are dealing with averages).

For the cosmic rays at the Earth we have

$$\kappa_e \sim 100 \quad , \quad \kappa_H \sim 1 \quad , \tag{15.17}$$

where we used the measured values $w_{c.r.} \sim 10^{-12}$ erg/cm^3 and $w_e \sim 10^{-14}$ erg/cm^3, and also the fact that on average in the Galaxy $H \sim 2$ to 5×10^{-6} Oe.

By virtue of the arguments already given, such as the discussions of the equipartition of energy under quasi-stationary conditions (whence would follow $\kappa_H \sim 1$), it is very reasonable to take the estimates (15.17) for the Galaxy as a whole (or, on average in the galactic space). One proceeds similarly with respect to extra-galactic sources and one reaches the above mentioned conclusion about the large role of cosmic rays in the Universe. Nonetheless we emphasized in Chapter 5 the importance of producing methods for measuring the quantities κ_e and κ_H; we shall also discuss this in Chapters 16 and 17. When applied to the Galaxy the estimates (15.17) become, possibly, invalid for the central region, and for the main galactic discrete sources (such as supernova shells) — in some stages of their evolution.

Using the values (15.17) for the Galaxy as a whole we can easily estimate the total cosmic ray energy in the radio-disc (with $w_{c.r.} \sim 10^{-12}$ erg/cm^3 and $H \sim 5 \times 10^{-6}$ Oe) and, somewhat arbitrarily,† in the whole of the Galaxy, including the radio-halo

$$W_{c.r.,disc} \sim w_{c.r.,disc} V_{disc} \sim 10^{55} \text{ erg} ,$$
$$W_{c.r.,halo} \sim w_{c.r.,halo} V_{halo} \sim 10^{56} \text{ erg} , \qquad (15.18)$$

where $V_{disc} \sim R^2 \cdot 2h \sim 10^{67}$ cm^3 is the volume of the radio-disc (see Fig.15.2) and $V_{halo} \sim 4\pi R^3/3 \sim 5 \times 10^{68}$ cm^3 the volume of the quasi-spherical radio-halo; the real halo is, apparently, flattened and, moreover, the cosmic ray energy density probably decreases when one gets away from the galactic plane; we have therefore taken as an estimate $W_{c.r.,halo} \sim 10^{56}$ erg. This quantity is impressive; one can, for instance, note that it corresponds to the rest mass energy of about 100 stars with solar masses ($M_\odot = 2 \times 10^{33}$ g , $M_\odot c^2 \sim 10^{54}$ erg). More important is, of course, that the energy $W_{c.r.}$ is comparable with or larger than the internal energy of the interstellar gas or the energy of the inter-

---

† The problem of the presence of a radio-halo of the Galaxy has been discussed already for more that 25 years, but remains unclear (however, see below). In our opinion (see Ginzburg and Ptuskin, 1976a, b) there must exist some kind of 'cosmic ray halo', that is, a region between the radio-disc and the intergalactic space in which the cosmic ray energy density is still appreciable (say, $w_{c.r.} \sim 10^{-13}$ to 10^{-14} erg/cm^3). It is, however, possible that because the magnetic field strength and the density of the electron component decrease, the radio-emission from that region (radio-halo) is very weak. The estimate (15.18) for $W_{c.r.,halo}$ clearly assumes the existence of such a cosmic ray halo.

stellar magnetic field. We note also that in powerful radio-galaxies the energy $W_{c.r.}$ reaches values 10^{60} or even 10^{61} erg, that is, of the order of $10^7 M_{\odot} c^2$.

What is the origin of the cosmic rays in general and the cosmic rays at the Earth in particular? Undoubtedly the fact itself of an efficient generation (acceleration) of particles up to relativistic and ultra-relativistic energies in the Universe is a reflection of the plasma nature of the corresponding regions, of plasma instabilities, and of the existence of cosmic outbursts (outbursts of galactic nuclei, supernova and nova outbursts, solar flares, and so on). It are therefore just the cosmic and subcosmic rays and the radiation (in particular, the X-ray and gamma-radiation) which serve as the practically necessary elements and at the same time indicators of particularly violent, active cosmic processes — this is one of the most important features of cosmic ray astrophysics (and in general of high-energy astrophysics) which determines its role for astronomy as a whole.

When we are dealing with the origin of the cosmic rays observed at the Earth (in what follows we shall in that case simply talk about the origin of the cosmic rays) one usually does not have in mind the acceleration mechanisms, but the construction of a model in which the cosmic ray sources must be indicated, as well as the region occupied by the cosmic rays, and so on. For many years there has now been competition between basically two models — galactic and metagalactic models. In the metagalactic models one assumes that by far the largest part of the cosmic rays (the proton-nuclear component) arrives in the Galaxy from outside — from the intergalactic (metagalactic) space. One can see† therefore that the cosmic ray energy density in the Metagalaxy w_{Mg} (at least in the region bordering upon the Galaxy) must be of the order of $w_{c.r.,G} \sim 10^{-12}$ erg/cm^3. This means, by the way, that in the Metagalaxy $\kappa_e \gg 10^2$, as the electron component of the cosmic rays observed at the Earth certainly are of galactic origin; indeed, by virtue of the large Compton

† The cosmic rays in the Metagalaxy, as in the Galaxy, must have a high degree of isotropy, as a pronounced cosmic ray anisotropy is destroyed by plasma instabilities (see Ginzburg and Syrovatskii, 1966b; Ginzburg, 1969a, 1975a; Kaplan and Tsytovich, 1973; Ginzburg, Ptuskin, and Tsytovich, 1973; Wentzel, 1974; Tsytovich, 1977; and also below). Under quasi-stationary conditions one can see from considering the motion of particles in a magnetic field that the intensity, and thus also the cosmic ray energy density in the Metagalaxy and in the Galaxy must be approximately the same (for details see Ginzburg and Syrovatskii, 1966b; Ginzburg, 1969a, 1975a).

losses (vide infra) relativistic electrons cannot wander long in the Metagalaxy or reach the Galaxy from radio-galaxies or quasars. Unfortunately, there is as yet no reliable estimate of the density w_{Mg}, but from a number of considerations one may assume that

$$w_{Mg} \ll w_{c.r.,G} \sim 10^{-12}\ \text{erg/cm}^3\ , \qquad (15.19)$$

and, most probably, $w_{Mg} \leqslant 10^{-15}$ to 10^{-16} erg/cm^3 (see Ginzburg, 1969a, 1975). If one can reliably confirm the inequality (15.19), which is possible in principle (see Chapter 17), the metagalactic models would turn out to be convincingly refuted. So far, however, they are, in our opinion (not shared by everybody) only very improbable, in contrast to the galactic models. In the latter (again in the opinion of many, in particular, of the author, but not all) the main cosmic ray sources are supernovae, including pulsars† and possibly, outbursts in the nucleus of the Galaxy. In the disc galactic model the characteristic volume occupied by the cosmic rays is the radio-disc or the region close to it ($V_{disc} \sim 10^{67}$ cm^3). In that case the characteristic lifetime of the cosmic rays in the Galaxy is $T_{c.r.,disc} \sim 10^7$ yr. In the galactic model with a halo the occupied volume is the cosmic ray halo ($V_{halo} \sim 10^{68}$ cm^3) and the characteristic lifetime $T_{c.r.,halo} \sim 10^8$ yr (we give below an estimate of the lifetime). Hence and from the values (15.18) for $W_{c.r.}$ it is clear that to retain a quasi-stationary regime the cosmic ray sources in the Galaxy in both models mentioned must emit (accelerate) cosmic rays with a power of the order of

$$U_{c.r.} \sim \frac{W_{c.r.,halo}}{T_{c.r.,halo}} \sim \frac{W_{c.r.,disc}}{T_{c.r.,disc}} \sim 10^{40} - 10^{41}\ \text{erg/s}\ . \qquad (15.20)$$

Both supernovae and outbursts of the galactic nucleus are able to inject cosmic rays with such a power.

The choice between galactic models has not yet been made; it is, for instance, connected with the measurement of the characteristic cosmic ray lifetime $T_{c.r.}$ and to do this there are several possibilities (Ginzburg and Ptuskin, 1976a,b). The same could be achieved by solving the problem of the existence or not of a radio-halo. The elucidation of the role of some sources or other of cosmic

---

† In most and even in all supernovae there is probably a rotating magnetized neutron star — a pulsar (see, for instance, Ginzburg, 1971). Particles may be accelerated in the supernova outburst itself, in the supernova shell, and close to the pulsar. The acceleration of particles by pulsars is thus one of three possibilities, and its share is insufficiently clear.

rays is also possible by various methods. As an example we note that the galactic nucleus cannot serve as a source for the electron component of the cosmic rays in the high-energy region $E \geqslant 1$ to $10\,\text{GeV}$. The fact is that on the path from the centre of the Galaxy to the solar system high-energy electrons lose most of their energy as a result of synchrotron and Compton losses (see Chapter 4).†

We shall restrict ourselves here to the remarks we have made about the origin of the cosmic rays as our main aim is to illustrate a number of physical processes and mechanisms which are of interest in high-energy astrophysics. The following processes and mechanisms must be analyzed.

The processes for the acceleration of the proton-nuclear and electron components of the cosmic rays in various cosmic conditions and various regions (stellar outbursts, turbulent plasma in supernova shells, acceleration in interstellar space, acceleration near pulsars, acceleration in solar flares, and so on).

Energy loss mechanisms of various kinds of fast particles. Transformation of nuclei in collisions.

Diffusion and isotropization mechanisms for cosmic rays, in particular taking plasma effects into account.

Processes and mechanisms for the generation (production) by cosmic and subcosmic rays of photons with various energies and their application to radioastronomy, optical, X-ray, and gamma-astronomy. The problem of the absorption and scattering of photons in all wavebands also belongs to this class of problems.

---

† We have given here the state of the problem of the origin of cosmic rays as it was in 1975-76. As this problem is studied on a very wide front (which is also true of X-ray and gamma-astronomy) the situation changes comparatively fast and some remarks and estimates in Chapters 15 to 17 turned out to be obsolete at the time of the publication of the English edition of the present book. We must, however emphasize once again that the aim of these chapters is not to give a survey of the state of high-energy astrophysics, but mainly to discuss some physical problems relevant to this topic. Recent data in cosmic rays and gamma-astronomy are contained in the proceedings of the International Conference on Cosmic Rays which occur every year (see Plovdiv, 1977 for the picture in 1977). In our opinion the most important achievement of recent days has been the confirmation of the galactic model with a halo (Ginzburg, 1978; see also Ginzburg and Ptuskin, 1976a,b).

Furthermore, there is, clearly, the problem of constructing a quantitative theory for the origin of cosmic rays in the Galaxy, taking into account losses, diffusion, transformation of the chemical composition, and so on. To do this one must, of course, give a more detailed model (make precise the region filled by the cosmic rays, the distribution of the sources, the parameters of the interstellar space, and so on). The state of the problem is such that the 'trial and error' method is unavoidable — one must work out various models and choose the best of them by comparison with the observational data (see Ginzburg and Syrovatskii, 1964a, 1966b, 1973; Ginzburg, 1969a, 1975a; Bulanov, Dogel', and Syrovatskii, 1972a,b; Ptuskin, 1972, 1974; Ginzburg and Ptuskin, 1976a,b).

It is completely obvious that the whole of the corresponding set of problems is extremely extensive. We have already mentioned this in relation to synchrotron emission (see Chapters 5 and 9). In the foregoing we have also touched upon several other processes which are of interest in astrophysics, but the largest part of the problems enumerated has not yet been elucidated. Unfortunately, it is totally impossible to do this in any detail or fully in the framework of the present book. Below (in this chapter and in Chapters 16 and 17) we restrict our discussion to that of energy losses, to an exposition of the general scheme of cosmic ray diffusion taking the transformation of their chemical composition into account and (in the case of the electron component) taking losses into account; after that we shall consider mechanisms for producing X-rays and gamma-rays and make a few remarks about related problems. We shall not consider particle acceleration mechanisms under cosmic conditions (see Ginzburg and Syrovatskii, 1964a; Dorman, 1972; Kaplan and Tsytovich, 1973; Toptygin, 1973; Tsytovich, 1977).

When charged particles pass through matter there occur several processes which together are usually called 'ionization energy losses'. If we assume the motion of the particle given (to be precise, uniform and rectilinear) and neglect changes in the mass and charge of the particle due to nuclear transformations and decay or 'stripping' of orbital electrons (we have here in mind the motion of an atomic nucleus) the ionization of the atoms in the medium, their excitation, and Cherenkov radiation all contribute to the ionization losses.

It is true that a sharp division of the action of a charged particle on a medium into those three kinds of processes is not always possible, especially not in a dense medium. Moreover, in the case of a plasma one must speak about the transfer of energy to the electrons and ions in the plasma and not about the ionization and excitation proceeding in a gas of neutral atoms or molecules. For sufficiently slow particles charge transfer plays a role. When there is a beam of particles present rather than separate particles collective effects may occur — beam instabilities and so on. The problems of the formation of δ-electrons (recoil electrons) in a medium and of multiple scattering when particles pass through a given layer of a substance are bordering on the problem of ionization losses. Sometimes one must also take into account fluctuations in ionization losses and a spread in mean free paths.

Even from this short list it is clear how wide-ranging the problems of ionization losses are; one might well devote a special set of lectures to them. Of course, there is an extensive literature devoted to this field, but we restrict ourselves to referring to Bohr's classical paper (1948; see also Heitler, 1947; Landau and Lifshitz, 1960, 1977; Ginzburg and Syrovatskii, 1964a; Hayakawa, 1969). In what follows we only give a number of formulae which one must use for calculating ionization losses in a gas or plasma, and we make a few remarks on that account.

The basis for the calculation of ionization losses for fast particles is the formula (sometimes called the Bethe-Bloch formula)

$$-\left(\frac{dE}{dx}\right)_i = -\frac{1}{v}\left(\frac{dE}{dt}\right)_i = \frac{2\pi NZ^2e^4}{mv^2}\left\{\ln\frac{2mv^2W_{max}}{\mathcal{I}^2(1-\beta^2)} - 2\beta^2 + f\right\}, \qquad (15.21)$$

where N is the electron density in the matter, m the electrom mass, $\beta = v/c$, v the velocity of the fast particle considered with charge Ze, $\mathcal{I}$ the average ionization energy of the atoms in the medium, W_{max} the maximum energy transferred by the particle to the atomic electrons, and f a correction for the 'density effect'. The basis for obtaining Eqn.(15.21) is, indeed, the classical Rutherford formula which determined the cross-section for the scattering of a particle of charge Ze, mass M and initial velocity v by a particle at rest of charge e, mass m (interaction energy Ze^2/r). In the collision the particle initially at rest (to be precise, an electron) acquires some energy W and the incident particle loses the same energy (elastic collision). The cross-section, expressed in terms of W, equals

$$d\sigma = 2\pi\frac{Z^2e^4}{mv^2}\frac{dW}{W^2}$$

(the derivation is given by Landau and Lifshitz (1976, § 19) and it is therefore inappropriate to repeat it). For the energy lost by the incident particle we have

$$dE = \int_{W_{min}}^{W_{max}} W\, d\sigma = \frac{2\pi Z^2 e^4}{mv^2} \ln \frac{W_{max}}{W_{min}} .$$

After multiplying by the electron density N we are from this led to a formula like (15.21) and the real problem is to give a more precise expression for the logarithmic factor taking relativistic effects into account (so far we have clearly used the non-relativistic formula), as well as the binding of the electrons in the atoms, and so on.

The qualitative meaning of the logarithmic term in (15.21) becomes clear if we take into account that it essentially has the form const. $\times \ln (p_{max}/p_{min})$, where p is the impact parameter. For close collisions ($p \sim p_{min}$) δ-electrons are formed with an energy reaching W_{max}. On the other hand, the contribution from distant collisions ($p \sim p_{max}$) increases as $\ln[1/(1-\beta^2)] = \ln(E/Mc^2)^2$ by virtue of the compression of the field of the particle as $v \to c$ (for that reason the Fourier component of the field with a given frequency $\omega \sim \mathcal{I}/\hbar$ corresponds for increasing energy, roughly speaking to ever further distances). However, when p_{max} decreases there are between the particle and the electron to which the energy is transferred more and more particles in the medium. The latter screen the field of the particle and ceteris paribus this screening is, of course, the larger the denser the medium. The effect of the screening (or the 'density effect') is just taken into account by the term f in (15.21). In the ultra-relativistic case (for more details see below) the term f has a universal character

$$f = \ln(1-\beta^2) + \ln \frac{\mathcal{I}^2}{\hbar^2 \omega_p^2} + 1 \quad , \quad \omega_p^2 = \frac{4\pi N e^2}{m} .$$

As a result Eqn.(15.21) takes the form

$$-\left(\frac{dE}{dx}\right)_i = \frac{2\pi N Z^2 e^4}{mc^2} \left\{ \ln \frac{m^2 c^2 W_{max}}{2\pi N \hbar^2 e^2} - 1 \right\} . \tag{15.22}$$

The independence of the given expression for f of the properties of the medium (apart from the electron density N) is connected with the fact that in the case discussed of rather high energies the properties of the medium at high frequencies are important, and then we have for any medium

$$\varepsilon = 1 - \frac{\omega_p^2}{\omega^2} = 1 - \frac{4\pi N e^2}{m\omega^2} .$$

The ionization losses of ultrarelativistic electrons ($E \gg mc^2$) in atomic hydrogen are, according to (15.21) equal to

$$-\left(\frac{dE}{dt}\right)_i = \frac{2\pi N e^4}{mc^2}\left\{\ell n \frac{E^3}{mc^2 \mathcal{J}^2} - 2\right\} =$$

$$= 1.22 \times 10^{-20}\, N\left\{3\, \ell n \frac{E}{mc^2} + 18.8\right\} \text{erg/s} =$$

$$= 7.62 \times 10^{-9}\, N\left\{3\, \ell n \frac{E}{mc^2} + 18.8\right\} \text{eV/s} =$$

$$= 2.54 \times 10^{-19}\, N\left\{3\, \ell n \frac{E}{mc^2} + 18.8\right\} \text{eV/cm} =$$

$$= 1.53 \times 10^{5} \left\{3\, \ell n \frac{E}{mc^2} + 18.8\right\} \text{eV.cm}^2/\text{g} \qquad (15.23)$$

(we have given here the values of the ionization losses in various units for the sake of convenience when one wants to use them in different cases).

In (15.23) N is the density of hydrogen atoms, and the effective ionization energy $\mathcal{J}$ was put equal to 15 eV ; in this formula we have taken into account the contribution from all recoil electrons with energies reaching $W_{max} = \frac{1}{2}E$ as because of the indistinguishability of the electrons this is just the maximum transferred energy W_{max} in electron-electron collisions.† We note that we have in (15.21) to (15.23) taken into account all processes — ionization (in particular, the formation of fast δ-electrons), excitation, and Cherenkov radiation. The fraction of the latter is even for hydrogen comparatively small (of the order of 15%). The energy $\mathcal{J}$ in (15.23) has not yet been calculated exactly, as far as we can establish; there is thus under the logarithm sign in (15.23) an undetermined factor of the order unity (this leads to an inaccuracy of Eqn.(15.23) of not more than a few per cent). Moreover, we neglected the density effect in (15.23), that is, we dropped the term f in (15.21). This is allowable as long as $v/c < 1/\sqrt{\varepsilon(0)}$, where v is the particle velocity and $\varepsilon(0)$ the dielectric permittivity of the medium at a frequency $\omega = 0$. In atomic hydrogen

$$\varepsilon(0) = 1 + 4\pi N\alpha \quad , \quad \alpha = \frac{9}{2}\left(\frac{\hbar^2}{me^2}\right)^3 \sim 10^{-24}\ \text{cm}^3 \ ,$$

† An electron with an energy larger than $\frac{1}{2}E$ we shall assume to be the scattered electron, rather than a δ-electron. We note also that Eqn.(15.21) is not completely exact in the case of electrons. Under the conditions of (15.23) we must replace the term −2 in the first pair of braces in (15.23) by $\frac{1}{8} - \ell n\, 2 = -0.57$. In practice this refinement is unimportant, in particular, because of the approximate definition of the energy $\mathcal{J}$ (vide infra).

and the density effect can be neglected when

$$\frac{E}{mc^2} < (4\pi N\alpha)^{\frac{1}{2}} \sim \frac{3\times 10^{11}}{\sqrt{N}} .$$

Even when $N \sim 10^2\ cm^{-3}$ this means that Eqn.(15.23) is valid for electrons with energies $E < 10^{16}$ eV.

For light non-hydrogen atoms the ionization losses are also given by Eqn. (15.23) to a first approximation, if we understand by N the density of all atomic electrons. It is clear that in interstellar space (which, contains, say, 10% helium atoms) the ionization losses are only approximately larger by 10% than in pure hydrogen (for the same total density of atoms). In a competely ionized plasma (N the electron density) the ionization losses for ultra-relativistic electrons equal

$$-\left(\frac{dE}{dt}\right)_i = \frac{2\pi N e^4}{mc}\left\{\ell n\,\frac{m^2c^2E}{4\pi N\hbar^2 e^2} - \frac{3}{4}\right\}$$

$$= 7.62\times 10^{-9}\,N\left\{\ell n\,\frac{E}{mc^2} - \ell n\,N + 73.4\right\}\ \text{eV/s} . \qquad (15.24)$$

This formula is obtained from (15.22) for $W_{max} = \frac{1}{2}E$; moreover, we have introduced a more precise numerical value of the logarithmic factor in agreement with Tsytovich's calculations (1962b): we have replaced the -1 in (15.22) by $-\frac{3}{4}$ in (15.24); of course, this refinement is not particularly important. Equations (15.23) and (15.24) usually give results which do not differ as to order of magnitude. For instance, for $N = 0.1\ cm^{-3}$ and $E = 5\times 10^8$ eV the losses (15.24) are twice those of (15.23). The losses (15.24) proceed forming δ-electrons (that is, transferring energy to the plasma electrons) and through Cherenkov radiation of plasma waves.† The necessity to apply Eqn.(15.22) for a plasma, which takes into account the density effect, is completely clear from what we have said earlier: for a rarefied plasma $\varepsilon = 1 - \omega_p^2/\omega^2$ for all frequencies and it is just using that expression which led to (15.22).

---

† We assume the plasma to be isotropic, that is, there to be no magnetic field; we know that under those circumstances a particle cannot emit transverse Cherenkov waves in the plasma. We emphasize also that we understand by plasma waves longitudinal waves which can propagate not only in a plasma, but also in any medium provided $\varepsilon(\omega) = 0$. The peculiarity of a plasma in this respect is merely the weak damping of sufficiently long-wavelength plasma waves. In a condensed medium an appreciable part of the ionization losses can also be connected just with the generation of plasma waves.

Equation (15.-3) refers to the case of ultra-relativistic electrons. If, however, the condition $E \gg mc^2$ is not satisfied and, in particular for non-relativistic electrons (but with a velocity $v \gg v_a$, where v_a is the velocity of the atomic electrons; in the case of hydrogen this means that the kinetic energy of the electron $E_k = E - mc^2 \gg 15$ eV), one can use Eqn. (15.21), replacing W_{max} by $\frac{1}{2} E_k$, for calculations with errors not exceeding a few per cent.

For particles of total energy E and mass $M \gg m = 9.1 \times 10^{-28}$ g, that is, for mesons, protons, and nuclei, Eqn. (15.21) leads to the following results. Let

$$E \ll \frac{M}{m} Mc^2 . \tag{15.25}$$

The maximum energy transferred to the electron is then equal to

$$W_{max} = 2mv^2 \left(\frac{E}{mc^2}\right)^2 . \tag{15.26}$$

When condition (15.25) is satisfied Eqn. (15.21) gives for losses in atomic hydrogen ($\mathcal{I}$ = 15 eV)

$$\left.\begin{aligned} -\left(\frac{dE}{dt}\right)_i &= 7.62 \times 10^{-9} Z^2 N \left(\frac{2Mc^2}{E_k}\right)^{\frac{1}{2}} \left\{\ln \frac{E_k}{Mc^2} + 11.8\right\} \text{ eV/s} , \\ E_k &= E - Mc^2 \approx \tfrac{1}{2} Mv^2 \ll Mc^2 , \end{aligned}\right\} \tag{15.27}$$

$$-\left(\frac{dE}{dt}\right)_i = 7.62 \times 10^{-9} Z^2 N \left\{4 \ln \frac{E}{Mc^2} + 20.2\right\} \text{ eV/s} , \quad E \gg Mc^2. \tag{15.28}$$

For protons condition (15.25) has the form $E \ll 2 \times 10^{12}$ eV and Eqn. (15.28) is thus in practice suitable for $2 \times 10^9 < E < 10^{12}$ eV.

If

$$E \gg \frac{M}{m} Mc^2 , \tag{15.29}$$

we have

$$W_{max} = E . \tag{15.30}$$

Equation (15.21) then becomes

$$\begin{aligned} -\left(\frac{dE}{dt}\right)_i &= \frac{2\pi N Z^2 e^4}{mc} \left\{\ln \frac{2mc^2}{\mathcal{I}^2} \cdot \frac{E^3}{(Mc^2)^2} - 2\right\} \\ &= 7.62 \times 10^{-9} Z^2 N \left\{3 \ln \frac{E}{mc^2} + \ln \frac{M}{m} + 19.5\right\} \text{ eV/s} , \end{aligned} \tag{15.31}$$

where the last expression refers to atomic hydrogen. The density effect has not been taken into account in (15.31); this is allowable in atomic hydrogen as long as $E/Mc^2 < 3 \times 10^{11}/\sqrt{N}$ (vide supra).

In a fully ionized plasma with electron density N we have in the non-relativistic case

$$-\left(\frac{dE}{dt}\right)_i = \frac{2\pi NZ^2e^4}{mv}\,\ln\frac{m^3v^4}{\pi Ne^2\hbar^2} =$$

$$= 7.62\times10^{-9}\,Z^2\,N\left(\frac{2Mc^2}{E_k}\right)^{\frac{1}{2}}\left\{\ln\frac{E_k}{Mc^2}-\frac{1}{2}\ln N+38.7\right\}\ \text{eV/s.} \qquad (15.32)$$

This formula is obtained from (15.21) for $E_k = \frac{1}{2}Mv^2 \ll Mc^2$, $W_{max} = 2mv^2$ (see (15.26)) and $\mathcal{I} = \hbar\omega_p = \hbar(4\pi Ne^2/m)^{\frac{1}{2}} = 3.7\times10^{11}\sqrt{N}$ eV. This replacing of $\mathcal{I}$ by the plasmon energy $\hbar\omega_p$ (or, put differently, replacing the frequency $\omega = \mathcal{I}/\hbar$ by the plasma frequency ω_p) is completely natural and is confirmed by a more consistent calculation (see Landau and Lifshitz, 1960).

In the ultra-relativistic case $E \gg Mc^2$ we must use for a plasma Eqn.(15.22) or the formula, which differs by a small factor ,

$$-\left(\frac{dE}{dt}\right)_i = \frac{2\pi NZ^2e^4}{mc}\,\ln\frac{m^2c^2W_{max}}{4\pi Ne^2\hbar^2} =$$

$$= 7.62\times10^{9}\,Z^2\,N\left\{\ln\frac{W_{max}}{mc^2}-\ln N+74.1\right\}\ \text{eV/s}\ , \qquad (15.33)$$

where we must use for W_{max} the value (15.26), when (15.25) holds, and the value (15.30) in the case (15.29). The numerical values are given in eV/s. For relativistic particles with $v \approx c$ one finds the expression for the losses in eV/cm by dividing by $c = 3\times10^{10}$ cm/s; the losses in hydrogen in eV/(g.cm^{-2}) are obtained from the losses in eV/cm by $6\times10^{23}/N = 1/(1.67\times10^{-24}\,N) = 1/M_pN$.

Let us now discuss the problem of the formation of δ-electrons. It is clear that the corresponding losses are completely taken into account in (15.21) and the other formulae in which already appears the maximum energy transferred to an electron W_{max}. Losses connected with the formation of δ-electrons with energies from W_{max} down to some value W_{min} in which we are interested (of course, $W_{min} \gg \mathcal{I}$, where $\mathcal{I}$ is the average binding energy of the electron, which occurs in (15.21)) can be obtained from (15.21) and is equal to

$$-\left(\frac{dE}{dt}\right)_i = \frac{2\pi NZ^2e^4}{mv}\,\ln\frac{W_{max}}{W_{min}}\ , \qquad (15.34)$$

where Ze is the charge and v the velocity of the incident particle, N the electron (not the atomic) density in the matter, and e and m the electron charge and mass. We have already earlier given (see (15.26), (15.30), and the explanations of Eqn.(15.23)) the values of W_{max} in various cases.

The probability for a particle with energy E to transfer an energy $W \gg \mathcal{I}$, lying in the range $W, W+dW$ to an electron in a layer of the matter of thickness 1 cm is equal to

$$P_\delta(E,W)\,dW = \frac{2\pi NZ^2e^4}{mv^2}\,\frac{dW}{W^2}\,F(E,W)\ ,\quad F(E,W) = \left(\frac{E}{E-W} - \frac{W}{E}\right)^2\ , \tag{15.35}$$

where the function F is written down for the case of ultra-relativistic electrons with $E \gg mc^2$ (in that case, of course, $Z = 1$). The differential cross-section per electron corresponding to the probability (15.35) equals

$$d\sigma_\delta = \frac{2\pi Z^2e^4F(E,W)}{mv^2W^2}\,dW\ .$$

Before we briefly discuss also other energy losses (besides the ionization losses) we make a few remarks about collective effects occurring when particles move through a medium. To be precise we consider a plasma in which a stream (beam) of fast charged particles moves. The problem of the processes in such a system (a beam in a plasma) is to some extent close to the problem of ionization losses.

We shall, however, start with a more general problem. In a sufficiently rarefied gas one can consider processes such as the emission of photons, the production of various other particles (for instance, ions) or the ionization and excitation of atoms without taking the effect of the medium into account. In other words, if the distance between particles (say, atoms) in the medium is sufficiently large all processes proceed as if there existed only the colliding particles (the particle flying past, the atom, the collision 'products'). However, it is clear (and well known) that when the density of the medium increases one must, in general, take into consideration the mutual effect of the particles in it and in that sense once can speak of collective effects. For instance, if one takes into account the effect of the refractive index of the medium on synchrotron radiation (see Chapter 6) one is led to such a collective effect. The Cherenkov radiation of both transverse and longitudinal (plasma) waves is, of course, also a collective effect and in this case the process does not occur at all without a medium (in vacuo). We have already discussed this kind of collective effects, whenever this should be done. Another category of collective effects refers to a 'collective' of incident particles. To be precise, energy losses by a particle beam when it passes through a medium are only in the simplest case equal to the sum of the losses suffered by the separate particles in the beam when there are no other particles in the beam. Such conditions are by far not always satisfied.

Without trying to classify the effects we divide nevertheless all collective effects into two classes connected with the emitting ('incident') particles themselves. For the first class the spatial inhomogeneity in the distribution of the emitting particles is important. For instance, it is clear from Eqn. (15.21) and from the substance of the effect itself, that ionization losses and the intensity of the Cherenkov radiation are proportional to the square of the charge Ze of the fast particle considered.† At the same time it is completely clear that the losses of a single particle of charge Ze will be equal to the losses of Z particles of charge e only if the latter fly together, forming a rather compact bunch with total charge Ze. The bunch can, clearly, be assumed to be small in the afore-mentioned sense if its size ℓ is small compared to the characteristic size p which, figuratively speaking, is responsible for the losses considered. For close encounters leading to the formation of δ-electrons the impact parameter is small and for a bunch one does not have to speak about collective effects.‡ On the other hand, Cherenkov radiation of wavelength λ is produced in regions of size of the order of λ; in that case bunches of particles can very well turn out to be sufficiently small and, hence, the intensity of the radiation can no longer be simply proportional to the number of particles in the beam.

Another possibility for the occurrence of collective effects (their second class) is connected with the reabsorption of radiation, beam instabilities, and so on. The spatial inhomogeneity of the particle distribution is for this class of collective effects, in general, unimportant (at least, when we calculate the absorption or amplification coefficient of the waves in the linear approximation). One such kind of process is the reabsorption of synchrotron radiation, discussed in Chapter 9. Here the picture is very simple: a single particle radiates and the other particles of the same collection can absorb this radiation and as a result the absorption coefficient depends on the

---

† Above we have already emphasized that the Cherenkov losses are contained in the complete expression for the ionization losses and, hence, that they are, as the latter, proportional to Z^2. All forms of ionization losses are proportional to Z^2, as the field of the particle $\mathbf{E}$ is proportional to Ze; the losses along the path of the particle are equal to the work $Ze v E \propto e^2 Z^2$ which the field produced by the particle performs on itself (just this method of calculation – evaluation of the work $Ze v E$ – was used for the determination of the ionization losses by Landau and Lifshitz (1960, Chapter 12).

‡ We have here in mind bunches of particles which are uncorrelated. If, on the other hand, the particles are correlated such, as for instance, the protons in a nucleus, the nature of the close collisions is usually determined by the charge Ze of the bunch.

density of the radiating particles. The instability of a beam of particles in a plasma which is connected with the occurrence of longitudinal waves in the beam is indeed the same process — in this case we are dealing with negative Cherenkov absorption (reabsorption) of plasma waves.

We have already touched upon this problem in Chapter 7 but now we shall discuss it in somewhat more detail as the problem is important both as regards methodology and as regards practical relations.

Let us consider a non-relativistic beam of particles of mass M, charge e, and density N_s moving in the 'parent' (ambient) plasma with density N and temperature T. The velocity distribution function of the particles in the beam is denoted by $f_s(\mathbf{v})$. As a rather typical example we use the distribution

$$f_s(\mathbf{v}) = N_s \left(\frac{M}{2\pi k_B T_s}\right)^{\frac{3}{2}} \exp\left\{-\frac{M(\mathbf{v}-\mathbf{v}_s)^2}{2K_B T_s}\right\} . \tag{15.36}$$

It is clear that we are here dealing with a beam moving with an average velocity $\mathbf{v}_s$; the spread in velocities around $\mathbf{v}_s$ is Maxwellian in shape with a temperature T_s. As we shall assume the parent plasma to be in equilibrium we can write for its electrons

$$f_0(\mathbf{v}) = N\left(\frac{m}{2\pi k_B T}\right)^{\frac{3}{2}} \exp\left(-\frac{mv^2}{2k_B T}\right) . \tag{15.37}$$

If there is no beam (or, if we can neglect its influence) in the collisionless isotropic plasma considered (we assume that there is no external magnetic field) electromagnetic transverse waves can propagate:

$$\left.\begin{aligned} &\mathbf{E} = \mathbf{E}_0 e^{i(\mathbf{k}\cdot\mathbf{r})-i\omega t} , \quad (\mathbf{k}\cdot\mathbf{E}) = 0 , \quad \mathbf{H} = \frac{c}{\omega}[\mathbf{k}\wedge\mathbf{E}] , \\ &n_\perp \equiv n_{1,2} = \frac{ck}{\omega} = \sqrt{\varepsilon} = \sqrt{\left(1-\frac{\omega_p^2}{\omega^2}\right)} , \quad \omega_p^2 = \frac{4\pi Ne^2}{m} , \quad \omega^2 = \omega_p^2 + c^2k^2 , \end{aligned}\right\} \tag{15.38}$$

as well as longitudinal waves

$$\left.\begin{aligned} &\mathbf{E} = \mathbf{E}_0 e^{i(\mathbf{k}\cdot\mathbf{r})-i\omega t} , \quad \mathbf{H} = 0 , \quad (\mathbf{k}\cdot\mathbf{E}) = kE , \\ &\omega^2 \approx \omega_p^2 + 3\frac{k_B T}{m}k^2 , \quad n_\parallel \equiv n_3 = \frac{ck}{\omega} \approx \frac{1-\omega_p^2/\omega^2}{3k_B T/mc^2} . \end{aligned}\right\} \tag{15.39}$$

The longitudinal waves propagate without specific collisionless damping only provided $k\sqrt{k_B T/m} \ll \omega_p$ (this means that $kr_D = k\sqrt{(k_B T/8\pi Ne^2)} \ll 1$, or $\lambda = 2\pi/k \gg r_D$, where $r_D = \sqrt{(k_B T/8\pi Ne^2)}$ is the Debye radius). For the sake of convenience we have repeated here what was said in Chapter 11.

In the independent particle approximation (a sufficiently rarefied beam) each particle in the beam moves, scatters, and radiates independently of the other particles. In that case one can, in general, consider the scattering and bremsstrahlung due to the particles in the parent plasma† as the result of binary collisions; the same applies also to the formation of δ-electrons and the part of the ionization losses which is connected with it which is due to close collisions. However, we have already emphasized that Cherenkov radiation is essentially a collective effect. In the case (15.38) the refractive index $n_\perp < 1$ and, hence, the phase velocity of the transverse waves $v_{ph\perp} = c/n_\perp > c$. It is clear that the condition for Cherenkov radiation, $\cos\theta = c/n(\omega)v$ (see (6.53)) cannot be satisfied when $n < 1$ (v is here the particle velocity and θ the angle between $\mathbf{v}$ and $\mathbf{k}$, the wavevector of the emitted wave). On the other hand, for the longitudinal waves (15.39) the Cherenkov condition (6.53) can clearly be satisfied and a beam particle is thus able to generate plasma (longitudinal) waves. The total power of this Cherenkov radiation is

$$-\left(\frac{dE}{dt}\right)_{Ch} = \frac{2\pi Ne^4}{mv}\,\ln\left(\frac{2v^2}{3k_BT/m}\right) = \frac{e^2\omega_p^2}{2v}\,\ln\left(\frac{2mv^2}{3k_BT}\right). \qquad (15.40)$$

The structure of Eqn.(15.40) is the same as that of all formulae (see, for instance, (15.21) for $\beta^2 \ll 1$) for the ionization losses (Eqn.(15.40) is part of those losses). As regards the logarithmic factor which is determined only approximately (in that sense the factor $\frac{2}{3}$ under the logarithm sign is purely approximative and Eqn.(15.40) does not really differ from Eqn.(7.33) given earlier), one can determine it only as a result of a more detailed calculation (see, for instance, Pines and Bohm, 1952; Zheleznyakov, 1970; essentially the necessary fact when taking into account the Cherenkov condition (6.53) with $n = n_\parallel$ is the condition $kr_D \leqslant 1$ for the existence of not strongly damped longitudinal waves).

All particles in the beam for which $v\cos\theta = c/n_\parallel(\omega)$ will emit waves of frequency ω at a given angle θ. In the radiation at the angle θ we have thus contributions from all particles with given values of the component of $\mathbf{v}$ along $\mathbf{k}$ and arbitrary values of $\mathbf{v}_\perp$ (that is, the component perpendicular to $\mathbf{k}$). In that connection we shall be interested not in the distribution function $f_s(\mathbf{v})$ itself, but in the function

† We are dealing here with electrons and ions; the density of the latter N_i is for $Z = 1$ equal to the electron density N (or, when there is a beam present, such as would guarantee quasi-neutrality of the system).

$$f_s(v_k) = \int f_s(\mathbf{v})\, d^2\mathbf{v}_\perp = N_s \left(\frac{M}{2\pi k_B T_s}\right)^{\frac{1}{2}} \exp\left\{-\frac{M(v_k - v_s\cos\theta)^2}{2k_B T_s}\right\}, \qquad (15.41)$$

where we have used the distribution (15.36). A function such as (15.41) was sketched in Fig. 7.2.

When the particle density N_s in the beam increases we must take into account the reabsorption (or amplification) of the Cherenkov waves, that is, their absorption and induced emission by other particles of the same beam. The fact that such processes (absorption and induced emission) are possible is at once clear if we use quantum language. In this language† the generation of plasma waves (in particular, their Cherenkov generation) is the emission of plasmons of energy $\hbar\omega$ and momentum $\hbar\mathbf{k} = (\hbar\omega/c)n_{\|}(\mathbf{k}/k)$ (we have made some provisos in this respect in Chapter 12 and there is no need to repeat them). When the plasmon is emitted a particle of energy $E_2 = \frac{1}{2}Mv_2^2$ and momentum $\mathbf{p}_2 = m\mathbf{v}_2$ changes into a state of energy $E_1 = \frac{1}{2}Mv_1^2 = E_2 - \hbar\omega$ and momentum $\mathbf{p}_1 = M\mathbf{v}_2 - \hbar\mathbf{k}$. It is completely clear that the reverse process is also possible as the modulus of the matrix elements for direct and inverse transitions are equal to one another. In such a reverse process a plasmon $(\hbar\omega, \hbar\mathbf{k})$ is absorbed by a particle $(E_1, \mathbf{p}_1)$ and as a result of this the energy and momentum of the particle become equal to $E_2 = E_1 + \hbar\omega$ and $\mathbf{p}_2 = \mathbf{p}_1 + \hbar\mathbf{k}$. The probability for induced emission is equal to the probability for absorption and, thus, if the state 2 lies 'higher' (as we assumed above), when there is a plasmon $(\hbar\omega, \hbar\mathbf{k})$ present, the system (particle) undergoes an induced transition $E_2 \to E_1$, $\mathbf{p}_2 \to \mathbf{p}_1$ with the emission of yet another plasmon $(\hbar\omega, \hbar\mathbf{k})$. The 'true' (resulting) absorption is given by the difference of the numbers of particles N_1 and N_2 in the states 1 and 2. In the case of Cherenkov radiation, clearly, only the component along $\mathbf{k}$ of the particle velocity $\mathbf{v}$, that is, the value of v_k, changes, and $Mv_{k,2} = Mv_{k,1} + (\hbar\omega/c)n_{\|}$. Moreover

$$1 - \frac{N_2}{N_1} = 1 - \frac{f_s(v_{k,2})}{f_s(v_{k,1})} = -\frac{1}{f_s}\frac{df_s}{dv_k}\frac{\hbar\omega}{Mc}n_{\|},$$

where the distribution function f_s and its derivative must be taken at the point $v_k = \omega/k \approx v_{k,1} \approx v_{k,2}$ (one sees easily that in the classical case $\hbar k \ll Mv_k$). The reabsorption coefficient is thus equal to

---

† As we have done several times before, we are talking here about 'language' as we have in mind a classical problem which can also be described completely in classical terms, but it turns out to be more convenient or more translucent to use quantum concepts.

$$\mu = -A\left(\frac{df_s(v_k)}{dv_k}\right)_{v_k = \omega/k} . \tag{15.42}$$

The field of a monochromatic wave propagating along the z-axis changes as

$$E = E_0 \exp\left[i\omega\left(\frac{n}{c}z - t\right)\right]\exp\left(-\frac{\omega}{c}\kappa z\right),$$

where κ is the absorption index. The absorption coefficient $\mu = 2\omega\kappa/c$ determines the change in intensity $I \propto |E|^2 \propto e^{-\mu z}$. Another statement of the problem is, however, possible and often encountered, in which the wavevector **k** is taken to be real and the frequency to be complex. In that case

$$E = E_0 \exp\left[i\omega'\left(\frac{n}{c}z - t\right)\right] e^{-\gamma t}$$

where $\omega = \omega' - i\gamma$ (here $\omega' = \mathrm{Re}\,\omega$).

In that case the intensity $I \propto e^{-2\gamma t}$. For a weakly absorbing (or weakly amplifying) medium one can rigorously show that, as is also immediately clear from intuitive considerations, we have

$$2\gamma = \mu v_{gr} , \tag{15.43}$$

where $v_{gr} = d\omega/dk$ is the group velocity of the waves; in the case (15.39)

$$v_{gr} = \frac{3k_B T}{mv_{ph}} \quad \text{and} \quad v_{ph} = \frac{\omega}{k} = \left(\frac{3k_B T/m}{1 - \omega_p^2/\omega^2}\right)^{\frac{1}{2}}.$$

One can find the coefficient A in Eqn.(15.42) as a result of calculations which we omit here (see Zheleznyakov, 1970; Kaplan and Tsytovich, 1973; Tsytovich, 1977, and for a Maxwellian plasma also Ginzburg, 1970b; Ginzburg and Rukhadze, 1975; and Chapter 11). As a result we get for γ (see also (15.43))

$$\gamma = -\frac{2\pi^2 e^2 \omega_p}{Mk^2}\left(\frac{df_s(v_k)}{dv_k}\right)_{v_k = \omega/k} . \tag{15.44}$$

As far as its meaning is concerned $f_s(v_k)$ is in (15.44) the total distribution function for an electron beam, taking into account both the presence of the beam and the existence of the parent plasma. If we nonetheless denoted the distribution function by f_s (the index s corresponds to the beam) we have merely in mind the application to a real case when close to the value $v_k = \omega/k$ the contribution from the particles of the Maxwellian parent plasma can be neglected. It is clear from what we have said that Eqn.(15.44) with $M = m$ can also be applied to a purely Maxwellian plasma without a beam when it leads to an expression for γ under collisionless damping conditions (see Chapter 11). As we said, there is then just damping ($\gamma > 0$) and this is connected with the fact that for the Maxwell distribution $df_s/dv_k < 0$ (see (15.37) or (15.41) with $v_s = 0$). From the above it is particularly clear that the nature of the collisionless or Landau damping in an isotropic plasma — we are dealing here with the inverse Cherenkov effect, that is, with Cherenkov absorption (such an absorption of a plasma wave is realized just for particles with $v_k = v \cos\theta = c/n_{\parallel}(\omega)$; we repeat here what was said in Chapters 7 and 11).

If the distribution function is such that in some region or other $df_s/dv_k > 0$, we have instead of damping an amplification of the waves or, put differently, there is negative absorption or an instability. For any beam which is 'submerged' in the plasma there is, as is clear, for instance, from Fig. 7.2, a region (region I), where $df_s/dv_k > 0$. Waves with a phase velocity $v_{ph} = \omega/k$ which lies in that region of v_k values which corresponds to region I in Fig. 7.2 will thus be amplified. As a result the amplitude of the waves increases and is bounded only due to non-linear effects. The amplification of the waves in this case is of the same nature as in quantum amplifiers or generators (masers and lasers). Indeed, the condition $df_s/dv_k > 0$ simply means that there are more particles in the upper than in the lower levels as a result of which induced emission dominates over absorption.

Substituting the function (15.41) into (15.44) we get

$$\left.\begin{aligned} \gamma &= \sqrt{\tfrac{1}{8}\pi}\,\frac{\omega_s^2\omega_p(v_{ph} - v_s\cos\theta)}{k^2 v_{T_s}^3}\exp\left\{-\frac{(v_{ph} - v_s\cos\theta)^2}{2\,v_{T_s}^2}\right\} \\ \omega_s^2 &= \frac{4\pi N_s e^2}{M}\,,\quad \omega_p^2 = \frac{4\pi N e^2}{m}\,,\quad v_{ph} = \frac{\omega}{k}\,,\quad v_{T_s}^2 = \frac{k_B T_s}{M} \end{aligned}\right\} \qquad (15.45)$$

In the region where $v_{ph} < v_s \cos\theta$, the waves grow ($\gamma < 0$). It is clear that the maximum value of γ for given k equals

$$|\gamma_{max}| \sim \frac{\omega_s^2\omega_p}{k^2 v_{T_s}^2}\,. \qquad (15.46)$$

The velocity v_{ph} occurring here is the phase velocity of the waves in the parent plasma. In order that Cherenkov radiation be possible this velocity should not exceed c. Therefore, in (15.46) $k_{min} \approx \omega_p/c$ and, hence,†

---

† If $v_{T_s} \ll v_s$, one can assume that the condition for Cherenkov radiation is $v_{ph} < v_s$ and $k_{min} \approx \omega_p/v_s$; hence $|\gamma_{max}| \leqslant \omega_s^2 v_s^2/\omega_p v_{T_s}^2$.

$$|\gamma_{max}| \leqslant \frac{\omega_s^2 c^2}{\omega_p v_{T_s}^2} \gg \frac{\omega_s^2}{\omega_p} . \tag{15.47}$$

In the classical approach to the problem one uses a kinetic equation for the distribution function f_s and one chooses as the initial distribution in the beam, for instance, the distribution (15.36). One then determines, say, the frequency $\omega = \omega' - i\gamma$ for a wave with real wavevector $\mathbf{k}$ (or complex wavevector $\mathbf{k}$ for real ω). As a result one obtains, of course, the same result (15.44) or, more precisely, (15.45). Just this identity of results indicates the complete equivalence of the classical and quantal approaches in the problem under discussion (see, for instance, Ginzburg and Zheleznyakov, 1969a, 1965 in this connection) where we have in mind when talking about the quantal approach the method using the Einstein coefficients for the transition probabilities†. The region of applicability of this method is restricted, in particular, in connection with the condition $|\gamma| \ll \omega \sim \omega_p$. However, in the region where it is applicable the Einstein coefficient method is very fruitful as we have already demonstrated in Chapter 9.

It is clear from what we have said that when there is a particle beam present in the plasma (average velocity of the particles in the beam $v_s \gg v_T = \sqrt{k_B T/m}$) this beam is unstable – longitudinal (plasma) waves in it grow. The growth rate γ is proportional to the particle density N_s in the beam (see (15.45) and also use the fact that $\omega_s^2 \propto N_s$). From this it follows already that in a beam of a sufficiently low density the growth of the waves due to negative absorption of Cherenkov waves is rather small (small over a time characterizing the process; small along the whole path of the beam, and so on); on the other hand, the collective effect (instability) in beams may in completely realistic cases be very important. As a result the energy losses in a beam and its spreading out may proceed much faster (or along a shorter path) than for separate particles. The solution of the problem about the losses and scattering (isotropization) of a beam is rather complicated as it is here not possible to limit ourselves to the linear approximation and we must develop a

---

† Only in this sense or a similar one could the quantal and classical approaches confront one another. If, on the other hand, we have in mind the very possibility of solving any classical problem using the equations from quantum theory, that possibility is obvious as classical mechanics and classical electrodynamics are limiting cases of the appropriate quantum-mechanical constructions.

non-linear theory (see Kaplan and Tsytovich, 1973; Tsytovich, 1977; and the literature cited there).

Why have we dwelled on the instability of beams in a chapter devoted to cosmic rays? At first sight this looks the stranger where we have earlier emphasized the isotropy of the cosmic rays due to which there are absolutely no conditions for the appearance of a beam instability. Moreover, because of the extreme rarefaction of the cosmic plasma (electron density $N \leqslant 1\ \mathrm{cm}^{-3}$ in interstellar space and $N \leqslant 10^{-5}\ \mathrm{cm}^{-3}$ in intergalactic space) plasma effects should turn out to be completely unimportant in cosmic ray astrophysics.

However, this last argument must of course not be taken to be serious as absolute values of the density N and of other quantities cannot play a role – one should compare them with the appropriate values which are important for the processes considered. As to the isotropy of the cosmic rays, one of the most important problems is to establish its cause, and also to elucidate the conditions under which there is no isotropy. An analysis of plasma effects in cosmic ray astrophysics is thus, indeed, necessary. Moreover, there is no doubt that these effects can be very important.

Let us, for instance, consider the 'outflow' of cosmic rays from a region with a magnetic field H_1 in which the cosmic rays are isotropic into a surrounding region with a magnetic field $H_2 \ll H_1$. Such a situation is fully realistic, say, when cosmic rays leave a supernova shell for the interstellar space or when they flow from a galaxy (or its core) into intergalactic space.

When a charged particle moves in a regular magnetic field the adiabatic invariant

$$\frac{p_\perp^2}{H} = \frac{p^2 \sin^2 \chi}{H} = \text{constant} \tag{15.48}$$

is conserved;† here p is the particle momentum and χ the angle between $\mathbf{p}$

† It is more correct to say that the left-hand side of (15.48) is an adiabatic invariant when a particle moves in a magnetic field. An adiabatic invariant remains constant when the parameters of the problem change slowly, in the given case when the field H changes slowly. This means that the field can vary only inappreciably over distances of the order of the radius of curvature r_H and times of the order of $1/\omega_H^*$, $\omega_H^* = (ZeH/Mc)(Mc^2/E)$. Under cosmic conditions such requirements are well satisfied in many cases. One must, however, bear in mind that we have assumed here not only that there is no electrical field (apart from the field connected with the change of H with time), but we have also neglected losses. The latter may, of course, lead to the adiabatic invariant not being constant. For instance, magneto-brems losses, particularly important for electrons, lead for $H = \text{const}$ to a decrease in the angle χ.

and **H**. In a constant, that is, time-independent, magnetic field the particle energy $E = \sqrt{(M^2c^4 + c^2p^2)}$ and its momentum p are unchanged and hence

$$\frac{\sin^2\chi}{H} = \text{constant}\ , \quad \frac{\partial H}{\partial t} = 0\ . \tag{15.49}$$

When the particle moves into a region with a smaller field the angle χ decreases, as one can see from (15.49). Hence it follows that when $H_2 \ll H_1$ in region 2 with a weak field the distribution of the particles over directions must become steeply anisotropic; they will move practically along the field lines (that is, for them the angle $\chi \ll 1$). This produces a particle beam.

In the Galaxy the cosmic ray density $N_{c.r.} \sim 10^{-10}$ particles/cm^3 (see (15.9)); for an estimate of the density of the particles in the 'beam' which enters interstellar space from shells, or which leaves the Galaxy we also take the estimate $N_s \sim N_{c.r.} \sim 10^{-10}$ particles/cm^3 whence we get for protons

$$\omega_s^2 = 4\pi N_s e^2/M \sim 0.3\ \text{s}^{-2}\ ;$$

this estimate is valid up to energies $E \sim Mc^2 \sim 10^9$ eV, that is, for the majority of the cosmic rays (in the relativistic case $\omega_s^2 = (4\pi N_s e^2/M)(Mc^2/E)$). At the same time for the parent plasma in the Galaxy

$$\omega_p = \sqrt{(4\pi N e^2/m)} = 5.64 \times 10^4 \sqrt{N} \leqslant 5 \times 10^4\ \text{s}^{-1} \quad (N \leqslant 1\ \text{cm}^{-3})$$

and in the Metagalaxy

$$\omega_p \leqslant 10^2\ \text{s}^{-1} \quad (N \leqslant 10^{-5}\ \text{cm}^{-3})\ .$$

Hence we get for the growth rate of the plasma waves due to the beam instability (see (15.47) with $v_{T_s} \sim c$)

$$\left.\begin{aligned} |\gamma_{max}| &\leqslant \frac{\omega_s^2}{\omega_p} \sim 10^{-5}\ \text{s}^{-1} \quad \text{(Galaxy)}\ ; \\ |\gamma_{max}| &\leqslant 10^{-3}\ \text{s}^{-1} \quad \text{(Metagalaxy)}\ . \end{aligned}\right\} \tag{15.50}$$

For the shortest-wavelength waves which may play a role in the problem

$$\gamma \sim \gamma_{min} \sim (v_T/c)^2\gamma_{max} \sim (k_B T/mc^2)\ \gamma_{max} \sim 10^{-6}\ \gamma_{max}$$

(for $T \sim 10^4$ °K) and $\gamma_{min} \sim 10^{-4}\ \gamma_{max}$ (for $T \sim 10^6$ °K which probably corresponds to the metagalactic gas). However, even for a value of $|\gamma| \sim 10^{-10}\ \text{s}^{-1}$ the plasma waves grow considerably over a time† $T \sim 1/|\gamma| \sim 10^{10}$ s ~ 300 yr, that is, over a time which is negligibly small compared to the characteristic

---

† The time and the temperature are here denoted by the same symbol T but this should not lead to any confusion.

time for the evolution of the Galaxy $T_G \sim 10^9$ to 10^{10} yr and the lifetime of the cosmic rays in the Galaxy $T_{c.r.} \sim 10^7$ to 10^8 yr. We shall not multiply the number of similar examples and estimates as their only aim was here to demonstrate that in cosmic ray astrophysics plasma effects are, in general, important (Ginzburg and Syrovatskii, 1966b; Ginzburg, 1966, 1969a, 1975a; Kaplan and Tsytovich, 1973; Ginzburg, Ptuskin, and Tsytovich, 1973; Wentzel, 1974; Tsytovich, 1977). The fact is simply that the frequencies ω_p of the cosmic plasma and the frequencies ω_s in possible cosmic ray beams are small only as compared to 'laboratory' frequencies; the same applies to the growth rates of various instabilities which must clearly be compared with the quantities $1/T$, where T is a characteristic time in the problem (time for spreading out of a supernova shell or the age of the cosmic rays, and so on).

The growth rate γ is different for different instabilities and, for instance, when plasma waves or magnetohydrodynamic waves are excited by a particle beam. Moreover, γ depends on the parameters of the parent plasma and on the characteristics of the beam itself, in particular, on the degree of anisotropy of the particles in the beam (for a beam with the distribution (15.36) we are talking about the ratio k_BT_s/Mv_s^2). For instance, for a beam of cosmic rays with a small anisotropy $\delta \ll 1$ only the excitation of magnetohydrodynamic waves plays a role (Ginzburg, Ptuskin, and Tsytovich, 1973; Wentzel, 1974). The growth rate $|\gamma|$ is thus largest and thereby, in general, most important under different conditions for different kinds of instabilities, different kinds of waves, and so on. The proviso made here of 'in general' is connected with two points. Firstly, the growth rate γ characterizes, say, the growth of plasma waves only in the initial, linear stage. However, the nature of the stationary state is determined by non-linear processes. The instability with the largest growth rate can thus in actual fact in the non-linear stage lead to less important perturbations than some more slowly developing instability. Secondly, we are usually interested not in the instabilities themselves or in the intensity of the waves which appear, but in some action or other of these waves (perturbations). For instance, if we are dealing with the radiation as the result of instabilities of electromagnetic (transverse) waves it is not less important to take into account the mutual transformation of different kinds of waves than to determine the intensity of the plasma waves. In cosmic ray astrophysics we are usually particularly interested in the problem of the feedback of the waves which are generated and of other perturbations on the beam itself and on the magnetic field in which it moves. As a result of the development of instabilities which lead to the appearance of various waves and other perturbations (say, aperiodic distortions of the magnetic field) the beam is, in general, blurred out and made isotropic and the magnetic field changes from a regular one into an altogether random, turbulent one. This is a general and rather obvious tendency but to analyze it in any detail requires large efforts and has as yet not been accomplished by far (Ginzburg, Ptuskin, and Tsytovich, 1973; Kaplan and Tsytovich, 1973; Tsytovich, 1977). This refers in particular to the cosmic conditions where an appreciable fraction of the lack of determination is connected with our insufficient knowledge of the parameters of the problem (characteristics of the beams and of the magnetic field, let alone even the parameters of the interstellar plasma). However, even the estimates given here for the beam instability allow us to reach a few conclusions which confirm the evaluation of the contribution of magnetohydrodynamic waves, the analysis of the role of non-linear processes, and so on (see Ginzburg, Ptuskin, and Tsytovich, 1973, and the literature cited there).

It is just the occurrence of plasma instabilities which leads in the Universe to an efficient generation of various waves and perturbations which, in turn, scatter the cosmic rays. As a result of this any steeply anisotropic cosmic ray distribution will relax rather fast and in the Galaxy or in the Metagalaxy the cosmic rays can have only a small degree of anisotropy $\delta \ll 1$. At the same time the scattering of the cosmic rays by the inhomogeneities and waves in conjunction with the already mentioned perturbation of the magnetic field leads to the well known turbulence of the magnetic field in the Galaxy and the mixing of the cosmic rays. Unfortunately, the quantitative side of the problem is as yet not very clear. In particular, in various energy ranges the importance of the scattering and isotropization of the cosmic rays by the waves which are themselves produced by the cosmic rays, or by waves of another provenance and by various static or rather quasi-static magnetic field inhomogeneities is not clear (Ginzburg, Ptuskin, amd Tsytovich, 1973; Wentzel, 1974).

We cannot develop here this theme, but it seemed useful to us, as elsewhere in this chapter, to discuss, albeit in general terms, a class of topics which stand in the forefront of cosmic ray astrophysics — at this moment we have in mind the taking into account of plasma effects. Hereby must also become more intelligible the approach which dominates at the present all attempts to discuss quantitatively this or that model of the origin of cosmic rays — the use of the diffusion approximation and the transfer equations.

We shall assume that the cosmic rays are locally isotropic — this means that anisotropy can occur only due to the spatial inhomogeneity of the particle densities $N_i(\mathbf{r}, t, E)$, where i is the kind of particles (the number of particles in an element $d^3\mathbf{r}\,dE$ at time t equals $N_i d^3\mathbf{r}\,dE$). The general transfer equation for N_i has in the approximation discussed here (for details see Ginzburg and Syrovatskii, 1964a; there is, of course, here no summation over the index i which occurs twice in (15.51))

$$\frac{\partial N_i}{\partial t} - \text{div}\,(D_i \nabla N_i) + \frac{\partial}{\partial E}(b_i N_i) = Q_i - P_i N_i + \mathcal{P}_i . \qquad (15.51)$$

In a moment we shall, of course, say what all terms in Eqn. (15.51) mean but we start with the first two — if we retain only those we get the diffusion equation:

$$\frac{\partial N_i}{\partial t} - \text{div}\,(D_i \nabla N_i) = 0 , \qquad (15.52)$$

where $D_i(\mathbf{r}, E)$ is the diffusion coefficient.

It is not at all clear that one can use for a description of the motion of the cosmic rays in magnetic fields the diffusion approximation (15.51) or (15.52). The fact that the field has a strongly expressed irregular, random component is no good reason for the validity of this approximation as also in that case there is a strong tendency for the motion of particles along the magnetic field lines, even though they are rather entangled. However, in the Galaxy, for instance, one must also take into account the fact that as a result of the differential rotation of the Galaxy and the motion of gas clouds and spiral arms the magnetic field lines are 'mixed up' all the time. Finally, we shall usually be interested in the picture which is averaged not only over rather large regions of space (say, regions of tens or hundreds of parsecs) but also over rather long periods. For instance, to estimate the mean gradients of cosmic rays and their life time in the Galaxy $T_{c.r.}$ it is sufficient to know N_i averaged over a time $t \ll T_{c.r.} \sim 10^7$ to 10^8 yr, that is, the averaging time may well be as large as 10^5 years.

Taking all that into account the diffusion approximation turns out to be acceptable, especially when we choose the coefficient D_i as a free parameter. It is true that in that way we do not abandon the possibility of evaluating D_i from a more detailed consideration (for instance, taking plasma instabilities into account) or, mainly, the possibility of confirming our assumptions about the validity of the diffusion mechanism by a comparison of the observational data with those calculated in the diffusion approximation (using an equation such as (15.51)), for instance, of the anisotropy, chemical composition, and other quantities characterizing all the cosmic rays or their various components.

In the diffusion picture the resulting cosmic ray flux equals

$$F_{D,i} = 2\pi \int_0^{\pi} J(\theta) \cos\theta \sin\theta \, d\theta = D_i \, |\nabla N_i| = -D_i \frac{dN_i}{dr} , \qquad (15.53a)$$

where the last expression has been written down assuming an appropriate symmetry of the problem; moreover, we have chosen the polar axis along the direction of the flux $\mathbf{F}_{D,i}$. Putting $J(\theta) = J_0 + J_1 \cos\theta$ for all cosmic rays we easily get for the degree of anisotropy the expression

$$\delta = \frac{J_{max} - J_{min}}{J_{max} + J_{min}} = \frac{J_1}{J_0} = \frac{3F_D}{4\pi J_0} = \frac{3D}{c} \frac{1}{N_{c.r.}} \left| \frac{dN_{c.r.}}{dr} \right| , \qquad (15.53b)$$

where we have also used the relation $J \approx J_0 = (4\pi/v) N_{c.r.} = (4\pi/c) N_{c.r.}$ (we consider ultra-relativistic particles); of course, one can write down similar expressions for particles of any kind. In the quasi-spherical picture

$$\left| \frac{dN_{c.r.}}{dr} \right| \sim \frac{N_{c.r.}}{R} ,$$

where R is a characteristic distance; in the case of the Galaxy we put $R \sim 10^{22}$ to 10^{23} cm (the distance of the Sun from the centre of the Galaxy $R = 3 \times 10^{22}$ cm) and $\delta \leqslant 10^{-4}$ (vide supra). Hence $D \sim \frac{1}{3} \delta cR \leqslant 10^{28}$ to 10^{29} cm^2/s. Other more reliable† estimates are based upon calculations of the chemical composition of the cosmic rays (see Ginzburg and Syrovatskii, 1964a; Ptuskin, 1972; Ginzburg and Ptuskin, 1976a,b; and the following). They give

$$D_{disc} \sim 3 \times 10^{26} \text{ cm}^2/\text{s} , \quad D_{halo} \sim 7 \times 10^{27} \text{ cm}^2/\text{s} , \qquad (15.54)$$

where the values D_{disc} and D_{halo} correspond, respectively, to the disc model and the model with a halo.

The diffusion coefficient in a gas is $D = \frac{1}{3} v \ell$ where ℓ is the mean free path and v the particle velocity. Applying the same relation and assuming that the velocity of the cosmic ray motion along the field $v \sim 10^{10}$ cm/s, we can estimate from (15.54) the effective mean free path: $\ell_{disc} \sim 10^{17}$ cm and

† The fact is not only that the degree of anisotropy δ has not yet been measured (only an upper limit has been determined and even that not very reliably). Not less important is that the anisotropy may reflect 'local' conditions in the neighbourhood of the solar system and not characterize the average gradient of the cosmic rays in the Galaxy.

$\ell_{halo} \sim 2 \times 10^{18}$ cm. In the diffusion picture the mean square distance $\overline{z^2}$ traversed by a particle in the z-direction during a time T is equal to $\overline{z^2} = 2DT$. We can use this formula to estimate the lifetime of the cosmic rays in the Galaxy taking for $\overline{z^2}$ the square of a characteristic size L of the system. One usually puts in the disc model $L \sim 3 \times 10^{20}$ cm (half-thickness of the gas disc) and in the model with a halo $L \sim R \sim 10^{22}$ cm (radius or, rather, half-thickness of the halo). Using (15.54) we then get

$$T_{c.r.,disc} \sim 5 \times 10^6 \text{ yr} \ , \ T_{c.r.,halo} \sim 2 \times 10^8 \text{ yr} . \qquad (15.55)$$

The crudeness of the estimates given here is obvious but, even taking this into account it is not yet clear why we have taken as the size L in the disc model the half-thickness of the gas disc rather than the half-thickness of the radio-disc $h \sim 10^{21}$ cm. The fact is that the interstellar gas is concentrated just in the gas disc, as its name indicates. For an estimate of the time the cosmic rays stay in the gaseous medium (for the gas disc $N \sim 1\ cm^{-3}$) we must therefore give $T_{c.r.,disc}$ just for the gas disc. The arbitrariness of such a time $T_{c.r.}$ is, however, thereby made particularly clear. We need hardly explain that when earlier putting $T_{c.r.,disc} \sim 10^7$ yr and $T_{c.r.,halo} \sim 10^8$ yr we did not go beyond the limits of accuracy of the estimates (15.55).

The diffusion coefficient D_i in (15.50) and (15.51) may depend on the coordinate $\mathbf{r}$ and the particle energy E (we assume the system to be stationary so that the coefficients in the transfer equation, in particular, the coefficient D_i, are time-independent; one needs special considerations for diffusion and other processes in non-stationary conditions). However, in practice one solves the problem either with a coefficient D_i which is constant in space, or for some regions in each of which the coefficient D_i is constant (at the boundary between these regions the component normal to the boundary of the flux $-D_i \nabla N_i$ must be continuous, and also the density N_i). As to the dependence of D_i on the particle energy E, the approximate constancy of the chemical composition of the cosmic rays indicates the approximate constancy of D_i in the energy range of about $\epsilon \leqslant 10^{12}$ eV/nucleon, and possibly up to energies $E \sim E_c \sim 1$ to 3×10^{15} eV. When $E \sim E_c$ the index γ in the cosmic ray spectrum changes (see (15.11) and (15.12)); it is natural to assume that this is connected with an appreciable E-dependence of D_i and to be precise with an increase in the diffusion coefficient with increasing E for $E > E_c$. Relatively recently there has been the indication mentioned already of a possible weak E-dependence of D_i also in the energy range $E < 10^{12}$ eV (see Ptuskin, 1974; Júliusson, 1974).

It is important to emphasize here that it is very well possible to take this dependence into account in the framework of the diffusion approximation.

Let us now discuss the other terms (apart from the first two) in the transfer Eqn.(15.51) which have the meaning of conservation laws for the number of particles in coordinate and energy space. This remark makes it already possible to understand that the quantity b_iN_i is the particle flux of kind i in 'energy space' where b_i is the velocity in energy space, that is, the change in particle energy per unit time

$$\frac{dE}{dt} = b_i(E) . \tag{15.56}$$

Therefore, $\frac{\partial}{\partial E}(b_iN_i)$ is, indeed, the divergence of a flux. We must then bear in mind that the change in the particle energy considered must be smooth and continuous (at least, within the limits of the accuracy of the approximation used). If we are talking about energy losses, of course, $b_i < 0$; as an example of such practically continuous losses we may mention the ionization losses discussed earlier (clearly, $\frac{dE}{dx} = \frac{1}{v}\frac{dE}{dt}$) or the magneto-brems losses discussed in Chapter 4. When the particles are accelerated $b_i > 0$. It is necessary also to emphasize that as in the case of losses, also in the case of particle acceleration there may be (often important) fluctuating energy changes as well as a regular average energy change over some time interval. As a result of such fluctuations the energy distribution of the particles changes, even if the average particle energy remains constant.† When there are such fluctuations in energy one must under certain conditions add to the left-hand side of Eqn.(15.51) a term

$$-\frac{1}{2}\frac{\partial^2}{\partial E^2}(d_iN_i) , \quad \text{where} \quad d_i(E) = \frac{d}{dt}\overline{(\Delta E)^2} ,$$

$\overline{(\Delta E)^2}$ is the mean square change in energy as a result of fluctuations (for details see Ginzburg and Syrovatskii, 1964).

---

† As an example we may mention the acceleration of particles in an electrical field with potential differences V which are well-defined as far as their absolute magnitude is concerned, but under condition where the sign of V changes randomly (that is, the particle moves sometimes with the field and sometimes against it as might happen if it were incident on a 'capacitor' from different sides). The average energy of all the particles together remains then constant, as $\overline{V} = 0$, but some of the particles may 'win' in such a way that they acquire a large energy as a result of being predominantly incident in regions with the field parallel to the direction of the particle momentum. In other words, we are talking about 'diffusion in energy' so that $\overline{(\Delta E)^2} \neq 0$, as $\overline{V^2} \neq 0$.

The term $Q_i(\mathbf{r},t,E)$ in (15.51) is the power of the 'external' particle sources — their number entering the system per unit time in the neighbourhood $d^3\mathbf{r}\,dE$ of the 'point' $\mathbf{r},E$ is equal to $Q_i\,d^3\mathbf{r}\,dE$. The term $-P_iN_i$ in (15.51) takes into account 'catastrophic' processes leading to particles of the kind i leaving the element considered $d^3\mathbf{r}\,dE$. The particle so to speak vanishes from that element and its neighbourhood. As an example we may mention the transformation of nuclei when a nucleus of the kind i vanishes altogether, changing into nuclei (and in principle also into other particles) of the kinds k, ℓ, m. A second example are brems (radiative) losses when an electron collides with other particles and emits a rather hard photon.

If σ_i is the cross section for the collision of particles of the kind i, v_i the velocity of those particles, $N_{gas} \equiv N$ the density of particles, say, of nuclei in the interstellar gas with which the collisions take place, we have

$$P_i = \sigma_i v_i N = \frac{v_i}{\ell_i} = \frac{1}{T_i}. \tag{15.57}$$

Clearly P_i has the meaning of the number of collisions (see Chapter 11), $\ell_i = 1/\sigma_i N$ is the mean free path, and T_i the average 'lifetime' or the mean flight time.

The last term $\mathcal{P}_i$ in Eqn. (15.51) takes into account the entry of particles (also as the result of 'catastrophic' collisions) into the range considered $d^3\mathbf{r}\,dE$. One can, for instance, write

$$\mathcal{P}_i = \sum_k \int P_i^k(E',E)\,N_k(\mathbf{r},t,E')\,dE', \tag{15.58}$$

where P_i^k is the probability for the process where a particle of the kind k changes into a particle of the kind i (included is also the case $i=k$) from the energy range E' into the energy range E.

The transfer Eqn. (15.51) is rather complicated and this makes it natural to consider various particular cases. As an example we mention the transition to the diffusion Eqn. (15,52) where all other terms have been dropped. Although not so extensive, a significant simplification can usually also be introduced for the analysis of the chemical composition of the nuclei. In nuclear transformations in the interstellar medium (if we neglect inelastic collisions involving meson production, and so on) the energy per nucleon $\epsilon = E/A$ is conserved. It is therefore expedient to change from the variable E to the variable ϵ for which

$$P_i^k(E'-E) = P_i^k\,\delta(\epsilon-\epsilon') \quad \text{and} \quad \mathcal{P}_i = \sum_{k<i} P_i^k N_k(\mathbf{r},t,\epsilon)$$

(see (15.58)). The index $k < i$ indicates here that a nucleus of kind i can appear only due to the disintegration of heavier nuclei for which the index k customarily is assumed to be less than i. Moreover, for relativistic nuclei the energy losses are relatively small (we are mainly talking here about ionization losses) and we can neglect them. As a result we are led to equations which are widely used for an analysis of the chemical composition of the cosmic rays :

$$\frac{\partial N_i}{\partial t} - \text{div}\,(D_i \nabla N_i) = Q_i(\mathbf{r}, t) - P_i N_i + \sum_{k<i} P_i^k N_k , \qquad (15.59)$$

where in Q_i we have dropped the variable ϵ and we can do the same in respect to $N_i(\mathbf{r}, t, \epsilon)$; of course, if we take continuous losses into account by adding to the left-hand side of (15.59) a term $\frac{\partial}{\partial E}(b_i N_i)$ we must, in general, take $Q_i = Q_i(\mathbf{r}, t, \epsilon)$ and $N_i = N_i(\mathbf{r}, t, \epsilon)$.

The mean free path ℓ_i (see (15.57)) can conveniently be expressed in g/cm^2, where $\ell_i = 1/\sigma_i N$ cm $= M/\sigma_i$ g/cm^2, where $M = \rho/N$ is the average mass of the nuclei in the interstellar gas of density ρ and density of nuclei (or atoms) N. Usually one assumes that in the interstellar gas hydrogen takes up 90% in number and helium 10%, and one can neglect the other nuclei. We give in Table 15.2 the values of σ_i and ℓ_i for the motion of nuclei of the kind (group) i in hydrogen and in the interstellar gas of the above-mentioned composition (see also Table 15.1)

Table 15.2

Group of nuclei	Average atomic weight $\bar{A}$	Cross-section $\sigma_i \times 10^{26}$, cm^2		Mean free path ℓ_i, g/cm^2	
		hydrogen	interstellar gas	hydrogen	interstellar gas
p	1	2.3	3	74	72
α	4	9.3	11	18	20
L	10	23	25	7.3	8.7
M	14	29	31	5.8	6.9
H	31	48	52	3.5	4.2
Fe	56	73	78	2.3	2.8

We must bear in mind that the mean free path ℓ_i characterizes a nucleus of the kind i leaving the flux of such particles independently of into which nucleus it is transformed. When we divide the nuclei into groups we must take into account the transformation of nuclei which retains them within the limits of the group. The corresponding effective mean free path is $\lambda_i = \ell_i/(1 - P_i^i)$, where P_i^i is the probability of forming nuclei of group i from other nuclei of the same group. As a result, we have, for instance, for nuclei of the M group in the interstellar gas $\lambda_i = 7.8\ \mathrm{g/cm^2}$ for $\ell_i = 6.9\ \mathrm{g/cm^2}$. However, one should not attach too much weight to these quantities as the data in Table 15.2 are only of value for orientation purposes.

Relativistic protons lose in collisions with nuclei of the interstellar gas (with cross section σ_i) on average about $\frac{1}{3}$ of their energy; the energy of the protons decreases therefore by a factor $e = 2.72$ over a path length

$$\lambda_E = 1/\sigma_E N \approx 180\ \mathrm{g/cm^2} \approx 10^{26}/N\ \mathrm{cm} \quad (\sigma_E \sim 10^{-26}\ \mathrm{cm^2}),$$

In a rough approximation we can say that the energy of the relativistic protons changes as (N is the density of nuclei in the interstellar gas)

$$-\left(\frac{dE}{dt}\right)_{\mathrm{nucl}} = \frac{cE}{\lambda_E} = \sigma_E cNE \approx 3\times 10^{-16}\ NE\ . \qquad (15.60)$$

Let us compare these losses in hydrogen with the ionization losses also in hydrogen (see (15.28) with $Z = 1$). Clearly

$$\eta_{\mathrm{nucl},i} = \frac{(dE/dt)_{\mathrm{nucl}}}{(dE/dt)_i} \sim 40\,\frac{E/Mc^2}{4\,\ell n\,(E/Mc^2) + 20.2}\ . \qquad (15.61)$$

The nuclear losses are thus appreciably larger than the ionization losses already for $E \sim 10\,Mc^2 \sim 10^{10}$ GeV. For the majority of the cosmic rays (but not in the low energy region $\epsilon_k \leqslant M_p c^2 \sim 10^9$ eV) one can neglect the ionization losses; the nuclear losses, on the other hand, are taken into account in (15.59) by the term $-P_i N_i$ (these losses belong more correctly to the category of 'catastrophic' losses and one can use an equation such as (15.60) only for estimating the average losses over a considerable time).

We have elsewhere (Ginzburg and Syrovatskii, 1964a, Chapter 5; see also Ptuskin, 1972; Ginzburg and Syrovatskii, 1973; Ginzburg and Ptuskin, 1976a,b) discussed methods for solving the set (15.59) and we shall not dwell upon them here. We merely note that when we solve the problem we usually introduce a whole number of further simplifications: we assume the problem to be stationary (we drop the derivative $\partial N_i/\partial t$), we put the diffusion coefficient D_i to be constant and we consider several regions in space and energy ranges, 'joining up' the solutions at the boundaries, and we write the power of the sources Q_i in the form $Q_i(\mathbf{r}, t, \epsilon) = q_i\,\chi(\mathbf{r}, t, \epsilon)$, and so on. We must make the model even more precise — we must give the spatial distribution of the power of the sources Q_i, the region where the cosmic rays are 'trapped' (disc, halo), and so on.

As a limiting case of an ever further simplified problem we can take a uniform model (also called the leaky-box model), which is often used to determine the chemical composition of the cosmic rays. In such a model one assumes that the diffusion takes place rather fast and that therefore the cosmic ray density is constant in the whole system (Galaxy). We must then, of course, give, on the other hand some lifetime of the cosmic rays in the system which determines how fast they leave the system. In other words, we replace the terms $\partial N_i/\partial t - \operatorname{div}(D_i \nabla N_i)$ in (15.59) by $N_i/T_{c.r.,i}$ (one arrives at this substitution most simply by dropping the diffusive term and putting $\partial N_i/\partial t = N_i/T_{c.r.,i}$). The set (15.59) can then be written (uniform model)

$$\frac{N_i}{x} = q_i - \sigma_i N_i - \sum_{k<i} \sigma_{ik} N_k \,, \tag{16.62}$$

where $x = c\rho T_{c.r.}$ is the thickness of interstellar gas traversed by the cosmic rays (we assume the particles to be relativistic so that their velocity $v = c$; for the sake of simplicity we assumed that the times $T_{c.r.,i} = T_{c.r.}$, that is, that they are independent of the kind of particle i), and σ_i and σ_{ik} are the appropriate cross-sections (see (15.57) and the definition of the quantities $P_i^k = \sigma_{ik} v N$; if we defined, as is commonly done, the thickness x_i in g/cm^2 and therefore introduce the density $\rho = MN\, g/cm^3$, the cross-sections in (15.62) are the normal cross-sections divided by the mass M of the 'average nucleus' in the gas).

As the q_i in (15.62) characterize the power of the cosmic ray sources, it is clear that $q_i \geq 0$; moreover, for nuclei with small source densities (especially nuclei of the group L, that is, for Li, Be, and B) we can assume that $q_i = 0$. The set (15.62) is algebraic and can be solved rather simply; all difficulties are connected with insufficient information about the cross-sections σ_i and σ_{ik} and even more with insufficiently accurate data about the chemical (and even more the isotopic) composition of the cosmic rays at the Earth. At present the uniform model (15.62) describes fairly well the chemical composition of cosmic rays with energies $\epsilon_k \gtrsim 1$ to 2 GeV/nucleon for $x_i = x \approx 5$ to $7\ g/cm^2$ (see Ptuskin, 1972; Ginzburg and Syrovatskii, 1973; Ginzburg and Ptuskin, 1976a,b).

Thus

$$T_{c.r.} = \frac{x}{\rho c} = \frac{x}{cMN} \approx \frac{4 \times 10^6}{N}\ \text{yr} \,, \tag{15.63}$$

where N is the gas density (of nuclei with an average mass $M \sim 2 \times 10^{-24}$ g) in the region occupied by the cosmic rays. In the disc model the average value of $N \sim 0.3$ to $1\ cm^{-3}$ and in the model with a halo $N \sim 1$ to $3 \times 10^{-2}\ cm^{-3}$; as a result the corresponding values do not contradict the estimates (15.55).† More important is something else — the chemical composition is first of all determined by the thickness x and it is therefore impossible to find the time $T_{c.r.}$ from data about the chemical composition. It is true that in more refined models which take diffusion into account the dependence of the chemical composition on $T_{c.r.}$ determined by the diffusion coefficient $D_i(E)$ is more important, but the accuracy of all data is still insufficient to solve the problem. More reliable for success promises to be another method — taking into account the role of radio-active nuclei in the cosmic ray composition (the best known example is the ^{10}Be nucleus for which the average lifetime $\tau = 2.2 \times 10^6\ E/Mc^2$ yr, where the factor E/Mc^2 takes into account the relativistic slowing-down of the time). The possibility of radio-active decay is not taken into account in (15.62) and if one does so for radio-active nuclei, there should be on the left-hand side the sum $(N_i/x_i) + (N_i/c\rho\tau_i)$. Determining the density of radio-active nuclei N_i in comparison with the densities of a number of stable nuclei one can in principle find both $x_i = c\rho T_{c.r.,i}$ and $c\rho\tau_i$ and thereby determine $T_{c.r.,i}$ and the average gas density ρ in the region occupied by these cosmic rays (for details see Ginzburg and Ptuskin, 1976a,b; Plovdiv, 1977; Ginzburg, 1978).

We now make more precise the transfer Eqns.(15.51) for an application to electrons and positrons. In that case we must, clearly, assume that in (15.51) $N_i = N_e(\mathbf{r}, t, E)$ or separately N_{e-} (electrons) and N_{e+} (positrons). Simplifications occur when we assume that the problem is stationary (we drop the derivative $\partial N_e/\partial t$) and when we neglect the 'catastrophic' energy losses. We then have

$$- \operatorname{div}(D_e \nabla N_e) + \frac{\partial}{\partial E}(b_e(E) N_e) = Q_e(\mathbf{r}, E)\ . \tag{15.64}$$

---

† The characteristic nuclear lifetime for the proton component of the cosmic rays is (see (15.60))

$$T_{nucl} \sim \frac{E}{|(dE/dt)_{nucl}|} \sim \frac{3 \times 10^{15}}{N}\ s\ . \tag{15.60a}$$

Even for the gaseous disc with $N \sim 1\ cm^{-3}$, the time $T_{nucl} \sim 10^8$ yr $\gg T_{c.r.,disc}$, and for $N \sim 10^{-2}\ cm^{-3}$ we have already $T_{nucl} \sim 10^{10}$ yr $\gg T_{c.r.,halo} \sim 1$ to 3×10^8 yr. Hence it follows that the cosmic ray lifetime $T_{c.r.}$ is, indeed, determined by their leaving the Galaxy and not by the losses (this is, in general, not the case for the rather heavy nuclei; see Table 15.2).

The term Q_e must take into account the appearance of electrons and (or) positrons not only as a result of their acceleration, but also due to various decays of unstable particles ($\mu^{\pm}$, and so on) which are formed in nuclear collisions with cosmic rays (we can refer here to δ-electrons and electron-positron pairs produced by gamma rays). In the equation like (15.64) for the positrons we must also introduce a term accounting for their annihilation.

The most important difference arising when we consider the electron component of the cosmic rays as compared to the proton-nuclear component consists in the necessity to take, in general, the energy losses of the electrons into account. The variable E in Eqn.(15.64) therefore does not remain a parameter as was the case in Eqns.(15.59) and (15.62).

Integrating Eqn.(15.64) enables us to find the electron spectrum $N_e(\mathbf{r},t,E)$. Knowing this spectrum for the whole Galaxy we can evaluate the intensity of the synchrotron radio-emission received on Earth. The same holds, of course, also for the radio-emission from supernova shells, radio-galaxies, and so on. We shall not give here the corresponding calculations (see Ginzburg and Syrovatskii, 1964a; Bulanov, Dogel', and Syrovatskii, 1972a,b; and Ginzburg and Ptuskin, 1976a,b) restricting ourselves to the physical processes which must be taken into account. The acceleration of the electrons in the sources and the generation of 'secondary' electrons and positrons by the proton-nuclear component of the cosmic rays determines the power of the sources $Q_i(\mathbf{r},E)$. The electrons lose energy as a result of ionization, brems (radiative), magneto-brems, and Compton losses; these all contribute to the coefficient $b_e(E)$ in (15.64).

We have already considered the ionization losses (see (15.23)), the magneto-brems losses were discussed in Chapter 4, and the brems and Compton losses will be considered in Chapter 16. It is, however, expedient to give all these losses for ultra-relativistic electrons in one place.

Ionization losses. In atomic hydrogen

$$-\left(\frac{dE}{dt}\right)_i = \frac{2\pi N e^4}{mc}\left(\ln\frac{E^3}{mc^2\mathcal{I}^2} - 0.57\right) =$$

$$= 7.62\times10^{-9}\, N\left\{3\ln\frac{E}{mc^2} + 20.2\right\} \text{ eV/s} . \qquad (15.23a)$$

In an ionized gas

$$-\left(\frac{dE}{dt}\right)_i = \frac{2\pi N e^4}{mc}\left(\ln\frac{m^2c^2E}{4\pi N e^2\hbar^2} - \frac{3}{4}\right) =$$

$$= 7.62\times10^{-9}\, N\left\{\ln\frac{E}{mc^2} - \ln N + 73.4\right\} \text{ eV/s} . \qquad (15.24)$$

Brems (radiative) losses. In the atomic interstellar gas (see (16.48))

$$-\left(\frac{dE}{dt}\right)_r = 10^{-15} N_a E \text{ eV/s} = 5.1\times 10^{-10} N_a \frac{E}{mc^2} \text{ eV/s}.$$

In a fully ionized gas (see (16.46))

$$-\left(\frac{dE}{dt}\right)_r = 7\times 10^{-11} N\left(\ell n \frac{E}{mc^2} + 0.36\right)\frac{E}{mc^2} \text{ eV/s}.$$

($N = N_a$ is the nuclear or electron density).

Magneto-brems and Compton losses.

$$-\left\{\left(\frac{dE}{dt}\right)_m + \left(\frac{dE}{dt}\right)_C\right\} = \frac{32\pi}{9}\left(\frac{e^2}{mc^2}\right)^2 c\left(\frac{H^2}{8\pi} + w_{ph}\right)\left(\frac{E}{mc^2}\right)^2 =$$

$$= 1.65\times 10^{-2}\left(\frac{H^2}{8\pi} + w_{ph}\right)\left(\frac{E}{mc^2}\right)^2 \text{ eV/s}. \qquad (15.65)$$

The part of this expression which corresponds to the magneto-brems losses is proportional to H^2 and is obtained from Eqn. (4.39), assuming that the direction of the magnetic field changes randomly so that $H_\perp^2 = \frac{2}{3}H^2$. The second part of expression (15.65) corresponds to Compton losses in an isotropic radiation field with energy density w_{ph} (the quantity $H^2/8\pi + w_{ph}$ is in (15.65) measured in erg/cm^3), and we have here considered only the region of electron energies $E \ll (mc^2/\epsilon_{ph})mc^2$, where ϵ_{ph} is the average photon energy (see Chapter 16 for details).

We have, for instance, for an ionized gas

$$\eta_{r,i} = \frac{(dE/dt)_r}{(dE/dt)_i} = \frac{1.8\times 10^{-8}\{\ell n(E/mc^2) + 0.36\}E}{\ell n(E/mc^2) - \ell n N + 73.4}. \qquad (15.66)$$

Even in the intergalactic gas $N \sim 10^{-5}$ to 10^{-6} cm^{-3} and, hence, $|\ell n N| < 15$. Therefore $\eta_{r,i} \leqslant 1$ when $E \leqslant 7\times 10^8$ eV. When $E > 10^9$ eV the brems losses dominate the ionization losses.

The ratio of the magneto-brems and Compton losses to the brems losses (16.46) equals

$$\eta_{mC,r} = \frac{(dE/dt)_m + (dE/dt)_C}{(dE/dt)_r} \approx \frac{3\times 10^7}{N}\left(\frac{H^2}{8\pi} + w_{ph}\right)\frac{E}{mc^2}. \qquad (15.67)$$

In the Galaxy (in the disc) $H^2/8\pi \sim 10^{-12}\text{ erg/cm}^3$ and $w_{ph} \sim 10^{-12}\text{ erg/cm}^3$ (for the black-body background radiation of temperature $T = 2.7\ {}^\circ K$ alone we have $w_{ph} = 4\times 10^{-13}\text{ erg/cm}^3$; in the Galaxy, especially in the disc there are also many optical photons emitted by the stars). Therefore in the gas disc (for $N \sim 1\text{ cm}^{-3}$) $\eta_{mC,r} \sim 3\times 10^{-5} E/mc^2 \gtrsim 1$ for $E \gtrsim 10^{10}$ eV, and in the halo (with $N \sim 10^{-2}\ cm^{-3}$) $\eta_{mC,r} \geqslant 1$ for $E \geqslant 10^{8}$ eV. Even in the radio-disc, let alone in the halo, the main role is therefore played for the electron component in the most interesting energy range $E \geqslant 10^{8}$ to 10^{9} eV by the magneto-brems and Compton losses. We must add to this that the brems losses in fact belong to the 'catastrophic' losses — they are mainly accompanied by the emission of photons of energies $\hbar\omega \sim E$. As a result the electron simply 'goes out of play'. The average characteristic time T_r for such losses (see (16.48) below) equals

$$T_r \sim \frac{E}{|(dE/dt)_r|} \sim \frac{10^{15}}{N}\ s\,. \tag{15.68}$$

Even when $N \sim 1\ cm^{-3}$ the time $T_r \sim 3\times10^{7}$ yr, which is less than the time the electrons move about in the gas disc. When $N \sim 10^{-2}\ cm^{-3}$ the time T_r is already so large (compared to $T_{c.r.} \leqslant 1$ to 3×10^{8} yr) that the brems losses do not play a role. For evaluating the electron spectrum in the Galaxy one therefore usually takes into account only the magneto-brems and Compton losses.

With this we conclude the present chapter which was devoted to a few problems in cosmic ray astrophysics. Here, in contrast to other chapters we paid considerable attention to descriptive, essentially astrophysical, material which diminished the space devoted to theoretical problems. We did this as the corresponding astrophysical information is practically completely absent in general physics and theoretical physics courses. However, if one does not use it and does not take it into account, one cannot deal at all with cosmic ray astrophysics and one is merely left with purely physical results which one could especially apply to cosmic rays. In that case, however, the choice of material remains undetermined and, in the main, loses all astrophysical features. We, on the other hand, wanted to retain the astrophysical aspects. This tendency will also be seen, although not quite so strongly, in the next two chapters.

Chapter XVI

X-RAY ASTRONOMY

Processes leading to the formation of X-rays and gamma-rays.
Definition of the quantities used in X-ray and gamma-astronomy.
X-ray brems-emission by a non-relativistic gas (plasma).
Bremsstrahlung by relativistic electrons and brems (radiative) energy losses.
Scattering of relativistic electrons by photons (inverse Compton effect).
Compton energy losses. X-ray synchrotron emission.
Remarks about the comparison between theory and observations.

X-ray and gamma rays 'by themselves' (that is, without considering their interaction with matter) differ not only merely in wavelength but are also 'neighbours' in the electromagnetic wave spectrum.† It is therefore expedient to start the discussion of the processes leading to the appearance of cosmic X-ray and gamma-radiation without splitting the bands in more detail. We shall therefore first of all list the processes which lead to the production of both X-rays and gamma rays.

It is true that it is opportune to note beforehand that the mean free path of the absorption coefficient of even hard gamma-rays, let alone those of the softer photons, does not exceed approximately $100\ \mathrm{g/cm^2}$. Hence it is clear that the cosmic gamma- and X-ray radiation which reaches the Earth cannot come from regions with an extra-ordinarily high density, for instance, from the interior of neutron stars. It is from this also clear that the photon emission and absorption processes which we shall encounter in X-ray and gamma-astronomy have, so to speak, the usual character of those in atomic or nuclear physics. In other words, we do not need to consider here some new, not yet known emission or absorption mechanisms. The specific features which arise in X-ray and gamma-astronomy are connected in the first place with the fact that

---

† Customarily we shall call photons with energies $100 < E_X < 10^5$ eV (wavelength $\lambda \approx 12400/E_X$ (eV) Å approximately between 0.1 and 100 Å) X-rays. However, radiation emitted by atomic nuclei is usually called γ-rays even when $E_X \equiv E_\gamma < 10^5$ eV. We have denoted the energy of X-ray and gamma-photons by E_X and E_γ, but sometimes we shall understand by E_γ the energy of any hard photon (in the X and γ ranges).

in the laboratory one is usually dealing with the scattering of hard photons by slow electrons, while in the Universe the scattering of high-energy electrons by optical and radio-photons plays a larger role. There are, of course, also other peculiarities but in all known cases they refer to an actual problem or to parameters which characterize the problem and not to the matter itself of the elementary processes discussed. Hence, when we are talking about elementary processes which are important for X-ray and gamma astronomy we may assume that the picture is rather obvious.

The following processes lead to the production of X-ray and gamma photons:

1. Bremsstrahlung of electrons and positrons apart from certain exceptions (we shall in what follows not mention positrons separately).

We have here in mind collisions of electrons with various nuclei and also with other electrons in which both the incident and the scattered electron have a continuous spectrum while, apart from recoil, the scattering particle does not change its state.

A particle with kinetic energy E_k can emit a brems photon only with an energy $E_{X,\gamma} \leq E_k$ (for the sake of simplicity we have in mind only collisions with rather heavy particles at rest; when recoil is taken into account $E_{X,\gamma} < E_k$). From this it is clear that non-relativistic electrons can produce only X-rays as the result of the brems mechanism. Relativistic electrons can also give gamma photons. The intensity of the bremsstrahlung at nuclei is for relativistic protons or nuclei of mass M less than for electrons with the same total energy by a factor $(M/m)^2 \geq 3.4 \times 10^6$. There is thus every reason to limit the discussion to bremsstrahlung by electrons. One can also take the radiation accompanying the appearance of electrons and positrons in the $\pi^{\pm} \rightarrow \mu^{\pm} \rightarrow e^{\pm}$ decay or that arising in the formation of δ-electrons (recoil electrons) to be bremsstrahlung.

We do not intend in what follows to dwell upon all aspects of the theory of bremsstrahlung. We shall consider only two cases: bremsstrahlung of an equilibrium non-relativistic plasma and bremsstrahlung of relativistic electrons.

2. Recombination and characteristic X-ray radiation occurring when an electron makes a transition from a level in the continuous spectrum to an atomic level or when it goes from one atomic level to another.

In astrophysical terminology we are dealing, respectively, with free-bound and bound-bound electron transitions, whereas bremsstrahlung corresponds to free-free transitions. Type 2 processes will be touched upon only in passing in the present chapter.

3. Compton scattering of relativistic electrons by X-ray, optical, and radio-photons.

We have already mentioned this process before (see, for instance, Chapter 15). Various relations are possible between the energies of the incident and scattered particles, but we shall consider only the case

$$E > E_{\gamma} \gg \epsilon_{ph} , \tag{16.1}$$

where E, E_{γ}, and ϵ_{ph} are the energies (in the 'laboratory' frame, that is, the frame fixed in the Earth or the Galaxy) of the primary electron, the scattered γ- or X-ray-photon and the primary photon, respectively; in the majority of cases the energy ϵ_{ph} refers to the region of the optical ($\epsilon_{ph} \sim 1$ eV) or the thermal black-body background radio-radiation ($\epsilon_{ph} \sim 10^{-3}$ eV, $\lambda \sim 1$ mm), but we are also interested in the scattering of electrons by cosmic X-rays (in that case $\epsilon_{ph} \sim 10^2$ to 10^4 eV and γ-rays are formed as the result of the scattering) and by radio-photons, say, of synchrotron origin (in that case $\epsilon_{ph} \sim 10^{-5}$ to 10^{-7} eV, $\lambda \sim 10$ cm to 10 m and X-ray and optical photons are produced by the scattering).

The scattering of relativistic protons and nuclei by photons is appreciably less efficient (in comparison with the scattering of electrons, there occurs a factor $(m/M)^2 \leqslant 3 \times 10^{-7}$) and in practically all known cases we can neglect it. Compton radiation will be considered in the present chapter.

4. Synchrotron radiation.

For the sake of convenience we write down once again Eqn. (5.40a) for the characteristic frequency ν_m emitted by an electron of energy $E \gg mc^2$ in a magnetic field

$$\nu_m = 1.2 \times 10^6 H_{\perp} \left(\frac{E}{mc^2}\right)^2 = 4.6 \times 10^{-6} H_{\perp} (E(\mathrm{eV}))^2 \ \mathrm{Hz} . \tag{16.2}$$

Hence one sees easily that in fields $H_{\perp} \leqslant 10^{-3}$ Oe radiation is emitted with a frequency $\nu_m \sim 10^{18}\ s^{-1}$ ($\lambda_m = c/\nu_m \sim 3$ Å) for $E \geqslant 10^{13}$ eV; emission with $\lambda \leqslant 0.1$ Å (γ-rays) is emitted only when $E \geqslant 10^{14}$ eV. In galaxies, radio-galaxies, and most supernova shells synchrotron X-ray and gamma-radiation can thus arise only if there are electrons of very high energies present. This fact clearly limits the possibility of applying the synchrotron mechanism to hard photons. It is sufficient to say that electrons of energy $E \geqslant 10^{13}$ eV in a field $H_\perp \sim 10^{-3}$ Oe lose half their energy in a time

$$T_m = 5.1 \times 10^8 \, mc^2/H_\perp^2 E \text{ s} \leqslant 1 \text{ yr}$$

(see (4.42)). In stars, quasars (close to their nucleus) and, possibly, in some regions of supernova shells (particularly near pulsars) there may exist rather strong magnetic fields. Clearly, under such conditions X-ray synchrotron radiation may be produced already by electrons of lower energies. For instance, for $H_\perp \sim 10^2$ Oe the frequency $\nu_m \sim 10^{18}$ s^{-1} occurs for electrons with energies $E \sim 5 \times 10^{10}$ eV; in that case, however, $T_m \sim 1$ s.

From what we have said it seems that the appearance of cosmic synchrotron X-ray and gamma-radiation (forgetting about pulsars, or, in general, the vicinity of compact sources — white dwarfs, neutron stars, 'black holes') is relatively improbable. Nonetheless. this conclusion is in fact rather conditional as there are circumstances under which the acceleration of electrons or their injection into an extended region with a rather strong field can be very efficient. As an example we may mention the Crab Nebula for which the X-ray emission has a synchrotron nature (as was shown relatively recently by measuring the polarization of the radiation). The pulsar PSR 0531 which is situated in the nebula plays here (directly or indirectly) the role of an efficient electron injector. Cosmic synchrotron X-ray emission is thus observed and, undoubtedly, plays an important role (and one should note that as data are accumulated this role becomes ever more important). We shall return to synchrotron X-ray emission later on.

5. Decay of neutral pions into two γ-photons ($\pi^0 \to \gamma + \gamma$).

The rest energy of a π^0 equals $m_\pi c^2 = 135$ MeV and, hence, the π^0 are produced only by cosmic rays. Their production occurs mainly in p-p, p-α, and α-p collisions (p is a proton, and α a helium nucleus).

However, at sufficiently high energies π^0's can also be produced through photo-production when cosmic rays collide with (radio-, optical, or X-ray) photons which occur in space. The general expression for the threshold E_{min} for photoproduction of particles of rest mass m_π when a nucleus (total energy $E \gg Mc^2$, rest mass $M = AM_p$) collides with photons of energy ϵ_{ph} has the form[†]

$$E_{min} = \frac{2M + m_\pi}{4\epsilon_{ph}} m_\pi c^4 = \epsilon_{ph,0} \frac{Mc^2}{2\epsilon_{ph}}, \tag{16.3}$$

[†] See footnote on next page

where

$$\epsilon_{ph,0} = \frac{2M + m_\pi}{2M} m_\pi c^2 \approx m_\pi c^2 \qquad (16.4)$$

is the threshold for pion photoproduction at a nucleus of mass M at rest. The π^0 photoproduction threshold $\epsilon_{ph,0}$ for nucleons at rest is approximately 150 MeV and, hence, the energy of cosmic protons which generate π^0's, for instance, on optical phonons with $\epsilon_{ph} \sim 1$ eV must exceed an energy

$$E_{min} \approx 150\, M_p c^2/2\epsilon_{ph}\, \text{MeV} \sim 10^{17} \text{ eV}.$$

When $\epsilon_{ph} \sim 10^{-3}$ eV (background radiation) $E_{min} \sim 10^{20}$ eV. It is just such photoproduction processes which will, in general, lead to the 'cut-off' of the cosmic ray spectrum for $E \gtrsim 10^{19}$ to 10^{20} eV.

Gamma-rays are, of course formed not only in the π^0 decay but also in several other decay processes and we shall discuss these processes in Chapter 17.

6. Electron and positron annihilation $(e^+ + e^- \to \gamma + \gamma)$.

There are always some positrons in the universe, as they are formed in the $\pi^+ \to \mu^+ \to e^+$ decay and in a number of other processes. One must distinguish the annihilation of relativistic, or at any rate fast, positrons in flight

---

† It is well known that one obtains expressions for the threshold for the production of particles from energy and momentum conservation laws. Nowadays the corresponding calculations are normally simplified by using four-dimensional vectors. For instant, we obtain Eqn. (16.3) by denoting the appropriate four-vectors for the incident particle, for the pion, and for the photon by

$$p_i = \{\mathbf{p}, iE/c\}, \quad \pi_i = \{\boldsymbol{\pi}, iE_\pi/c\}, \quad k_i = \{\mathbf{k}, i\epsilon_{ph}/c\},$$

$$p_i^2 = p^2 - E^2/c^2 = -M^2c^2, \quad \pi_i^2 = -m_\pi^2 c^2, \quad k_i^2 = 0.$$

(in a different notation, for instance, $p^i = (E/c, \mathbf{p})$; see Landau and Lifshitz, 1975). The energy-momentum conservation law for the photoproduction of a pion has the form $k_i + p_{1,i} = p_{2,i} + \pi_i$. Squaring this relation and using the above notation we get $M^2c^2 + 2k_i p_{1,i} = (p_{2,i} + \pi_i)^2$. At the threshold of photoproduction, however, we have $(p_{2,i} + \pi_i)^2 = -(M + m_\pi)^2 c^2$, as we can use for the evaluation of this quantity any frame of reference, and in the centre of mass frame at the threshold for the production the particle and the pion are at rest. Moreover, if the photon and the particle collide head-on

$$2k_i p_i = -2(\epsilon_{ph}/c)|p_1| - 2\epsilon_{ph}E/c^2 \approx -4\epsilon_{ph}E_1/c^2,$$

where the last expression refers to the case $E_1 \gg Mc^2$. We thus get at once Eqn. (16.3) for $E_1 = E_{min}$. For photoproduction on a particle at rest $2k_i p_i = -2\epsilon_{ph} M$ and we get Eqn. (16.4) for $\epsilon_{ph} = \epsilon_{ph,0}$.

and the annihilation of (slow) positrons at rest. In the first case gamma-rays are formed with a continuous, or at least a very wide, spectrum. In the second case (annihilation of positrons which have been stopped) the gamma-radiation is monochromatic ($E_\gamma = mc^2 = 0.51$ MeV) and because of this feature it can, in principle, be distinguished above the background of the continuous spectrum.

The gamma-radiation arising from the annihilation of anti-protons by protons or of any other particles by their anti-particles can practically play no role unless one talks about regions where matter and anti-matter come into contact. We do not think that such a possibility has a great probability and at any rate there are no sufficiently well defined indications for its realization (for details see Stecker, 1971; Stecker and Trombka, 1973).

7. Nuclear gamma-rays arising from radiative transitions in atomic nuclei.

In stellar atmospheres and in outbursts (such as supernova outbursts) nuclei are excited in nuclear reactions and as a result of collisions with fast particles, and this can lead to gamma-radiation. The exciting agent in interstellar and intergalactic space are the cosmic and subcosmic rays. It is important to emphasize that the spectrum of the nuclear gamma-radiation may be either continuous, or discrete (we refer here to the presence of more or less sharp lines). The latter case occurs for nuclear reactions in which slow particles (nuclei in interstellar space) are excited. However, if a nucleus which is part of the cosmic rays is excited in some collision, it usually has a large velocity and when we take the contributions from cosmic rays with different energies into account, its gamma-radiation produces a continuous spectrum.

We now remind the reader of some basic definitions and notations (we follow here and in other places in this chapter Ginzburg and Syrovatskii, 1965 and Ginzburg, 1969b).

In observations one measures one of the following quantities:

the intensity $J_\gamma(E_\gamma)$ and the flux $F_\gamma(E_\gamma)$ in photon number, or
the intensity $I_\gamma(E_\gamma)$ and the flux $\Phi_\gamma(E_\gamma)$ in energy;

$$\left.\begin{aligned} F_\gamma(E_\gamma) &= \int_\Omega J_\gamma(E_\gamma)\, d^2\Omega\,, \quad I_\gamma(E_\gamma) = E_\gamma J_\gamma(E_\gamma)\,, \\ \Phi_\gamma(E_\gamma) &= E_\gamma F_\gamma(E_\gamma) = E_\gamma \int_\Omega J_\gamma(E_\gamma)\, d^2\Omega\,. \end{aligned}\right\} \qquad (16.5)$$

The intensities and fluxes given here are differential quantities; for instance,

$J_\gamma(E_\gamma)\,dE_\gamma$ is the number of photons with energies in the range E_γ, $E_\gamma + dE_\gamma$ crossing unit area (normal to the photon momentum) per unit time and per unit solid angle. The corresponding integral quantities have the form

$$\left.\begin{aligned} J_\gamma(>E_\gamma) &= \int_{E_\gamma}^{\infty} J_\gamma(E'_\gamma)\,dE'_\gamma\,, \\ I_\gamma(>E_\gamma) &= \int_{E_\gamma}^{\infty} I_\gamma(E'_\gamma)\,dE'_\gamma = \int_{E_\gamma}^{\infty} E'_\gamma J_\gamma(E'_\gamma)\,dE'_\gamma\,, \\ F_\gamma(>E_\gamma) &= \int_{E_\gamma}^{\infty} F_\gamma(E'_\gamma)\,dE'_\gamma = \int_{\Omega}\int_{E_\gamma}^{\infty} J_\gamma(E'_\gamma)\,dE'_\gamma\,d^2\Omega\,, \\ \Phi_\gamma(>E_\gamma) &= \int_{E_\gamma}^{\infty} \Phi_\gamma(E'_\gamma)\,dE'_\gamma\,. \end{aligned}\right\} \quad (16.6)$$

Let in a volume element dV a source of X-rays or gamma-rays produce per unit time $q(E_\gamma)\,dE_\gamma\,dV\,d^2\Omega$ photons moving off into an element of solid angle $d^2\Omega$, with energies within the range E_γ, $E_\gamma + dE_\gamma$. The quantity $q(E_\gamma)$ is called the emittance (emissive power) in photon number. The emittance ε_ν which we used earlier (see, for instance, (5.52)) is connected with $q(E_\gamma)$ through the obvious relation $E_\gamma\,q(E_\gamma)dE_\gamma = \varepsilon_\nu\,d\nu$, whence $q(E_\gamma) = \varepsilon_\nu/h^2\nu$. If the emission is isotropic, it is convenient to use also the emittance in all directions

$$\tilde{q}(E_\gamma) = 4\pi\,q(E_\gamma) = \frac{4\pi\,\varepsilon_\nu}{h^2\nu} \quad (16.7)$$

In gamma-astronomy one uses mainly the emittance $q(E_\gamma)$ whereas in X-ray astronomy the application of the emittance ε_ν (and in general the energy-scale quantities) is not less wide-spread. If the X-rays or gamma-rays are formed by cosmic rays (or by any other particles) with an isotropic intensity $J(E)$, we have

$$\tilde{q}(E_\gamma)\,dE_\gamma = 4\pi q(E_\gamma)\,dE_\gamma = 4\pi N(\mathbf{r})\,dE_\gamma \int_{E_\gamma}^{\infty} \sigma(E_\gamma, E)\,J(E)\,dE\,. \quad (16.8)$$

Here $N(\mathbf{r})$ is the density of the atoms (or, say, electrons, soft photons, and so on) in the source and

$$\sigma(E_\gamma, E)\,dE_\gamma = dE_\gamma \int \sigma(E_\gamma, E, \Omega')\,d^2\Omega' \quad (16.9)$$

is the cross-section, integrated over the angles at which the photons leave, for the production of photons (with energies in the range E_γ, $E_\gamma + dE_\gamma$) by particles of energy E.

Let the source be at a distance R from the observer. In that case the radiation flux from the source into a solid angle $d^2\Omega$ equals

$$dF_\gamma(E_\gamma) = J_\gamma(E_\gamma)\, d^2\Omega = d^2\Omega \int_0^L \frac{\tilde{q}(E_\gamma)}{4\pi R^2} R^2 dR = d^2\Omega \int_0^L q(E_\gamma)\, dR \qquad (16.10)$$

and

$$J_\gamma(E_\gamma) = \int_0^L q(E_\gamma)\, dR = \tilde{N}(L) \int_{E_\gamma}^\infty \sigma(E_\gamma, E)\, J(E)\, dE\ , \qquad (16.11)$$

where

$$\tilde{N}(L) = \int_0^L N(R)\, dR \qquad (16.12$$

is the number of atoms (or other particles with which the cosmic rays which produce the gamma-radiation collide) along the line of sight; in the case of soft photons

$$\tilde{N}_{ph}(L) = \int_0^L N_{ph}\, dR$$

along the line of sight. In (16.11) we have assumed the cosmic ray intensity to be constant along the whole path L. One can easily drop that restriction. In the case of gamma-rays one often calls $J_\gamma(E_\gamma)$ and $J_\gamma(> E_\gamma)$ the differential and integral gamma-ray energy spectra.

For discrete sources (especially when their dimensions are small) one normally uses the following expressions for the flux :

$$F_\gamma(E_\gamma) = \int_\Omega J_\gamma(E_\gamma)\, d^2\Omega = \frac{\int q(E_\gamma)\, d^3\mathbf{r}}{R^2} \approx \frac{\tilde{N}_V}{R^2} \int_{E_\gamma}^\infty \sigma(E_\gamma, E)\, J(E)\, dE\ , \qquad (16.13)$$

where the integration is over the solid angle under which the source is seen; the source is at a distance R from the observer; in (16.13)

$$\tilde{N}_V = R^2 \int_\Omega \tilde{N}(L)\, d^2\Omega \approx \int N(\mathbf{r})\, d^3\mathbf{r} \qquad (16.14)$$

is the total number of particles (or soft photons) in the source.

We now turn to a discussion of the mechanisms of X-ray emission and we start with X-ray bremsstrahlung of a hot non-relativistic gas (plasma).

A hot gas which is partially or completely ionized is a source of brems-, recombination, and line-spectrum (characteristic) X-ray emission. At sufficiently high temperatures (we shall make clear what this means in what follows) bremsstrahlung plays the main role. Moreover, if we are dealing with a hydrogen or hydrogen-helium plasma, one does not consider the line-spectrum X-ray emission at all. X-ray bremsstrahlung is observed in X-ray 'stars' and also in the solar spectrum.

We give below the basic formulae with which one is dealing when discussing X-ray bremsstrahlung (for a more general approach see Heitler, 1947; Bethe and Salpeter, 1957; Hayakawa, 1969; Blumenthal and Gould, 1970; Berestetskii, Lifshitz, and Pitaevskii, 1971). Apart from for X-ray astronomy problems, this kind of radiation is also of interest for thermo-nuclear investigations and in research about the use of a hot plasma in laboratories as a powerful source of X-rays.

For sufficiently fast, but still non-relativistic electrons one may assume that the following conditions hold:

$$e^2 Z/\hbar v \ll 1 \quad , \quad \tfrac{1}{2} mv^2 \ll mc^2 \, . \tag{16.15}$$

As $e^2/\hbar c = 1/137$ the first condition (16.15) will, of course, not be satisfied for very heavy elements, but we have in mind the case of light elements. The total energy emitted in a single collision equals (see Heitler, 1947, § 25)

$$W = \int E_\gamma \sigma(E_\gamma, E)\, dE_\gamma = \frac{16 Z^2 E^6}{3 mc^3 \hbar} = \frac{16}{3} \alpha r_e^2 Z^2 mc^2 \; , \tag{16.16}$$

where, as before, we have denoted the photon energy by E_γ and $r_e = e^2/mc^2$. Brems (radiative) losses of a single electron per unit time are equal to

$$-\left(\frac{dE}{dt}\right)_r = W N_a v = \frac{16 Z^2 e^6 N_a v}{3 mc^3 \hbar} = 2.5 \times 10^{-33} \, Z^2 N_a v \text{ erg/s} \; , \tag{16.17}$$

where N_a is the density of nuclei in the medium. When $E \sim mc^2$ and $v \approx c$ Eqn.(16.17) and Eqn.(16.46) given below for the relativistic region give for $Z = 1$ approximately the same value $-(\frac{dE}{dt})_r \approx 8 N_a e^6/mc^2 \hbar$. We note that in the non-relativistic approximation the radiation from electron-electron collisions is appreciably weaker than for electron-proton collisions. The point is that when identical particles collide there is no dipole radiation due to the momentum conservation law, while the quadrupole radiation is weaker than the dipole radiation by a factor of the order of $(v/c)^2$.

For an equilibrium plasma the density of electrons with velocities in the range $v, v + dv$ equals

$$dN = N(v)\, dv = 4\pi N \left(\frac{m}{2\pi k_B T}\right)^{\frac{3}{2}} v^2 \exp\left(-\frac{mv^2}{2 k_B T}\right) dv \; , \quad \int N(v)\, dv = N \, . \tag{16.18}$$

The total power of the radiation from unit volume of the plasma is thus

$$4\pi \varepsilon = \int \left|\frac{dE}{dt}\right|_r dN = \int_0^\infty W N_a N 4\pi \left(\frac{m}{2\pi k_B T}\right)^{\frac{3}{2}} \exp\left(-\frac{mv^2}{2 k_B T}\right) dv^3 \, dv = \frac{32\sqrt{2}\, Z^2 e^6 N_a N (k_B T/m)^{\frac{1}{2}}}{3\sqrt{\pi}\, mc^3 \hbar} \, . \tag{16.19}$$

The factor 4π is used here in order that the integral emittance $\varepsilon = \int \varepsilon_\nu d_\nu$ is defined per unit solid angle. Due to the quasi-neutrality condition, which is usually well satisfied, we have for a completely ionized plasma of one kind of atoms $N = ZN_a$. Thus, we have from (16.19) for a hydrogen plasma

$$4\pi\varepsilon = 1.57 \times 10^{-27} N^2 \sqrt{T} \text{ erg/cm}^3.\text{s} , \tag{16.20}$$

where the temperature is measured in degrees absolute (T has the meaning of the electron temperature); clearly,

$$T(°K) = (k_B/1.6 \times 10^{-12})^{-1} T(\text{eV}) = 1.6 \times 10^4 T(\text{eV})$$

and N is the electron density in cm^{-3}. Somewhat more exact calculations which take into account electron-electron collisions and relativistic corrections lead to the expression

$$4\pi\varepsilon = 1.6 \times 10^{-27} N^2 \sqrt{T} (1 + 4.4 \times 10^{-10} T) . \tag{16.21}$$

Equations (16.19) and (16.20) are applicable only when conditions (16.15) hold which for $Z = 1$ give $v \gg 3 \times 10^8$ cm/s, or,

$$T \sim \frac{mv^2}{3k_B} \gg \frac{me^4}{\hbar^2 k_B} \sim 10^5 \text{ °K} . \tag{16.22}$$

On the other hand, from the condition that relativistic corrections be small we have

$$T \ll \frac{mc^2}{k_B} \sim 10^{10} \text{ °K} . \tag{16.23}$$

Of course, not only the integral emittance ε, but also the differential emissivity ε_ν, introduced earlier, is of interest. By definition

$$\varepsilon = \int_0^\infty \varepsilon_\nu d\nu = \int_0^\infty dE_\gamma E_\gamma \int_{E_\gamma}^\infty \sigma(E_\gamma, E) v(E) N(E) dE =$$

$$= h^2 \int_0^\infty \nu d\nu \int_{\sqrt{2h\nu/m}}^\infty \sigma(h\nu, E) vN(v) dv ,$$

where $E_\gamma = h\nu = \hbar\omega$ is the photon energy and $E = \frac{1}{2}mv^2$ the electron energy; in thermal equilibrium $N(v) dv$ is given by Eqn. (16.18).

The cross-section $\sigma(E_\gamma, E)$ depends only weakly on E (see Heitler, 1947; Blumenthal and Gould, 1970) and to a first approximation we can put

$$\sigma(E_\gamma, E) = \text{const}/E_\gamma = \text{const}/\nu ,$$

whence

$$\varepsilon_\nu = \text{const} \times \exp(-h\nu/k_B T) .$$

One can easily determine the constant from the condition

$$\int_0^\infty \varepsilon_\nu \, d\nu = \varepsilon$$

and we have thus

$$\varepsilon_\nu = \varepsilon \frac{h}{k_B T} \exp\left(-\frac{h\nu}{k_B T}\right) = \frac{7.7 \times 10^{-38} N^2}{4\pi\sqrt{T}} \exp\left(-\frac{h\nu}{k_B T}\right) \frac{\text{erg}}{\text{cm}^3.\text{s.Hz}} . \tag{16.24}$$

We now discuss the other limiting case of small energies (low temperatures) when

$$\frac{Ze^2}{\hbar v} \gg 1 . \tag{16.25}$$

When condition (16.25) holds one can perform the calculations classically. Indeed, the electron wavelength $\lambda = h/mv = 2\pi\hbar/mv$ while the smallest distance to which the electron approaches the nucleus is found from the condition $Ze^2/r_{min} = \frac{1}{2}mv^2$ and is equal to $r_{min} = 2Ze^2/mv^2$. It is clear that when inequality (16.25) holds $r_{min} \gg \lambda/\pi$ and one can describe the electron motion classically. One can also describe the radiation classically, but one must in the integration over frequencies take a quantal element into consideration — one integrates only up to the frequency $\nu = mv^2/2h$, where v is the electron velocity before the emission.† Landau and Lifshitz (1975, § 70) give a detailed account of the classical evaluation of the radiation of a particle moving in a Coulomb field. We are here interested in the quantity

$$dW = \int \tilde{W}(p)\, 2\pi p \, dp ,$$

where $\tilde{W}(p)$ is the energy emitted in the range ω, $\omega + d\omega$ when a particle of charge e passes a nucleus of charge Ze at a distance p (the mass of the particle which we assume to be an electron is equal to m; we neglect the recoil of the nucleus). If

$$\omega \gg \frac{mv^3}{Ze^2} \quad \left(\text{that is, } \frac{Ze^2}{\hbar v} \gg \frac{mv^2}{\hbar\omega}\right), \tag{16.26}$$

we have

$$dW = \frac{16\pi Z^2 e^6}{3\sqrt{3}\, v^2 m^2 c^3} d\omega = \frac{32\pi^2 Z^2 e^6}{3\sqrt{3}\, v^2 m^2 c^3} d\nu . \tag{16.27}$$

One can, of course, obtain the same formula quantum-mechanically, but in this case the classical calculation is completely sufficient (the condition (16.25) is at once the condition for the applicability of the quasi-classical approximation for a Coulomb field).

---

† To be more precise, a classical consideration is applicable provided $h\nu \equiv \hbar\omega \ll \frac{1}{2}mv^2$; in a number of cases one can use the classical formulae approximately also when $h\nu \le \frac{1}{2}mv^2$.

The total energy emitted in the collision is equal to

$$W = \int_0^{h\nu_{max} = \frac{1}{2}mv^2} dW = \frac{16\pi^2 Z^2 e^6}{3\sqrt{3}\,mc^3 h} \qquad (16.28)$$

The inaccuracy of the initial expression (16.27) caused by condition (16.26) is unimportant when we integrate over the frequency. More important is the restriction connected with the use of Eqn.(16.27) up to the frequency $\omega = mv^2/2\hbar$ (vide infra). For a hydrogen plasma with a Maxwellian distribution we have

$$4\pi\varepsilon_\nu = \int_0^{\sqrt{2h\nu/m}} \frac{dW}{d\nu} N^2 4\pi \left(\frac{m}{2\pi k_B T}\right)^{\frac{3}{2}} v^3 \exp\left(-\frac{mv^2}{2k_B T}\right) dv =$$

$$= \frac{32\pi\sqrt{2\pi}\,N^2 \exp(-h\nu/k_B T)}{3\sqrt{3}\,m^{\frac{3}{2}}(k_B T)^{\frac{1}{2}}\,c^3} = 4\pi\varepsilon \frac{h\exp(-h\nu/k_B T)}{k_B T} =$$

$$= 6.8\times 10^{-38}\frac{N^2}{\sqrt{T}}\exp\left(-\frac{h\nu}{k_B T}\right)\frac{\text{erg}}{\text{cm}^3.\text{s.Hz}}. \qquad (16.29)$$

In deriving this formula we used the fact that a photon of energy $h\nu$ can be emitted only by an electron of energy $\frac{1}{2}mv^2 \geq h\nu$. The integral emittance ε is equal to

$$4\pi\varepsilon = \int 4\pi\varepsilon_\nu d\nu = \int W\,dN = \frac{16\sqrt{2\pi}\,N^2 e^6 (k_B T/m)^{\frac{1}{2}}}{3\sqrt{3}\,mc^3\hbar} =$$

$$= 1.42\times 10^{-27} N^2\sqrt{T}\ \frac{\text{erg}}{\text{cm}^3.\text{s}}. \qquad (16.30)$$

The value (16.30) differs from (16.20) merely by a factor $1.57/1.46 = 1.1$ which in applications to astrophysical problems can usually be assumed to be equal to unity. Therefore, although one should use Eqns.(16.29) and (16.30) in the temperature range (see (16.25) with $Z = 1$)

$$T \sim \frac{mv^2}{3k_B} \ll \frac{me^4}{\hbar^2 k_B} \sim 10^5\ {}^\circ\text{K}, \qquad (16.31)$$

one can in fact always use Eqns.(16.20), (16.21), and (16.24). To be more precise, the use of (16.20), (16.21), and (16.24) at all temperatures is allowable as long as we are talking about accuracies of the order of tens of percents (however, (16.24) is inaccurate when $h\nu/k_B T \ll 1$). In the high-temperature region given by (16.22) and (16.23) Eqn.(16.21) is clearly very accurate. Even at high temperatures Eqn.(16.24) is rather accurate only when

$h\nu/k_BT \geqslant 1$ (when $h\nu/k_BT \ll 1$ one must multiply ε_ν by $(\sqrt{3}/\pi)\ \ln(4k_BT/1.781\,h\nu)$; vide infra). Equations (16.28) to (16.30) are approximate in connection with the use of classical expressions up to the frequency $\nu = mv^2/2h$. The exact formulae differ from (16.29) and (16.30) by the presence of an additional factor $g(\nu, T)$ which is usually called the Gaunt factor. Greene (1959) and Karzas and Latter (1961) give the appropriate expressions and curves for $g(\nu, T)$. The factor $g \sim 1$ and this is just the reason, as we have already mentioned, that when we are concerned with calculations with an accuracy of tens of percents one can usually employ Eqns.(16.20), (16.21), and (16.24) for all temperatures. This conclusion, like the majority of similar ones, must be treated with caution, as the accuracy depends, of course, on the energy range of the X-ray considered. For instance, for a plasma with $k_BT = 6$ keV in the photon energy range 2 to 20 keV the Gaunt factor changes from 1.35 to 0.55, that is, by a factor of about 2.5. Such a change, especially in conditions of an ever increasing accuracy of measurements in the X-ray astronomy region, is rather important and should not be ignored (Margon, 1973). We must also bear in mind that at low temperatures recombination occurs and it is usually not possible to assume the plasma to be completely ionized.

The maximum power of recombination radiation $4\pi\varepsilon_{rec,max} \approx 10^{-21}\ N^2/\sqrt{T}$ and hence

$$\frac{\varepsilon_{rec}}{\varepsilon_{brems}} \leqslant \frac{8 \times 10^5}{T}\ . \tag{16.32}$$

Thus for $T \leqslant 10^6$ °K we must take the recombination radiation into account (that is, the free-bound and the bound-bound transitions; see Bethe and Salpeter, 1957). When $T \gg 10^6$ °K a hydrogen plasma emits practically only bremsstrahlung (see (16.21) and (16.24)). By virtue of the presence of recombination radiation the inaccuracy of Eqns.(16.21) and (16.24) is usually even less important in the range $T \leqslant 10^6$ °K.

The absorption coefficient for the radiation which is connected with free-free transitions (that is, the process which is the inverse of bremsstrahlung) is under the conditions (16.26) equal to†

---

† We have elsewhere (Ginzburg, 1970b, § 37) given a detailed derivation (by the Einstein coefficient method) of Eqn.(16.33a) which follows below. Equation (16.33) is obtained in exactly the same way, but with Eqn.(37.7) of the above reference replaced by (37.7a) and the difference $N_2 - N_1 = Nh\nu/k_BT$ replaced by $N_2 - N_1 = N[1 - \exp(-h\nu/k_BT)]$ as now we do not use the condition $h\nu \ll k_BT$.

$$\mu = \mu_\nu = \frac{16\pi^2 Z^2 e^6 N N_a}{3\sqrt{3}\,h\,c(2\pi m)^{\frac{3}{2}}(k_B T)^{\frac{1}{2}}\,\nu^3}\left[1 - \exp\left(-\frac{h\nu}{k_B T}\right)\right] =$$

$$= \mu_{\nu,0}\left[1 - \exp\left(-\frac{h\nu}{k_B T}\right)\right]$$

$$= 3.68 \times 10^8\,\frac{N^2}{\sqrt{T}\,\nu^3}\left[1 - \exp\left(-\frac{h\nu}{k_B T}\right)\right]\ \mathrm{cm}^{-1}\ , \qquad (16.33)$$

where the last expression refers to a hydrogen plasma $(Z = 1\ , N = N_a)$. We note that in the factor $[1 - \exp(-h\nu/k_BT)]$ the induced emission has been taken into account which leads to a decrease in the observed absorption. When $h\nu \ll k_BT$ Eqn. (16.33) becomes

$$\mu_\nu = \frac{0.018\ N^2}{T^{\frac{3}{2}}\ \nu^2}\ . \qquad (16.34)$$

However, in this limiting case Eqn. (16.33) and, hence, also (16.34) is inaccurate as in its derivation we have assumed the frequency to be rather large (condition (16.26)). When $h\nu \ll k_BT$ the following formula holds :

$$\mu = \frac{8 N^2 e^6}{3\sqrt{2\pi}\,(mk_BT)^{\frac{3}{2}}\,c\,\nu^2}\ \ln\left[\frac{(2k_BT)^{\frac{3}{2}}}{2.115 \times 2\pi e^2 m^{\frac{1}{2}}\,\nu}\right]$$

$$\approx \frac{10^{-2}\ N^2}{T^{\frac{3}{2}}\,\nu^2}\left[17.7 + \ln\left(\frac{T^{\frac{3}{2}}}{\nu}\right)\right]\ , \qquad (16.33a)$$

or, for $T \gg 10^5\ {}^\circ K$ the same formula but with the substitution of

$$\ln\frac{4\,k_BT}{1.781\ h\nu}\quad \text{for}\quad \ln\frac{(2k_BT)^{\frac{3}{2}}}{2.115 \times 2\pi e^2 m^{\frac{1}{2}}\,\nu}\ ,$$

or, in the last expression in (16.33a) by replacing $\ln(T^{\frac{3}{2}}/\nu)$ by $\ln(10^3 T^{\frac{3}{2}}/\nu)$. If we use for μ Eqn. (16.33a) with the logarithmic term replaced by $\ln(4k_BT/1.781\,h\nu)$, as should be done under the conditions (16.22), through the method indicated below we are led to Eqn. (16.24) with an additional factor $(\sqrt{3}/\pi)\ln(4k_BT/1.781\,h\nu)$. This fact had already been noted earlier (we are talking about making Eqn. (16.24) more accurate for $h\nu \ll k_BT$). If we do not consider the logarithmic term, Eqns. (16.33a) and (16.34) depend in the same way on $N = N_e$, T, and ν and give the same result, as far as order of magnitude is concerned. The region of applicability and the accuracy of Eqn. (16.33) is the same as (16.29). Moreover, one of them follows from the other (this is just the reason why we have gone into this problem in detail).

Indeed, according to Kirchhoff's theorem (law) the emittance is in a state of thermal equilibrium equal to

$$\varepsilon_\nu = B_\nu \mu_\nu = \frac{2h\nu^3}{c^2} \mu_{\nu,0} \exp\left(-\frac{h\nu}{k_B T}\right), \tag{16.35}$$

where

$$B_\nu = \frac{2h\nu^3}{c^2} \frac{1}{\exp(h\nu/k_B T) - 1}$$

is the black-body spectral density, per unit volume and unit solid angle, and where we have used the notation $\mu_\nu = \mu_{\nu,0}[1 - \exp(-h\nu/k_B T)]$. Substituting expression (16.29) into (16.35) we arrive at Eqn. (16.33) and vice versa. At first sight this might indicate a particular case as Kirchhoff's theorem, strictly speaking, refers to a state of local thermodynamic equilibrium. However, in actual fact the region of applicability of Eqn. (16.35) is appreciably wider by virtue of the weak interaction between electromagnetic radiation and matter. For this reason, for instance, bremsstrahlung by a gas is, in general, independent of the state of the radiation field and will be equilibrium (thermal) radiation even when there is no equilibrium with the radiation. In somewhat more detail and more precisely one can state this as follows. The condition for the applicability of Eqn. (16.35) consists in the presence of thermal equilibrium for the electrons (in other words, the velocity distribution function of the electrons must be Maxwellian). Under conditions of a weak interaction between the radiation and the particles† the state of the radiation field does usually or at least in a whole number of cases not affect the electron distribution function and, provided the latter remains an equilibrium one, the emittance also remains the equilibrium one.

In the foregoing we have everywhere calculated and discussed expressions for the emissivity (the quantities ε_ν and $\varepsilon = \int \varepsilon_\nu d\nu$). In order to determine the radiation intensity or flux emitted from a source we must, in general, solve a transfer equation which takes into account the absorption of the radiation and, in particular, its reabsorption (see Chapter 9). The result, however, is at once known for two limiting cases — for a 'thick' and for a 'thin' layer. If the layer of radiating plasma is 'thick', that is, if it is

---

† Formally, the weakness of the electromagnetic interaction manifests itself in the fact that the fine structure constant $\alpha = e^2/\hbar c \approx 1/137$ is small. For a meson field with $g^2/\hbar c \geqslant 1$ the analogue of Kirchhoff's law would, in general, turn out to be applicable for a non-equilibrium meson field. On the other hand, even in the electromagnetic case there are situations possible in which the non-equilibrium character of the radiation leads to a violation of Kirchhoff's law (the result depends both on the intensity of the radiation field and on the probability for certain radiative transitions in the system).

completely non-transparent, it emits as a black body — the spectral density of the radiation is proportional to $\nu^3/[\exp(h\nu/k_BT) - 1]$. In the case of a completely transparent ('thin') layer, its emission is simply proportional to $\varepsilon_\nu L$ (L is the thickness of the layer which is assumed to be uniform). It then follows from (16.24) and (16.29) that the spectrum is exponential — it has the form $\exp(-h\nu/k_BT)$ (it is true that when $h\nu \ll k_BT$ the spectrum is not constant but depends logarithmically on the frequency; see (16.33a)). It is clear that one can consider the layer to be thin as long as $\mu L \ll 1$.

In the case of X-ray sources of a brems nature (clouds or atmospheres of a hot plasma) it is particularly convenient to use Eqns. (16.21) and (16.24) for estimates. For instance, for a uniform source of volume V the X-ray luminosity (radiation power) is

$$L_X = 1.6 \times 10^{-27}\, N^2 \sqrt{T}\, V\,. \tag{16.36}$$

Here we have, of course, assumed that the source is 'thin'. In the case of a quasi-uniform density distribution the average value of $N \sim \sqrt{(\overline{N^2})}$ and, hence, for a gas mass $M = M_p NV$ and an internal energy $W_T = \frac{3}{2} k_B TNV$ in the source we have

$$M \sim 2 \times 10^{-24}\, V\sqrt{(\overline{N^2})} = \frac{(L_X V)^{\frac{1}{2}}}{2 \times 10^{10}\, T^{\frac{1}{4}}}\ \text{g}\,,$$

$$W_T \sim \sqrt{(\overline{N^2})}\, k_B T V \sim 3 \times 10^{-3} (L_X V)^{\frac{1}{2}}\, T^{\frac{3}{4}}\ \text{erg}\,. \tag{16.37}$$

One can use these simple formulae to reach some conclusions about the 'brems model' of X-ray emission of various sources.

We now consider the bremsstrahlung of relativistic electrons. The bremsstrahlung of relativistic electrons (and positrons) can be an important mechanism for energy losses (radiative losses). From that point of view we have already mentioned bremsstrahlung in Chapter 15. Here, on the other hand, we shall be interested in the bremsstrahlung itself, but only in the case of ultra-relativistic electrons and of rather high energies of the emitted photons (gamma-ray region) †.

Let us, thus, consider bremsstrahlung by ultra-relativistic electrons and let us denote the initial and final electron energies by $E \equiv E_1$ and E_2, respectively, and the energy of the emitted brems photon by E_γ. In the case of

† This problem is nonetheless discussed in the present chapter and not in Chapter 17, as bremsstrahlung by relativistic electrons should logically be discussed at once after bremsstrahlung by non-relativistic electrons.

bremsstrahlung when electrons are scattered by nuclei (we restrict ourselves now to this case) one can normally assume the nucleus to be fixed. Under such circumstances the nucleus only receives momentum, but one can neglect its kinetic energy, that is, the electron energy after the emission equals $E_2 = E - E_\gamma$.

If

$$E \gg mc^2 \quad , \; E_2 = E - E_\gamma \gg mc^2 \quad , \tag{16.38}$$

the photons fly mainly away within angles $\theta \sim mc^2/E$ with the direction of the momentum of the incident electron $\mathbf{p} \equiv \mathbf{p}_1$. The cross-section then equals

$$\sigma_r(E_\gamma, E)\, dE_\gamma = 4\frac{e^2}{\hbar c} Z^2 \left(\frac{e^2}{mc^2}\right)^2 \frac{dE_\gamma}{E_\gamma} \left[\left\{1 + \left(1 - \frac{E_\gamma}{E}\right)^2\right\} \Phi_1 + \left\{1 - \frac{E_\gamma}{E}\right\} \Phi_2\right] , \tag{16.39}$$

where Z is the charge of the nucleus of the scattering atom, while the functions Φ_1 and Φ_2 are given below. By definition the probability for the emission per unit time by an electron of energy $E \gg mc^2$ of a photon of energy E_γ in the range E_γ, $E_\gamma + dE_\gamma$ is equal to $P(E_\gamma , E)\, dE_\gamma = \sigma(E_\gamma , E)\, F\, dE_\gamma$, where F is the flux of electrons through unit area. If we are dealing with a 'bare' nucleus, that is, with scattering by a Coulomb centre,

$$\Phi_1 = \ln\left(\frac{2E}{mc^2} \frac{E - E_\gamma}{E_\gamma}\right) - \frac{1}{2} \quad , \quad \Phi_2 = -\frac{2}{3}\, \Phi_1 \; . \tag{16.40}$$

In the scattering by atoms the electrons in the atomic shells screen the nuclear charge, and as a result the cross-section changes. The amount of screening is determined by the parameter

$$\xi = \frac{\hbar c}{e^2} \frac{mc^2}{E} \frac{E_\gamma}{E - E_\gamma} Z^{-\frac{1}{3}} \; .$$

The meaning of this parameter becomes clear if we take into account that when an electron is scattered by a nucleus the latter acquires a momentum $\Delta p \sim \hbar/r$, where r is the effective distance at which the electron passed the nucleus (for details see, for instance, Heitler, 1947, § 25). Moreover, it follows from the conservation laws that

$$\Delta p = \frac{1}{2} \frac{mc^2}{E} \frac{E_\gamma}{E - E_\gamma} mc$$

and thus

$$r \sim \frac{\hbar}{mc} \frac{E(E - E_\gamma)}{mc^2\, E_\gamma} \; .$$

On the other hand, in the statistical model of the atom the radius of the atom $a \sim a_0 Z^{-\frac{1}{3}}$, where $a_0 = \hbar^2/me^2 = (\hbar/mc)(e^2/\hbar c)^{-1} = 5.3 \times 10^{-9}$ cm. The parameter ξ introduced a moment ago is clearly of the order of the ratio a/r. The harder the emitted photon, the closer the electron must pass the nucleus and the less the screening will be; in that case, if the parameter $\xi \gg 1$, Eqns. (16.40) are valid. Softer photons are emitted when the electron passes at appreciably larger distances from the nucleus. When $\xi \ll 1$ the screening is large and for heavy atoms we have

$$\Phi_1 = \ln(191 \times Z^{-\frac{1}{3}}) \ , \ \Phi_2 = -\tfrac{2}{3}\ln(191 \times Z^{-\frac{1}{3}}) + \tfrac{1}{9} \ . \qquad (16.41)$$

It is clear from Eqns. (16.39) to (16.41) that the E-dependence of $\sigma(E_\gamma, E)$ is a weak one and that the E_γ-dependence is given by the factor $1/E_\gamma$ (this remark is especially valid for complete screening, (16.41)). Moreover, Eqn. (16.41) is inaccurate for light elements.

If we are talking about errors which are not less than a few percent the cross-section for bremsstrahlung under conditions of complete screening can be written in the form

$$\sigma_r(E_\gamma, E)\, dE_\gamma = \frac{M dE_\gamma}{t_r E_\gamma} \ , \qquad (16.42)$$

where t_r is the radiative unit length (in g/cm^2) in a gas of atoms of mass M; under the conditions of (16.41)

$$\frac{1}{t_r} = 4\,\frac{e^2}{\hbar c}\, Z^2 M^{-1} \ln(191 \times Z^{-\frac{1}{3}})$$

and for hydrogen ($Z = 1$, $M = 1.67 \times 10^{-24}$ g) we would have $t_r \approx 73\ g/cm^2$. In fact, because of the inaccuracy of Eqn. (16.41) for light elements the value of t_r for hydrogen is somewhat lower even when we take into account the contribution from electron-electron collisions. Detailed calculations (Dovchenko and Pomanskii, 1964) lead to the following values for t_r : $t_H = 62.8$, $t_{He} = 93.1$, $t_C = 43.3$, $t_N = 38.6$, $t_O = 34.6$, $t_{Fe} = 13.9\ g/cm^2$ (we have dropped here the index r and replaced it by the symbol for the relevant element). For the un-ionized interstellar medium (about 90% H and roughly 10% He) we can with good accuracy put $M = 2 \times 10^{-24}$ g and $t_r = 66\ g/cm^2$.

The intensity of brems gamma-photons is equal to

$$J_{\gamma,\mathrm{brems}}(E_\gamma) = \int_0^L dR \int_{E_\gamma}^{\infty} Na(\mathbf{R})\, \sigma_r(E_\gamma, E)\, J_e(E, \mathbf{R})\, dE \ , \qquad (16.43)$$

where J_e is the intensity of the electron component which produces the gamma-rays and $N_a(\mathbf{R})$ is the density of atoms in the interstellar medium. Using Eqn. (16.42) and assuming that the intensity J_e is constant along the line of sight we get

$$J_{\gamma,\mathrm{brems}}(E_\gamma) = \frac{M\tilde{N}(L)}{t_r} \frac{J_e(>E_\gamma)}{E_\gamma} = 1.5\times 10^{-2}\,\tilde{M}(L)\,\frac{J_e(>E_\gamma)}{E_\gamma}, \qquad (16.44)$$

where $\tilde{M}(L) = M\tilde{N}(L)$ is the mass of gas along the line of sight (in g/cm^2) and

$$J_e(>E_\gamma) = \int_{E_\gamma}^{\infty} J_e(E)\,dE\ .$$

In un-ionized hydrogen with $E_\gamma \leqslant E$ the parameter $\xi = 10^2 mc^2/E$ and $\xi \ll 1$ for $E \gg 5\times 10^7$ eV; under those conditions one can use Eqns.(16.42) and (16.44). In a fully ionized medium it is practically always permissible to neglect screening. Indeed, in that case the screening radius is the Debye radius $r_D = \sqrt{(k_BT/8\pi Ne^2)}$ which is, for instance, for $T\sim 10^4\ {}^\circ K$, $N\sim 0.1\ cm^{-3}$ of the order of 10^3 cm. In that example $r_D \sim r \sim (\hbar/mc)(E/mc^2)$ (vide supra) only when $E/mc^2 \sim 3\times 10^{13}$, that is, $E \sim 10^{19}$ eV. In an ionized gas we have therefore usually $r \ll r_D$, and screening is unimportant. When there is no screening one must under the conditions (16.38) use Eqns.(16.39), (16.40), and (16.43). We note that the bremsstrahlung is then not taken into account which arises in the collisions of the incident electron with the atomic electrons (and in general with the electrons in the medium, for instance, in a plasma). To a rough approximation one can take the effect of the electron-electron collisions on the cross-section (16.39) into account by replacing the factor Z^2 by $Z(Z+1)$. The meaning of that substitution consists clearly in the fact that the cross-section $\sigma(E_\gamma,E)$ is for electron-electron collisions approximately the same as for electron-proton collisions; moreover, we have, of course, used the fact that there are Z electrons in the atom. We emphasize that this approximate way of taking the electron-electron collisions into account is integrated over the angles. Just because of this the contribution to the integral cross-section $\sigma(E_\gamma,E)$ from processes connected with transferring a large momentum to the atomic electron turns out to be insignificant. The value $t_r = 66\ g/cm^2$ given above for the interstellar medium was obtained taking into account the bremsstrahlung in electron-electron collisions (see Dovzhenko and Pomanskii, 1964). Below we shall also take the electron-electron collision into account by replacing Z^2 by $Z(Z+1)$ or by the appropriate choice of the value of t_r.

Due to bremsstrahlung the electrons lose energy — the corresponding losses are, as we have already mentioned, called brems or radiative losses. We emphasize that radiative losses occur mainly in large amounts (that is, they belong to the class of 'catastrophic' losses; see Chapter 15). It is, for instance, clear from (16.42) that the transferred energy is

$$\int E_\gamma \sigma(E_\gamma, E)\, dE_\gamma \sim \int_0^E \text{const.}\, dE_\gamma \sim \text{const.}\, E ,$$

that is, it is determined by the emission of photons of energy $E_\gamma \sim E$. The radiative losses therefore fluctuate wildly. We restrict ourselves, however, to evaluating the average losses per unit path length

$$-\left(\frac{dE}{dx}\right)_r = \int_0^E N_a E_\gamma \sigma(E_\gamma, E)\, dE_\gamma , \qquad (16.45)$$

where N_a is the density of atoms and where we have used the fact that the cross-section $\sigma(E_\gamma, E)$ is normalized to unit electron flux (moreover, by virtue of (16.38) we have replaced the upper limit of the integral $E - mc^2$ by E); for the ultra-relativistic electrons considered here the losses $(\frac{dE}{dt})_r$ per unit time are obtained simply by multiplying the value (16.45) by $c = 3 \times 10^{10}$ cm/s.

In a completely ionized gas (plasma) or when there is no screening we have from (16.39), (16.40), and (16.45)

$$\begin{aligned}
-\frac{1}{E}\left(\frac{dE}{dt}\right)_r &= \frac{4Z(Z+1) N_a e^6}{m^2 c^4 \hbar}\left\{\ln\frac{2E}{mc^2} - \frac{1}{3}\right\} = \\
&= 1.37 \times 10^{-16} N_a \left\{\ln\frac{E}{mc^2} + 0.36\right\} \text{ s}^{-1} \\
&= 2.74 \times 10^{-3} \left\{\ln\frac{E}{mc^2} + 0.36\right\} \text{ g}^{-1}\,\text{cm}^2 \\
&= 4.6 \times 10^{-27} N_a \left\{\ln\frac{E}{mc^2} + 0.36\right\} \text{ cm}^{-1} , \qquad (16.46)
\end{aligned}$$

where we have put $Z = 1$ (hydrogen) in going over to the last three expressions.†

For complete screening for heavy elements we find

$$\begin{aligned}
-\frac{1}{E}\left(\frac{dE}{dt}\right)_r &= \frac{4Z(Z+1) N_a e^6}{m^2 c^4 \hbar}\left\{\ln(191 \times Z^{-\frac{1}{3}}) + \frac{1}{18}\right\} \\
&= 7.26 \times 10^{-16} N_a \text{ s}^{-1} , \qquad (16.47)
\end{aligned}$$

where the numerical values refer to hydrogen $(Z = 1)$ when Eqn. (16.47) is

---

† We explain once more that the losses evaluated per g/cm^2 are obtained from the losses per unit path length by replacing the density of the atoms N_a by $1/M$, where M is the mass of an atom.

already inaccurate. We gave nonetheless the numerical result in order to compare Eqns.(16.46) and (16.47) for hydrogen. It is clear that the losses (16.46) and (16.47) become the same for $E/mc^2 = 140$; in un-ionized hydrogen Eqn.(16.46) must be used for $E/mc^2 \leqslant 10^2$, and (16.47) for $E/mc^2 \gg 10^2$. In the latter case (complete screening) one obtains a more exact value for the interstellar medium if we take the radiative unit length t_r to be equal to $66\,g/cm^2$. We then get directly from (16.42) and (16.45)

$$-\frac{1}{E}\left(\frac{dE}{dt}\right)_r = \frac{M\,c\,N_a}{t_r}\,s^{-1} = \frac{1}{t_r}\,g^{-1}\,cm^2 = 1.5\times 10^{-2}\,g^{-1}\,cm^2 \approx$$

$$\approx 10^{-15}\,N_a\,s^{-1} \approx 3\times 10^{-26}\,N_a\,cm^{-1}\,. \qquad (16.48)$$

By virtue of (16.48) an electron loses energy on average according to the law $E = E_0 \exp(-L/66)$, where E_0 is its initial energy and L the path traversed in g/cm^2. As the radiative losses are mainly 'catastrophic' in nature (the energy loss in a single act $\Delta E \sim E$) one may assume that to a rough approximation the electron has a chance equal to $\exp(-L/66)$ to traverse a path L (in g/cm^2) without any radiative losses.

Apart from gamma-photons ultra-relativistic electrons can produce electron-positron pairs e^+, e^- when being scattered, and also other particles (for instance, μ^+, μ^- pairs). The corresponding total cross-section for e^+, e^- pair production in electron-photon or electron-electron collisions is of the order of

$$\sigma_{pair} \sim \frac{1}{\pi}\left(\frac{e^2}{\hbar c}\right)^2 \left(\frac{e^2}{mc^2}\right)^2 \ln^3\left(\frac{E}{mc^2}\right) \qquad (16.49)$$

(to be more precise, for electron-proton collisions one should have a factor $E/M_p c^2$ under the logarithm sign, where M_p is the proton mass). Moreover, when there is no screening (as in (16.49)) the cross-section for bremsstrahlung in hydrogen (see (16.39) and (16.40)) is equal to

$$\sigma_r = \int \sigma(E_\gamma, E)\,dE_\gamma \sim 4\left(\frac{e^2}{\hbar c}\right)\left(\frac{e^2}{mc^2}\right)^2 \ln\frac{E}{mc^2}\,. \qquad (16.50)$$

Thus

$$\sigma_{pair} \sim \frac{1}{4\pi}\frac{e^2}{\hbar c}\,\sigma_r \ln^2\left(\frac{E}{mc^2}\right) \sim 10^{-3}\,\sigma_r \ln^2\left(\frac{E}{mc^2}\right)$$

and if there is no screening $\sigma_{pair} \ll \sigma_r$ as long as $E/mc^2 \ll 10^{12}$ to 10^{13} (or $E \ll 10^{18}\,eV$).

Let us now consider the radiation arising when electrons and positrons are formed as a result of the $\pi^\pm \to \mu^\pm \to e^\pm$ decay. The intensity of such radiation is usually very small compared to the other kinds of radiation (bremsstrahlung by electrons, gamma-rays emitted in π^0-decay). We shall therefore restrict ourselves only to some estimates.

When particles of charge e and energy

$$E = \frac{mc^2}{\sqrt{(1 - v^2/c^2)}} \gg mc^2$$

are produced we have in the classical approximation the emission of the following amount of energy (see Chapter 7 and Landau and Lifshitz, 1975, § 69)

$$\left.\begin{aligned} dW_\gamma &= \frac{\alpha}{\pi}\left(\frac{c}{v}\,\ell n\,\frac{c+v}{c-v} - 2\right) dE_\gamma \approx \frac{2\alpha}{\pi}\left(\ell n\,\frac{2E}{mc^2} - 1\right) dE_\gamma\,, \\ \alpha &= \frac{e^2}{\hbar c} \approx \frac{1}{137}\,. \end{aligned}\right\} \quad (16.51)$$

For estimates this expression is adequate also when $E_\gamma \leq E$ (the quantum constant $\hbar = h/2\pi$ does, in fact, not appear in (16.51) as $dE_\gamma = hd\nu$). The number of photons produced equals dW_γ/E_γ and, hence, the intensity is

$$\begin{aligned} J_{\gamma,prod}(E_\gamma) &= \frac{L}{4\pi}\int_{E_\gamma}^{\infty} \frac{dW_\gamma}{dE_\gamma}\,\frac{q_e(E)\,dE}{E} \approx \\ &\approx 4\times 10^{-4}\left(\ell n\,\frac{2\bar{E}_\gamma}{mc^2} - 1\right)\frac{Q_e(E > E_\gamma)}{E_\gamma}\,, \end{aligned} \quad (16.52)$$

where

$$Q_e(E > E_\gamma) = L\int_{E_\gamma}^{\infty} q_e(E)\,dE$$

is the number of electrons with energies $E > E_\gamma$ produced per sec along a path length L along the line of sight while $\bar{E}_\gamma$ is some average value. Actual estimates for the Galaxy indicate that the intensity $J_{\gamma,prod}$ is, for instance, for $E_\gamma = 5\times 10^7$ eV several orders of magnitude smaller than the bremsstrahlung intensity $J_{\gamma,brems}$. As to a comparison of the intensity $J_{\gamma,prod}$ (we are talking here about the formation of the decay produces of $\pi^\pm$ mesons) with J_{γ,π^0}, the intensity of gamma-rays from π^0 decay, an estimate is possible without knowing the spectrum and the intensity of the proton-nuclear and the electron components of the cosmic rays. Indeed, π^0-mesons are produced in numbers approximately half of those of $\pi^\pm$-mesons. From this it is clear that in the $\pi^0 \to 2\gamma$ decay approximately as many gamma-photons are produced as electrons and positrons from the $\pi^\pm \to \mu^\pm \to e^\pm$ decay. Moreover, the probability for the emission of a photon when an electron (positron) is produced contains an additional factor $\alpha = e^2/\hbar c = 1/137$. Hence, the intensity of gamma-photons accompanying the $\pi^{\pm} \rightarrow \mu^{\pm} \rightarrow e^{\pm}$ decay is smaller by two orders of magnitude than the gamma-ray intensity due to π^0-decay.

The next and, moreover, an especially important, process to which we now turn is the scattering of relativistic electrons by photons, which is often called the inverse Compton effect.†

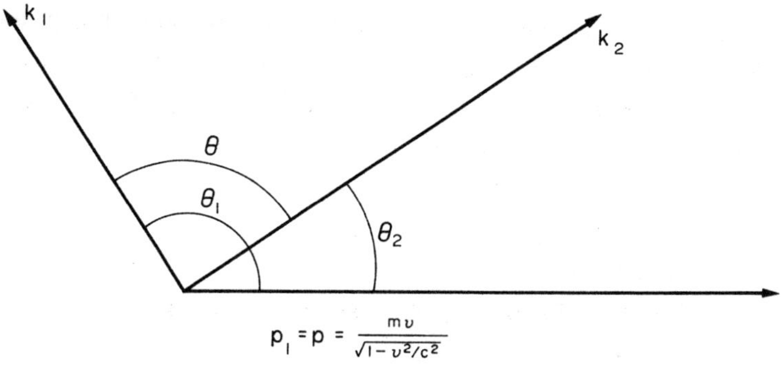

Fig. 16.1 Scattering of a photon of momentum $\hbar\mathbf{k}$ and energy $\epsilon_{ph,1} \equiv \epsilon_{ph}$ by an electron with momentum $\mathbf{p}_1 \equiv \mathbf{p}$ and energy $E_1 \equiv E$.
The scattered photon has a momentum $\hbar\mathbf{k}_2$ and energy $\epsilon_{ph,2} \equiv E_\gamma$. The vectors $\mathbf{k}_1$, $\mathbf{k}_2$, and $\mathbf{p}$ do in general not lie in one plane.

Let there be in the laboratory frame of reference (in the case of interest to us this is a reference frame fixed in the Earth or some other astronomical frame of reference) an electron of momentum

$$\mathbf{p}_1 = \frac{m\mathbf{v}_1}{\sqrt{(1 - v_1^2/c^2)}}$$

and energy

$$E_1 = \frac{mc^2}{\sqrt{(1 - v_1^2/c^2)}},$$

and also a photon of momentum $\hbar\mathbf{k}_1$ and energy $\epsilon_{ph,1} = \hbar\omega$ ($k = 2\pi/\lambda$, $\omega = 2\pi c/\lambda = ck$); let, moreover, there be no medium or let it not exert any influence; this is admissible if $\omega \gg \omega_0 = \sqrt{(4\pi N e^2/m)} = 5.64 \times 10^4 \sqrt{N}$. When

† This terminology arose because the Compton effect was discovered and is usually studied in laboratories where the gamma-photon is scattered by an electron at rest or a slow electron. In what follows we talk in all cases, independent of the photon and electron energies, simply about the Compton effect.

they scatter one another, the electron and photon exchange energy and momentum so that in the final state the corresponding quantities are equal to $\mathbf{p}_2$, E_2, $\hbar\mathbf{k}_2$, and $\epsilon_{ph,2}$. As $E = \sqrt{(m^2c^4 + c^2p^2)}$ and $\epsilon_{ph} = \hbar ck$, overall there are in the scattering problem for given $\mathbf{p}_1$ and $\mathbf{k}_1$ six unknown quantities (the momenta $\mathbf{p}_2$ and $\hbar\mathbf{k}_2$). The energy and momentum conservation laws give four relations and we can thus find the energy of the scattered photon, $\epsilon_{ph,2}$, only by giving two other parameters. For those one normally takes the angle θ_2 between $\mathbf{k}_2$ and $\mathbf{p}_1$ and the angle θ between $\mathbf{k}_2$ and $\mathbf{k}_1$ (in that way the direction of $\mathbf{k}_2$ is fixed; see Fig. 16.1). For the sake of convenience, bearing in mind applications to the case of hard scattered photons (photons number 2) we shall usually use the following notation:

$$\mathbf{p}_1 \equiv \mathbf{p} = \frac{m\mathbf{v}}{\sqrt{(1 - v^2/c^2)}}, \quad E_1 \equiv E, \quad \epsilon_{ph,1} = \epsilon_{ph}, \quad \epsilon_{ph,2} = E_\gamma, \tag{16.53}$$

θ_1 the angle between $\mathbf{k}_1$ and $\mathbf{p}$, θ_2 the angle between $\mathbf{k}_2$ and $\mathbf{p}$ and θ the angle between $\mathbf{k}_1$ and $\mathbf{k}_2$.

The energy of the scattered photon equals

$$E_\gamma = \frac{\epsilon_{ph}(1 - (v/c)\cos\theta_1)}{1 - (v/c)\cos\theta_2 + (\epsilon_{ph}/E)(1 - \cos\theta)} = \phi(\epsilon_{ph}, E, \theta_1, \theta_2, \theta). \tag{16.54}$$

If

$$E > E_\gamma \gg \epsilon_{ph}, \tag{16.55}$$

we have approximately

$$E_\gamma = \frac{\epsilon_{ph}(1 - (v/c)\cos\theta_1)}{1 - (v/c)\cos\theta_2 + (\epsilon_{ph}/E)(1 - (v/c)\cos\theta_1)\cos\theta_2}. \tag{16.56}$$

It is convenient to write the cross-section for the scattering of unpolarized particles in invariant form (see Berestetskii, Lifshitz, and Pitaevskii, 1971)

$$\left.\begin{aligned}\sigma_C(\mathbf{k}_1, \mathbf{k}_2, \mathbf{v})\, d^2\Omega_2 &= 2\left(\frac{e^2}{mc^2}\right)^2 \frac{E_\gamma^2}{m^2c^4\kappa_1^2}\left\{4\left(\frac{1}{\kappa_1} + \frac{1}{\kappa_2}\right)^2\right.\\ &\left.\quad - 4\left(\frac{1}{\kappa_1} + \frac{1}{\kappa_2}\right) - \left(\frac{\kappa_1}{\kappa_2} + \frac{\kappa_2}{\kappa_1}\right)\right\} d^2\Omega_2,\\ \kappa_1 = -\frac{2}{m^2c^4}\epsilon_{ph}E\left(1 - \frac{v}{c}\cos\theta_1\right), &\quad \kappa_2 = \frac{2}{m^2c^4}E_\gamma E\left(1 - \frac{v}{c}\cos\theta_2\right),\end{aligned}\right\} \tag{16.57}$$

where $d^2\Omega_2$ is an element of solid angle corresponding to the direction of $\mathbf{k}_2$ (the solid angle Ω_1 used below refers to the direction of $\mathbf{k}_1$).

If in the initial state the electron is at rest (that is, $\mathbf{p} = \mathbf{p}_1 = 0$) we get from (16.54) and (16.57) the well known expressions

$$E_\gamma = \frac{\epsilon_{ph}}{1 + (\epsilon_{ph}/mc^2)(1 - \cos\theta)} , \tag{16.58}$$

$$\sigma_C \, d^2\Omega_2 = \frac{1}{2}\left(\frac{e^2}{mc^2}\right)^2 \left(\frac{E_\gamma}{\epsilon_{ph}}\right)^2 \left(\frac{\epsilon_{ph}}{E_\gamma} + \frac{E_\gamma}{\epsilon_{ph}} - \sin^2\theta\right) d^2\Omega_2 . \tag{16.59}$$

In the non-relativistic limit when in (16.59) $\epsilon_{ph} \ll mc^2$ we can put $E_\gamma = \epsilon_{ph}$ (this is, of course, also clear from (16.58)) and then

$$\sigma_C \, d^2\Omega_2 = \frac{1}{2}\left(\frac{e^2}{mc^2}\right)^2 (1 + \cos^2\theta) \, d^2\Omega_2 . \tag{16.60}$$

Integrating the cross-section (16.59) over the solid angle $d^2\Omega_2 = 2\pi \sin\theta \, d\theta$ (we can in this case identify the angle θ_2 with the angle θ betweeen $\mathbf{k}_1$ and $\mathbf{k}_2$) we get the Thomson cross-section

$$\sigma_T = \int \sigma_C \, d^2\Omega_2 = \frac{8\pi}{3}\left(\frac{e^2}{mc^2}\right)^2 = 6.65 \times 10^{-25} \text{ cm}^2 . \tag{16.61}$$

We emphasize once again that we assumed the electrons and photons in the initial state to be unpolarized and the cross-section given by (16.57), (16.59), or (16.60) is summed over the polarizations in the final state. The development of X-ray and gamma-astronomy leads, undoubtedly, to the very important and interesting possibility of measuring the polarization of cosmic X-rays and gamma-rays (this is important for elucidating the nature of the radiation; for instance, the X-ray synchrotron radiation from the Crab Nebula must be rather strongly polarized which has been observed, whereas the bremsstrahlung from a hot gas is unpolarized; for details see below). However, for the sake of simplicity we shall assume in what follows that the scattered radiation (photons number 2) is not polarized. This should be the case if the soft radiation (photons number 1) are unpolarized (see Berestetskii, Lifshitz, and Pitaevskii, 1971, § 87 for polarization effects in Compton scattering).

We remind ourselves that according to the definition of the cross-section (see, for instance, Blumenthal and Gould, 1970) $\sigma_C \, d^2\Omega_2 \, F$ is the number of photons scattered per unit time into a solid angle $d^2\Omega_2$ while $F = N_{ph}(c - v\cos\theta_1)$ is the flux density of the photons scattered by electrons (N_{ph} is the photon density, $v \equiv v_1$ is the velocity of the scattering electrons, θ_1 the angle between $\mathbf{k}_1$ and $\mathbf{v}$); the cross-section $\sigma_C d^2\Omega_2$ is a relativistic invariant. We shall now assume that photons of energy ϵ_{ph} are isotropically distributed in direction and we are interested in scattered photons of energy E_γ also independently of the direction of their propagation. Under such conditions we must evaluate the cross-section (see (16.54) and (16.57)):

$$\sigma(E_\gamma, \epsilon_{ph}, E) = \frac{1}{4\pi}\int\left(1-\frac{v}{c}\cos\theta_1\right)\sigma_C(\mathbf{k}_1, \mathbf{k}_2, \mathbf{v}) \times$$

$$\times\ \delta(E_\gamma - \phi(\epsilon_{ph}, E, \theta_1, \theta_2\ \theta))\, d^2\Omega_1\, d^2\Omega_2\ . \qquad (16.62)$$

Indeed, we denote the density of soft photons 1 with arbitrary propagation direction and energies in the range ϵ_{ph}, $\epsilon_{ph} + d\epsilon_{ph}$ by $N_{ph}(\epsilon_{ph})\, d\epsilon_{ph}$. The number of hard photons 2 produced per unit time as a result of the scattering of soft photons by an electron is equal to $c\int\sigma(E_\gamma, \epsilon_{ph}, E)\, N_{ph}(\epsilon_{ph})\, d\epsilon_{ph}$. Let

$$4\epsilon_{ph}\frac{E}{mc^2} \ll mc^2 \qquad (16.63)$$

and let, moreover, condition (16.55) be satisfied. Using then Eqns.(16.56) and (16.57) we can evaluate the cross-section (16.62) and after that also the quantity $\sigma_t = \int\sigma(E_\gamma, \epsilon_{ph}, E)\, dE_\gamma$. Under the condition (16.63) we get

$$\sigma(E_\gamma, \epsilon_{ph}, E) = \tfrac{1}{4}\pi\left(\frac{e^2}{mc^2}\right)^2\frac{(mc^2)^4}{\epsilon_{ph}^2 E^3}\left\{2\frac{E_\gamma}{E} - \frac{(mc^2)^2E_\gamma^2}{\epsilon_{ph}E^3} + \right.$$

$$\left. + 4\frac{E_\gamma}{E}\ln\frac{(mc^2)^2E_\gamma}{4\epsilon_{ph}E^2} + \frac{8\epsilon_{ph}E}{(mc^2)^2}\right\}, \qquad (16.64)$$

where E_γ lies within the range $\epsilon_{ph} \le E_\gamma \le 4\epsilon_{ph}(E/mc^2)^2$; in fact, the range of applicability of Eqn.(16.64) is somewhat wider. Moreover,

$$\sigma_t = \int\sigma(E_\gamma, \epsilon_{ph}, E)\, dE_\gamma = \frac{8\pi}{3}\left(\frac{e^2}{mc^2}\right)^2 = \sigma_T\ . \qquad (16.65)$$

We did not discuss in detail the corresponding calculations as the result (16.65) is very translucent.† In the frame of reference fixed in the electron the energy of photon 1 equals $\epsilon'_{ph,1} = (E/mc^2)\,\epsilon_{ph}(1 - (v/c)\cos\theta_1)$. For isotropic radiation $\overline{\epsilon'_{ph}} = (E/mc^2)\,\overline{\epsilon_{ph}}$ and moreover when $\overline{\epsilon'_{ph}} \ll mc^2$ Eqns.(16.60) and (16.61) are valid — in other words, one can assume the scattering to be classical. However, the condition $\overline{\epsilon'_{ph}} \ll mc^2$ is in view of what we have said

---

† We make only one elucidation regarding the dimensionality of Eqn.(16.62). It is equal to area × energy^{-1} or, to be precide cm^2/erg as the dimension of σ_C is cm^2, while that of $\delta(E_\gamma)$ is $1/E_\gamma$, as $\int\delta(E_\gamma)\,dE_\gamma = 1$. Hence it is clear that the quantity σ_t is, indeed, a cross-section (we also use the therm cross-section for $\sigma(E_\gamma, \epsilon_{ph}, E)$ although it would more correct to speak of the energy density of the cross-section).

equivalent to the condition $\overline{\epsilon_{ph}}(E/mc^2) \ll mc^2$ which is essentially the same as (16.63).

Thus, provided

$$E \ll \frac{1}{4}\,\frac{mc^2}{\epsilon_{ph}}\,mc^2 \approx \frac{6\times 10^{10}}{\epsilon_{ph}(eV)}\ eV \tag{16.66}$$

the scattering of photons by moving electrons is classical and the total cross-section equal to σ_T. When electrons are scattered by optical photons, $\overline{\epsilon_{ph}} \sim 1$ eV (thermal radiation from stars) and condition(16.66) has the form $E \ll 10^{11}$ eV; if, however, we are dealing with the scattering by radio-photons, condition (16.66) is, of course, even weaker. In the majority of situations encountered in astronomy condition (16.63) or (16.66) is satisfied and we restrict ourselves in what follows mainly to that case (however, we shall give approximate expressions also for another limiting case and also indicate a realistic example when condition (16.66) is violated).

We evaluate the average losses (Compton losses) which are suffered by an electron of energy E as a result of scattering by photons.

When the electrons are scattered by sufficiently soft photons when condition (16.66) is satisfied we have

$$-\left(\frac{dE}{dt}\right)_C = c\int \sigma(E_\gamma, \epsilon_{ph}, E)\, N_{ph}(\epsilon_{ph})\, E_\gamma\, dE_\gamma\, d\epsilon_{ph} =$$

$$= \frac{8\pi}{3}\left(\frac{e^2}{mc^2}\right)^2\left(\frac{E}{mc^2}\right)^2 c\times\frac{4}{3}\int \epsilon_{ph}\, N_{ph}(\epsilon_{ph})\, d\epsilon_{ph}$$

$$= cN_{ph}\,\sigma_T\times\frac{4}{3}\overline{\epsilon_{ph}}\left(\frac{E}{mc^2}\right)^2 = \frac{32\pi}{9}\left(\frac{e^2}{mc^2}\right)^2 c\, w_{ph}\left(\frac{E}{mc^2}\right)^2, \tag{16.67}$$

where the total photon density, their energy density, and the average photon density are determined by the relations

$$N_{ph} = \int N_{ph}(\epsilon_{ph})\, d\epsilon_{ph}, \quad w_{ph} = \int \epsilon_{ph}\, N_{ph}(\epsilon_{ph})\, d\epsilon_{ph} = \overline{\epsilon_{ph}}\, N_{ph}. \tag{16.68}$$

It is clear from (16.65) and (16.67) that the average energy of a scattered photon (hard photon 2) equals

$$E_\gamma = \frac{4}{3}\overline{\epsilon_{ph}}\left(\frac{E}{mc^2}\right)^2. \tag{16.69}$$

An exact calculation requires here (see (16.67)) using the cross-section $\sigma(E_\gamma, \epsilon_{ph}, E)$ which is given by Eqn.(16.64). However, if we forget about the factor $\frac{4}{3}$ the relation (16.69) can easily be established on the basis of an elementary calculation performed using the energy and momentum conservation laws. We shall not give here that calculation (see, for instance, Ginzburg and Syrovatskii, 1964a) as one can obtain the result (16.69) even more simply using a wave representation or an actual formula for the Doppler effect. Indeed, we saw in Chapter 5 that taking the Doppler effect into account leads to the appearance of a factor of the order $(E/mc^2)^2$ or, in other words, a frequency ν_0, evaluated without taking the Doppler effect into account, becomes a frequency $\nu \sim \nu_0(E/mc^2)^2$; for magnetobremsstrahlung $\nu_0 \sim eH_\perp/2\pi mc$ and $\nu \sim (eH_\perp/2\pi mc)(E/mc^2)^2$ (see (5.7) and (5.40)). In the case of scattering, on the other hand, $\nu_0 \sim \epsilon_{ph}/h$, $\nu \sim (\epsilon_{ph}/h)(E/mc^2)^2$ and for $E_\gamma = h\nu$ we get an expression such as (16.69).

Comparing the synchrotron losses (4.39) for a randomly directed magnetic field (in such a field $H_\perp^2 = \frac{2}{3}H^2$) with the Compton losses (16.67) we find for the total losses the expression

$$-\left(\frac{dE}{dt}\right)_{mC} = -\left\{\left(\frac{dE}{dt}\right)_m + \left(\frac{dE}{dt}\right)_C\right\} = \frac{32\pi}{9}\left(\frac{e^2}{mc^2}\right)^2 c\left(\frac{H^2}{8\pi} + w_{ph}\right)\left(\frac{E}{mc^2}\right)^2 =$$

$$= 2.65\times10^{-14}\left(\frac{H^2}{8\pi} + w_{ph}\right)\left(\frac{E}{mc^2}\right)^2 \text{ erg/s}, \qquad (16.70)$$

where $H^2/8\pi + w_{ph}$ is measured in erg/cm^3; we gave this formula already in Chapter 15 (see (15.65)).

Such a result, the equivalent of the synchrotron and Compton energy losses for identical energy densities in, respectively, (on average) isotropic magnetic fields and radiation is, of course, not accidental. The fact is that in the classical region (and Eqn.(16.67) refers solely to the classical region (16.66)) the radiation power is determined by the acceleration of the charge and, hence, by the force acting upon it. However, in an electromagnetic field in vacuo $E = H$ and for an ultra-relativistic particle the Lorentz force $(e/c)[\mathbf{v} \wedge \mathbf{H}]$ equals for $\mathbf{v} \perp \mathbf{H}$ the force produced by the electric field $e\mathbf{E}$. Moreover, as $v \to c$ it is just the acceleration at right angles to the velocity which is important and as a result the total losses are in the isotropic case determined by the energy density $(E^2 + H^2)/8\pi$ of the electromagnetic field independently of the spectral composition of the field (this spectral composition determines, however, the spectral composition of the resulting radiation; in the case discussed that of the synchrotron radiation and of the scattered electromagnetic waves).

It is important to emphasize that when condition (16.66) is violated the Compton losses increase weakly with energy. Indeed, in the limiting case

$$E \gg \frac{(mc^2)^2}{\epsilon_{ph}} \tag{16.71}$$

the energy of the photons 1 satisfies in the frame of reference fixed in the electron the condition $\epsilon'_{ph} = (E/mc^2)\,\epsilon_{ph} \gg mc^2$. In that case the total cross-section for scattering, which one can obtain from (16.59), has the form (see Heitler, 1947; Berestetskii, Lifshitz, and Pitaevskii, 1971)

$$\sigma_t = \frac{3}{8}\,\sigma_T\,\frac{mc^2}{\epsilon'_{ph}}\left\{\ell n\,\frac{2\epsilon'_{ph}}{mc^2} + \frac{1}{2}\right\}. \tag{16.72}$$

If (16.71) holds, there is emitted a gamma-photon† of energy $E_\gamma \sim E$ in each collision corresponding to the cross-section (16.72). In the scattering an electron therefore loses energy at the rate

$$-\left(\frac{dE}{dt}\right)_C \sim \sigma_t N_{ph} cE \approx \frac{3}{8}\,c\,\sigma_T w_{ph}\left(\frac{mc^2}{\bar{\epsilon}_{ph}}\right)^2\left\{\ell n\,\frac{2E\bar{\epsilon}_{ph}}{m^2c^4} + \frac{1}{2}\right\}$$

$$\approx 10^{-14}\left(\frac{mc^2}{\bar{\epsilon}_{ph}}\right)^2 w_{ph}\,\ell n\,\frac{2E\bar{\epsilon}_{ph}}{m^2c^4}\ \text{erg/s}, \tag{16.73}$$

where we have used the fact that $\overline{\epsilon'_{ph}} \sim (E/mc^2)\,\bar{\epsilon}_{ph}$ and $w_{ph} = N_{ph}\,\overline{\epsilon_{ph}}$. When $(E/mc^2)(\overline{\epsilon_{ph}}/mc^2) \sim 1$ the losses (16.67) and (16.73) have the same order of magnitude, as should be the case. Compton losses thus increase proportional to the square of the energy E^2 only as long as $E \leqslant (mc^2)^2/4\epsilon_{ph}$. On the other hand, in the limiting case (16.71) the Compton losses are practically constant.

In the majority of cases which one so far has encountered in astrophysics condition (16.66) is satisfied (as we indicated, for scattering by optical photons with $\overline{\epsilon_{ph}} \sim 1$ eV this condition has the form $E \ll 10^{11}$ eV, while the relativistic electrons corresponding to the observed cosmic radio-emission have usually energies $E \leqslant 10^{10}$ eV). We can nevertheless already indicate some important exceptions. Firstly, in a number of sources (Crab Nebula, Virgo A, quasars) optical synchrotron emission is observed which corresponds to high-energy electrons $(E \gg 10^{10}$ eV). Therefore, even for $\overline{\epsilon_{ph}} \sim 1$ eV condition (16.66)

---

† When electrons are scattered by photons electron-positron pairs may be formed apart from photons, when (16.71) holds. The corresponding energy losses differ from the losses (16.73) by a factor of the order of $2\times 10^{-3}\,\ell n\,(E\bar{\epsilon}_{ph}/m^2c^4)$. This factor is in the majority of cases we meet with appreciably smaller than unity.

will be violated for those electrons. Secondly, powerful X-ray emission has been observed from galactic and extra-galactic sources. In those sources and close to them there is a large energy density of X-ray photons with $\overline{\epsilon_{ph}} \sim 3$ to 5×10^3 eV ($\overline{\epsilon_{ph}} \sim k_B T$, $T \sim 5 \times 10^7$ °K). For the scattering by such photons condition (16.66) has the form $E \ll 10^7$ eV and is violated already for the electrons corresponding to cosmic rays. The limitation to the classical region (16.66) in X-ray and gamma-astronomy has thus by no means an ubiquitous character.

We now turn to the problem of the energy spectrum of the scattered (hard) photons. As always we denote by $J_e(E)\,dE$ the intensity of relativistic electrons, that is the number of electrons passing per unit time through unit area (at right angles to the area) into unit solid angle with energies in the range $E, E + dE$. The intensity of the X-rays is then equal to (see (16.11))

$$J_\gamma(E_\gamma) = \int_0^L dR \int_{E_\gamma}^\infty J_e(E, \mathbf{R})\,dE \int_0^L \sigma(E_\gamma, \epsilon_{ph}, E)\, N_{ph}(\epsilon_{ph}, \mathbf{R})\, d\epsilon_{ph} =$$

$$= \tilde{N}_{ph}(L) \int_{E_\gamma}^\infty \sigma(E_\gamma, E)\, J_e(E)\, dE\,, \tag{16.74}$$

where in going over to the last expression we have assumed that J_e is independent of $\mathbf{R}$ along the line of sight (the integration over dR is along the line of sight) and

$$\tilde{N}_{ph}(L)\,\sigma(E_\gamma, E) = \int \sigma(E_\gamma, \epsilon_{ph}, E)\, N_{ph}(\epsilon_{ph}, \mathbf{R})\, d\epsilon_{ph}\, dR\,. \tag{16.75}$$

For a mono-energetic electron spectrum $J_e(E) = J_0\,\delta(E - E_0)$ and the intensity $J_\gamma(E_\gamma)$ is determined by Eqn.(16.64) for $\sigma(E_\gamma, \epsilon_{ph}, E_0)$. More often one is dealing with a power-law spectrum

$$J_e(E) = K_J\, E^{-\gamma}\,. \tag{16.76}$$

For an isotropic distribution of relativistic electrons with density $N(E) = K_e E^{-\gamma}$ it is clear that

$$J_e(E) = \frac{c}{4\pi} N(E)\,.$$

In order not to perform complicated calculations (it is quite obvious how to perform them) we use for the evaluation of $J_\gamma(E_\gamma)$ for the power-law spectrum (16.76) to begin with as the cross-section $\sigma(E_\gamma, E)$ averaged over the spectrum of the thermal (soft) photons the expression

$$\left.\begin{aligned}\sigma(E_\gamma, E) &= \frac{1}{N_{ph}}\int_0^\infty \sigma(E_\gamma, \epsilon_{ph}, E)\, N_{ph}(\epsilon_{ph})\, d\epsilon_{ph} = \\ &= \sigma_T\, \delta\left(E_\gamma - \tfrac{4}{3}\overline{\epsilon_{ph}}\left(\frac{E}{mc^2}\right)^2\right), \\ \sigma_T &= \frac{8\pi}{3}\left(\frac{e^2}{mc^2}\right)^2, \quad N_{ph} = \int N_{ph}(\epsilon_{ph})\, d\epsilon_{ph}\ .\end{aligned}\right\} \tag{16.77}$$

In other words, we assume that all soft photons have the average energy (6.69) and according to (16.77) the total cross-section for scattering equals σ_T. The cross-section (16.77) leads thus necessarily to the correct expressions for $\int \sigma_T(E_\gamma, E)\, dE_\gamma$ and $\int E_\gamma\, \sigma(E_\gamma, E)\, dE_\gamma$.

Substituting (16.77) into (16.74) we get for the case of a uniform distribution of all quantities along the length L (so that $\tilde{N}_{ph}(L) = N_{ph} L$)

$$\begin{aligned}J_\gamma(E_\gamma) &= \sigma_T N_{ph} L \int J_e(E)\, \delta\left(E_\gamma - \tfrac{4}{3}\overline{\epsilon_{ph}}\left(\frac{E}{mc^2}\right)^2\right) dE = \\ &= \frac{\sqrt{3}\, N_{ph} L\, \sigma_T mc^2}{4\, (\overline{\epsilon_{ph}}\, E_\gamma)^{\frac{1}{2}}}\, J_e\left(mc^2 \sqrt{\{3E_\gamma / 4\overline{\epsilon_{ph}}\}}\right) = \\ &= \tfrac{1}{2} N_{ph} L \sigma_T (mc^2)^{1-\gamma}\, (\tfrac{4}{3}\overline{\epsilon_{ph}})^{\frac{1}{2}\gamma - \frac{1}{2}} K_J\, E_\gamma^{-\frac{1}{2}(\gamma+1)}\ .\end{aligned} \tag{16.78}$$

The calculation given here is completely analogous to the one used in Chapter 5 in the application to synchrotron emission (see (5.51)). In an exact calculation (Ginzburg and Syrovatskii, 1964b, 1965; Ginzburg, 1969b) there appears a numerical factor $f(\gamma)$. This factor equals 0.84, 0.86, 0.99, and 1.4 for $\gamma = 1$, 2, 3, and 4, respectively.

In the case of thermal radiation $\overline{\epsilon_{ph}} = 2.7\, k_B T$, where T is the temperature of the radiation. As an example we mention that for $T = 5000$ °K ($\overline{\epsilon_{ph}} = 1.2$ eV)

$$\begin{aligned}J_\gamma(E_\gamma) &= 2.8 \times 10^{-25} (7.9 \times 10^{-2})^{\gamma-1} \times \\ &\times f(\gamma)\, L w_{ph} K_J\, E_\gamma^{-\frac{1}{2}(\gamma+1)}\ \frac{\text{photons}}{\text{cm}^2.\text{sterad.s.GeV}},\end{aligned} \tag{16.79}$$

where

$$w_{ph} = N_{ph}\, \overline{\epsilon_{ph}},$$

E_γ is measured in GeV and K_J in units $(\text{GeV})^{\gamma-1}\ (\text{cm}^2.\text{s.sterad})^{-1}$.

If we are dealing with the intensity in number of particles rather that the intensity in the energy scale, $I_\gamma(E_\gamma) = E_\gamma J_e(E_\gamma)$.

In the case of the power-law spectrum (16.76) clearly $I_\gamma(E_\gamma) \propto E_\gamma^{-\frac{1}{2}(\gamma-1)} \propto \nu^{-\alpha}$, $\alpha = \frac{1}{2}(\gamma-1)$. This result is the same as the behaviour (5.50) for synchrotron radiation, as one should expect for Compton radiation when (16.66) holds — we have seen that under that condition synchrotron and Compton radiation are related to one another.

We have already mentioned that synchrotron (magneto-brems) X-ray emission under cosmic conditions is 'atypical' in a certain sense, but it is observed and, probably, its specific weight in the course of further studies will increase. It is completely obvious that synchrotron emission can have arbitrarily high frequencies while it can be described classically as long as

$$h\nu \ll E . \tag{16.80}$$

Cosmic synchrotron radiation most often lies in the radio-band, as the magnetic field strength in the corresponding regions is small ($H \leqslant 10^{-3}$ Oe) while the electron energy E is also not very large. To be precise, the intensity of the synchrotron radiation from an electron of total energy $E \gg mc^2$ is a maximum at a frequency (see (5.40) and (16.2))

$$\nu_m = 1.2 \times 10^6 H_\perp \left(\frac{E}{mc^2}\right)^2 = 4.6 \times 10^{-6} H_\perp (\text{Oe})[E(\text{eV})]^2 \text{ Hz} . \tag{16.81}$$

Moreover, for instance, for $H_\perp = H \sin\chi = 10^{-3}$ Oe and $E = 10^{10}$ eV the frequency $\nu_m = 4.6 \times 10^{11}$ Hz ($\lambda_m = c/\nu_m \sim 6.5 \times 10^{-2}$ cm). The cosmic optical and X-ray synchrotron radiation therefore arises either when there are electrons present with very high energies $E > 10^{11}$ eV (Crab Nebula, Virgo A), or when the magnetic field strength is very large ($H \geqslant 10$ to 100 Oe) and when there are electrons present with energies reaching 10^9 to 10^{10} eV (such a situation exists clearly for quasars; we do not yet mention pulsars with their enormous magnetic fields).

To estimate the fields and energies for which waves of various frequencies are emitted it is convenient to use the formula

$$\frac{\nu_2}{\nu_1} = \frac{H_{\perp,2}}{H_{\perp,1}} \frac{E_2^2}{E_1^2} , \tag{16.82}$$

which follows from (16.81).

Let, for instance, $\nu_1 = 3 \times 10^8$ Hz ($\lambda_1 = c/\nu_1 = 1$ m) in a field $H_{\perp,1} = 3 \times 10^{-6}$ Oe which is typical for the Galaxy. According to (16.81), the energy of the emitting electrons is then $E_1 \sim 5 \times 10^9$ eV. In the same field $H_{\perp,2} = H_{\perp,1}$ waves with optical frequencies $\nu_2 \sim 10^{14}$ to 10^{15} Hz ($\lambda_2 = 0.3$ to $3\ \mu$m) can be emitted only by electrons of energies $E_2 \sim 5 \times 10^{12}$ eV. For X-rays $\nu_2 \sim 10^{18}$ Hz and, hence, for an unchanged magnetic field electrons must have an energy $E_2 \sim 3 \times 10^{14}$ eV.

We must bear in mind that synchrotron losses are proportional to $H_\perp^2 E^2$ (see (4.39)) and therefore particles with very high energies or when they move in a strong field are slowed down fast. An estimate of the energy and of the 'lifetime' in a magnetic field can conveniently be found by using Eqns.(4.41) and (4.42). We can then express in Eqn.(4.42) the electron energy in terms of the characteristic frequency of its radiation (16.81) and, thus, obtain immediately a relation between the observed frequency and the characteristic lifetime (the time over which the energy reduces by a factor two) of the emitting electrons:

$$T_m = \frac{5 \times 10^8}{H_\perp^2} \frac{mc^2}{E} \text{ s} \approx \frac{5.5 \times 10^{11}}{H_\perp^{\frac{3}{2}} \nu^{\frac{1}{2}}} \text{ s} \approx \frac{1.8 \times 10^4}{H_\perp^{\frac{3}{2}} \nu^{\frac{1}{2}}} \text{ yr.} \qquad (19.83)$$

Here $H_\perp$ is measured in Oersted and ν in Hertz. The time T_m expressed in terms of the frequency has, of course, a somewhat arbitrary character as we have chosen for ν the frequency corresponding to the maximum in the radiation spectrum of mono-energetic electrons.

In a field $H_\perp = 3 \times 10^{-6}$ Oe the time T_m for electrons with energies 5×10^9, 5×10^{12}, and 3×10^{14} eV is, respectively, 2×10^8, 2×10^5, and 3×10^3 year. For our Galaxy and, in general, for normal galaxies for which the value $H = 3 \times 10^{-6}$ Oe may be assumed to be typical a characteristic time T_m of the order of 10^5 year and, even more, one of the order of 10^3 year is very short and it is therefore natural that the optical and X-ray synchrotron radiation will be weak. The position can change only when there is a powerful injection of high-energy electrons into the interstellar space from some kind of sources, for instance, from supernova shells.

As we have indicated, the optical and X-ray synchrotron radiation is completely described by the formulae given earlier (see Chapter 5; condition (16.80) is assumed to be satisfied). Moreover, there occurs even a simplification which is connected with the fact that at high frequencies one can neglect the fact that the refractive index $\tilde{n}(\omega)$ differs from unity, as well as the reabsorption and the rotation of the polarization plane in the cosmic plasma. One must, however, solely take into account the absorption of the radiation on its path to the Earth or in the source itself (by gas, dust).

For the sake of convenience we shall nevertheless give a few expressions which are useful for calculations. In the X-ray region and sometimes also in the optical region one often uses instead of the energy flux the particle number (photon) flux or intensity which we have denoted by F_ν and J_ν. The transition is clearly obtained by dividing the energy-expressions by the photon energy $h\nu$. The photon number intensity is thus according to (5.48) equal to

$$J(\nu) = \frac{I_\nu}{h\nu} = 3.26\times 10^{-15}\, a(\gamma)\, L K_e H^{\frac{1}{2}(\gamma+1)} \times \times \left(\frac{6.26\times 10^{18}}{\nu}\right)^{\frac{1}{2}(\gamma+1)} \frac{\text{photons}}{\text{cm}^2.\text{s.sterad.Hz}}, \quad (16.84)$$

or, if we change from the frequency ν to the photon energy $\epsilon_{ph} = h\nu$, expressed in eV,

$$J(\epsilon_{ph}) = J(\nu)\frac{d\nu}{d\epsilon_{ph}} = 0.79\, a(\gamma)\, L K_e H^{\frac{1}{2}(\gamma+1)} \times \times \left(\frac{2.59\times 10^{4}}{\epsilon_{ph}}\right)^{\frac{1}{2}(\gamma+1)} \frac{\text{photons}}{\text{cm}^2.\text{s.sterad.MeV}}; \quad (16.85)$$

here L is measured in cm, K_e in $\text{erg}^{\gamma-1}.\text{cm}^{-3}$, H in Oe, and ϵ_{ph} in eV. Similarly, the photon flux from a discrete source (see (5.59)) equals

$$F(\nu) = \frac{\Phi(\nu)}{h\nu} = 3.26\times 10^{-15}\, a(\gamma)\, \frac{V K_e H^{\frac{1}{2}(\gamma+1)}}{R^2} \times \times \left(\frac{6.26\times 10^{18}}{\nu}\right)^{\frac{1}{2}(\gamma+1)} \frac{\text{photons}}{\text{cm}^2.\text{s.Hz}}, \quad (16.86)$$

or, in terms of the photon energy $\epsilon_{ph} = h\nu = 4.14\times 10^{-15}\,\nu$ eV,

$$F(\epsilon_{ph}) = 0.79\, a(\gamma)\, \frac{V K_e H^{\frac{1}{2}(\gamma+1)}}{R^2} \left(\frac{2.59\times 10^{4}}{\epsilon_{ph}}\right)^{\frac{1}{2}(\gamma+1)} \frac{\text{photons}}{\text{cm}^2.\text{s.eV}}. \quad (16.87)$$

Moreover, if we can assume the electron spectrum to be the same in the whole of the source, it is convenient to use the following expression for the ratio of the radiative fluxes ar different frequencies ν_1 and ν_2 (see (5.59) or (16.86))

$$\frac{\Phi_2(\nu_2)}{\Phi_1(\nu_1)} = \frac{V_2}{V_1}\left(\frac{H_2}{H_1}\right)^{\frac{1}{2}(\gamma+1)} \left(\frac{\nu_2}{\nu_1}\right)^{\frac{1}{2}(\gamma-1)}. \quad (16.88)$$

We have assumed here that the radiation at the frequency ν_1 arises in a source of volume V_1 in which the magnetic field strength is H_1 and the radiation at the frequency ν_2 comes from a volume V_2 with a field H_2. In that case, if we are dealing with radiation by electrons with the same energy $E_2 = E_1$, the frequencies ν_1 and ν_2 are related through the Eqn.(16.82), while the ratio of the fluxes equals

$$\frac{\Phi_2(\nu_2)}{\Phi_1(\nu_1)} = \frac{V_2H_2}{V_1H_1} . \qquad (16.89)$$

Equations (16.88) and (16.89) are useful if in a small region V_2 of the source with total volume V_1 the field $H_2 \gg H_1$ while there is in the electron spectrum a break at the high-energy side so that the electrons in the volume V_1 do not emit at frequencies $\nu_2 \gg \nu_1$ while the emission from the volume V_2 at the frequency ν_1 is small due to the fact that that volume is small. The observed flux ratio at the frequencies ν_1 and ν_2 from the whole source will then be determined by the flux ratio from the volumes V_1 and V_2. Such a situation can occur, for instance, in the case of nebulae in the central part of which there is a region (say, surrounding a pulsar) with a very strong magnetic field.

A characteristic feature of the synchrotron radiation is the fact that in an ordered field this radiation is polarized to a large degree. For instance, for a power-law electron spectrum $N_e(E) = K_e E^{-\gamma}$ in a uniform field the degree of polarization equals (see (5.46))

$$\Pi_0 = \frac{I_{max} - I_{min}}{I_{max} + I_{min}} = \frac{\gamma + 1}{\gamma + \frac{7}{3}} . \qquad (16.90)$$

As we have mentioned, there are no depolarization factors in the X-ray region which are caused by the presence of a medium (various forms of Faraday rotation). The degree of polarization for a given index γ reflects thus only the degree of the ordered field, and it reaches its maximum value (16.90) in a uniform field.

Bremsstrahlung is polarized only when the electron distribution function is anisotropic (or, strictly speaking, the distribution function for the relative velocity of the colliding particles). For instance, polarization of bremsstrahlung arises when there is a directed electron flux which is scattered in a cold plasma. Under cosmic conditions, forgetting the Sun and a few nonstationary regions, there are no special grounds to expect the existence of any strong anisotropy in the electron velocity distribution (see Chapter 15; moreover, when there are collisions, the relaxation of an anisotropic electron velocity distribution, that is, isotropization, will proceed even faster than in the collisionless case). When electromagnetic radiation is scattered by particles the degree of polarization can in principle be large. This is well known, for instance, for the case of scattering of light in gases or in a plasma (see Chapter 14). However, in the case of scattering of soft, unpolarized photons by relativistic electrons, forming hard (X-ray and gamma-ray) photons the degree of polarization of the latter is of the order of $(mc^2/E)^2$ that is, very small.

One should expect that under cosmic conditions (excluding possibly dense magnetospheres of white dwarfs, neutron stars, and so on) for all other radiation mechanisms, apart from synchrotron and synchro-Compton mechanisms (see end of Chapter 15) no appreciable polarization will occur either. The observation of polarization of cosmic X-ray or gamma-ray emission, as in the case of the cosmic radio-emission, allows us therefore to assume that the corresponding radiation is synchrotron (or synchro-Compton) emission. In particular, the synchrotron (and not brems) nature of the X-ray emission of the Crab Nebula was finally recognized only as the result of the observation of the polarization of the X-ray emission.

If the observation of the polarization of the X-ray emission indicates its synchrotron nature, the opposite conclusion is, of course, invalid — it is sufficient to say that synchrotron radiation in a random magnetic field is not polarized.† It is thus always difficult to establish the nature of the cosmic X-ray emission. The main criterion (apart from polarization) is the shape of the spectrum. Bremsstrahlung from a hot plasma has an exponential spectrum (see, for instance, (16.24)) and, moreover, in the case of a hot plasma one can observe, in principle, characteristic X-ray lines from heavy elements (first of all iron). Compton X-ray emission is produced by relativistic electrons which normally have a power-law spectrum (16.76) with an index γ so that for the X-ray emission $J_\gamma(E_\gamma) \propto E_\gamma^{-\beta}$, $\beta = \frac{1}{2}(\gamma+1)$ (see (16.78)) and $I_\gamma(E_\gamma) \propto E_\gamma^{-\alpha}$, $\alpha = \frac{1}{2}(\gamma-1)$. Moreover, for a given known object the relativistic electrons may also produce synchrotron radiation, the spectrum of which enables us to determine the index γ (we remind ourselves once again that for synchrotron radiation, as for Compton radiation, $I(\nu) \propto \nu^{-\alpha}$, $\alpha = \frac{1}{2}(\gamma-1)$; see Chapter 5). We have already mentioned that one can in this way for a known radiation field in which Compton scattering occurs find also the magnetic field in the same emitting region. Unfortunately, in practice all this is not

---

† One must, moreover, bear in mind that polarimetric measurements in the X-ray region, particularly for low degrees of polarization and for relatively weak cosmic X-ray emission tluxes, are very difficult.

so simple: first of all we must note that the spectra are not strictly power laws and the same relativistic electrons produce synchrotron and Compton radiation with a given α only in limited and a priori unknown frequency intervals $\Delta\nu$ (in the radio- and X-ray bands). One may nevertheless expect in the future a great deal of progress along the line of combined (complex) studies of the radiation in a very wide range of (radio- , optical , and X-ray-) frequencies in conjunction with polarimetric measurements and also with measurements with a high angular resolution. This last requirement is connected with the fact that for a low angular resolution (so far characteristic for the majority of observations in X-ray astronomy) one cannot only not distinguish the structure of the discrete X-ray sources but the nature of the X-ray background also remains unclear — it may partially (and in principle even almost completely) be determined by a set of unresolved discrete sources.

The present state of X-ray astronomy which has already produced brilliant successes (the observation of powerful X-ray sources — 'X-ray stars', including pulsars, and so on) is on the whole in a stage of being established and of fast development and it seems particularly inappropriate to discuss here the available observational data (Gratton, 1970; Schnopper and Delvaille, 1972; Friedman, 1973; Stecker and Trombka, 1973; Culhane, 1978; Cooke, Lawrence, and Perola, 1978). We conclude therefore the present chapter only for a certain orientation with a few remarks on this subject.

The observed discrete X-ray sources (X-ray stars) belong to a few types. Within the Galaxy mainly two kinds of sources are observed: supernova shells (these are in fact extended sources) and X-ray stars in the strict sense of the word — practically point sources which are bright in the X-ray band. In the majority of cases and, possibly, almost always the latter are part of binary system; the X-ray pulsars and 'fluctuars' (which are, possibly, 'black holes') belong to this class; the pulsar PSR 0532 in the Crab Nebula is a well known exception — it is an X-ray 'point' source (neutron star with its magnetosphere) which is not part of a binary system. The X-ray emission from supernova shells has either a synchrotron character (Crab Nebula) or it is basically bremsstrahlung from a hot plasma (characteristic temperature 10^6 to 10^8 °K corresponding to an average particle energy of 10^2 to 10^4 eV). The appearance of powerful X-ray sources in tight binary systems is very understandable — in such a situation there occurs strong accretion, that is, transfer of plasma from the lighter to the heavier star. In that case, especially for a compact star (white dwarf, neutron star) the plasma fluxes when they approach the stellar photosphere reach high velocities and when they are decelerated ('fall' on the star) the plasma is greatly heated up ($T \sim 10^7$ to 10^9 °K). If the star has a sufficiently strong magnetic field, one may expect that not only the brems- but also the magneto-bremsstrahlung will turn out to be significant.

The power (X-ray luminosity) of galactic X-ray source L_X reaches 10^{37} to 10^{38} erg/s † which exceeds the solar luminosity $L_\odot = 3.86 \times 10^{33}$ erg/s by 4 to 5 orders of magnitude. For $L_X \sim 10^{38}$ erg/s and an isotropic emission the X-ray flux at the Earth equals

$$\Phi = \frac{L_X}{4\pi R^2} \sim \frac{L_X/10^{38}}{[R(\mathrm{pc})]^2} \sim \frac{1}{[R(\mathrm{pc})]^2} \frac{\mathrm{erg}}{\mathrm{cm^2.s}} , \qquad (16.91)$$

where $R(\mathrm{pc})$ is the distance to the source in parsec.

According to this estimate we get for the Crab Nebula ($R \approx 2000$ pc) in the whole of the X-ray band $\Phi_X \approx 2 \times 10^{-7}$ erg/cm^2.s (the flux of photons of energy 2 to 10 keV is $F_X \approx 2$ photons/cm^2.s). For a comparison we note that a black body of temperature T (in °K) emits from unit surface a flux

$$\Phi_0 = \sigma T^4 , \quad \sigma = \frac{\pi^2 k_B^2}{60 \hbar^3 c^2} = 5.67 \times 10^{-5} \frac{\mathrm{erg}}{\mathrm{cm^2.s.(°K)^4}} , \quad \Phi(R) = \Phi_0 \frac{r^2}{R^2} , \qquad (16.92)$$

where $\Phi(R)$ is the flux from a black body sphere of radius r observed at a distance R.

The flux of the solar radiation is at the Earth equal to

$$\Phi_\odot = \frac{L_\odot}{4\pi R^2} = 1.4 \times 10^6 \frac{\mathrm{erg}}{\mathrm{cm^2.s}}$$

($R \approx 1.5 \times 10^{13}$ cm, $L_\odot = 4\pi r_\odot^2 \Phi_\odot$, $T_\odot \approx 5700$ °K), but the radiation is concentrated in the optical part of the spectrum. The X-ray flux from the quiet Sun equals $\Phi_{\odot,X} \sim 10^{-4}$ to 10^{-5} erg/cm^2.s and only during powerful flares does it reach values $\Phi_{\odot,X} \sim 1$ erg/cm^2.s. Hence it follows that the X-ray emission from the closest stars ($R \sim 4 \times 10^{18}$ cm), if they emit like the Sun, will be very weak: $\Phi_X \sim 10^{-11}\, \Phi_{\odot,X} \leqslant 10^{-11}$ erg/cm^2.s. This was just the reason why the observation in 1962 of a bright X-ray 'star' in Scorpio (Sco X-1) for which $\Phi_X \sim 10^{-6}$ erg/cm^2.s was unforeseen. As for Sco X-1 the distance $R \leqslant 500$ pc, the total X-ray luminosity of this source does, apparently, not exceed that of the Crab Nebula and is, possibly, weaker by an order of

† Undoubtedly, our Galaxy is in that respect not at all exceptional. Similar sources have already been observed in the Magellanic Clouds and clearly exist also in other galaxies.

magnitude. Of course, a large X-ray luminosity is connected with a relatively large electron energy and, to be precise, for thermal sources with a high electron temperature. For instance, a star of the size of the Sun and a surface temperature $T \sim 6 \times 10^6$ °K would have the enormous luminosity $L_X \sim 3 \times 10^{45}$ erg/s, just in the X-ray region (the maximum of the intensity in the black-body spectrum occurs at a wavelength $\lambda_m = hc/4.965\,k_B T \approx 3 \times 10^7/T$ (°K) Å).

An ordinary star cannot have such a high temperature (we are talking here about the photosophere which emits approximately as a black body) just because of the large, practically unsuppliable energy losses through radiation. For a neutron star of radius $r \sim 7 \times 10^5$ cm $\sim 10^{-5} r_\odot$ and $T \sim 6 \times 10^6$ °K we have already $L_X \sim 3 \times 10^{35}$ ergs/s which is allowable for a certain time. As regards thermal sources which are not so compact, they are 'thin' sources (or, as one says, optically thin; we avoid using that term as in the X-ray band it may lead to confusions). If we, therefore, use Eqns.(16.36) and (16.37) we see easily that for $T \sim 6 \times 10^6$ °K a luminosity $L_X \sim 10^{38}$ erg/s is, for instance, reached by a plasma cloud of volume $V \sim 10^{30}$ cm^3 and with a comparatively small mass $M \sim 10^{22}$g $\sim 10^{-11}$ $M_\odot$ (the solar mass $M_\odot = 2 \times 10^{33}$ g) the average electron density in the cloud is in that case $N \sim \sqrt{(N^2)} \sim 10^{16}$ cm^{-3}.

Discrete extragalactic X-ray sources are galaxies (in particular, radio-galaxies), quasars, and clusters of galaxies. The X-ray luminosity of a normal galaxy (such as our Galaxy) does not exceed 10^{39} to 10^{40} erg/s. The flux of such a galaxy at a distance of $R \sim 10^7$ pc (the distance from the radio-galaxy Virgo A ≡ NGC 4486 ≡ M87) would be $\Phi_X \sim 10^{-12}$ to 10^{-13} erg/cm^2.s (see (16.91)). However, the radio-galaxy M 87 emits an appreciably more powerful X-ray radiation so that for it $L_X \sim 10^{43}$ to 10^{44} erg/s. Both in this case and in the case of other powerful X-ray sources — galaxies, quasars, and clusters of galaxies — the emission does clearly not reduce to a system of radiation from 'X-ray stars' and the X-ray source is provided by the relativistic electrons (synchrotron and Compton mechanisms) or the hot plasma (brems mechanism) which fill the galaxy, the cluster, or the 'corona' of the quasar.

In principle the situation is particularly 'simple' in the case of the hot plasma in an extended source such as a cluster of galaxies. For instance, at a temperature $T \sim 6 \times 10^6$ °K a plasma of volume $V \sim 3 \times 10^{73}$ cm^3 will have a luminosity $L_X \sim 10^{44}$ erg/s for $N \sim 10^{-3}$ cm^{-3} which corresponds to a mass $M \sim 2 \times 10^{-24}$ NV $\sim 10^{13}$ $M_\odot$; such a mass is fully acceptable for a cluster of galaxies. However, even this example shows that it is not easy to explain the enormous luminosities of powerful extra-galactic sources or it is more correct to say that such an explanation is accompanied by far-reaching assumptions which must be checked (and in principle can fully be checked) in a number of ways.

From this it is also clear how exceptionally useful X-ray astronomy can be, for instance, for a study of the hot plasma in the Universe (see also Zel'dovich, 1975).

Apart from the discrete sources, one observes an X-ray background, that is, radiation coming from all directions without on the celestial sphere having any pronounced 'granular' structure. It is not excluded that such a background is nonetheless connected (partially or even completely) with a system of discrete sources which our instruments do not resolve. It is at the same time fully possible and even probable that there exists some (true) X-ray background which is produced in interstellar and particularly in intergalactic space. The background arises as the result of the bremsstrahlung of the hot intergalactic gas and/or of the Compton scattering of relativistic electrons by the relict thermal radiation (in intergalactic space) and in the Galaxy, moreover, by other kinds of radiation (we have in mind, in particular, Compton scattering by infra-red and optical photons).

The spectrum of the X-ray background (in the energy range $E_X > 1$ keV, $\lambda < 10$ Å), as that of the discrete sources is a decreasing one — the radiative intensity decreases with increasing energy of the photons $E_X = h\nu_X$. So far the shape of the spectrum has not been established reliably, but for orientation purposes we give here as an example such a spectrum ($20 < E_X < 1000$ keV):†

$$J_X(E_X) = 25 \left(\frac{E_X}{1\ \text{keV}}\right)^{-2.1} \frac{\text{photons}}{\text{cm}^2.\text{s.sterad.keV}}\ , \qquad (16.93)$$

where the photon energy is measured in keV (this is, strictly speaking, already reflected in the way we have written the formula. but we repeated it here to avoid misunderstanding.

There is a widespread discussion in the literature of the problem of the shape of the spectrum for $E_X < 20$ to 40 keV; for instance, there were indications

---

† We have here clearly already included the soft gamma-ray region and we could with the same effort denote $J_X(E_X)$ by $J_\gamma(E_\gamma)$. We note that in the literature such a differential intensity in terms of photon numbers is also denoted as dN/dE, $N(E)$, and so on.

of a 'break' in the spectrum for $E_X \approx 40$ keV, and the spectrum itself could for $0.2 < E_X < 20$ keV be written in the form

$$J_X(E_X) = 15\ (E_X/1\ \text{keV})^{-1.8}\ \text{photons/cm}^2\text{.s.sterad.keV}\ .$$

Another approximation of the spectrum is (see Gratton, 1970; Schnopper and Dalvaille, 1972; Friedman, 1973)

$$J_X(E_X) = 12.4(E_X/1\ \text{keV})^{-1.7\pm0.2}\ \text{photons/cm}^2\text{.s.sterad.keV},\quad 1 < E_X < 40\ \text{keV}$$

$$J_X(E_X) = 20(E_X/1\ \text{keV})^{-2}\ \text{photons/cm}^2\text{.s.sterad.keV},\quad E_X > 60\ \text{keV}\ .$$

The mechanism leading to the appearance of the X-ray background is as yet not clear and we have already indicated the various possibilities.

X-ray astronomy, that is, the study of the cosmic X-ray emission and the comparison of the appropriate data with the theory and with all other astronomical information opens up exceptionally wide and for astronomy as a whole important possibilities for studying hot cosmic plasmas and relativistic cosmic electrons. One can hardly overestimate the value of this astronomical method.

Chapter XVII

GAMMA ASTRONOMY

Gamma-radiation produced by the proton-nuclear cosmic ray component.
Example of the Magellanic Clouds and their central region of the Galaxy.
Absorption of X-rays and gamma-rays.

We have already considered in Chapter 16 two important mechanisms for gamma-radiation, namely, the brems- and Compton mechanisms. The same is true for synchrotron radiation which is of less interest in the gamma-ray region as in practice it can turn out to be important only in regions with a very strong field (for instance, near pulsars). As to bremsstrahlung of relativistic electrons and especially their Compton scattering, their contribution may be noticeable and even play a decisive role in the whole gamma-band. For instance, in the case of scattering by optical photons ($\epsilon_{ph} \sim 1$ eV) the classical region (16.66) extends up to electron energies $E \sim 5 \times 10^{10}$ eV and the photons produced in that case have energies $E_\gamma \sim \epsilon_{ph}(E/mc^2)^2 \leqslant 10^{10}$ eV (see (16.69)). In the region of even higher electron energies, especially in the quantum region (16.71), the energy of the Compton photons $E_\gamma \sim E$. True, at very high energies the 'existence condition' for electrons due to their large losses is less favourable than for protons or nuclei. Moreover, electrons are in the Universe accelerated less efficiently than protons and in any case in the Galaxy the intensity of the electron component is two orders of magnitude smaller than that of the proton component (see Chapter 15). We can thus assume that with increasing energy and, quite possibly, already for $E > 50$ to 100 MeV (vide infra) the cosmic gamma-radiation is mainly produced by the proton-nuclear component of the cosmic rays. In the experimental plane there is here no clarity, and in general in the field of observations gamma-astronomy is as yet only taking the first steps (for a survey of the available data see Gratton, 1970; Stecker, 1971; Schnopper and Delvaille, 1972; Gal'per, Kirillov-Ugryumov, Luchnov, and Prilutskii, 1972; Stecker and Trombka, 1973; Friedman, 1973; Gal'per, Kirillov-Ugryumov, and Luchkov, 1974; Plovdiv, 1977; Ginzburg, 1978).

All mechanisms for cosmic gamma-emission and all energy ranges are of interest but such a rather trivial statement is made here only to isolate one, particularly interesting and important possibility. For us, namely, of most importance is the possibility to use gamma-astronomy to obtain reliable conclusions about the proton-nucleus component of the cosmic rays far from the Earth. We have already emphasized in Chapter 5 and 15 that the absence of the corresponding direct data presents us with difficulties as a matter of principle when we develop the astrophysics of cosmic rays (formally the question is the lack of knowledge of the coefficients $\kappa_e = w_{c.r.}/w_e$ and $\kappa_H = w_H/w_{c.r.}$; see (15.16)).

Let us discuss this problem in detail without being afraid of repeating ourselves. Moreover, we shall use observational data which will already be obsolete by the time this book appears. However, our aim is not just the discussion of actual results but an illustration of the nature of the estimates and arguments which are used in high-energy astrophysics.

Protons and nuclei which form part of the cosmic rays undergo collisions with the protons and nuclei of the intergalactic or interstellar gas. As a result of nuclear collisions π^0 mesons and Σ^0 hyperons are formed which decay fast, producing gamma-rays. The π^0 decay goes with a probability of 98.8% (that is, practically always) through the $\pi^0 \to 2\gamma$ channel, so that the energy of the gamma-rays from the decay of a π^0 at rest equals $E_\gamma = \frac{1}{2} m_\pi c^2 = 67.5$ MeV; the average lifetime of a π^0 is 0.84×10^{-16} s. For the Σ^0 decays (with practically a probability of 100%) through the channel $\Sigma^0 \to \Lambda + \gamma$, the energy $E_\gamma \approx 77$ MeV, and the average lifetime of the Σ^0 is less than 10^{-14} s. Apart from the direct π^0 production in nuclear collisions, they are also formed as a result of the decay of various mesons and hyperons ($K^\pm \to \pi^\pm + \pi^0$, $\Lambda \to n + \pi^0$, and so on), as a result of which γ-rays are emitted. The probabilities and kinematics of all possible reactions are rather well known (Ginzburg and Syrovatskii, 1965; Ginzburg, 1969b; Stecker, 1971) and this enables us to evaluate the gamma-ray spectrum with an accuracy which is fully sufficient for the astrophysical applications discussed here. It is in that case important that the cosmic gamma-ray flux is generated, of course, not by monoenergetic particles, but by the cosmic rays which are isotropic in direction and have an intensity $J_{c.r.}(E)$. There is thus an averaging over the spectrum and, to be precise, the intensity of gamma-rays of energy E_γ is equal to

$$J_\gamma(E_\gamma) = \tilde{N}(L) \int_{E_\gamma}^{\infty} \sigma(E_\gamma, E)\, J_{c.r.}(E)\, dE \,, \tag{17.1}$$

where σ is the appropriate cross-section averaged using the chemical composition of the cosmic rays and of the gas (and also, of course, using the fact that two photons are produced in the π^0 decay), while

$$\tilde{N} = \int_0^L N(R)\, dR$$

is the number of particles in the gas along the line of sight (we have assumed in Eqn.(17.1), which is the same as (16.11), and also in the formulae given below that the intensity $J_{c.r.}$ does not depend on position). The integral intensity equals

$$J_\gamma(>E_\gamma) = \int_{E_\gamma}^{\infty} J_\gamma(E_\gamma)\, dE_\gamma\ .$$

For the gamma-ray flux from a discrete source we have

$$F_\gamma(>E_\gamma) = \int_\Omega J_\gamma(>E_\gamma)\, d^2\Omega \approx \frac{\overline{(\sigma I_{c.r.})}\, N(V)}{R^2}$$

$$\approx \frac{5\times 10^{23}\,\overline{(\sigma I_{c.r.})}\, M}{R^2}\ \frac{\text{photons}}{\text{cm}^2.\text{s}}\ , \qquad (17.2)$$

where Ω is the solid angle, R the distance from the source (in cm), $N(V) = NV$ the number of particles (nuclei) in the source (V is the volume, N the average gas density), and $M = 2\times 10^{-24}\, N(V)$ the mass of gas in the source in grammes (we have assumed the chemical composition of the source to correspond to the average distribution of elements and thus the mass of the 'average' nucleus of the gas is assumed to equal 2×10^{-24} g, due to taking especially the He nuclei into account). We show in Fig.17.1 the value of

$$\overline{\sigma J_0(>E_\gamma)} = \int_{E_\gamma}^{\infty} dE_\gamma \int_{E_\gamma}^{\infty} dE\, \sigma(E_\gamma, E)\, J_{c.r.,0}(E)\, dE\ ,$$

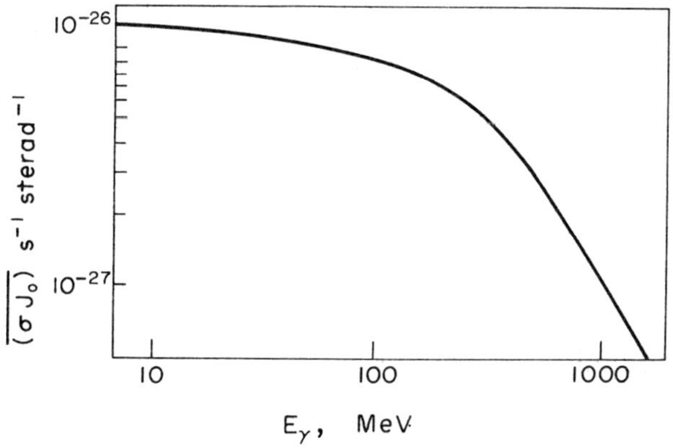

Fig. 17.1 The quantity $\overline{\sigma J_0}\,(>E_\gamma)$, averaged over the galactic cosmic rays, as function of the energy of the gamma-rays produced.

taken from Stecker's book (1971) for the spectrum of the cosmic rays at the Earth (intensity $J_{c.r.,0}(E) \equiv J_0(E)$). Below we shall use the value $\overline{\sigma J_0(E_\gamma > 100\,\text{MeV})} = 10^{-26}\ \text{s}^{-1}.\text{sterad}^{-1}$, and, hence,

$$F_\gamma(E_\gamma > 100\,\text{MeV}) = \frac{10^{-26}\,NVw_{c.r.}/w_0}{R^2} = \frac{5\times 10^{-3}\,Mw_{c.r.}/w_0}{R^2}\,\frac{\text{photons}}{\text{cm}^2.\text{s}}, \tag{17.3}$$

where $w_{c.r.}$ is the cosmic ray energy density in the source, and where we have assumed that the shape of their spectrum is the same as at the Earth (so that $w_{c.r.}/w_0 = J_{c.r.}/J_{c.r.,0}$, where $w_{c.r.,0} \equiv w_0 \sim 10^{-12}\ \text{erg/cm}^3$ is the cosmic ray energy density at the Earth; see (15.9)). Within the limits of the approximations made for sources such as the Galaxy where the neutral hydrogen atoms dominate, $M \approx 1.2\,M_{HI}$, where M_{HI} is the mass of neutral hydrogen; the accuracy of the calculations can be increased as one can immediately find the ratio M_{HI}/R^2 from the hydrogen line ($\lambda = 21$ cm) data. We shall in what follows not aim at such a refinement, which is as yet not pressing.

We note, finally, that the gamma-ray spectrum of nuclear provenance about which we are talking is for obvious reasons concentrated in the energy range $E_\gamma \geqslant 50$ to $100\,\text{MeV}$ (the cosmological red shift is here, of course, not taken into account and we are therefore considering sources which are not too far away). What we have said is clear from Fig.17.1 and more precisely from the following example (Fichtel, Hartman, Kniffen, and Sommer, 1972); for gamma-rays from π^0 decay we have the ratio

$$\xi = \frac{F_\gamma(E_\gamma > 50\,\text{MeV}) - F_\gamma(E_\gamma > 100\,\text{MeV})}{F_\gamma(E_\gamma > 100\,\text{MeV})} = 0.12\,.$$

Moreover, for relativistic electrons with a spectrum $J_e(E) = K_e E^{-2.6}$ in the case of brems gamma-rays $\xi = 2.03$ and for gamma rays of synchrotron provenance or emitted in the inverse Compton effect $\xi = 0.74$. The spectral measurements of the gamma-ray flux thus allow us in principle to establish relatively easily their 'nuclear' nature. If this is done, we get at once from measuring the flux $F_\gamma(E_\gamma > 100\,\text{MeV})$ or the corresponding intensity J_γ the ratio $w_{c.r.}/w_0$ in the source, that is, the basic parameter which is now lacking. In that case we make, of course, an assumption about the similarity of the cosmic ray spectra in the source and at the Earth. There are, however, good grounds for such an assumption and, moreover, under realistic conditions it can, apparently, lead only to the appearance of a numerical coefficient of the order of unity.

At any rate, even such a determination of the energy density or the total density $W_{c.r.} = w_{c.r.} V$ of the cosmic rays in the sources would be an important step forward.

We can make what we have said more concrete using the Magellanic Clouds and the central region of the Galaxy as examples.

It is of interest by itself of course, to consider the Magellanic Clouds. However, this example is even more important in connection with attempts to answer the question: how to elucidate most convincingly the fate of the metagalactic model of the origin of the cosmic rays. It is clear from what we said in Chapter 15 that for this it is sufficient to determine the energy density w_{Mg} in the region surrounding the Galaxy. If it turns out that $w_{Mg} \ll w_{c.r.,G} \sim 10^{-12}\ \mathrm{erg/cm^3}$, the metagalactic model can be dismissed. The best solution from all possible ways to solve this problem which we know is at once the measurement of the gamma-ray flux from the Magellanic Clouds. For these Clouds (the Large Magellanic Cloud — LMC and the Small Magellanic Cloud — SMC) the distances from the Sun and the mass of the neutral hydrogen are, respectively, equal to

$$R(GMC) = 55\ \mathrm{kpc}\ , \quad R(SMC) = 63\ \mathrm{kpc}\ ,$$

$$M_{HI}(LMC) = 1.1 \times 10^{42}\ \mathrm{g}\ , \quad M_{HI}(SMC) = 0.8 \times 10^{42}\ \mathrm{g}\ .$$

Therefore we have, from (17.3) for $w_{c.r.} = w_0$

$$\left. \begin{aligned} F_{\gamma,LMC}(E_\gamma > 100\,\mathrm{MeV}) &\approx 2 \times 10^{-7}\ \mathrm{photons/cm^2.s}\ , \\ F_{\gamma,SMC}(E_\gamma > 100\,\mathrm{MeV}) &\approx 1 \times 10^{-7}\ \mathrm{photons/cm^2.s}\ . \end{aligned} \right\} \qquad (17.4)$$

As we have already mentioned these fluxes can also be calculated more exactly. Here something else is important — the fluxes (17.4) are directly obtained for any known metagalactic models as in those models the role of the intrinsic cosmic ray sources in the Magellanic Clouds, as in the Galaxy, is small so that $w_{Mg} \approx w_{c.r.,G} \approx w_{LMC} \approx w_{SMC}$. On the other hand, there is no ground for expecting the observation of such an equality in galactic models. Even for equal cosmic ray source activity it is very probable that $w_{c.r.G} > w_{LMC} > w_{SMC}$ because of the smaller size of the Clouds and the corresponding faster departure from them of the cosmic rays.

Thus, for a convincing refutation of the metagalactic models of the origin of cosmic rays† it would be sufficient, for instance, to establish that from the

---

† see footnote on next page

two clouds taken together $F_{\gamma,MC}(E_\gamma > 100\,\text{MeV}) \ll 3\times10^{-7}$ photons/cm^2.s or that $F_{\gamma,SMC} \ll \frac{1}{2} F_{\gamma,LMC}$. When the calculations have been made more accurate one can in principle replace the $\ll$-sign by a $<$-sign. It is in this case important that any contribution to the gamma-ray flux connected with relativistic electrons leads only to an increase in the fluxes F_γ and, hence, in no way affects the interpretation given, say, of the result $F_{\gamma,MC}(E_\gamma > 100\,\text{MeV}) \ll 3\times10^{-7}$ photons/cm^2.s. Such a result we call a negative one and we note that there is a certain asymmetry, as often is the case, when interpreting positive and negative results of an experiment. For instance, if the measurement of the gamma-ray flux from the Magellanic Clouds would indicate the presence of an appreciable flux $F_{\gamma,MC} \geqslant 3\times10^{-7}$ photons/cm^2.s (such a result we would customarily call positive) this would not yet prove the validity of metagalactic models, as in principle such a flux could also be generated by the cosmic rays (and also by the relativistic electrons) which are accelerated in the Clouds themselves. Unfortunately, the measurement of the gamm-ray flux from the Magellanic Clouds is a difficult problem and its solution lies in the future. It is appreciably simpler, but also very interesting, to study the gamma-emission from the central region of the Galaxy. Such radiation (with $E_\gamma > 100\,\text{MeV}$) has already been observed and the intensity of the corresponding ('equivalent') linear source is 1 to 2.5×10^{-4} photons/cm^2.s.radian (see Fichtel, Hartman, Kniffen, and Sommer, 1972; Kraushaar, Clark, Garmire, Borken, Higbie, Leong, and Thorson, 1972; Ginzburg, 1973a). If we multiply this value by the angular resolution (about $\pi/6$) we get for the flux from the central galactic source

$$F_\gamma(E_\gamma > 100\,\text{MeV}) = 3 \text{ to } 10\times10^{-5} \text{ photons/cm}^2\text{.s} . \qquad (17.5)$$

Doubts have been expressed about the reality of the obtained result and the validity of its interpretation as evidence for the presence of an extended gamma-ray source in the direction of the centre of the Galaxy, but it is now clear (Stecker and Trombka, 1973; Puget and Stecker, 1974) that the

---

† It is important that we are talking here about all known metagalactic models, whereas a measurement of the isotropic background of gamma-rays generated in the intergalactic space (see Ginzburg and Syrovatskii, 1964a, 1965; Ginzburg, 1969; Stecker, 1971; Gal'per, Kirillov-Ugryumov, Luchkov, and Prilutskii, 1972; Stecker and Trombka, 1973; Gal'per, Kirillov-Ugryumov, and Luchkov, 1974) can serve as a refutation only of those models in which the cosmic rays fill a very large volume, in particular, the whole of intergalactic space (moreover, the density of the metagalactic gas has not yet been established).

corresponding gamma-emission exists although its source is not concentrated in the galactic centre itself (this was assumed earlier; see Fichtel et al., 1972; Kraushaar et al., 1972; Ginzburg, 1973a). We shall assume below that the value (17.5) is real. In that case, both on the basis of spectral measurements and also from a series of indirect considerations it is probable that we are dealing with gamma-rays generated by cosmic rays (that is, mainly with π^0 decay products). Accepting this interpretation we draw a few conclusions.†

Substituting the value (17.5) into Eqn. (17.3) we reach the conclusion that in the central galactic source (we indicated the corresponding quantities by an index GC) cosmic rays are concentrated with a total energy

$$W_{GC} = w_{GC} V_{GC} \sim 3 \text{ to } 10 \times 10^{66} \frac{w_0}{N_{GC}} \sim 3 \times 10 \frac{10^{54}}{N_{GC}} \text{ erg}, \qquad (17.6)$$

as $w_0 \equiv w_{c.r.,0} \sim 10^{-12}$ erg/cm^3, while the distance from the Sun to the central source $R = 10$ kpc. If we assume that the central source is not small (characteristic size L_{GC} less than or of the order of 300 pc) the gas density N_{GC} cannot be put larger than approximately unity (for $L_{GC} \sim 10^{21}$ cm, the volume $V_{GC} \sim 10^{63}$ cm^3 and the gas mass $M_{GC} \sim 2 \times 10^{-24} N_{GC} V_{GC} \sim 10^6 N_{GC} M_\odot$; for $N_{GC} \sim 10$ cm^{-3} we have already $M_{GC} \sim 10^7\ M_\odot$ which is, probably, the limit for a region with the volume chosen).

When $N_{GC} \sim 1$ cm^{-3} we get from (17.6) the estimate $W_{GC} \sim 3$ to 10×10^{54} erg which is only one order of magnitude smaller than the total energy of all cosmic rays in the Galaxy (see (15.18)). On the other hand, a value of the order of 10^{55} erg is just obtained from an analysis of astronomical data which give evidence for an outburst of the galactic nucleus about 10^7 years ago.

If the size of the central gamma-ray source turned out to be less than 200 to 300 pc, a value $N_{GC} \gtrsim 10$ cm^{-3} would not be excluded. The energy W_{GC} then, of course, becomes smaller (see (17.6)), but the intensity of the cosmic rays $J_{c.r.,GC} \equiv J_{GC}$ is, in general, not changed. For instance, for $N_{GC} \sim 10$ cm^{-3}

---

† As the result of more recent measurements (Plovdiv, 1977; Ginzburg, 1978) it has, apparently, become clear that an appreciable part of the gamma-emission from the region of the galactic disc is connected not only with π^0 decay, but also with discrete sources and, perhaps, with bremsstrahlung from relatively soft electrons ($E_e \gtrsim 100$ MeV). We must therefore treat the actual estimates and values given in the present chapter for the galactic gamma-emission as having merely an illustrative value.

we get from (17.6)

$$W_{GC} \sim 3 \text{ to } 10 \times 10^{53} \text{ erg}, \; J_{GC}/J_{c.r.,0} = w_{GC}/w_0 \sim 300 \text{ to } 3000.$$

It is very difficult to contain cosmic rays for 10^7 years in a smaller region (for instance, for $T_{GC} = 3 \times 10^{14}$ s the diffusion path is $\sqrt{2DT_{GC}} \sim 10^{21}$ cm for $D \sim 10^{27}$ cm^2/s which corresponds to a very small value of $\ell \sim 0.03$ pc for the effective mean free path $\ell \sim D/v$, $v \sim 10^{10}$ cm/s). The value $W_{GC} \sim 3 \times 10^{53}$ erg therefore seems to us to be the smallest possible one and one should rather have $W_{GC} \sim 3 \times 10^{54}$ erg. In that case the presence of a central cosmic ray source would have an important value for the whole energy balance of the cosmic rays in the Galaxy (average power of injection $U_{GC} \sim W_{GC}/T_{GC} \gtrsim 10^{40}$ erg/s which is of the same order of magnitude as the total power of injection in galactic models; see (15.20)).

If the value (17.5) is valid the total flux from the central source equals $\mathfrak{F}_\gamma(E_\gamma > 100\,\text{MeV}) = 4\pi R^2 F_\gamma \sim 3$ to 10×10^{41} photons/s, corresponding to a luminosity $L_\gamma \sim \overline{E_\gamma}\mathfrak{F}_\gamma \sim 10^{38}$ erg/s. Moreover, the whole Galaxy, if it were filled uniformly with cosmic rays would emit $\mathfrak{F}_{\gamma,G}(E_\gamma > 100\,\text{MeV}) = 4 \times 10^{-3}\, 4\pi M_G \sim 10^{41}$ photons/s as the total mass of gas in the Galaxy is $M_G \sim 3 \times 10^{42}$ g.

This is not the end of the story, as one can (and must) compare the gamma-astronomy data with radio-astronomy results. Unfortunately, in the case of an extended source of non-thermal radio-emission in the central region of the Galaxy we dispose only of obsolete data (for references see Ginzburg, 1973a). According to those data the flux from an extended source of size $1° \times 3°$ (probable volume $V \sim 10^{63}$ cm^3) is $\Phi_\nu = 3 \times 10^{-20}$ erg/cm^2.s.Hz at a frequency $\nu = 85.5$ MHz with a spectral index $\alpha = 0.7$ ($\gamma = 2a + 1 = 2.4$). Substituting those parameters into Eqns.(5.70) and (5.71) we arrive for the central radio-source at the values (the index r indicates that we are here using radio-data)

$$W_{GC,r} = \kappa_e W_{e,r} \sim 9 \times 10^{51} \kappa_H^{-\frac{3}{7}} \kappa_e^{\frac{4}{7}} \text{ erg}, \; H \sim 10^{-5}(\kappa_H \kappa_e)^{\frac{2}{7}} \text{ Oe}, \tag{17.7}$$

where we have assumed that the radio-emission has a spectral index $\alpha = 0.7$ in the range 10^7 to 10^9 Hz; the estimates are not very sensitive to the choice of this range or, within certain limits, to that of the other parameters. If we now put, as at the Earth, $\kappa_e \sim 100$ and $\kappa_H \sim 1$ we have (for $V \sim 10^{63}$ cm^3)

$$W_{GC.r.} \sim 3 \times 10^{52} \text{ erg}, \; W_{e,r} \sim 3 \times 10^{50} \text{ erg},$$

$$H \sim 3 \times 10^{-5} \text{ Oe} \sim 10\, H_G, \tag{17.8}$$

$$w_{GC,r} = \frac{W_{GC,r}}{V} \sim 3\times 10^{-11}\ \text{erg/cm}^3 \sim 30\ w_{c.r.,G}$$
$$w_{e,r} = \frac{W_{e.r}}{V} \sim 3\times 10^{-13}\ \text{erg/cm}^3 \sim 30\ w_{e,G}\ .$$

Even with all inaccuracies of the initial data (and therefore also of the estimates to a certain extent) the value of $W_{GC,r}$ given here is appreciably smaller than the value $W_{GC} \sim 3\times 10^{53}$ to 10^{55} erg determined above from the gamma-data.

One might find a rather natural explanation for such a discrepancy (Ginzburg, 1973a) but we have very strong arguments for not discussing this problem. The fact is that the estimates given above to some extent are based upon the assumption that the gamma-radiation sources with an enhanced intensity which are observed in the direction of the galactic centre are concentrated in some region close to the centre itself. Just this interpretation (Fichtel et al., 1972; Kraushaar et al., 1972; Ginzburg, 1973a) was thought to be the most probable one until the appearance of measurements which became known in 1973/4 (Stecker and Trombka, 1973; Puget and Stecker, 1974). According to these later data the observed radiation does not come from the central galactic region itself, but from a very extended section situated between the Sun and the centre of the Galaxy. One possible interpretation of the results consists in assuming that the cosmic ray intensity (perhaps, only as far as particles with energies $E \sim 1$ to $10\,\text{GeV}$ are concerned) in a toroidal region at a distance between 4 and 5 kpc from the galactic centre is about an order of magnitude higher than that near the Earth. The increased cosmic ray intensity might in turn be explained to be due to their additional acceleration and trapping in that region (Puget and Stecker, 1974). It is, however, possible that the cosmic ray intensity in that region is the same as at the Earth but that the gas density there is appreciably higher due to the presence of molecular hydrogen (Fichtel et al., 1975). However, as we are dealing with preliminary data and hypotheses, it is inappropriate to go into details, the more so as we here have another aim. Namely, we wanted not only to state, but also to illustrate by an actual example how one can use data on cosmic gamma-rays in the energy range $E_\gamma > 50\,\text{MeV}$, which are formed in the decay of π^0-mesons and other particles which are produced in the gas by the cosmic rays (see also Ginzburg and Ptuskin, 1976a,b; Plovdiv, 1977; Ginzburg, 1978).

Above we concentrated our attention on gamma-rays of nuclear origin, that is, those which are formed in the gas during nuclear collisions of cosmic rays.

Moreover, in that case we were talking only about energies $E_\gamma \geqslant 50$ to 100 MeV. It is at the same time well known, of course, that there are a whole number of other possibilities. For instance, the energy range $E_\gamma = 1$ to 50 MeV is of interest; this is the range of energies of gamma-rays emitted in the π^0 decay in regions with a large red-shift parameter z (to be precise, we are talking about gamma-radiation in the Lemaitre cosmological model, about the observation of matter-antimatter annihilation with $z \gg 1$, and so on).[†] Of course, when $E_\gamma < 50$ MeV, the competing processes, other than π^0 decay are much more important than for $E_\gamma > 50$ MeV. There are undoubtedly certain possibilities to perform measurements in the energy range $E_\gamma \sim 1$ to 50 MeV (see Gal'per et al., 1972, 1974; Stecker and Trombka, 1973). One can say the same about observations of gamma-rays with energies $E_\gamma > 10^{11}$ eV which are performed by the method of Cherenkov fluorescence in the atmosphere (Gal'per et al., 1972, 1974). Particular mention deserve gamma-rays emitted by excited nuclei which appear in nuclear reactions and the gamma-radiation from positron-electron annihilation (if we neglect the red shift the energy of those gamma-rays must be concentrated near an energy of 0.51 MeV). We have already mentioned these gamma-radiation processes at the start of Chapter 16. Ginzburg and Syrovatskii (1965; Ginzburg, 1969b), for instance, have given a scheme for calculating the corresponding intensities (see also the books by Stecker (1971; Stecker and Trombka, 1973; and the literature quoted by Ginzburg, 1973a and by Gal'per et al., 1972, 1974). It is very well possible that the various possibilities for gamma-astronomy enumerated here, or most of them, will be developed in the future. Nevertheless at the present time it seems that none of them has such a potentially wide and universal character as the gamma-astronomical study of the proton-nuclear cosmic ray component. This was the reason why we paid here so much attention to it; we shall not discuss in detail any other, particular possibilities (see also Ginzburg and Ptuskin, 1976a,b; Plovdiv, 1977, Ginzburg, 1978).

Concluding this chapter we shall touch upon the problem of the absorption of gamma-rays (and also partially of X-rays) as this problem is one of interest because of the principles involved.

---

† For a source with a distance from us characterized by the parameter z the energy observed on the Earth is $E_\gamma = E_{\gamma,0}/(1+z)$, where $E_{\gamma,0}$ is the photon energy in the source.

To evaluate the absorption coefficient μ for gamma- and X-rays it is usually important to take into account how the primary flux of the radiation is attenuated, that is, to pay attention to absorption and to scattering. By definition, the quantity μ occurs in the equation†

$$\frac{dJ}{dz} = -\mu J \ , \quad \mu = \sigma N \ , \tag{17.9}$$

where σ is the total cross-section for absorption and scattering and N the particle (atom, electron) density responsible for the absorption and scattering. In the geometric optics approximation (it is always applicable in the cases of interest to us), and this is just the case to which Eqn. (17.9) refers, we have for the intensity $J = J_0 e^{-\tau}$ and for the optical length $\tau = \int_0^L \mu \, dR$ (or in a uniform medium $\tau = \mu L$).

We assumed in (17.9), moreover, that there is no emission of photons along the line of sight. If, on the other hand, emission takes place, the transfer equation becomes

$$\frac{dJ(\nu)}{dz} = q(\nu) - \mu(\nu) \, J(\nu) \ , \tag{17.10}$$

where $q(\nu)$ is the emittance (in the photon number scale) at the frequency considered $\nu = E_\gamma / h$ (see (16.7)).

There are, in principle, very many processes which contribute to σ, namely:

1. Photoeffect (ionization of atoms).
2. Compton scattering.
3. Transitions in the continuous spectrum (free-free absorption).
4. Transitions between atomic levels (excitation of atoms).
5. Formation of e^+, e^- pairs in the medium.
6. Formation of e^+, e^- pairs on thermal and, in general 'soft' photons (the $\gamma + \gamma' \to e^+ + e^-$ process, where γ' is a soft photon).
7. Absorption by nuclei (nuclear photoeffect and excitation of nuclei.
8. Production of $\pi^\pm$ and π^0 mesons at protons and nuclei. Production of other particles.

We have already considered some of the processes listed here. For instance, the absorption coefficient for free-free transitions (process 3) in a hydrogen plasma is given by Eqn. (16.33). Absorption due to bound-bound transitions

---

† We do not consider here the possibility of induced absorption or scattering; as far as one can judge this assumption is justified in the X-ray and gamma bands, but on the whole this problem requires a more detailed analysis.

(process 4) in the X-ray region can play a role only for comparatively heavy elements for the simple reason that the ionization potential for light elements, even for the K-shell, is too small (for instance, for Al atoms with $Z = 13$, it equals approximately 1500 V, which for the edge of the K-band corresponds to absorption at a wavelength of 8 Å). The cross-section for Compton scattering (process 2) was given in Chapter 16.

For low energies the main part in the absorption is played by the photoeffect, while with increasing energy Compton scattering starts to dominate. A consideration of the photoeffect (process 1) requires, in general, taking into account the actual chemical composition of the medium and its degree of ionization. We shall not dwell upon this process which determines the absorption of not too hard X-rays (see Bell and Kingston, 1967; Vainshtein, Kurt, and Sheffer, 1968; Brown and Gould, 1970; Fireman, 1974), but we emphasize that a detailed study of the absorption of soft X-rays in the interstellar and intergalactic medium is of exceptional interest. Possibly, that is the way to obtain useful information about the density, composition, and degree of ionization of the gas in regions about which we know very little at the moment (particularly in the intergalactic medium). However, this is a special problem and it is not possible to elucidate it here in the necessary manner.

When the energy increases absorption due to the photoeffect decreases and in air the contributions from Compton scattering and from the photoeffect are equal at an energy $E_\gamma \approx 25$ keV. At $E_\gamma = 50$ keV photo-absorption is already about five times smaller than Compton absorption. For X-rays with $E_\gamma > 50$ keV and up to energies $2mc^2 = 1$ MeV when e^+, e^- pairs start to be formed, one therefore needs take into account only Compton scattering. For

$$E_\gamma = h\nu \ll mc^2 \approx 5 \times 10^5 \text{ eV}$$

the total cross-section for scattering σ_C which occurs in (17.9) is, when we neglect other processes, equal to the Thomson cross-section

$$\sigma_T = \frac{8}{3}\pi (e^2/mc^2)^2 = 6.65 \times 10^{-25} \text{ cm}^2 .$$

When the frequency increases the cross-section diminishes and for $h\nu = mc^2$ we have already $\sigma_C = 0.43\,\sigma_T$. Therefore with the usually (but not always!) encountered 'astrophysical accuracy' one can put $\sigma_C \sim \sigma_T$ for all gamma-ray energies $E_\gamma \leqslant 1$ MeV. When $E_\gamma \gg mc^2$ one must use Eqn. (16.72) and, for instance, for $E_\gamma = 10^3 mc^2 = 5 \times 10^8$ eV we have $\sigma_C = 3 \times 10^{-3}\sigma_T$. One can find more detailed formulae and also tables with σ_C in § 36 of Heitler's book

(1947). We note also that when Compton scattering is taken into account we must in (17.9) understand by N the total electron density in the medium. If we put $\sigma = \sigma_C = \sigma_T$, we have in the interstellar medium (N is the total density of all electrons)

$$\mu_C = \sigma_T N = 6.65 \times 10^{-25} N \approx 0.4 \text{ cm}^2/\text{g} . \qquad (17.11)$$

In the energy range $E_\gamma < 10^8$ eV Compton scattering in the interstellar medium makes the dominating contribution to μ. Pair production (process 5) is responsible for the absorption of the gamma-rays in the range $E_\gamma < 10^8$ eV. In a neutral gas in the energy range $E_\gamma > 10^8$ eV pair production occurs to a first approximation under conditions of complete screening. The corresponding value of the absorption coefficient in the interstellar medium equals

$$\mu_{pair} = 1.2 \times 10^{-2} \text{ cm}^2/\text{g} = 2 \times 10^{-26} N_a \text{ cm}^{-1} ; \qquad (17.12)$$

we have used here the value of the t-unit length equal to 66 g/cm^2 (see Chapter 16) and N_a is the density of atoms. In a plasma (a completely ionized gas) one can neglect screening in the cases of interest to us and

$$\mu_{pair} = \frac{4Z(Z+1)e^2}{\hbar c}\left(\frac{e^2}{mc^2}\right)^2 N_a \left\{\frac{7}{9} \ln \frac{2E_\gamma}{mc^2} - \frac{109}{54}\right\} =$$

$$= 3.6 \times 10^{-27}\left(\ln \frac{E_\gamma}{mc^2} - 1.9\right) N_a \text{ cm}^{-1}$$

$$= 2.1 \times 10^{-3}\left(\ln \frac{E_\gamma}{mc^2} - 1.9\right) \text{cm}^2/\text{g} , \qquad (17.13)$$

where the numerical values are given for hydrogen $(Z = 1)$. We have restricted ourselves here to merely giving the results, as we have already discussed the role of screening of Chapter 16 (see Heitler's book (1947) or the book by Berestetskii, Lifshitz, and Pitaevskii (1971) for details about the pair production process). The values (17.12) and (17.13) are approximately the same for $E_\gamma \sim 10^9$ eV.

The Compton 'absorption' (17.11) is appreciably larger than the absorption due to pair production (at least as long as $\ln(E_\gamma/mc^2) \ll 100$). However, as we have already emphasized this coefficient refers only to the energy range $E_\gamma \ll 10^6$ eV. The coefficients $\mu_C = \sigma_C N$ and μ_{pair} are equal for $E_\gamma \sim 10^8$ eV. Already for $E_\gamma = 5 \times 10^8$ eV the coefficient $\mu_C \sim 2 \times 10^{-27} N \sim 0.1\, \mu_{pair}$ (see (17.12)).

In the direction to the centre of the Galaxy $\tilde{N}(L) \approx NL \sim 3 \times 10^{22}$ cm^{-2} and the corresponding gas mass is $\tilde{M}(L) = 2 \times 10^{-24} \tilde{N}(L) \approx 6 \times 10^{-2}$ g/cm^2; in the

Metagalaxy $\tilde{M}(L) \sim 0.1\ g/cm^2$ (for $L = R_{ph} \sim 10^{28}$ cm). It is thus at once clear from (17.12) that under the conditions discussed the absorption of gamma-rays is small (for $E_\gamma \geqslant 10^8$ eV); for instance, for $\tilde{M}(L) \sim 0.1\ g/cm^2$ the optical depth $\tau \sim 10^{-3}$ and the factor $e^{-\tau} \approx 1 - \tau$ can be assumed to equal unity with an accuracy of the order of 0.1%. This conclusion remains valid also when we take into account the absorption of gamma-rays by nuclei (process 7).

One must consider especially process 6 — the absorption of gamma-rays connected with the production of e^+, e^- pairs by thermal photons (Nikishov, 1962; Berezinskii, 1970). In a frame of reference in which the total momentum of the two photons vanishes, pair production starts at an energy $E'_\gamma = mc^2$. In the laboratory frame in which there is a gamma-photon of energy E_γ and a thermal photon of energy ϵ_{ph} the threshold for pair production corresponds to an energy†

$$E_{\gamma,0} = \frac{mc^2}{\epsilon_{ph}} mc^2 = 5 \times 10^5 \frac{mc^2}{\epsilon_{ph}} \text{ eV} . \tag{17.14}$$

For optical photons $\epsilon_{ph} \sim 1$ eV and $E_{\gamma,0} \sim 2 \times 10^{11}$ eV; for photons of the relict metagalactic background with a temperature $T \sim 3\ ^\circ K$ the average energy $\overline{\epsilon_{ph}} \sim 10^{-3}$ eV (for thermal radiation $\overline{\epsilon_{ph}} = 2.7\ k_B T$) and $E_{\gamma,0} \sim 2 \times 10^{14}$ eV. Only in the case of pair production on X-ray phtons with $\epsilon_{ph} \sim 10^3$ to 10^4 eV the energy $E_{\gamma,0} \sim 10^7$ to 10^8 eV and the corresponding absorption can be important for comparatively soft γ-rays. However, on average in the Galaxy and in the Metagalaxy the energy density of the X-ray radiation is very small (in the Galaxy $\bar{w}_{ph,X} \sim 10^{-6}\ eV/cm^3$); it is appreciable only in the sources for cosmic X-rays. For optical photons in metagalactic space the energy density $w_{ph,0} \sim 10^{-2}\ eV/cm^3$ and for the relict metagalactic radiation $w_{ph,T} \sim 0.3$ eV/cm^3. The absorption coefficient caused by the optical photons was calculated by Nikishov (1962) and by Berezinskii (1970); the corresponding value

---

† This result can, apparently, be obtained most simply without even transforming from one frame of reference to another. Indeed, at threshold for the production of pairs in the laboratory we have according to the energy and momentum conservation laws

$$E_\gamma + \epsilon_{ph} = \frac{2mc^2}{\sqrt{(1 - v^2/c^2)}}, \quad E_\gamma - \epsilon_{ph} = \frac{2mcv}{\sqrt{(1 - v^2/c^2)}}$$

where v is the velocity of the pair which is produced. Hence

$$E_\gamma = \frac{mc^2(1 + v/c)}{\sqrt{(1 - v^2/c^2)}}, \quad \frac{v}{c} = \frac{E_\gamma - \epsilon_{ph}}{E_\gamma + \epsilon_{ph}}$$

and when $E_\gamma \gg \epsilon_{ph}$ we at once get (17.14).

of μ is a maximum for $E_\gamma = 10^{12}$ eV :

$$\mu_{max} \sim 7 \times 10^{-26} \, w_{ph} \text{ cm}^{-1} , \tag{17.15}$$

where w_{ph} is the energy density of the radiation in eV/cm^3. For thermal radiation with a temperature $T = 5800\ ^\circ K$ ($k_B T = 0.5$ eV) the magnitude of the ratio μ/w_{ph} is given in Table 17.1.

Table 17.1

Energy of gamma-photons, eV	$10^{26}\ \mu/w_{ph}$, $cm^{-1}(eV/cm^3)^{-1}$	Energy of gamma-photons, eV	$10^{26}\ \mu/w_{ph}$ $cm^{-1}(eV/cm^3)^{-1}$
10^{11}	0.05	5×10^{12}	4
5×10^{11}	5	10^{13}	2
10^{12}	7	5×10^{13}	0.7

For $w_{ph} \sim 10^{-2}\ eV/cm^3$ the optical thickness $\tau_{max} \sim 7 \times 10^{-28}\,L$ (cm) and at the photometric radius of the Metagalaxy $R_{ph} \sim 10^{28}$ cm we have already $\tau_{max} \sim 7$. The absorption by photons of the relict background is one and a half orders of magnitude larger, but reaches its maximum only for $E_\gamma \sim 10^{15}$ eV.

For gamma-rays with a very high energy $E_\gamma > 10^{11}$ eV and particularly $E_\gamma > 10^{14}$ eV the absorption of gamma-rays due to pair production on thermal photons (process 6) may thus be large. We note that for collisions of cosmic rays (protons and nuclei) of rather high energy with thermal photons photo-nuclear reactions occur (process 7). However, in this case (in the laboratory system) these processes are not directly related to the problem of gamma-ray absorption. Another fact is that they may play an important role when we consider the generation of gamma-rays as the result of π^0 decay and when we discuss the problem of the change in chemical composition of the cosmic rays and the problem of the 'cut-off' in their spectrum at $E \sim 10^{19}$ to 10^{20} eV (see Berezinskii and Zatsepin, 1971).

REFERENCES

A.I. Akhiezer, I.A. Akhiezer, R.V. Polovin, A.G. Sitenko, and K.N. Stepanov, 1975a, Plasma Electrodynamics, vol.1, Linear Theory, Pergamon Press, Oxford.

A.I. Akhiezer, I.A. Akhiezer, R.V. Polovin, A.G. Sitenko, and K.N. Stepanov, 1975b, Plasma Electrodynamics, vol.2, Non-linear Theory and Fluctuations, Pergamon Press, Oxford.

A.A. Abrikosov, L.P. Gor'kov, and I.E. Dzyaloshinskii, 1965, Quantum Field Theoretical Methods in Statistical Physics, Pergamon Press, Oxford.

V.M. Agranovich, 1968, Theory of Excitons, Nauka, Moscow.

V.M. Agranovich and V.L. Ginzburg, 1966, Spatial Dispersion in Crystal Optics and the Theory of Excitons, Wiley, New York. (A new Russian edition will appear in 1979.)

V.M. Agranovich and V.L. Ginzburg, 1971, Progress in Optics, **9**, 235.

V.M. Agranovich and V.L. Ginzburg, 1972, Sov. Phys. JETP, **34**, 664.

V.M. Agranovich and V.L. Ginzburg, 1973, Sov. Phys. JETP, **36**, 440.

V.M. Agranovich and V.I. Yudson, 1973, Optics Commun., **9**, 58.

V.A. Alekseev and I.I. Sobel'man, 1969, Sov. Phys. JETP, **28**, 991.

D.F. Alferov, Yu.A. Bashmakov, and E.G. Bessonov, 1975, Proc. Lebedev Inst. Phys., **80**, 100.

A.I. Alikhan'yan, E.S. Belyakov, G.M. Garibyan, M.P. Dorikyan, K.Zh. Markaryan, and K.K. Shikhlyarov, 1972, JETP Lett., **16**, 222.

J.L. Anderson and J.W. Ryon, 1969, Phys. Rev., **181**, 1765.

J. Arons, 1972, Astroph. J., **177**, 395.

F.R. Arutyunyan, A.A. Nazaryan, and A.A. Frangyan, 1971, Abstracts 12th Conf. Cosmic Rays, vol.6, 2469.

F.R. Arutyunyan, A.A. Nazaryan, and A.A. Frangyan, 1972, Sov. Phys. JETP, **35**, 1067.

Yu.S. Barash, 1975a, Radiophys. Qu. Electron., **16**, 836.

Yu.S. Barash, 1975b, Radiophys. Qu. Electron., **16**, 945.

Yu.S. Barash, 1978, Radiophys. Qu. Electron., **21**, in course of publication.

Yu.S. Barash and V.L. Ginzburg, 1972, JETP Lett., **15**, 403.

Yu.S. Barash and V.L. Ginzburg, 1975, Sov. Phys. Uspekhi, **18**, 305.

Yu.S. Barash and V.L. Ginzburg, 1976, Sov. Phys. Uspekhi, **19**, 263.

F.G. Bass and V.M. Yakovenko, 1965, Sov. Phys. Yspekhi, **8**, 420.

K.L. Bell and A.E. Kingston, 1967, Monthly Not. Roy. Astron. Soc. **136**, 241.

A.P. Belousov, 1939, Zh. Eksp. Teor. Fiz., **9**, 658.

V.B. Berestetskii, E.M. Lifshitz, and L.P. Pitaevskii, 1971, Relativistic Quantum Theory, Pergamon Press, Oxford.

V.S. Berezinskii, 1970, Sov. J. Nucl. Phys., **11**, 222.

V.S. Berezinskii and G.T. Zatsepin, 1971, Sov. J. Nucl. Phys., **13**, 453.

H.A. Bethe and E.E. Salpeter, 1957, Quantum Mechanics of One- and Two-electron Atoms, Berlin.

R.D. Blandford, 1972, Astron. Astrophys., **20**, 135.

N. Bloembergen, 1967, Am. J. Phys., **35**, 989.

G.R. Blumenthal and R.J. Gould, 1970, Rev. Mod. Phys., **42**, 237.

L.S. Bogdankevich, 1960, Sov. Phys. Tech. Phys., **4**, 992.
D. Bohm and E.P. Gross, 1949, Phys. Rev., **75**, 1851.
N. Bohr, 1948, Proc. Dan. Acad. Sc., **18**, No.8.
B.M. Bolotovskii, 1960, Adv. Phys. Sc., (vol.62), p.372.
B.M. Bolotovskii, 1962, Sov. Phys. Uspekhi, **4**, 781.
B.M. Bolotovskii and V.L. Ginzburg, 1972, Sov. Phys. Uspekhi, **15**, 184.
M. Born, 1909, Ann. Physik, **30**, 1.
M. Born, 1935, Optik, Berlin.
T.H. Boyer, 1970, Ann. Phys., **56**, 474.
V.L. Bratman and E.V. Suvorov, 1969, Sov. Phys. JETP, **28**, 740.
I. Brevik, 1970a, Proc. Dan. Acad. Sc., **37**, No.11.
I. Brevik, 1970b, Proc. Dan. Acad. Sc., **37**, No.13.
R.L. Brown and R.J. Gould, 1970, Phys. Rev., **D1**, 2252.
S.V. Bulanov, V.A. Dogel', and S.I. Syrovatskii, 1972a, Cosmic Research, **10**, 32.
S.V. Bulanov, V.A. Dogel', and S.I. Syrovatskii, 1972b, Cosmic Research, **10**, 721.
S.V. Bulanov, V.A. Dogel', and S.I. Syrovatskii, 1976, Astrophys. Space. Sc., **44**, 255.
H.B.G. Casimir, 1948, Proc. Roy. Dutch Acad. Sc. **51**, 793.
A. Cavalieri, P. Morrison, and L. Sartori, 1971, Science, **173**, 625.
S. Chandrasekhar, 1960, Radiative Transfer, Dover, New York.
S. Chen and M. Takeo, 1957, Rev. Mod. Phys., **29**, 20.
M.L. Cherry, D. Müller, and J.A. Prince, 1974, Nucl. Instr. Meth., **115**, 141.
D.M. Chitre and R.H. Price, 1972, Phys. Rev. Lett., **29**, 185.
B.A. Cooke, A. Lawrence, and G.C. Perola, 1978, Monthly Not. Roy. Astron. Soc., **182**, 661.
J.L. Culhane, 1978, Qu. J. Roy. Astron. Soc., **19**, 1.
R.C. Davidson, 1972, Methods in Nonlinear Plasma Theory, Academic Press, New York.
L. Davis and M. Goldstein, 1970, Astrophys. J., **159**, L81.
L.I. Dorman, 1972, Accelerator Processes in the Cosmos, VINITI.
A.G. Doroshkevich, I.D. Novikov, and A.G. Polnarev, 1973, Sov. Phys. JETP, **36**, 816.
O.I. Dovzhenko and A.A. Pomanskii, 1964, Sov. Phys. JETP, **18**, 187.
L. Durand, 1973, Astrophys. J., **182**, 417.
I.E. Dzyaloshinskii, E.M. Lifshitz, and L.P. Pitaevskii, 1961, Sov. Phys. Uspekhi, **4**, 153.
V.Ya. Eidman, 1958, Sov. Phys, JETP, **7**, 91 (errata:ibid. **9**, 947 (1959)).
V.Ya. Eidman, 1960, Radiofizika, **3**, 192.
V.Ya. Eidman, 1962, Sov. Phys. JETP, **14**, 1401.
V.Ya. Eidman, 1974a, Radiophys. Qu. Electron., **15**, 480.
V.Ya. Eidman, 1974b, Astrophys, **8**, 359.
A. Einstein, 1907, Ann. Physik, **23**, 371.
A. Einstein, 1910, Ann. Physik, **33**, 1275.
N.S. Erokin and S.S. Moiseev, 1973, Sov. Phys. Uspekhi, **16**, 64.
I.L. Fabelinskii, 1968, Molecular Scattering of Light, New York.
V.M. Fain, 1972, Quantum Radiophysics: Photons and Non-Linear Media, Soviet Radio Publishing House.
F.I. Fedorov, 1973, Sov. Phys. Uspekhi, **15**, 849.
E.L. Feinberg, 1966, Sov. Phys. JETP, **23**, 132.
E.L. Feinberg, 1972, Tamm Memorial Volume, p.248.
C.E. Fichtel, R. Hartman, D. Kniffen, and M. Sommer, 1972, Astrophys. J., **171**, 31.
C.E. Fichtel, R.C. Hartman, D.A. Kniffen, D.J. Thompson, G.F. Bignami, H. Ögelman, M.E. Özel, and T. Tümer, 1975, Astrophys. J., **198**, 163.

E.L. Fireman, 1974, Astrophys. J., **187**, 57.
E.S. Fradkin, 1950, Zh. Eksp. Teor. Fiz., **20**, 211.
I.M. Frank, 1942, Izv. Akad. Nauk. SSSR, **6**, 3.
I.M. Frank, 1952, Vavilov Memorial Volume, p.173.
I.M. Frank, 1959, Nobel Lecture.
I.M. Frank, 1972, Tamm Memorial Volume, p.350.
I.M. Frank and V.L. Ginzburg, 1975, J. Phys. USSR, **9**, 35.
H. Friedman, 1973, AIAA Paper No.73-197.
T. Fulton and F. Rohrlich, 1960, Ann. Phys., **9**, 499.

A. Gailitis, 1964, Radiofizika, **7**, 646.
A.M. Gal'per, V.G. Kirillov-Ugryumov, and B.I. Luchkov, 1974, Sov. Phys. Uspekhi, **17**, 186.
A.M. Gal'per, V.G. Kirillov-Ugryumov, B.I. Luchkov, and O.F. Prilutskii, 1972, Sov. Phys. Uspekhi, **14**, 630.
F.F. Gardner and J.B. Whiteoak, 1966, Ann. Rev. Astron. Astrophys., **4**, 245.
G.M. Garibyan, 1960, Sov. Phys. JETP, **10**, 372,
G.M. Garibyan, 1970, Preprint EFI-TF-13, Erevan.
V.P. Gavrilov and A.A. Kolomenskii, 1971, JETP Lett., **14**, 431.
V.P. Gavrilov and A.A. Kolomenskii, 1972, JETP Lett., **16**, 19.
E. Gerlach, 1971, Phys. Rev., **B4**, 393.
M.E. Gertsenshtein, 1954, Zh. Eksp. Teor. Fiz., **27**, 180.
G.G. Getmantsev and V.L. Ginzburg, 1952, Dokl. Akad. Nauk. SSSR, **87**, 187.
V.L. Ginzburg, 1939a, C.R. de l'Acad. Sc. URSS, **23**, 774.
V.L. Ginzburg, 1939b, C.R. de l'Acad. Sc. URSS, **24**, 131.
V.L. Ginzburg, 1939c, Zh. Eksp. Teor. Fiz., **9**, 981.
V.L. Ginzburg, 1940a, J. Phys. USSR, **2**, 441.
V.L. Ginzburg, 1940b, J. Phys. USSR, **3**, 95.
V.L. Ginzburg, 1940c, J. Phys. USSR, **3**, 101.
V.L. Ginzburg, 1941, C.R. Acad. Sc. URSS, **30**, 399.
V.L. Ginzburg, 1944a, J. Phys. USSR, **8**, 33.
V.L. Ginzburg, 1944b, C.R. Acad. Sc. URSS, **42**, 168.
V.L. Ginzburg, 1945, Izv. Akad. Nauk SSSR, **9**, 174.
V.L. Ginzburg, 1946, Proc. Lebedev Inst., **3**, 195.
V.L. Ginzburg, 1947, C.R. Acad. Sc. URSS, **56**, 145.
V.L. Ginzburg, 1952a, Usp. Fiz. Nauk, **46**, 348.
V.L. Ginzburg, 1952b, Vavilov Memorial Volume, p.193.
V.L. Ginzburg, 1958, Sov. Phys. JETP, **7**, 1096.
V.L. Ginzburg, 1960, Sov. Phys. Uspekhi, **2**, 874.
V.L. Ginzburg, 1963, Sov. Phys. Uspekhi, **5**, 649.
V.L. Ginzburg, 1966, Sov. Astron. AJ, **9**, 877.
V.L. Ginzburg, 1969a, Origin of Cosmic Rays, Gordon and Breach, New York.
V.L. Ginzburg, 1969b, Elementary Processes of Cosmic Ray Astrophysics, Gordon and Breach, New York.
V.L. Ginzburg, 1970a, Contemporary Astronomy, Nauka, Moscow.
V.L. Ginzburg, 1970b, Propagation of Electromagnetic Waves in Plasmas, Pergamon, Oxford.
V.L. Ginzburg, 1970c, Sov. Phys. Uspekhi, **12**, 565.
V.L. Ginzburg, 1971, Sov. Phys. Uspekhi, **14**, 83.
V.L. Ginzburg, 1972a, Tamm Memorial Volume, p.192.
V.L. Ginzburg, 1972b, Short Communications Lebedev Inst., **2**, 40.
V.L. Ginzburg, 1972c, JETP Lett., **16**, 357.
V.L. Ginzburg, 1972d, Sov. Phys. Uspekhi, **15**, 114.
V.L. Ginzburg, 1973a, Sov. Phys. Uspekhi, **15**, 626.
V.L. Ginzburg, 1973b, Sov. Phys. Uspekhi, **15**, 839.
V.L. Ginzburg, 1973c, Sov. Phys, Uspekhi, **16**, 434.

V.L. Ginzburg, 1975a, Phil. Trans. Roy. Soc., **277**, 463.
V.L. Ginzburg, 1975b, Radiophys. Qu. Electron. **16**, 386.
V.L. Ginzburg, 1978, Sov. Phys. Uspekhi, **21**, No.1.
V.L. Ginzburg and V.Ya. Eidman, 1959a, Sov. Phys. JETP, **8**, 1055.
V.L. Ginzburg and V.Ya. Eidman, 1959b, Sov. Phys. JETP, **9**, 1300.
V.L. Ginzburg and V.Ya. Eidman, 1963, Sov. Pnys. JETP, **16**, 1316.
V.L. Ginzburg and V.M. Fain, 1957a, Sov. Phys. JETP, **5**, 123.
V.L. Ginzburg and V.M. Fain, 1957b, Radiotekh. Elektron., **2**, 780.
V.L. Ginzburg and V.M. Fain, 1959, Sov. Phys. JETP, **8**, 567.
V.L. Ginzburg and I.M. Frank, 1946, Zh. Eksp. Teor. Fiz., **16**, 15.
V.L. Ginzburg and I.M. Frank, 1947a, Dokl. Akad. Nauk. SSSR, **56**, 583.
V.L. Ginzburg and I.M. Frank, 1947b, Dokl. Akad. Nauk. SSSR, **56**, 699.
V.L. Ginzburg and V.V. Kelle, 1973, JETP Lett., **17**, 306.
V.L. Ginzburg and A.P. Levanyuk, 1974, Phys. Lett., **47A**, 345.
V.L. Ginzburg and L.M. Ozernoy, 1966, Astrophys. J., **144**, 599.
V.L. Ginzburg and V.S. Ptuskin, 1976a, Sov. Phys. Uspekhi, **18**, 931.
V.L. Ginzburg and V.S. Ptuskin, 1976b, Rev. Mod. Phys., **48**, 161,675.
V.L. Ginzburg, V.S. Ptuskin, and V.N. Tsytovich, 1973, Astrophys. Space Sc., **21**, 13.
V.L. Ginzburg and A.A. Rukhadze, 1975, Waves in a Magneto-active Plasma, Nauka, Moscow. (First edition in Handb. Phys., **49/4**, 395,1972.)
V.L. Ginzburg, V.N. Sazonov, and S.I. Syrovatskii, 1968, Sov. Phys. Uspekhi, **11**, 34.
V.L. Ginzburg and S.I. Syrovatskii, 1964a, Origin of Cosmic Rays, Pergamon Press, Oxford.
V.L. Ginzburg and S.I. Syrovatskii, 1964b, Sov. Phys. JETP, **19**, 1255.
V.L. Ginzburg and S.I. Syrovatskii, 1965, Sov. Phys. Uspekhi, **7**, 696.
V.L. Ginzburg and S.I. Syrovatskii, 1966a, Sov. Phys. Uspekhi, **8**, 674.
V.L. Ginzburg and S.I. Syrovatskii, 1966b, Sov. Phys. Uspekhi, **9**, 223.
V.L. Ginzburg and S.I. Syrovatskii, 1969, Ann. Rev. Astron. Astrophys., **7**, 375.
V.L. Ginzburg and S.I. Syrovatskii, 1973, see Hobart 1973, p.53.
V.L. Ginzburg and V.N. Tsytovich, 1974a, Sov. Phys. JETP, **38**, 65.
V.L. Ginzburg and V.N. Tsytovich, 1974b, Sov. Phys. JETP, **38**, 909.
V.L. Ginzburg and V.N. Tsytovich, 1978, Phys. Repts., in course of publication.
V.L. Ginzburg and V.A. Ugarov, 1976, Sov. Phys. Uspekhi, **19**, 94.
V.L. Ginzburg and V.A. Ugarov, 1977, Sov. Phys. Uspekhi, **20**, No.6.
V.L. Ginzburg and G.F. Zharkov, 1965, Sov. Phys. JETP, **20**, 1525.
V.L. Ginzburg and V.V. Zheleznyakov, 1958a, Radiofizika, **1**, No.1, p.59.
V.L. Ginzburg and V.V. Zheleznyakov, 1958b, Radiofizika, **1**, No.5/6, p.9.
V.L. Ginzburg and V.V. Zheleznyakov, 1959a, Sov. Astron. AJ, **2**, 653.
V.L. Ginzburg and V.V. Zheleznyakov, 1959b, Sov. Astron. AJ, **3**, 235.
V.L. Ginzburg and V.V. Zheleznyakov, 1965, Phil. Mag., **11**, 197 (errata p.876).
V.L. Ginzburg, V.V. Zheleznyakov, and V.Ya. Eidman, 1962, Phil. Mag., **7**, 451.
P. Goldsmith and J.V. Jelley, 1959, Phil. Mag., **4**, 836.
L.M. Gorbunov, 1973, Sov. Phys. Uspekhi, **16**, 217.
G.S. Gorelik, 1951, Usp. Fiz. Nauk, **44**, 33.
G.S. Gorelik and I.M. Suchchinskii, 1969, Sov. Phys. Uspekhi, **12**, 399.
R.J. Gould, 1972a, Physica, **60**, 145.
R.J. Gould, 1972b, Physica, **62**, 555.
R.J. Gould, 1974, Ann. Phys., **84**, 480.
R.J. Gould, 1975, Astrophys. J., **196**, 689.
I.S. Gradshteyn and I.M. Ryzhik, 1965, Tables of Integrals, Sums, Series, and Products, Academic Press, New York.
W.T. Grandy Jr., 1970, Nuovo Cim., **65A**, 738.
L. Gratton (ed.), 1970, IAU Symposium No.37.

J. Greene, 1959, Astrophys. J., **130**, 693.
J.E. Gunn and J.P. Ostriker, 1971, Astrophys. J., **165**, 523.
A.V. Gurevich and A.B. Shvartsburg, 1973, Non-Linear Theory of the Propagation of Radio-Waves in the Ionosphere, Nauka, Moscow.

D. ter Haar, 1971, Elements of Hamiltonian Mechanics, Pergamon Press, Oxford.
D. ter Haar, 1972, Phys. Repts., **3**, 57.
S. Hayakawa, 1969, Cosmic Ray Physics, Wiley, New York.
W. Heitler, 1947, Quantum Theory of Radiation, Oxford University Press.
Hobart, 1973, Proceedings of 12th International Conf. on Cosmic Rays.

S. Ichimaru, 1973, Basic Principles of Plasma Physics, Benjamin, New York.

J.M. Jauch and K.M. Watson, 1948a, Phys. Rev., **74**, 950.
J.M. Jauch and K.M. Watson, 1948b, Phys. Rev., **74**, 1485.
J.V. Jelley, 1958, Cherenkov Radiation and its Applications, Pergamon Press, Oxford.
E. Júliusson, 1974, Astrophys. J., **191**, 331.

N.G. van Kampen, B.R.A. Nijboer, and K. Schram, 1968, Phys. Lett., **26A**, 307.
S.A. Kaplan and V.N. Tsytovich, 1973, Plasma Astrophysics, Pergamon Press, Oxford.
W.J. Karzas and R. Latter, 1961, Astrophys. J. Suppl., **6**, 167.
D.A. Kirzhnitz, 1976, Sov. Phys. Uspekhi, **19**, 530.
A.A. Korchak and S.I. Syrovatskii, 1962, Sov. Astron., **5** , 678.
A. Kovetz and G.E. Tauber, 1969, Am. J .Phys., **37**, 382.
H.A. Kramers, 1944, Ned. Tijds. Natuurk., **11**, 134 (see: Collected Papers, North Holland, Amsterdam, 1956, p.838).
J.D. Kraus, 1966, Radio-Astronomy, New York.
W.L. Kraushaar, G.W. Clark, G.P. Garmire, R. Borken, P. Higbie, C. Leong, and T. Thorsen, 1972, Astrophys. J., **177**, 341.

L.D. Landau, 1946, J. Phys. USSR, **10**, 25. (Collected Papers, Pergamon Press, 1965, p.445.)
L.D. Landau and E.M. Lifshitz, 1959, Fluid Dynamics, Pergamon Press, Oxford.
L.D. Landau and E.M. Lifshitz, 1960, Electrodynamics of Continuous Media, Pergamon Press, Oxford.
L.D. Landau and E.M. Lifshitz, 1969, Statistical Physics, Pergamon Press, Oxford.
L.D. Landau and E.M. Lifshitz, 1975, Classical Theory of Fields, Pergamon Press, Oxford.
L.D. Landau and E.M. Lifshitz, 1976, Mechanics, Pergamon Press, Oxford.
L.D. Landau and E.M. Lifshitz, 1977, Quantum Mechanics, Pergamon Press, Oxford.
C. Leibovitz and A. Peres, 1963, Ann. Phys., **25**, 400.
M. Leontovich, 1931, Zs. Physik, **72**, 247.
M. Leontovich (ed), 1965, Rev. Plasma Phys., (Continuing series.), Academic Press, New York.
A.P. Levanyuk, 1974, Sov. Phys. JETP, **39**, 1111.
E.M. Lifshitz, 1956, Sov. Phys. JETP, **2** , 73.
N.D. Lubart, 1974, Phys. Rev., **D9**, 2717.

L.I. Mandel'shtam, 1947, Collected Papers, **2**, 334.
B. Margon, 1973, Astrophys. J. **184**, 323.
M.A. Markov, 1946, Zh. Eksp. Teor. Fiz., **16**, 800.
A.I. Markushevich, 1961, Course of Analytical Functions, Fizmatgiz.
R. McCray, 1966, Science, **154**, 1320.
N.C. McGill, 1968, Contemp. Phys. **9**, 33.
J.P. McTague and G. Birnbaum, 1971, Phys. Rev., **A3**, 1376.

D.B. Melrose, 1968, Astrophys. Space Sc., **2**, 171.
F.C. Michel and H.C. Goldwire, 1970, Astrophys. Lett., **5**, 21.
C. Møller, 1972, Theory of Relativity, Oxford University Press.
A.G. Molchanov, 1966, Sov. Phys. Solid State, **8**, 922.
E.J. Moniz and D.H. Sharp, 1977, Phys. Rev., **D15**, 2850.
V.V. Musakhanyan and A.I. Nikishov, 1974, Sov. Phys. JETP, **39**, 615.

A.I. Nikishov, 1962, Sov. Phys. JETP, **4**, 393.
A.I. Nikishov and V.I. Ritus, 1969, Sov. Phys. JETP, **29**, 1093.
A.I. Nikishov and V.I. Ritus, 1970, Sov. Phys. JETP, **13**, 303.
B.W. Ninham, V.A. Parsegian, and G.H. Weiss, 1970, J. Stat. Phys., **2**, 323.
J.F. Nye, 1957, Physical Properties of Crystals, Oxford University Press.
H. Nyquist, 1928, Phys. Rev., **32**, 110.

J.P. Ostriker and J.E. Gunn, 1969, Astrophys. J., **157**, 1395.
L.M. Ozernoi, O.F. Prilutskii, and I.L. Rozental', 1978, High Energy Astrophysics, Pergamon Press, Oxford.
L.M. Ozernoy and V.N. Sazonov, 1969, Astrophys. Space Sc., **3**, 395.

A.G. Pacholczyk, 1970, Radio Astrophysics, Freeman, San Francisco.
V.E. Pafomov, 1957, Sov. Phys. JETP, **5**, 307.
V.E. Pafomov, 1959, Sov. Phys. JETP, **9**, 1321.
W. Pauli, 1958, Theory of Relativity, Pergamon Press, Oxford.
D. Pines, 1963, Elementary Excitations in Solids, Benjamin, New York.
D. Pines and D. Bohm, 1952, Phys. Rev., **85**, 338.
I.Ya. Pomeranchuk, 1939, Zh. Eksp. Teor. Fiz., **9**, 915.
Plovdiv, 1977, Papers at the 15th International Cosmic Ray Conference, Plovdiv, Bulgaria.
V.S. Ptuskin, 1972, Cosmic Research, **10**, 351.
V.S. Ptuskin, 1974, Astrophys. Space Sc., **28**, 3.
J.L. Puget and F.W. Stecker, 1974, Astrophys. J., **191**, 323.

R. Ramaty, 1968, J. Geophys. Res., **73**, 3573.
M.J. Rees, 1967, Monthly Not. Roy. Astron. Soc., **135**, 345.
M.J. Rees, 1971a, Nature, **229**, 312.
M.J. Rees, 1971b, Nature Phys. Sc., **230**, 55.
M.J. Rees and M. Simon, 1968, Astrophys. J., **152**, L145.
M.I. Riazanov, 1957, Sov. Phys. JETP, **5**, 1013.
V.I. Ritus, 1972a, Nucl. Phys., **B44**, 236.
V.I. Ritus, 1972b, Ann. Phys., **69**, 555.
F.N.H. Robinson, 1973, Macroscopic Electromagnetism, Pergamon Press, Oxford.
F.N.H. Robinson, 1975, Phys. Repts., **16**, 313.
F. Rohrlich, 1965, Classical Charged Particles, Addison-Wesley.
M. Ryle and M.S. Longair, 1967, Monthly Not. Roy. Astron. Soc., **136**, 123.
S.M. Rytov, 1966, Introduction to Statistical Radiophysics, Nauka, Moscow.
Yu.A. Ryzhov, 1959, Radiofizika, **2**, 869.

V.N. Sazonov, 1969, Sov. Phys. JETP, **29**, 578.
V.N. Sazonov, 1970, Sov. Astron. AJ, **13**, 797.
V.M. Sazonov, 1973, Sov. Astron. AJ, **16**, 971.
V.N. Sazonov and V.N. Tsytovich, 1968, Radiophys. Qu. Electron. **11**, 731.
H.W. Schnopper and J.P. Delvaille, 1972, Sc. American, **227**, (1), 26.
C.S. Shen, 1970, Phys. Rev. Lett., **24**, 410.
W.A. Shurcliff, 1962, Polarized Light, Harvard Univ. Press, Cambridge, Mass., USA.
V.P. Silin, 1971, Introduction to the Kinetic Theory of Gases, Nauka, Moscow.
V.P. Silin, 1973, Sov. Phys. Uspekhi, **15**, 742.
V.P. Silin and A.A. Rukhadze, 1961, Electromagnetic Properties of a Plasma and Plasma-like Media, Gosatomizdat.

A.G. Sitenko, 1967, Electromagnetic Fluctuations in a Plasma, Academic Press, New York.
A.G. Sitenko, 1977, Fluctuations and Non-linear Interactions of Waves in a Plasma, Kiev.
D.V. Skobel'tsyn, 1973, Sov. Phys. Uspekhi, **16**, 381.
B.M. Smirnov, 1972, Physics of a Weakly Ionized Gas, Nauka, Moscow.
Ya.A. Smorodinskii and V.A. Ugarov, 1972, Sov. Phys. Uspekhi, **15**, 340.
I.I. Sobel'man, 1972, Introduction to the Theory of Atomic Spectra, Pergamon Press, Oxford.
A. Sommerfeld, 1904a,b, Gött. Nachr., pp.99, 363.
A. Sommerfeld, 1905, Gött. Nachr., p.201.
A. Sommerfeld, 1914, Ann. Physik, **44**, 177.
A. Sommerfeld, 1964, Optics, Academic Press, New York.
L. Spitzer, Jr., 1962, Physics of Fully Ionized Gases, Interscience, New York.
V.S. Starunov and I.L. Fabelinskii, 1970, Sov. Phys. Uspekhi, **12**, 463.
F.W. Stecker, 1971, Cosmic Gamma Rays, NASA, Washington.
F.W. Stecker and J.I. Trombka (Eds), 1973, Gamma-Ray Astrophysics, Washington.
T.H. Stix, 1962, The Theory of Plasma Waves, McGraw-Hill, New York.
M.M. Sushchinskii, 1969, Raman Spectra of Molecules and Crystals, Nauka, Moscow.
E.V. Suvorov and Yu.V. Chugunov, 1973, Astrophys. Space Sci., **23**, 189.

I.E. Tamm, 1939, J. Phys. USSR, **1**, 459.
I.E. Tamm, 1959, Nobel Lecture.
I.E. Tamm, 1976, Basic Electricity Theory, Nauka, Moscow.
I.E. Tamm and I.M. Frank, 1937, C.R. Acad. Sc. USSR, **14**, 109.
V.V. Tamoykin, 1972, Astrophys. Space Sci., **16**, 120.
M.L. Ter-Mikaelyan, 1972, High-Energy Electromagnetic Processes in Condensed Media, Wiley, N.Y.
W.B. Thompson, 1964, Introduction to Plasma Physics, Pergamon, Oxford.
I.N. Toptygin, 1973, Astrophys. Space Sci., **20**, 351.
B.A. Trubnikov, 1958, Sov. Phys. Doklady, **3**, 136.
V.N. Tsytovich, 1962a, Sov. Phys. JETP, **15**, 320.
V.N. Tsytovich, 1962b, Sov. Phys. JETP, **15**, 561.
V.N. Tsytovich, 1964, Sov. Phys. Tech. Phys., **8**, 599.
V.N. Tsytovich, 1977, Theory of a Turbulent Plasma, Plenum, N.Y.
V.N. Tsytovich and S.A. Kaplan, 1974, Astrophysics, **8**, 260.

L.A. Vainshtein, 1957, Electromagnetic Waves, Soviet Radio Publ. House.
L.A. Vainshtein, 1976, Sov. Phys. Uspekhi, **19**, 189.
L.A. Vainshtein, V.G. Kurt, and E.K. Sheffer, 1968, Sov. Astron. AJ, **12**, 189.

G.B. Walker and D.G. Lahoz, 1975a, Nature, **253**, 339.
G.B. Walker and D.G. Lahoz, 1975b, Can. J. Phys., **53**, 2577.
C.L. Wang, G.F. Dell Jr, H. Uto, and L.C.L. Yuan, 1972, Phys. Rev. Lett., **29**, 814.
K.M. Watson and J.M. Jauch, Phys. Rev., **75**, 1249.
T.C. Weekes, 1969, High Energy Astrophysics, Chapman and Hall, London.
V.F. Weisskopf, 1960, Physics Today, **13**, (9), 24.
D.G. Wentzel, 1974, Ann. Rev. Astron. Astrophys., **12**, 71.
K.C. Westfold, 1959, Astrophys. J., **130**, 241.
A.W. Wolfendale, 1970, Cosmic Ray Neutrinos, Lecture to British Association, September 1970.

G.B. Yodh, X. Artru, and R. Ramaty, 1973, Astrophys. J., **181**, 725.

Ya.B. Zel'dovich, 1975, Sov. Phys. Uspekhi, **18**, 79.
V.V. Zheleznyakov, 1959, Radiofizika, **2**, 14.

V.V. Zheleznyakov, 1967a, Sov. Astron. AJ, **11**, 33.
V.V. Zheleznyakov, 1967b, Sov. Phys. JETP, **24**, 381.
V.V. Zheleznyakov, 1968, Astrophys. Space Sc., **2**, 417.
V.V. Zheleznyakov, 1970, Radioemission of the Sun and Planets, Pergamon Press, Oxford.
V.V. Zheleznyakov, 1974, Sov. Astron. AJ, **18**, 142.
V.V. Zheleznyakov and E.V. Suvorov, 1968, Sov. Phys. JETP, **27**, 335.
V.V. Zheleznyakov and E.V. Suvorov, 1972, Astrophys. Space Sci., **15**, 24.
V.P. Zrelov, 1968, Vavilov-Cherenkov Radiation and its Application in High Energy Physics, Atomizdat.

INDEX

Theoretical Physics and Astrophysics

This is the first textbook on theoretical physics to cover astrophysical applications and covers many new advances in X-ray, γ-ray and radio astronomy. It covers various topics in theoretical physics in depth, especially topics related to electrodynamics, and their application to different aspects of astrophysics. Over 300 references.

Contents

ISBN 0 08 023066 0 f